b14062859

NORTH DAKOTA
STATE UNIVERSITY

OCT 27 2008

SERIALS DEPT.
LIBRARY

D1787465

WITHDRAWN
NDSU

Sulfur: A Missing Link between Soils, Crops, and Nutrition

Sulfur: A Missing Link between Soils, Crops, and Nutrition

Joseph Jez, Editor

Book and Multimedia Publishing Committee
David Baltensperger, Chair
Kenneth Barbarick, ASA Editor-in-Chief
Craig Roberts, CSSA Editor-in-Chief
Sally Logsdon, SSSA Editor-in-Chief
Mary Savin, ASA Representative
Hari Krishnan, CSSA Representative
April Ulery, SSSA Representative

Managing Editor: Lisa Al-Amoodi

Agronomy Monograph 50

Copyright © 2008 by American Society of Agronomy, Inc.
Crop Science Society of America, Inc.
Soil Science Society of America, Inc.

ALL RIGHTS RESERVED. No part of this publication may be reproduced or transmitted in any form or by any means, electronic or mechanical, including photocopying, recording, or any information storage and retrieval system, without permission in writing from the publisher.

The views expressed in this publication represent those of the individual Editors and Authors. These views do not necessarily reflect endorsement by the Publisher(s). In addition, trade names are sometimes mentioned in this publication. No endorsement of these products by the Publisher(s) is intended, nor is any criticism implied of similar products not mentioned.

American Society of Agronomy, Inc.
Crop Science Society of America, Inc.
Soil Science Society of America, Inc.
677 South Segoe Road, Madison, WI 53711-1086 USA

ISBN: 978-0-89118-168-2

Library of Congress Control Number: 2008934217

Cover design: Patricia Scullion

Cover photos: Joseph Jez, The Danforth Center; Silvia Haneklaus, Elke Bloem, and Ewald Schnug, Julius-Kühn Institute

Copy editor: Nicholas H. Rhodehamel

Printed in the United States of America.

Contents

1
Sulfur Forms and Cycling Processes in Soil and Their Relationship to Sulfur Fertility   1
    Jeff J. Schoenau
    Sukhdev S. Malhi

2
Sulfur Nutrition of Crops in the Indo-Gangetic Plains of South Asia   11
    M. P. S. Khurana
    U. S. Sadana
    Bijay-Singh

3
Soil Sulfur Cycling in Temperate Agricultural Systems   25
    Jørgen Eriksen

4
History of Sulfur Deficiency in Crops   45
    Silvia Haneklaus
    Elke Bloem
    Ewald Schnug

5
Availability of Sulfur to Crops from Soil and Other Sources   59
    Warren A. Dick
    David Kost
    Liming Chen

6
Sulfur and Cysteine Metabolism   83
    Rainer Hoefgen
    Holger Hesse

7
Sulfur Response Based on Crop, Source, and Landscape Position   105
    Dave Franzen
    Cynthia A. Grant

8
Sulfur Management for Soybean Production   117
Kiyoko Hitsuda
Kazunobu Toriyama
Guntur V. Subbarao
Osamu Ito

9
Sulfur in a Fertilizer Program for Corn   143
George W. Rehm
John G. Clapp

10
Sulfur Nutrition and Wheat Quality   153
Hamid A. Naeem

11
Sulfur and Marketable Yield of Potato   171
Alexander D. Pavlista

12
Sulfur, Its Role in Onion Production and Related Alliums   183
George E. Boyhan

13
Sulfur and the Production of Rice in Wetland and Dryland Ecosystems   197
Richard W. Bell

14
Evaluation of the Relative Significance of Sulfur and Other Essential Mineral Elements in Oilseed Rape, Cereals, and Sugar Beet Production   219
Ewald Schnug
Silvia Haneklaus

15
Improving the Sulfur-Containing Amino Acids of Soybean to Enhance its Nutritional Value in Animal Feed   235
Hari B. Krishnan

16
Methionine Metabolism in Plants   251
Rachel Amir
Yael Hacham

17
Plant Sulfur Compounds and Human Health   281
      Joseph M. Jez
      Naomi K. Fukagawa

18
A Future Crop Biotechnology View of Sulfur
and Selenium   293
      Muhammad Sayyar Khan
      Rüdiger Hell

Subject Index   313

Foreword

The American Society of Agronomy, the Crop Science Society of America, and the Soil Science Society of America are pleased to publish this interesting, comprehensive, and very timely treatise on the role of sulfur in plant nutrition. In addition to the comprehensive coverage that readers of our monographs have come to expect, this volume includes a link to human health, which is increasingly critical given the many nutritional problems facing the world today.

As pointed out by the editor, Dr. Joseph Jez, sulfur is an essential mineral nutrient that is often overshadowed by nitrogen, phosphorus, and potassium. This book provides a current snapshot of the relationships between sulfur and nutrition of crops, animals, and humans. It brings a unique perspective of the interrelationships between sulfur and the dietary needs of animals and humans.

The Societies certainly appreciate the efforts of Dr. Joseph Jez, who chose an outstanding group of authors, and who skillfully and carefully guided the development of the book. We anticipate this excellent work will be a highly valued resource in the scientific community.

Kenneth Moore, President of the American Society of Agronomy
William Wiebold, President of the Crop Science Society of America
Gary A. Peterson, President of the Soil Science Society of America

Preface

Sulfur is an essential mineral nutrient, although it is often overshadowed by nitrogen, phosphorus, and potassium. Because of its central role in soil condition, plant growth, and nutrition, understanding how plants utilize sulfur is critical for optimizing crop yield and quality. Moreover, sulfur incorporated into methionine and cysteine in plants directly impacts the nutritional value of human food and livestock feeds. The goal of this book is to provide an overview of sulfur's importance as a link between soil, plants, and nutrition by bringing together a group of authors with backgrounds that span from the molecular level to the field. The following chapters cover how sulfur cycles in the environment, the requirements for this element in soil and plants, the metabolism of sulfur compounds in plants, the importance of sulfur in specific crops, and the role sulfur plays in livestock and human health. As editor, I would like to thank all of the authors who made this book possible and the administrative help of Marti Shafer.

Joseph Jez, Editor
Donald Danforth Plant Science Center, St. Louis, MO

Contributors

Amir, R.	Dep. of Plant Science, MIGAL, Galilee Technology Center, P.O. Box 831, Kiryat-Shmona, 11016, Israel (Rachel@migal.org.il)
Bell, R.W.	School of Environmental Science, Murdoch University, 90 South St., Murdoch, WA 6150, Australia (R.Bell@murdoch.edu.au)
Bijay-Singh	Dep. of Soils, Punjab Agricultural University, Ludhiana 141 004, India (bijaysingh20@hotmail.com)
Bloem, E.	Institute for Crop and Soil Science, Julius-Kühn Institute, Federal Research Centre for Cultivated Plants (JKI), Bundesallee 50, 38116 Braunschweig, Germany (elke.bloem@jki.bund.de)
Boyhan, G.E.	University of Georgia, Horticulture Dep., Southeast Extension Center, P.O. Box 8112, GSU, Statesboro, GA 30460-8112 (gboyhan@uga.edu)
Chen, L.	School of Environment and Natural Resources, The Ohio State University Agricultural Research & Development Center, Wooster, OH 44691-4096 (chen.280@osu.edu)
Clapp, J.G.	Tessenderlo Kerley Inc., 310 Clapp Farms Road, Greensboro, NC 27405 (jclapp@tkinet.com)
Dick, W.A.	School of Environment and Natural Resources, The Ohio State University Agricultural Research & Development Center, Wooster, OH 44691-4096 (dick.5@osu.edu)
Eriksen, J.	Dep. of Agroecology and Environment, Faculty of Agricultural Sciences, University of Aarhus, P.O. Box 50, 8830 Tjele, Denmark (Jorgen.Eriksen@agrsci.dk)
Franzen, D.	Extension Soil Specialist, NDSU, Box 5758, Fargo, ND 58105-5758 (david.franzen@ndsu.edu)
Fukagawa, N.K.	Dep. of Medicine and General Clinical Research Center, University of Vermont College of Medicine, Burlington, VT 05405 (Naomi.Fukagawa@uvm.edu)
Grant, C.	AAFC Brandon Research Centre, Box 1000A RR#3, Brandon, MB, Canada R7A 5Y3 (CGrant@agr.gc.ca)
Hacham, Y.	Dep. of Plant Science, MIGAL, Galilee Technology Center, P.O. Box 831, Kiryat-Shmona, 11016, Israel (Yaelh@migal.org.il)
Haneklaus, S.	Institute for Crop and Soil Science, Julius-Kühn Institute, Federal Research Centre for Cultivated Plants (JKI), Bundesallee 50, 38116 Braunschweig, Germany (silvia.haneklaus@jki.bund.de)
Hell, R.	Heidelberg Institute of Plant Sciences, University of Heidelberg, Im Neuenheimer Feld 360, 69120 Heidelberg, Germany (rhell@hip.uni-hd.de)
Hess, H.	Max-Planck-Institut für Molekulare Pflanzenphysiologie, Dep. Molecular Physiology, Wissenschaftspark Golm, Am Mühlenberg 1, D-14476 Potsdam, Germany (hesse@mpimp-golm.mpg.de)
Hitsuda, K.	Consulting Division, Japan Development Service Co. Ltd., Toranomon Bldg. 4F, 1-1-12 Toranomon, Minato-ku, Tokyo 105-0001 Japan (hitsuda@jds2)

Hoefgen, R.	Max-Planck-Institut für Molekulare Pflanzenphysiologie, Dep. Molecular Physiology, Wissenschaftspark Golm, Am Mühlenberg 1, D-14476 Potsdam, Germany (hoefgen@mpimp-golm.mpg.de)
Ito, O.	Crop Production & Environment Division, JIRCAS, 1-1 Ohwashi, Tsukuba, Ibaraki 305-8686 (osamuito@affrc.go.jp)
Jez, J.M.	Donald Danforth Plant Science Center, 975 N. Warson Rd., St. Louis, MO 63132 (jjez@danforthcenter.org)
Khan, M.S.	Heidelberg Institute of Plant Sciences, University of Heidelberg, Im Neuenheimer Feld 360, 69120 Heidelberg, Germany (skhan@hip.uni-heidelberg.de)
Khurana, M.P.S.	Dep. of Soils, Punjab Agricultural University, Ludhiana 141 004 India (khuranamps1@rediffmail.com)
Kost, D.	School of Environment and Natural Resources, The Ohio State University Agricultural Research & Development Center, Wooster, OH 44691-4096 (kost.2@osu.edu)
Krishnan, H.B.	Plant Genetics Research Unit, Agricultural Research Service-USDA, and Division of Plant Sciences, University of Missouri, Columbia, MO 65211 (KrishnanH@missouri.edu)
Malhi, S.S.	Agriculture and Agri-Food Canada, Melfort Research Farm, Melfort, SK, Canada S0E 1A0 (MalhiS@agr.gc.ca)
Naeem, H.A.	Dep. of Food Science, University of Manitoba, 250 Ellis Building, Winnipeg, MB, Canada R3T 2N2 (hamid.naeem@gmail.com)
Pavlista, A.D.	Dep. of Agronomy and Horticulture, University of Nebraska, Panhandle Research and Extension Center, Scottsbluff, NE 69361 (apavlista@unl.edu)
Rehm, G.W.	University of Minnesota, 8431 County 17 Blvd., Cannon Falls, MN 55009 (grehm@umn.edu)

Conversion Factors for SI and Non-SI Units

To convert Column 1 into Column 2 multiply by	Column 1 SI unit	Column 2 non-SI unit	To convert Column 2 into Column 1 multiply by
Length			
0.621	kilometer, km (10^3 m)	mile, mi	1.609
1.094	meter, m	yard, yd	0.914
3.28	meter, m	foot, ft	0.304
1.0	micrometer, µm (10^{-6} m)	micron, µ	1.0
3.94×10^{-2}	millimeter, mm (10^{-3} m)	inch, in	25.4
10	nanometer, nm (10^{-9} m)	Angstrom, Å	0.1
Area			
2.47	hectare, ha	acre	0.405
247	square kilometer, km^2 (10^3 $m)^2$	acre	4.05×10^{-3}
0.386	square kilometer, km^2 (10^3 $m)^2$	square mile, mi^2	2.590
2.47×10^{-4}	square meter, m^2	acre	4.05×10^3
10.76	square meter, m^2	square foot, ft^2	9.29×10^{-2}
1.55×10^{-3}	square millimeter, mm^2 (10^{-3} $m)^2$	square inch, in^2	645
Volume			
9.73×10^{-3}	cubic meter, m^3	acre-inch	102.8
35.3	cubic meter, m^3	cubic foot, ft^3	2.83×10^{-2}
6.10×10^4	cubic meter, m^3	cubic inch, in^3	1.64×10^{-5}
2.84×10^{-2}	liter, L (10^{-3} m^3)	bushel, bu	35.24
1.057	liter, L (10^{-3} m^3)	quart (liquid), qt	0.946
3.53×10^{-2}	liter, L (10^{-3} m^3)	cubic foot, ft^3	28.3
0.265	liter, L (10^{-3} m^3)	gallon	3.78
33.78	liter, L (10^{-3} m^3)	ounce (fluid), oz	2.96×10^{-2}
2.11	liter, L (10^{-3} m^3)	pint (fluid), pt	0.473
Mass			
2.20×10^{-3}	gram, g (10^{-3} kg)	pound, lb	454
3.52×10^{-2}	gram, g (10^{-3} kg)	ounce (avdp), oz	28.4
2.205	kilogram, kg	pound, lb	0.454
0.01	kilogram, kg	quintal (metric), q	100
1.10×10^{-3}	kilogram, kg	ton (2000 lb), ton	907
1.102	megagram, Mg (tonne)	ton (U.S.), ton	0.907
1.102	tonne, t	ton (U.S.), ton	0.907
Yield and Rate			
0.893	kilogram per hectare, kg ha^{-1}	pound per acre, lb $acre^{-1}$	1.12
7.77×10^{-2}	kilogram per cubic meter, kg m^{-3}	pound per bushel, lb bu^{-1}	12.87
1.49×10^{-2}	kilogram per hectare, kg ha^{-1}	bushel per acre, 60 lb	67.19
1.59×10^{-2}	kilogram per hectare, kg ha^{-1}	bushel per acre, 56 lb	62.71

Table cont.

To convert Column 1 into Column 2 multiply by	Column 1 SI unit	Column 2 non-SI unit	To convert Column 2 into Column 1 multiply by
1.86×10^{-2}	kilogram per hectare, kg ha^{-1}	bushel per acre, 48 lb	53.75
0.107	liter per hectare, L ha^{-1}	gallon per acre	9.35
893	tonne per hectare, t ha^{-1}	pound per acre, lb acre^{-1}	1.12×10^{-3}
893	megagram per hectare, Mg ha^{-1}	pound per acre, lb acre^{-1}	1.12×10^{-3}
0.446	megagram per hectare, Mg ha^{-1}	ton (2000 lb) per acre, ton acre^{-1}	2.24
2.24	meter per second, m s^{-1}	mile per hour	0.447
Specific Surface			
10	square meter per kilogram, m^2 kg^{-1}	square centimeter per gram, cm^2 g^{-1}	0.1
1000	square meter per kilogram, m^2 kg^{-1}	square millimeter per gram, mm^2 g^{-1}	0.001
Density			
1.00	megagram per cubic meter, Mg m^{-3}	gram per cubic centimeter, g cm^{-3}	1.00
Pressure			
9.90	megapascal, MPa (10^6 Pa)	atmosphere	0.101
10	megapascal, MPa (10^6 Pa)	bar	0.1
2.09×10^{-2}	pascal, Pa	pound per square foot, lb ft^{-2}	47.9
1.45×10^{-4}	pascal, Pa	pound per square inch, lb in^{-2}	6.90×10^3
Temperature			
1.00 (K − 273)	kelvin, K	Celsius, °C	1.00 (°C + 273)
(9/5 °C) + 32	Celsius, °C	Fahrenheit, °F	5/9 (°F − 32)
Energy, Work, Quantity of Heat			
9.52×10^{-4}	joule, J	British thermal unit, Btu	1.05×10^3
0.239	joule, J	calorie, cal	4.19
10^7	joule, J	erg	10^{-7}
0.735	joule, J	foot-pound	1.36
2.387×10^{-5}	joule per square meter, J m^{-2}	calorie per square centimeter (langley)	4.19×10^4
10^5	newton, N	dyne	10^{-5}
1.43×10^{-3}	watt per square meter, W m^{-2}	calorie per square centimeter minute (irradiance), cal cm^{-2} min^{-1}	698
Transpiration and Photosynthesis			
3.60×10^{-2}	milligram per square meter second, mg m^{-2} s^{-1}	gram per square decimeter hour, g dm^{-2} h^{-1}	27.8
5.56×10^{-3}	milligram (H$_2$O) per square meter second, mg m^{-2} s^{-1}	micromole (H$_2$O) per square centimeter second, μmol cm^{-2} s^{-1}	180
10^{-4}	milligram per square meter second, mg m^{-2} s^{-1}	milligram per square centimeter second, mg cm^{-2} s^{-1}	10^4
35.97	milligram per square meter second, mg m^{-2} s^{-1}	milligram per square decimeter hour, mg dm^{-2} h^{-1}	2.78×10^{-2}
Plane Angle			
57.3	radian, rad	degrees (angle), °	1.75×10^{-2}

Table cont.

Conversion Factors for SI and Non-SI Units

To convert Column 1 into Column 2 multiply by	Column 1 SI unit	Column 2 non-SI unit	To convert Column 2 into Column 1 multiply by
Electrical Conductivity, Electricity, and Magnetism			
10	siemen per meter, S m^{-1}	millimho per centimeter, mmho cm^{-1}	0.1
10^4	tesla, T	gauss, G	10^{-4}
Water Measurement			
9.73 × 10^{-3}	cubic meter, m^3	acre-inch, acre-in	102.8
9.81 × 10^{-3}	cubic meter per hour, m^3 h^{-1}	cubic foot per second, ft^3 s^{-1}	101.9
4.40	cubic meter per hour, m^3 h^{-1}	U.S. gallon per minute, gal min^{-1}	0.227
8.11	hectare meter, ha m	acre-foot, acre-ft	0.123
97.28	hectare meter, ha m	acre-inch, acre-in	1.03 × 10^{-2}
8.1 × 10^{-2}	hectare centimeter, ha cm	acre-foot, acre-ft	12.33
Concentration			
1	centimole per kilogram, cmol kg^{-1}	milliequivalent per 100 grams, meq 100 g^{-1}	1
0.1	gram per kilogram, g kg^{-1}	percent, %	10
1	milligram per kilogram, mg kg^{-1}	parts per million, ppm	1
Radioactivity			
2.7 × 10^{-11}	becquerel, Bq	curie, Ci	3.7 × 10^{10}
2.7 × 10^{-2}	becquerel per kilogram, Bq kg^{-1}	picocurie per gram, pCi g^{-1}	37
100	gray, Gy (absorbed dose)	rad, rd	0.01
100	sievert, Sv (equivalent dose)	rem (roentgen equivalent man)	0.01
Plant Nutrient Conversion			
	Elemental	Oxide	
2.29	P	P$_2$O$_5$	0.437
1.20	K	K$_2$O	0.830
1.39	Ca	CaO	0.715
1.66	Mg	MgO	0.602

1

Sulfur Forms and Cycling Processes in Soil and Their Relationship to Sulfur Fertility

Jeff J. Schoenau
University of Saskatchewan, Saskatoon, Canada

Sukhdev S. Malhi
Agriculture and Agri-Food Canada, Melfort, Saskatchewan, Canada

Abstract

Sulfur may be present in soil in a variety of organic and inorganic forms. In well-drained, upland agricultural soils, organic forms of sulfur dominate, while inorganic sulfate is the main inorganic sulfur form. Sulfate present in soil solution represents immediately plant-available sulfur. The microbial conversion of organic sulfur in the form of humus and crop residues to sulfate, termed mineralization, is a dominant mechanism for replenishment of available sulfur. Typically, 1 to 5% of the organic sulfur in a soil is mineralized to sulfate over a growing season. Warm, moist soils with large amounts of organic matter containing easily mineralized organic sulfates exhibit the highest mineralization rates. Microbial oxidation of sulfur is also an important process when reduced sulfur fertilizers such as elemental sulfur are added to soil. Like mineralization, the oxidation of reduced sulfur forms to sulfate is maximized when soils are warm and near field capacity moisture content. The conversion of elemental sulfur fertilizers to plant-available sulfate is increased when particle size is small and the particles are dispersed in the soil. Sulfate can be adsorbed to minerals and organic matter surfaces in soils of acid pH. A portion of the adsorbed sulfate is plant available and adsorption can be beneficial by reducing leaching losses in humid environments. In semiarid environments, sulfate salts can accumulate within the soil profile, especially when drainage is restricted. Sulfates found at depth in the soil profile can contribute to supplies of plant-available sulfur later in the growing season.

Optimizing the sulfur nutrition of crops is key to achieving top crop yields and quality (Tabatabai, 1984). Considerable progress has been made in the last few decades in identifying the nature and cause of sulfur deficiency in soils throughout the world and in recommendations for alleviating sulfur deficiency through the application of sulfur fertilizers. Significant crop responses to sulfur fertilization in the year of application and in subsequent years continue to be observed (Wen et al., 2003; Malhi et al., 2005a). The increased incidence of sulfur deficiency in many cropping systems has been attributed to reduced inputs of sulfur from the atmosphere because of curbing of sulfur emissions, increased

---

Copyright © 2008. American Society of Agronomy, Crop Science Society of America, Soil Science Society of America, 677 S. Segoe Rd., Madison, WI 53711, USA. *Sulfur: A Missing Link between Soils, Crops, and Nutrition.* Agronomy Monograph 50.

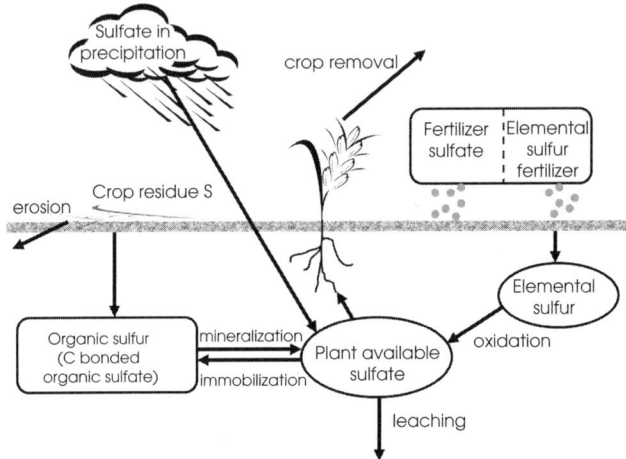

Fig. 1–1. Sulfur cycle showing important pools and transformations of sulfur in upland agricultural systems.

yields and crop removal, and greater acreages of high sulfur demanding crops like canola (*Brassica napus* L.). Ensuring proper sulfur nutrition of plants requires a sound understanding of the basic processes in the soil sulfur cycle that influence the ability of a soil to supply available sulfur over the short and long term. These processes include biological, chemical, and physical transformations of sulfur. The objective of this chapter is to introduce the basic forms and processes involved in the cycling of sulfur in upland, agricultural soils (Fig. 1–1) that closely control the availability of sulfur to plants. Processes that are covered include microbial mineralization–immobilization, microbial oxidation–reduction transformations, sulfate adsorption–precipitation phenomena, and leaching.

Microbial Sulfur Mineralization and Immobilization

Mineralization is the conversion of organically bound sulfur to plant-available inorganic sulfate in the soil, while immobilization is the opposite process in which inorganic sulfate is assimilated by the soil biota and converted to organic forms that cannot be absorbed by roots. In unfertilized soils, mineralization is often the dominant input to the plant-available sulfate pool over a growing season. As mineralization and immobilization processes are affected by organic matter quantity and quality, a review of soil organic sulfur is in order first.

Nature of Soil Organic Sulfur

In the top-soil of well-drained, well-aerated mineral soils, the majority (>90%) of the soil sulfur is contained in organic forms (Schoenau and Germida, 1992). These organic forms include humus as well as recent crop residues. The primary form of sulfur that plant roots absorb and therefore is considered immediately available is sulfate (SO_4^{-2}) dissolved in the soil solution. If present, thiosulfate ($S_2O_3^{-2}$) may also be absorbed. The sulfur contained in organic forms in residue and humus is rendered plant available by mineralization (conversion from organic form to inorganic sulfate).

Within the humus fraction, two main fractions of organic sulfur have been identified on the basis of susceptibility to reduction by hydriodic acid (Freney, 1961): (i) organic sulfates and (ii) carbon-bonded sulfur. The organic sulfate fraction consists of sulfate esters, thioglucosides, and sulfamates in which the sulfur is not directly bonded to carbon but is linked to carbon via an oxygen or nitrogen atom ($C-O-SO_3$, $C-N-SO_3$). The organic sulfates are easily hydrolyzed to inorganic sulfate by mild physical and chemical treatments as well as extracellular enzymes termed arylsulfatases. As such, the organic sulfate fraction is considered a labile, readily mineralizable fraction, and soils with high content of organic sulfur in organic sulfate form are considered to have more easily mineralizable humus (Bettany et al., 1979). Organic sulfates comprise between 30 and 70% of the total organic sulfur in soils. The rest of the organic sulfur is carbon-bonded sulfur, consisting of sulfur that is directly bonded to carbon in the form of sulfur-containing amino acids in proteins, sulfonic acids, and complex heterocyclic compounds (Biederbeck, 1978). Although amino acids may be readily mineralized by soil microorganisms (Maynard et al., 1985), much of the carbon-bonded sulfur appears to reside in high molecular weight forms that can be quite resistant to decomposition.

Organic sulfur in crop residues mainly take the form of carbon-bonded sulfur such as amino acids (cysteine and methionine) in proteins as well as other sulfur containing compounds including vitamins (thiamine and biotin) and glucoside oils. Decomposition of crop residue in the soil results in conversion of carbon-bonded sulfur compounds in the residue into microbial biomass and products of humification that are rich in organic sulfates (Roberts and Bettany, 1985). The sulfur in living microbial biomass in the soil represents about 1 to 4% of the total organic sulfur (Gupta, 1989). Factors controlling microbial activity and production of enzymes and the carbon to sulfur (C/S) ratio in the crop residue affect the release of plant-available sulfate by mineralization. These are described in the following section.

Factors Controlling Sulfur Mineralization–Immobilization

As organic matter quantity and quality greatly affect the release of available nitrogen through mineralization, similar controls exist for release of available sulfate from mineralization of organic sulfur in humus and plant residues. The sources of decomposable sulfur in the soil include three main groups (Freney, 1967): (i) fresh plant and animal residues, (ii) soil biomass including microbial cells and metabolites, and (iii) humus. Extracellular sulfatase enzymes released by soil microorganisms are capable of hydrolyzing peptide and ester bonds to plant-available inorganic sulfate (Tabatabai and Bremner, 1970). Proteinaceous components and macromolecules containing sulfate esters are broken down into smaller units. Microorganisms rapidly assimilate simple organic molecules such as low molecular weight ester sulfates and amino acids and also inorganic sulfate when the sulfur in the substrate is insufficient to meet microbial requirements. This results in immobilization of plant-available sulfate that temporarily reduces the supply of sulfate available to plants.

The C/S ratio in the substrate being decomposed is an important factor controlling whether there will be an initial net increase or decrease in supplies of plant-available sulfate when a residue is being decomposed. Whether net mineralization occurs is dependent on microbial and substrate C/S ratios as well as

Table 1–1. Nature of organic sulfur in the surface A horizon of native and cultivated pairs of grassland, boreal forest, and wetland soils in Saskatchewan, Canada.

Soil type	Organic sulfur concentration	C/S ratio	Percentage hydriodic acid-reducible sulfur
	mg sulfur kg^{-1}		%
Aridic Boroll (grassland)			
Native	453	66	60
Cultivated	268	62	60
Boralf (boreal forest)			
Native	73	133	29
Cultivated	144	124	34
Aquoll (wetland)			
Native	824	77	42
Cultivated	372	73	44

organic sulfate content (Gupta, 1989). Decomposition of crop residue with wide C/S ratio (>400:1) is often associated with initial microbial immobilization of plant-available sulfate, while sulfur-rich residues with narrow C/S ratios (<200:1) result in immediate net mineralization and increase in sulfate content. Manures and composts can be effective sources of available sulfur through mineralization, but some liquid manures stored under anaerobic conditions are reported to be low in available sulfur relative to nitrogen, such that supplemental addition of fertilizer sulfur was found to be beneficial to crop growth (Schoenau and Davis, 2006). The C/S ratio of soil humus is around 100:1 (Roberts et al., 1989) so that decomposition of soil humus typically results in net mineralization. The total organic sulfur content, C/S ratios, and percentage of total sulfur comprised of organic sulfate are shown for a range of soils including cultivated and native short-grass prairie, boreal forest, and wetland soils of Canada (Table 1–1). Soils containing humus with high content of organic sulfate compounds such as many soils formed under grassland generally have higher rates of sulfur mineralization than forest soils formed from litter containing higher C/S ratios and a lower proportion of organic sulfur comprised of organic sulfates (Roberts and Bettany, 1985). Loss of organic matter through cultivation reduces the sulfur fertility of the soil. However, adoption of cropping systems that reduce nutrient losses and increase soil organic matter such as forages and no-till have been shown to increase sulfur mineralization rates (Greer and Schoenau, 1992).

Predictions of sulfur mineralization in soils by mathematical models have been made from first order kinetic models that include measures of the amount of potentially mineralizable organic sulfur and rate constants that are adjusted for temperature and moisture (Ellert and Bettany, 1988; Coughenour et al., 1980). Rates of available sulfate release by mineralization are greatest when the soil is at or near field capacity and at temperatures approaching 35°C. Therefore, warm, moist soil conditions favor sulfur mineralization. Reported contributions to the plant-available sulfate pool from mineralization of soil organic sulfur over the growing season vary widely but typically represent from 1 to 5% of the total organic sulfur. Roberts (1985) reported approximately 4 to 13 kg sulfur ha^{-1} was mineralized over an 11-wk incubation in a range of cultivated soils in Saskatchewan, Canada, representing from 1 to 3% of the organic sulfur in these soils.

Microbial Oxidation and Reduction

The microbial oxidation of reduced inorganic sulfur forms such as sulfides, elemental sulfur, and thiosulfates to forms of higher oxidation state, including sulfate, is an important process in soils where reduced forms of sulfur are present. This may include parent material containing sulfides, sulfides that originated from previously flooded anoxic conditions, and where reduced forms of sulfur such as elemental sulfur are used as fertilizers. Microbial oxidation is performed by both autotrophic and heterotrophic microorganisms, including species of *Thiobacillus* (autotroph), heterotrophic bacteria like *Pseudomonas*, *Arthrobacter*, and *Bacillus* and some fungi. In some cases, heterotrophic sulfur oxidizers may dominate (Lawrence and Germida, 1988), especially in the rhizosphere (Grayston and Germida, 1990). In the oxidation of reduced sulfur compounds like elemental sulfur (S^0), acidity is produced along with sulfate as depicted in the following equation.

$$2\,S^0 + 3\,O_2 + 2H_2O \rightarrow 2\,H_2SO_4 \quad [1]$$

Reducing conditions in which redox is decreased as a result of poor aeration, flooding, and high oxygen consumption results in the microbial conversion of sulfates to sulfide after other electron acceptors including oxygen and nitrate are depleted. This process, termed dissimilatory sulfate reduction, is performed by species of *Desulfovibrio* and *Desulfotomaculum* bacteria. Hydrogen sulfide gas is retained in the soil by reaction with iron to form an iron sulfide mineral, depicted in the following equation.

$$SO_4 \rightarrow H_2S\,(g) + Fe \rightarrow FeS\ \text{mineral} \quad [2]$$

These sulfides may be oxidized back to sulfates if the soil becomes aerobic again. Since this chapter focuses on sulfur cycling processes in upland, aerobic agricultural soils, emphasis is placed on the sulfur oxidation process.

Influence on Sulfur Fertility

The progressive oxidation of sulfides contained in parent material can contribute large amounts of sulfate salts such that in soils with restricted drainage, the buildup of salts in the soil profile restricts plant growth. Saline soils in western Canada are dominated by inorganic sulfur in the form of sulfate salts, mainly sodium, magnesium, and calcium sulfates originating from oxidation of pyrite in the parent material (Mermut and Arshad, 1987).

A major role of oxidation in soil is the conversion of reduced sulfur fertilizer forms, especially elemental sulfur, to plant-available sulfate. The sulfate forms of sulfur fertilizer are immediately available for uptake, but the reduced forms including elemental sulfur and sulfides must be oxidized to sulfate (Table 1–2).

The agronomic effectiveness of any reduced sulfur fertilizer is directly related to oxidation rate that provides plant-available sulfate following application (Malhi et al., 2005a). The challenge is to predict accurately the rate of oxidation and release of plant-available sulfate. Generally, the oxidation of thiosulfate forms is quite rapid compared with elemental sulfur forms, with 56 to 70% of thiosulfate sulfur recovered as sulfate after only 25 d of incubation, representing 77 to 98% of the sulfate levels found in the corresponding sulfate treatment (Janzen and Bettany, 1986).

Elemental sulfur is an insoluble, hydrophobic particle that is dependent on microbial colonization of its surface and subsequent oxidation. Oxidation

Table 1–2. Sulfur fertilizer forms.

Fertilizer	Fertilizer analysis
Sulfate (SO_4^{-2}) forms:	%N–%P_2O_5–% K_2O–%S
Ammonium sulfate	21–0–0–24
Potassium sulfate	0–0–50–18
Gypsum (calcium sulfate)	0–0–0–19
Single superphosphate	0–20–0–14
Thiosulfate ($S_2O_3^{-2}$) forms:	
Ammonium thiosulfate (liquid ATS)	12–0–0–26
Potassium thiosulfate	0–0–25–17
Elemental(S^0) forms:	
sulfur-bentonites, micronized granulars	0–0–0–90 to 0–0–0–99
emulsions, suspensions	0–0–0–60 and others
mixtures of elemental sulfur and sulfate	various
Sulfide (S^{-2}) forms:	
Ammonium polysulfide	20–0–0–45
Potassium polysulfide	0–0–22–23

rates of elemental sulfur are slow in cold, dry soils and there are many reports of inadequate supplies of sulfate from elemental sulfur oxidation in the year of application for annual cropping systems in western Canada (Karamanos and Poisson, 2004; Malhi et al., 2005a). In three field experiments in northeastern Saskatchewan, there was little or no seed yield response of canola to granular elemental sulfur fertilizers (ES-90 or ES-95) compared with sulfate-sulfur fertilizers (potassium sulfate or ammonium sulfate) in the year of application (Malhi et al., 2005a, 2005b; S. S. Malhi, 1999, unpublished results). For example, average seed yields of three field experiments for the zero-sulfur control, granular ES-90, granular ES-95 and granular sulfate-sulfur on the severely sulfur-deficient soils were 29, 28, 35, and 1135 kg ha^{-1}, respectively, when adequate amounts of nitrogen fertilizer were applied. Seed yields of canola improved considerably when elemental sulfur fertilizers were surface broadcast or broadcast and incorporated into the soil as finely ground powder or as suspension (Malhi et al., 2005b). Seed yields of canola with powder or suspension elemental sulfur were comparable to sulfate-sulfur fertilizers.

The particle size of elemental sulfur greatly affects oxidation rates, with oxidation rate increasing with decreased particle size (Janzen and Bettany, 1986). A smaller particle size means increased surface area for colonization and microbial oxidation. Consequently, many studies have shown that dispersion of elemental sulfur into small micronized particles in the soil greatly enhances the oxidation rate. One study showed an 8- to 16-fold increase in oxidation when elemental sulfur granules were dispersed before application compared with application of intact granules (Table 1–3).

The manipulation of particle size of elemental sulfur is the most powerful tool in enhancing oxidation rate and consequently this is a factor to which much attention is given in modified elemental sulfur fertilizers (Ceccotti, 1994). It has been suggested that for annual cropping purposes, particle diameters smaller than 150 to 200 µm need to be used (Edmeades et al., 1994), and many products that granulate micrometer sized elemental sulfur particles into prills that are

Table 1–3. Percentage of elemental sulfur oxidized to sulfate from intact and dispersed elemental sulfur prills during a 10-wk incubation in a loamy Saskatchewan grassland soil.†

Temperature	Form	Percentage of applied elemental sulfur oxidized to sulfate
°C		%
5	Intact	1
	Dispersed	16.5
10	Intact	4
	Dispersed	31
20	Intact	7
	Dispersed	60

† Adapted from Wen et al. (2001).

suitable for field application have been developed. Combination of 82-μm sized elemental sulfur particles with sewage sludge (Cowell and Schoenau, 1995) and hydrated lime (Sulewski and Schoenau, 1998) was shown to increase oxidation rates over elemental sulfur alone. This was attributed to stimulation of heterotrophic sulfur oxidizer activity by the organic material and the ability to buffer extreme pH declines surrounding elemental sulfur surfaces. Further work is needed to develop products and models to predict more accurately availability of sulfur from reduced sulfur fertilizer forms to ensure an adequate supply of available sulfur to crops from oxidation.

Sulfate Adsorption, Precipitation, and Leaching

In well-drained, well-aerated surface soils of neutral to alkaline pH, inorganic sulfate exists mainly in the soil solution in the form of water-soluble salts of calcium, magnesium, and sodium (Schoenau and Germida, 1992), typically comprising 1 to 10% of the total soil sulfur pool. In acid soils, sulfate may be also be adsorbed to surfaces of amorphous iron and aluminum oxides and hydrous oxides as well as positively charged sites on edges of clay minerals and humus functional groups (Williams, 1975). Sulfate adsorption is insignificant at pH values greater than about 6. Like phosphate sorption, sulfate adsorption increases with decreasing pH and is greater in soils with high contents of amorphous iron and aluminum. As such it is a common phenomenon in many highly weathered, acidic, tropical soils. In soils that have large amounts of water percolating through the soil profile, sulfate adsorption is an important mechanism that helps to maintain sulfate in the soil profile and reduce losses by leaching (Bohn et al., 1986). The sulfate that is loosely held on colloid surfaces is in rapid chemical equilibrium with sulfate in solution and can replenish solution sulfate when depleted via plant uptake. Therefore, a significant portion of the adsorbed sulfate may be considered plant available. In soils of neutral and alkaline pH where sulfate is present mainly in solution, sulfate can be as mobile as nitrate, with losses of available sulfur by deep leaching of sulfate below the root zone (Bettany and Stewart, 1983).

Many Luvisolic soils (Boralfs), developed under boreal forest vegetation and subhumid climate in western Canada, have low concentrations of sulfate-sulfur throughout the soil profile (Table 1–4). The low content of sulfate reflects low supplies of available sulfur through mineralization in the surface horizons, limited retention by sulfate adsorption, and periodic deep leaching events. This con-

Table 1–4. Concentrations of soluble sulfate, and insoluble sulfate co-precipitated with calcium carbonate, in the genetic horizons of a Boralf and aridic Boroll in Saskatchewan, Canada.

Soil	Genetic horizon	Depth	Soluble sulfate	Insoluble sulfate
		cm	mg sulfur kg^{-1}	
Boralf	Ap	0–16	2.1	n.d.†
	Bt	16–65	0.9	n.d.
	Ck	65–95	1.2	30
Aridic Boroll	Ap	0–15	3.8	n.d.
	Bm	15–39	3.5	n.d.
	Cksa	39–50	2003	62

†n.d. = not detected.

tributes to a high incidence of sulfur deficiency in the Luvisols, especially when high sulfur demanding crops like canola are grown (Malhi et al., 2005a). Some of the soils also show evidence of slow transfer of soluble low molecular weight organic sulfur compounds from the A horizons to deeper B and C horizons as part of pedogenic processes operative during soil formation (Schoenau and Bettany, 1987). However, deep movement of sulfur in organic forms is not believed to be a significant process over the short term in most soils because of the tendency for the organic forms to react and precipitate in subsoils.

Under more arid conditions, limited downward leaching can result in significant accumulations of sulfate in the form of gypsum ($CaSO_4 \cdot 2H_2O$) and other soluble sulfate salts in subsurface horizons within the root zone. This is a common phenomenon in many prairie soils formed under semiarid conditions, especially with restricted drainage. Concentrations of sulfate in the gypsiferous, saline subsoils of soils like the Aridic Boroll in Table 1–4 are typically in the hundreds and thousands of milligrams sulfate-sulfur per kilogram of soil. As such, sulfur deficiency is not common, and plants will quickly recover from early deficiency in these soils, once the roots access the sulfate at depth.

Insoluble sulfate can also exist in soils as barium and strontium sulfate, sulfate associated with calcium carbonate, and iron and aluminum sulfates. Sulfate as a cocrystallized impurity in calcium carbonate is the most common form, and sulfate coprecipitated with carbonate minerals is reported to account for 40 to 50% of the total sulfur in calcareous horizons of soils in Australia (Williams, 1974) and Canada (Bettany and Stewart, 1983). Despite a relatively large content of sulfate coprecipitated with carbonate in the Ck horizon of the Boralf (Table 1–4), this sulfate is of very low solubility and makes little or no contribution to the plant-available sulfate pool over the short term (Williams and Steinbergs, 1964; Williams, 1974) because it would only be released by carbonate mineral dissolution.

Conclusions

The sulfur fertility of a soil is a product of the conditions under which the soil formed and its subsequent management. A host of factors in the soil sulfur cycle must be considered when assessing the ability of a soil to supply available sulfur for crop growth. The activity of the soil biota has a major influence on the supplies of plant-available sulfate through mineralization–immobilization and oxidation–reduction processes. In upland soils, biological replenishment of soil solution sulfate occurs through mineralization of organic sulfur along with oxidation of any reduced sulfur fertilizer forms added like elemental sulfur.

Management of soil to promote adequate supplies of substrate organic sulfur by building humus and returning crop residues, along with a diverse and active soil microbial pool to carry out mineralization and oxidation, will help to ensure good sulfur nutrition. The sensitivity of these biological sulfur transformations to temperature, moisture, substrate amount, and composition implies the need to build predictive models that take these factors into account when predicting contributions to available sulfur and the need for additional fertilizer sulfur amendments. Long-term supplies of available sulfur are dependent on balancing outputs from the system with sulfur inputs and reducing any losses.

Physico-chemical processes like sulfate adsorption remove sulfate from solution and can reduce the supply of available sulfur in the short term. However, sorption can also contribute to retention in high leaching environments. A measurement of the amount of soluble sulfate in a soil provides a good snapshot of the availability of sulfur at the start of the growing season. However, the balance between inputs and outputs from this pool of sulfur over time can have a large influence on what is actually available to the crop for uptake and utilization and should be known when making sulfur fertilizer recommendations.

References

Bettany, J.R., J.W.B. Stewart, and S. Saggar. 1979. The nature and forms of sulfur in organic matter fractions of soils selected along an environmental gradient. Soil Sci. Soc. Am. J. 43:481–485.

Bettany, J.R., and J.W.B. Stewart. 1983. Sulfur deficiency in the prairie provinces of Canada. p. 787–800. *In* A.I. More (ed.) Proceedings of international sulfur '82 conference. The British Sulphur Corporation Limited, London.

Biederbeck, B.P. 1978. Soil organic sulfur and fertility. p. 273–310. *In* M. Schnitzer and S. Khan (ed.) Soil organic matter. Elsevier, New York.

Bohn, H.L., N.J. Barrow, S.S. Rajan, and R.L. Parfitt. 1986. Reactions of inorganic sulfur in soils. p. 233–246. *In* M.A. Tabatabai (ed.) Sulfur in agriculture. Agron. Monogr. 27. ASA-CSSA-SSSA, Madison, WI.

Ceccotti, S.P. 1994. Sulphur fertilizers: An overview of commercial developments and technological advances. Sulphur Agric. 18:58–64.

Coughenour, M.B., W.J. Parton, W.K. Lauenroth, J.L. Dodd, and R.G. Woodmansee. 1980. Simulation of a grassland sulfur cycle. Ecol. Model. 9:179–213.

Cowell, L.E., and J.J. Schoenau. 1995. Stimulation of elemental sulfur oxidation by sewage sludge. Can. J. Soil Sci. 75:247–249.

Edmeades, D.C., A.G. Sinclair, J.H. Watkinson, S.F. Ledgard, A. Ghani, B.S. Thorrold, C.C. Boswell, A.C. Braithwaite, and M.W. Brown. 1994. Some recent developments in sulfur research in New Zealand agriculture. Sulphur Agric. 18:3–8.

Ellert, B.H., and J.R. Bettany. 1988. Comparison of kinetic models for describing net sulfur and nitrogen mineralization. Soil Sci. Soc. Am. J. 52:1692–1702.

Freney, J.R. 1961. Some observations on the nature of organic sulphur compounds in soil. Aust. J. Agric. Res. 12:424–432.

Freney, J.R. 1967. Sulfur containing organics. p. 220–259. *In* A.D. McLaren and G.H. Peterson (ed.) Soil biochemistry. Marcel Dekker, New York.

Grayston, S.J., and J.J. Germida. 1990. Influence of crop rhizosphere on populations and activity of heterotrophic sulfur-oxidizing microorganisms. Soil Biol. Biochem. 22:457–463.

Greer, K.J., and J.J. Schoenau. 1992. Soil organic matter content and nutrient turnover in Thin Black Oxbow soils after intensive conservation management. p. 167–173. *In* Proceedings of the Soils and Crops Workshop, Management of Agriculture Science, University Extension Press, University of Saskatchewan, Saskatoon, SK, Canada.

Gupta, V.V.S.R. 1989. Microbial biomass sulfur and biochemical mineralization of sulfur in soil. Ph.D. thesis. University of Saskatchewan, Saskatoon, SK, Canada.

Janzen, H.H., and J.R. Bettany. 1986. Release of available sulfur from fertilizers. Can. J. Soil Sci. 66:91–103.

Karamanos, R.E., and D.P. Poisson. 2004. Short and long term effectiveness of various sulfur products in prairie soils. Commun. Soil Sci. Plant Anal. 35:2049–2066.

Lawrence, J.R., and J.J. Germida. 1988. Most probable number procedure to enumerate sulfur oxidizing, thiosulfate producing heterotrophs in soil. Soil Biol. Biochem. 20:577–578.

Malhi, S.S., J.J. Schoenau, and C.A. Grant. 2005a. A review of sulfur fertilizer management for optimum yield and quality of canola in the Canadian Great Plains. Can. J. Plant Sci. 85:297–307.

Malhi, S.S., E.D. Solberg, and M. Nyborg. 2005b. Influence of formulation of elemental sulfur fertilizer on yield, quality and sulfur uptake of canola seed. Can. J. Plant Sci. 85:793–802.

Maynard, D.G., J.W.B. Stewart, and J.R. Bettany. 1985. The effects of plants on soil sulfur transformations. Soil Biol. Biochem. 17:127–134.

Mermut, A.R., and M.A. Arshad. 1987. Significance of sulfide oxidation in soil salinization in southeastern Saskatchewan, Canada. Soil Sci. Soc. Am. J. 51:247–251.

Roberts, T.L. 1985. Sulfur and its relationship to carbon, nitrogen and phosphorus in a climotoposequence of Saskatchewan soils. Ph.D. thesis, University of Saskatchewan, Saskatoon, SK, Canada.

Roberts, T.L., and J.R. Bettany. 1985. The influence of topography on the nature and distribution of soil sulfur across a narrow environmental gradient. Can. J. Soil Sci. 65:419–434.

Roberts, T.L., J.R. Bettany, and J.W.B. Stewart. 1989. A hierarchical approach to the study of organic C, N, P and S in western Canadian soils. Can. J. Soil Sci. 69:739–749.

Schoenau, J.J., and J.R. Bettany. 1987. Organic matter leaching as a component of carbon, nitrogen, phosphorus and sulfur cycles in a forest, grassland and gleyed soil. Soil Sci. Soc. Am. J. 51:646–651.

Schoenau, J.J., and J.G. Davis. 2006. Optimizing soil and plant responses to land-applied manure nutrients in the Great Plains of North America. Can. J. Soil Sci. 86:587–595.

Schoenau, J.J., and J.J. Germida. 1992. Sulfur cycling in upland agricultural systems. p. 261–277. In R. W. Howarth et al. (ed.) Sulfur cycling on the continents. John Wiley & Sons, New York.

Sulewski, G.D., and J.J. Schoenau. 1998. Can the plant availability of elemental sulfur be enhanced through its combination with sewage sludge and hydrated lime? Can. J. Soil Sci. 78:459–466.

Tabatabai, M.A. 1984. Importance of sulfur in crop production. Biogeochemistry 1:45–62.

Tabatabai, M.A., and J.M. Bremner. 1970. Arylsulfatase activity of soils. Soil Sci. Soc. Am. Proc. 34:225–229.

Wen, G., J.J. Schoenau, S.P. Mooleki, S. Inanaga, T. Yamamoto, K. Hamamura, M. Inoue, and P. An. 2003. Effectiveness of an elemental sulfur fertilizer in an oilseed–cereal–legume rotation on the Canadian Prairies. J. Plant Nutr. Soil Sci. 166:54–60.

Wen, G., J.J. Schoenau, T. Yamamoto, and M. Inoue. 2001. A model of oxidation of an elemental sulfur fertilizer in soils. Soil Sci. 166:607–613.

Williams, C.H.. 1974. The chemical nature of sulphur in some New South Wales soils. p. 16–23. In K.D. McLachlan (ed.) Handbook on sulfur in Australian agriculture. CSIRO, Melbourne, Australia.

Williams, C.H. 1975. The chemical nature of sulfur compounds in soil. p. 21–30. In K.D. McLachlan (ed.) Sulfur in Australian agriculture. Sydney Univ. Press, Sydney, Australia.

Williams, C.H., and A. Steinbergs. 1964. The evaluation of plant-available sulphur in soils. II. The availability of adsorbed and insoluble sulphates. Plant Soil 21:50–62.

2

Sulfur Nutrition of Crops in the Indo-Gangetic Plains of South Asia

M. P. S. Khurana, U. S. Sadana, and Bijay-Singh
Punjab Agricultural University, Ludhiana, India

Abstract

Sulfur is emerging as a major plant nutrient for crops grown in the Indo-Gangetic Plains spread over 13 million hectares in Pakistan, India, Nepal, and Bangladesh. The extent of sulfur deficiency in soils in the region is continuously increasing with the adoption of high-yield cultivars of rice (*Oryza sativa* L.), wheat (*Triticum aestivum* L.), maize (*Zea mays* L.), oilseeds, and pulses and because of the increased use of fertilizers lacking sulfur. Sufficient evidence is available from hundreds of experiments that sulfur fertilization of crops needs to be an integral component of balanced nutrition for producing optimum yield and quality of crops. Each kilogram of sulfur applied to oilseed crops could increase the production of edible oil by 3.0 to 3.5 kg, suggesting that sulfur is a master nutrient in oil production. With an integral role of sulfur in protein and oil production, sulfur nutrition of crops has a greater bearing on the nutrition of the vegetarian population in South Asia. Critical sulfur content in most crop plants in distinguishing sulfur deficiency falls between 0.20 and 0.25% sulfur. Sulfur can be supplied to crops through gypsum and fertilizers such as ammonium sulfate, potassium sulfate, or single superphosphate during seeding of crops or as pyrites and elemental sulfur 3 to 4 wk before seeding of crops. Substantial amounts of sulfur also become available to crop plants through irrigation with underground waters and recycling of organic manures, including green manures. Application of sulfur to a crop leaves a significant residual effect for subsequent crops grown in the rotation.

Among different regions in the world, Asia represents the region with the highest sulfur fertilizer requirement. Currently, India and China account for about 60% of the total estimated sulfur-deficit areas in Asia. Possibly, continuous mining of sulfur from soils has led to widespread sulfur deficiency and negative soil budget in India. For instance, of the total 400 districts of India, more than 200 districts have varying proportions of sulfur-deficient soils. In the mid-1990s, about 51 million hectares or about 30% of the total cultivated area in the country experienced varying degrees of sulfur deficiency (Tandon, 1995). Extensive soil surveys in India have revealed that sulfur deficiency varies from 5 to 83% with an overall average of 41% (Singh, 2001). Most of the alluvial soils of the Indo-Gangetic Plains (IGP) were found deficient with respect to plant available sulfur. Similarly, deficiencies of sulfur in rice-growing areas of Bangladesh have been

---

Copyright © 2008. American Society of Agronomy, Crop Science Society of America, Soil Science Society of America, 677 S. Segoe Rd., Madison, WI 53711, USA. *Sulfur: A Missing Link between Soils, Crops, and Nutrition.* Agronomy Monograph 50.

well documented (Sakai, 1980). It was estimated that about 2.8 million hectares under rice suffered from sulfur deficiency.

The deficiency of sulfur in soils in the IGP of South Asia is attributed to several factors (Pasricha and Aulakh, 1986, 1991; Tandon, 1991).

1. Adoption of high-yielding crop cultivars during the 1970s removed large quantities of nutrients in the harvested crop resulting in mining of nutrients from the soil, particularly those not applied in required quantities through fertilizers and manures.
2. Increased cropping intensity.
3. Decline in the use of sulfur-containing fertilizers like ammonium sulfate and single superphosphate, pesticides, and other agrochemicals.
4. Increased use of high-analysis sulfur-free fertilizers like urea and diammonium phosphate for supplying nitrogen and phosphorus to crops.
5. The reduced recycling of sulfur contained in crop residues and organic manures. In industrialized nations, annual sulfur deposition is between 3.7 to 25.2 kg sulfur ha^{-1} (Pasricha and Fox, 1993). In India, sulfate concentration in rainwater is usually less than 0.1 mg L^{-1}, which indicates small additions of sulfur from air pollution and acid rain (Singh, 1999).

Total removal of sulfur at the present level of crop production in India is more than 1.3 million Mg sulfur yr^{-1} (Aulakh, 2003). With total application of fertilizer sulfur stagnant around 0.6 million Mg sulfur yr^{-1} since the mid-1990s (FAI, 2006), sulfur mining of soils in India is expected to increase in the future. The picture in the IGP in India, Pakistan, and Bangladesh is worse since more than a1000 million people live in this region and there is very high cropping intensity. During the last three decades, a large number of investigations have been performed in the IGP of South Asia to study the role of sulfur in sustaining high production and quality levels of crops. This chapter provides an account of these studies.

The Indo-Gangetic Plains in South Asia

The IGP are spread over a vast area spanning from Punjab in Pakistan in the west to the Brahmputra floodplains of Bangladesh in the east. More than 85% of the rice–wheat systems practiced in South Asia are located in the IGP. It is a relatively homogeneous ecological region in terms of vegetation but can be subdivided into five broad transects based primarily on physiography, bioclimate, and social factors (Fig. 2–1) (Gupta et al., 2002). The trans-IGP (Transect 1 and 2) occupy large areas of Punjab in Pakistan and Punjab and Haryana in India. Upper and middle IGP (Transect 3 and 4) comprise of areas in the in west-central and eastern Uttar Pradesh, Bihar, and the Tarai in Uttaranchal in India and in Nepal. The lower parts of the IGP (Transect 5) are located in West Bengal in India and parts of Bangladesh.

The IGP have a continental monsoonal climate. There are wide variations in soil types that are generally coarser in the trans- and upper IGP and finer with the run of the river systems. Soils are primarily calcareous and micaceous alluviums with sandy loam to loam in texture in the upper reaches and to finer textured in the distal plains close to the mouth of the river systems (Gupta et al., 2002). Most soils in the IGP are deficient in nitrogen and phosphorus. Iron and

Fig. 2–1. The Indo-Gangetic plains in South Asia.

zinc deficiencies are also common. Sulfur and boron deficiencies are increasingly being reported (Velayutham et al., 1999).

Some farmers in the IGP grow mungbean (*Phaseolus aureus* Roxb.) during the transition phase between wheat and rice. To overcome herbicide resistance to isoproturon in *Phalaris minor* Retz., farmers replace wheat with Indian-mustard [*Brassica juncea* (L.) Czern. or *napus* L.], sugarcane (*Saccharum officinarum* L.), or berseem (*Trifolium alexandrinum* L.). In India and Pakistan, farmers intercrop or mix crop mustard with wheat (Hobbs et al., 1985). Many Nepalese farmers grow soybean [*Glycine max* (L.) Merr.] on the peripheral bunds of rice fields. In the middle IGP, many farmers grow three rice crops in a year. In this region, farmers also diversify the rice–wheat system more to cover risks of drought- and flood-prone agriculture. Many farmers replace wheat by oilseeds (*Brassica juncea* or *napus*), pulses (pea, *Pisum sativum* L.), grass pea (*Lathyrus* spp.), chickpea (*Cicer arietinum* L.), lentil (*Lens culinaris* Medikus), potato (*Solanum tuberosum* L.), or sugarcane, and occasionally rice by pigeonpea [*Cajanus cajan* (L.) Huth], maize, sunflower (*Helianthus annuus* L.), soybean, and sorghum [*Sorghum bicolor* (L.) Moench].

Sulfur Concentration in Plant Tissue and Uptake by Crops

Sulfur content in most plants is between 0.1 to 0.3%; although levels as high as 2% sulfur in leaves has been recorded. Roots invariably have the lowest amounts. Sulfur concentration among different plants and within different parts of the same plant varies widely. Generally, grains contain more sulfur than does straw. The crucifers show wider differences, whereas legumes show smaller differences in sulfur content in grain and straw. Sulfur content of cereal plants ranges from 0.16 to 0.25%; less than 0.2% is considered to be suboptimal (Aulakh et al., 1985).

Table 2–1. Uptake of sulfur vis-à-vis nitrogen, phosphorus, and potassium by different crops grown in the Indo-Gangetic plains of South Asia.

Crop	Yield	Total nutrient uptake†			
		S	N	P	K
Cereals			kg ha^{-1}		
Wheat	3900	12	137	26	137
Rice	2682	7	56	12	59
Maize	2132	7	61	10	51
Pulses					
Chickpea	1500	13	91	6	49
Lentil	2000	6	114	13	36
Mungbean	870	7	82	13	90
Blackgram (*Phaseolus mungo* L.)	600	7	45	8	90
Pigeonpea	1200	9	85	8	16
Oilseeds					
Rapeseed–mustard	2596	45	131	25	133
Groundnut	1900	15	121	19	43
Sunflower	2380	17	114	26	141
Sesame	1200	14	62	24	64
Other crops					
Sugarcane	87600	26	180	29	270
Forage grasses	19970	39	187	29	284

† Nutrient uptake includes grain, seed, tuber or cane, straw, and leaves. Source: Tandon and Messick (2002).

The average sulfur content in oilseed and leguminous crops is as follows: cruciferous oilseeds—1.19% in seeds and 0.13% in straw; sunflower and sesame—0.34% in seeds and 0.22% in straw; and legumes—0.24% in seeds and 0.20% in straw. A high sulfur requirement is also characteristic of protein rich crops (such as legumes), crucifers, and *Brassica*. The sulfur requirement of rapeseed–mustard crops (*Brassica campestris* L., *B. juncea, B. napus,* and *B. carinata* A. Braun) is nearly three times that of cereals (Aulakh et al., 1985). In plant species capable of synthesizing oils, sulfur is stored as organic sulfur rather than sulfate sulfur. This explains the relationship between sulfur supply and oil content.

Crucifers (cabbage—*B. oleracea* L., radish—*Raphanus sativus* L, rapeseed–mustard), legumes (alfalfa—*Medicago sativa* L. subsp. *sativa.*, soybean, groundnut—*Arachis hypogaea* L., onion—*Allium cepa* L., garlic—*A. sativum* L., cotton–*Gossypium* spp., sugarcane, and maize require large amounts of sulfur; whereas, cereals require relatively smaller amounts of the nutrient. For normal yields, the crops with high sulfur requirements need 20 to 45 kg sulfur ha^{-1}. Crops with medium sulfur requirements need 15 to 35 kg sulfur ha^{-1} (Aulakh et al., 1985). Uptake of sulfur for different crops grown in the IGP are listed in Table 2–1.

Sulfur uptake is generally 9 to 15% of the nitrogen uptake; although, it can range from 5to 30% (Tandon, 1991). In crucifers like mustard, sulfur uptake is among the highest and can approach one third of nitrogen uptake. As a rule of thumb, crops absorb nearly as much sulfur as phosphate. The sulfur uptake for 1 Mg grain production can be (Tandon and Messick, 2002) 3 to 4 kg for cereals (range 1–6), 8 kg for pulses (range 5–13), and 12 kg for oil seed (range 5–20). High crop yield may lead to high sulfur requirement of the crops. For example, sulfur content of maize varies from 0.034% in a sulfur-deficient soil to 0.16% under a sulfur sufficient condition. Similarly, sulfur requirements of a rice crop may vary from as low as 0.26 kg sulfur ha^{-1} to as high as 12.8 kg sulfur ha^{-1} (Aulakh et al., 1985).

Response of Crops to Sulfur Application

Application of sulfur and the associated response of crop plants has not received adequate attention, primarily because of substantial applications along with phosphatic fertilizers and through several other sources. In the IGP, phosphate needs of crops during the 1970s and 1980s were primarily met through applications of single superphosphate so that sulfur deficiencies were overlooked even though adequate amounts were not applied. During the late 1980s and later, a number of experiments were conducted to study the response of cereals, pulses, oilseeds, vegetables, forages, and other crops to sulfur application. The results have been reported in several reviews (Aulakh, 2003; Aulakh and Pasricha, 1986, 1988, 1997; Biswas and Tewatia, 1991; Aulakh and Chhibba, 1992; Pasricha and Aulakh, 1997; Tandon, 1991, 1995; Singh, 1999, 2001). According to Aulakh (2003), all type of crops in South Asia including cereals, pulses, oilseeds, forage crops, and vegetables responded to sulfur applications with yield increases ranging from 14 to 74%. Crop-wise average yield responses of different crops to application of sulfur in a large number of experiments conducted in the IGP as compiled by Tandon (1991) are listed in Table 2–2. Mean increase in yield due to sulfur application for cereals, pulses, and oilseed crops were in the range of 739 to 813, 137 to 340, and 144 to 560 kg ha^{-1}, respectively. For rice, wheat, and rapeseed–mustard for which considerable data are available, the percentage increase in average yield due to sulfur application was 17.1, 25.3, 30, and 31.7, respectively. Biswas and

Table 2–2. Ranges and mean crop yield increases due to sulfur application in different studies conducted under field conditions in India.

Crop	Number of studies	Yield without S application	Yield increase due to S			
			Range	Mean	Range	Mean
			kg ha^{-1}		%	
Wheat	32	3209	150–2120	813	4.5–109.5	25.3
Rice	27	4389	56–1720	752	0.7–39.5	17.1
Maize	3	1806	346–1511	739	16.6–60.4	40.9
Blackgram	7	787	130–200	153	17.0–26.3	19.9
Green gram	7	804	100–234	137	13.0–72.5	20.2
Chickpea	4	1851	315–513	340	15.9–29.6	18.4
Groundnut	23	1785	133–1480	566	8.2–106.6	31.7
Rapeseed–mustard	18	1122	83–839	335	10.1–92.8	30.0
Soybean	8	1426	202–698	361	14.2–35.6	25.3
Sunflower	6	1233	70–410	249	5.8–29.7	20.2
Sesame	3	674	50–279	144	9.8–32.7	21.4
Linseed	5	1571	47–459	246	3.2–31.3	15.7
Potato	3	14567	1661–4281	3080	8.1–63.9	21.1
Berseem†	3	12633	3250–7575	4562	19.6–49.7	36.1
Sugarcane	2	58750	10192–32292	21242	19.6–49.3	36.1
Onion	3	2480	80–1210	480	2.0–41.0	19.0

† *Trifolium alexandrinum*. Source: Tandon (1991).

Table 2–3. Summary of yield increases due to sulfur application in FAO sulfur network experiments in India.†

Crop	Number of studies	Yield with no S	Percentage increase due to application of			Response per kg S at 30 kg S ha^{-1}
			10 kg S ha^{-1}	20 kg S ha^{-1}	30 kg S ha^{-1}	
		kg ha^{-1}		%		kg
Wheat	20	3301	10.6	19.5	25.5	28.0
Rice	13	5617	3.8	8.2	15.2	23.3
Rapeseed–mustard	6	774	7.8	18.2	25.7	6.6
Groundnut	5	2224	6.5	22.4	26.2	19.4
Soybean	5	1244	15.8	25.6	28.1	11.6

† Source: Biswas and Tewatia (1991).

Tewatia (1991) documented results of FAO sulfur trial network in which responses of wheat, rice, groundnut, mustard, and soybean were studied at 125 sites in India, primarily in the IGP. The data from these experiments are summarized in Table 2–3. Average yield increase (kg ha^{-1}) due to applications of 30 kg sulfur ha^{-1} was 841 in wheat, 700 in rice, 582 in groundnut, and 109 kg in rapeseed–mustard.

Experiments conducted during the last decade continue to exhibit substantial responses of crops to applications of sulfur in the IGP. Singh (2001) compiled data from experiments conducted by Indian Council of Agricultural Research and found that the average response of sulfur application to different group of crops was low for pulses (357 kg ha^{-1}), moderate for oilseeds (570 kg ha^{-1}), and maximal for cereal crops (650 kg ha^{-1}). Application of 60 to 120 kg sulfur ha^{-1} increased the seed yield of chickpea by 323 to 839 kg ha^{-1} in alluvial soils of IGP (Singh, 2001). Khurana and Bansal (2007) studied the sulfur nutrition of a mungbean–raya (*Brassica juncea*) rotation and concluded that application of 20 kg sulfur ha^{-1} to each crop or 40 kg sulfur ha^{-1} to the first crop (mungbean) was adequate to obtain optimum yields of both the crops. In eight on-farm experiments conducted by Khurana et al. (2003) response of raya ranged from 4.5 to 28.5 kg grain kg^{-1} sulfur (mean 10.9 kg grain kg^{-1} S) when sulfur was applied at the rate of 20 kg sulfur ha^{-1}. In seven on-farm experiments with gobhi sarson (*Brassica napus*) conducted with a similar set up in the northwestern India, the responses ranged from 8.8 to 29.9 kg grain kg^{-1} sulfur with a mean value of 19.7 kg grain kg^{-1} sulfur. Significant yield responses of lentil to applications of sulfur up to 20 kg ha^{-1} were also reported by Khurana et al. (2002). Maity and Giri (2003) observed a 27% increase in pod yield of groundnut because of application of 30 kg sulfur ha^{-1}. Significant increases in yield of chickpea, mungbean, and mustard because of applications of sulfur have been reported by Rana and Rana (2003), Sital et al. (2007), and Singh and Meena (2004), respectively. Average responses of different crops to sulfur application based on several experiments conducted in the region are listed in Table 2–4.

Hoque and Hobbs (1980) reported that in Bangladesh, an average application of 34 kg sulfur ha^{-1} of ammonium sulfate increased the yield of rice by 100 to 1300 kg ha^{-1} and on farmers' fields by 300 to 2200 kg ha^{-1} over and above the yield obtained due to application of 60 kg nitrogen ha^{-1}. According to Frederick (1983), sulfur deficiency was increasingly recognized as a factor that limits rice

Table 2–4. Average crop responses to application of sulfur.†

Crop	Number of experiments	Average application rate	Response to S
		kg S ha^{-1}	kg grain kg^{-1} S
Wheat	35	39	19.0
Rice	23	42	16.6
Chickpea	6	85	5.3
Blackgram	9	30	5.4
Green gram	6	40	3.3
Groundnut	29	34	13.3 (pods)
Rapeseed–mustard	29	43	7.7
Sunflower	10	24	13.0
Linseed	5	26	9.5

† Source: Tandon and Messick, 2002.

production in Bangladesh. In 51 field trials using gypsum as a source of sulfur, rice yields increased by 860 kg ha^{-1}. In another set of 25 trials, the increase in yield ranged from 220 to 4200 kg ha^{-1} with a mean increase of 1120 kg ha^{-1}. Extent of response of wheat to applications of sulfur was similar to that of rice.

Sulfur Nutrition and Quality Aspects of Crops

Sulfur nutrition influences protein and amino acid contents in different crops and determines the oil content in oilseed crops. Because of its impact on protein yield and quality, sulfur deficiency in crop plants has serious implications on human nutrition. Das et al. (1975) observed that sulfur application influenced the content of essential amino acids and sulfur-containing amino acids in the grains of maize, wheat, and rice and thus helped in maintaining the quality of these cereals. It decreased methionine content in wheat, increased it in maize, and showed a variable response in rice. A significant increase in oil, protein, and methionine content in groundnut and mustard because of an application of sulfur was observed by Singh et al. (1970). Beneficial effect of sulfur application on protein as well as on yield of some pulses has been documented by Aulakh and Sharma (2005). Data reported in Tables 2–5 and 2–6 suggest that sulfur nutrition of crops has a significant bearing on the vegetarian population in the IGP of South Asia.

Applications of sulfur with balanced amounts of other nutrients increased the oil content in different oilseed crops. The increase was about 9% in groundnut, 5 to 6% in rapeseed–mustard, and 2% in linseed (*Linum usitatissimum* L.) and soybean. According to Aulakh (2003), increased oil content along with yield resulted in an overall oil yield increase from 15 to 30%. On the basis of results of several reports from India, average increase in the oil content of oilseeds because of sulfur application ranged from as low as 3.8% in sunflower to as high as 11.3% in groundnut (Fig. 2–2) (Tandon and Messick, 2002). Each kilogram of sulfur applied could increase the production of edible oil by 3.0 to 3.5 kg, suggesting that sulfur is a master nutrient in oil production. Applications of sulfur significantly improve various parameters of quality with in plants. In sesame (*Sesamum indicum* L.)–mustard, the application of 60 kg sulfur ha^{-1} resulted in an 11% increase in protein of sesame, as well as oil content, and a 5% increase in protein and 11%

Table 2–5. Increase in protein content of different crops due to sulfur application.

Crop	Protein content		Percentage increase	Reference
	−S	+S		
	%			
Chickpea	23.5	24.5	4	Tiwari (1989)
Green gram	24.4	26.3	8	Kamat et al. (1981)
Soybean	28.1	31.9	14	Aulakh et al. (1990)
Sunflower	13.9	16.6	19	Gangadharan et al. (1990)
Rapeseed–mustard	22.9	36.8	34	Pasricha et al. (1987)
Toria (*Brassica campestris*)	18.1	19.8	9	Singh and Ganga Saran (1987)

Table 2–6. Effect of sulfur application on the content of protein and sulfur containing amino acids in mungbean. †

Applied sulfur	Protein	Methionine	Cysteine
kg ha^{-1}	%	g 100 g^{-1} protein	
0	24.4	2.42	2.63
10	24.5	2.45	3.25
20	24.8	3.02	4.58
30	26.3	3.13	3.80

† Source: Kamat et al. (1981).

in oil content in the seeds of succeeding mustard crop (Singh and Tiwari, 1985). Aulakh et al. (1989) found that sulfur fertilization of linseed with an adequate supply of nitrogen and phosphorus accelerated the metabolic pathway of linolenic acid synthesis and resulted in a large decrease in the percentage of stearic, oleic, and linoleic acids with concurrent increase in the content of linolenic acid. Linseed oil with high linolenic acid and low oleic acid can be used in the manufacture of paints, oil cloths, and linoleum.

Managing Sulfur in Crops and Cropping Systems

Critical Values of Sulfur in Plants

Critical levels of sulfur in selected plant parts are used for diagnosing deficiency of sulfur in crops. The critical values of sulfur in some crops at different

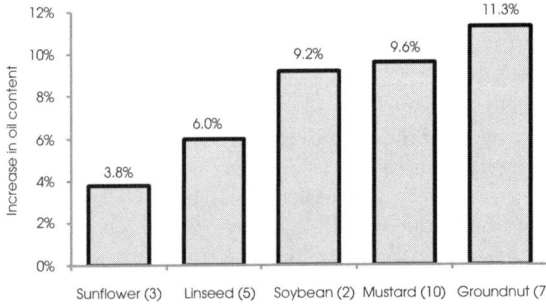

Fig. 2–2. Effect of sulfur application on increase in oil content of different oilseed crops in India. Data averaged for number of studies indicated in parentheses (Data source: Tandon and Messick, 2002).

Table 2–7. Critical values of S content in different plant parts and at different growth stages of crops grown in the Indo-Gangetic plain of South Asia for distinguishing S deficiency.

Crop	Critical level	Reference
Rice	0.16% S at tillering	Pillai and Singh (1975)
Wheat	0.20% S	Cheema and Arora (1984)
Wheat	0.22% S, N/S ratio of 16	Shinde (1988)
Maize	N/S ratio of 11 at 44 d	Shinde (1988)
Mungbean	0.23% S in whole shoot at 45 d	Khurana and Bansal (2007)
Rapeseed–mustard	0.18% S in whole shoot at 45 d, 0.39% S in grain	Khurana et al. (1995)
	0.26% S in whole shoot at 45 d growth	Khurana et al. (1998)
	0.21% S, N/S ratio of 15.5	Pasricha et al. (1988)
	0.55% S at 30 d, 0.45% S at 60 d	Shinde (1988)
Groundnut	0.20% S in plant	Cheema and Arora (1984)
Raya	0.37% S in whole shoot at 45 d	Khurana and Bansal (2007)
Soybean	0.14% S in tissue at 50 d	Shinde (1988)
Sunflower	0.24% S at pre bloom	Marok and Dev (1979)
	0.17–0.20% S in leaves or 0.20–0.21% S in flowers at 75% bloom stage	Gangwar and Parameswaran (1976)
Alfalfa (*Medicago sativa* L.)	0.16% at first cut, 0.21% S at second cut, N/S ratio of 16	Bansal et al. (1979)

stages of growth are given in Table 2–7. It is interesting to note that most of the values of critical limits fall within range of 0.20 to 0.25% sulfur irrespective of the crop. The wheat plant containing less than 0.2% sulfur is considered deficient, while 0.3 to 0.4% sulfur in the top leaves at ear emergence stage was found to be optimum in the Punjab in trans-IGP (Cheema and Arora, 1984). Bahl et al. (1990) observed that the sulfur concentration in plants of different cultivars of rape seed–mustard ranged between 0.30 to 0.40% sulfur at 30 d but was found to be adequate for obtaining 90% of the potential yield. But critical levels of sulfur at 70-d growth ranged between 0.11 to 0.18% sulfur, suggesting that some cultivars had higher sulfur requirements during earlier periods. Aulakh et al. (1977) reported that to get high yields of rapeseed–mustard, plants must contain 3.7% nitrogen and 0.27% sulfur at the preflowering stage. In Table 2–8, Singh (1999) has

Table 2–8. Deficiency and sufficiency levels of sulfur in 45- to 55-d-old groups of crop plants.†

Crops	S concentration in dry matter		
	Deficient	Moderately sufficient	Sufficient
	——————— % ———————		
Wheat, rice, maize, millets	0.10–0.20	0.20–0.30	>0.30
Groundnut, mustard, soybean, cowpea	0.10–0.25	0.25–0.40	>0.40
Sunflower, linseed, brinjal (*Solanum melongena* L.), French bean (*Phaseolus coccineus* L.), cucumber (*Cucumis sativus* L.)	0.25–0.35	0.35–0.55	>0.55
Chickpea, pea	0.15–0.45	0.45–0.75	>0.75
Potato, cauliflower (*Brassica oleracea* L.), spinach (*Spinacia oleracea* L.)	0.30–0.40	0.40–0.75	>0.75

† Source: Singh (1999).

summarized the critical levels of sulfur concentration at 45 to 55 d of growth in different groups of crops given ranges of deficiency and sufficiency.

Dose of Sulfur

With a gradual decrease in the use of sulfur-containing fertilizers like ammonium sulfate and single superphosphate in the IGP, precise recommendations for applications of sulfur to different crops are being developed for most of the crops showing response to sulfur. Under most field conditions, 20 to 40 kg sulfur ha^{-1} is suggested (Tandon and Messick, 2002). For different groups of crops the sulfur application rates are as follows: (i) cereals (wheat, maize, rice)—24 to 40 kg sulfur ha^{-1}; (ii) pulses (chickpea, lentil, green gram, blackgram)—10 to 40 kg sulfur ha^{-1}; (iii) oilseeds (groundnut, rapeseed–mustard, sunflower)—20 to 50 kg sulfur ha^{-1}; and (iv) fodders (grasses, sorghum, legumes)—25 to 50 kg sulfur ha^{-1}

Sulfur Sources and Management

In the IGP, sulfur is being supplied to crop plants from a variety of sources. These include chemical fertilizers such as gypsum, pyrites and elemental sulfur, and fertilizers supplying sulfur along with nitrogen, phosphorus, and potassium. Important ones in the latter category are single superphosphate, ammonium sulfate, potassium sulfate, and sodium sulfate. In general, different sources of sulfur are comparable. Tandon (1995) has summarized important management aspects of sulfur sources in Table 2–9.

Besides inorganic sources of sulfur, significant amounts of sulfur are being recycled through irrigation in ground waters and through application of organic manures, including green manure with leguminous crops. Cheema and Arora (1984) and Singh (1987) analyzed samples of ground water used for irrigation in two districts of northwestern India. On the basis of their sulfur content, they have observed that 1 to 6 kg sulfur ha^{-1} was added with every irrigation. Commonly grown leguminous green manure crops like sesban [*Sesbania sesban* (L.) Merr.], cowpea [*Vigna unguiculata* (L.) Walp.], and cluster bean (*Cyamopsis psoraloides* DC.) accumulate 10 to 15 kg sulfur ha^{-1} during 40 to 50 d growth (Aulakh, 2003) and when incorporated into the soil can prove to be valuable sources of sulfur for crops like wheat and rapeseed mustard. Using ^{35}S-labeled *Sesbania aculeata*

Table 2–9. Management aspects of sulfur containing fertilizers.†

Sulfur source	Management aspect
Ammonium sulfate	Supplies both nitrogen and sulfur, suitable for top dressing to correct sulfur-deficiency
Single superphosphate	Supplies sulfur along with phosphorus and calcium. Applied at planting in bands or mixed in soil after broadcast
Potassium sulfate	Supplies both potassium and sulfur and is applied to crops sensitive to chloride
Ammonium phosphate sulfate	Used for basal application of nitrogen, phosphorus, and sulfur
Elemental sulfur	Particularly suitable for fine textured calcareous soils. Applied 3–4 wk before planting in moist and aerated soils. Products with small particle size lead to improved availability of sulfur to plants
Iron pyrites	Suitable for alkaline soils. Applied 3–4 wk before planting in moist and aerated soils.
Gypsum	Suitable for crops which need both sulfur and calcium

† Adapted from Tandon and Messick (2002).

Poiret green manure, Dhillon et al. (2001) showed that sulfur contained in the green manure was released rapidly into plant available forms.

Time of Fertilizer Sulfur Application

Sulfur-containing fertilizers should generally be applied at the time of seeding of crops. Since physiological functions such as enzyme activities and photosynthesis, which are severely inhibited by sulfur deficiency, are restored to normal rates within 5 d of transfer of sulfur from soil solution (Friederich and Schrader, 1978). It was found that if application of sulfur was missed at seeding, this can still be corrected by an application during early-growth stages. Pasricha and Aulakh (1986) observed that application of sulfur at planting, pegging, or pod formation was equally effective in obtaining yield response of groundnut. Better performance of groundnut when gypsum was applied as a source of sulfur in split doses rather than full dose at seeding, could be due to combined advantage of providing sulfur and calcium during flowering and fruiting of groundnut (Chahal and Virmani, 1973; Dwivedi, 1981; Ahmad et al., 1999). As described in Table 2–9, when sulfur is applied as sources such as elemental sulfur or iron pyrites, these should be applied 3 to 4 wk ahead of seeding of crops.

Method of Application of Sulfur Fertilizers

Ideally sulfate-sulfur sources can be band placed or drilled in to soil at 10- to 15-depth at the time of crop seeding. When applied during crop growth, these can be surface broadcasted. However, pyrites or elemental sulfur should be broadcast and mixed with the soil because it ensures contact with a large volume of soil.

Management of Sulfur in Cropping Systems

Sulfur applications can benefit more than one crop grown in a sequence and produce a significant residual response. In FAO sulfur network trials conducted at several locations in India, spectacular direct and residual responses to sulfur fertilization were observed in different cropping systems (Table 2–10). It seems that crops responded equally to direct application and residual effects of sulfur. It has been observed that crops directly fertilized contributed 33 to 82% to the rotational response, while crops raised on residual contributed 18 to 67%.

Table 2–10. Direct and residual responses to sulfur in the FAO sulfur networks trials conducted in India.†

Cropping system	Experiments averaged	Rotational response	
		Direct	Residual
		kg ha^{-1} yr^{-1}	—— % ——
Rice–wheat	3	968	77	23
Rice–mustard	2	1066	82	18
Groundnut–wheat	3	1186	56	44
Groundnut–groundnut	1	299	47	53
Wheat–rice	1	799	33	67
Wheat–maize	1	756	36	64
Wheat–groundnut	1	933	80	20

† Source: Biswas and Tewatia (1991).

Aulakh and Pasricha (1979) reported that the residual effect of 40 kg sulfur ha^{-1} applied either to chickpea and lentil crops almost doubled the grain yield of the following crop of green gram [*Vigna radiata* (L.) R. Wilczek]. In wheat–groundnut cropping system in alluvial soils in the middle IGP, wheat following groundnut benefited more from residual sulfur (22% yield increase) than did wheat after rice, where the yield increase was 7% (Tiwari 1989). In the same cropping system in the trans-IGP, marked direct and residual effects were registered irrespective of the crop to which fertilizer sulfur was directly applied (Aulakh and Pasricha 1986). Sakal et al. (2001) indicated that sulfur application exerted its effect up to the fourth crop in rice–wheat cropping system being practiced on a ustifluvent.

References

Ahmad, A., Y.P. Abrol, and M.Z. Abdin. 1999. Effect of split application of sulphur and nitrogen on growth and yield attributes of brassica genotypes differing in their time of flowering. Can. J. Plant Sci. 79:175–180.

Aulakh, M.S. 2003. Crop responses to sulphur nutrition. p. 341–358. *In* Y.P. Abril and A. Ahmed (ed.) Sulphur in plants. Kluwer Academic Publishers, London.

Aulakh, M.S., and I.M. Chhibba. 1992. Sulphur in soils and responses of crops to its application in Punjab. Fert. News 37(9):33–45.

Aulakh, M.S., and N.S. Pasricha. 1979. Responses of gram (*Cicer arietinum* L.) and lentil (*Lens cultinaris* L.) to phosphorus as influenced by applied sulphur and its residual effect on moong (*Phaseolus aureus* L.). Bull. Indian Soc. Soil Sci. 12:433–438.

Aulakh, M.S., and N.S. Pasricha. 1986. Role of sulphur in the production of the grain legumes. Fert. News 31(9):31–35.

Aulakh, M.S., and N.S. Pasricha. 1988. Sulphur fertilization of oilseeds for yield and quality. Section II.3, p. 1–14. *In* Sulphur in Indian agriculture. The Sulphur Institute, Washington, DC, and Fertiliser Association of India, New Delhi, India.

Aulakh, M.S., and N.S. Pasricha. 1997. Role of balanced fertilization in oilseed based cropping systems. Fert. News 42(4):101–111.

Aulakh, M.S., N.S. Pasricha, and A.S. Azad. 1990. Phosphorus-sulphur interrelationships for soybeans on P and sulfur deficient soil. Soil Sci. 150:705–709.

Aulakh, M.S., N.S. Pasricha, A.S. Azad, and K.L. Ahuja. 1989. Response of linseed (*Linum usitatissimum* L.) to fertilizer nitrogen, phosphorus, and sulphur, and their effect on the removal of soil sulphur. Soil Use Manage. 5:194–198.

Aulakh, M.S., N.S. Pasricha, and G. Dev. 1977. Response of different crops to sulphur fertilization in Punjab. Fert. News 22(9):32–36.

Aulakh, M.S., and P. Sharma. 2005. Nutrient dynamics and their efficient use in pulse crops. p. 241–278. *In* Gurigbal Singh et al. (ed.) Pulses. Agrotech Publishing Academy, Udaipur, India.

Aulakh, M.S., B.S. Sidhu, B.R. Arora, and B. Singh. 1985. Content and uptake of nutrients by pulses and oilseed crops. Indian J. Ecol. 12:238–242.

Bahl, G.S., N.S. Pasricha, and H.S. Baddesha. 1990. Sulphur nutrition of three varieties of Indian mustard (*Brassica juncea*) at three stages of growth. Indian J. Agric Sci. 60:556–558.

Bansal, K.N., D.N. Sharma, and D. Singh. 1979. Evaluation of some soil test methods for measuring available sulfur in alluvial soils of Madhya Pradesh. J. Indian Soc. Soil Sci. 27:308–313.

Biswas, B.C., and R.K. Tewatia. 1991. Results of FAO sulphur trials network in India. Fert. News 36(4):11–35.

Chahal, R.S., and S.M. Virmani. 1973. Preliminary study on the effect of time of application of superphosphate on the yield and nutrient uptake by groundnut. Indian J. Agric. Sci. 43:731–733.

Cheema, H.S., and C.L. Arora. 1984. Sulphur status of soils, tubewell waters and plants in some areas of Ludhiana under groundnut-wheat cropping system. Fert. News 29:28–31.

Das, S.K., P. Chhabra, S.R. Chatterjee, Y.P. Abrol, and D.L. Deb. 1975. Influence of sulphur fertilization on yield of maize and protein quality of cereals. Fert. News 20(3):30–32.

Dhillon, S.K., K.S. Dhillon, and Bijay-Singh. 2001. Dynamics of sulphur in a calcareous loam soil amended with Sulphur-35 labelled *Sesbania aculeata* green manure. J. Indian Soc. Soil Sci. 49:259–265.

Dwivedi, R.S. 1981. Fertilizer use in groundnut-based cropping systems under different agro-climatic conditions. Fert. News 26(10):28–34.

FAI. 2006. Fertilizer statistics 2005–06. Fertilizer Association of India, New Delhi, India.

Frederick, M.T. 1983. Review of alternatives and recommendations for using phosphogypsum as an agricultural sulfur source for Bangladesh Agricultural Development Corporation. International Fertilizer Development Centre, Muscle Shoals, AL.

Friederich, J.W., and L.E. Schrader. 1978. Sulphur deprivation and nitrogen metabolism in maize seedlings. Plant Physiol. 61:900–903.

Gangadharan, G.A., M.H. Manjunathaiah, and T. Satyanaryana. 1990. Effect of sulphur on yield, oil content of sunflower and uptake of micronutrient by plants J. Indian Soc. Soil Sci. 39:692–695.

Gangwar, M.S., and P.M. Parameswaran. 1976. Phosphorus-sulphur relationship in sunflower. II. Studies on P and sulfur nutrition. Oilseed J. 6:33–37.

Gupta, R.K., R.K. Naresh, P.R. Hobbs, J. Zheng, and J.K. Ladha. 2002. Sustainability of post-green revolution agriculture: The rice–wheat cropping system of the Indo-Gangetic plains and China. p. 1–25. *In* J.K. Ladha et al. (ed.) Improving the productivity and sustainability of rice-wheat systems: Issues and impacts. ASA Spec. Publ. 65. ASA, Madison, WI.

Hobbs, P.R., A. Razzaq, N.I. Hashmi, M. Munir, and B.R. Khan. 1985. Effect of mustard grown as a mixed or intercrop on the yield of wheat. Pakistan J. Agric. Res. 6:241–247.

Hoque, M.Z., and P.R. Hobbs. 1980. Response of rice crop to added sulphur at BRRI station and nearby project area. p. 15–19. *In* Proceedings of the Workshop on Sulphur Nutrition in Rice. Publication No. 41. Bangladesh Rice Research Institute, Dhaka, Bangladesh.

Kamat, V.N., V.G. Kankute, R.B. Puranik, W.S. Kohadkar, and R.P. Joshi. 1981. Effect of sulfur and Mo application on yield, protein and S-amino acid contents of green gram. J. Indian Soc. Soil Sci. 29:225–227.

Khurana, M.P.S., and R.L. Bansal. 2007. Sulphur management in moong (*Phaseolus aureus* L.) and raya (*Brassica juncea* L.) crops. Acta Argon. Hung. 55(4):437–445.

Khurana, M.P.S., R.L. Bansal, and V.K. Nayyar. 2002. Effect of sulphur fertilization on yield and uptake of sulphur and micronutrient cations by lentil in alluvium derived soil. Ann. Agric. Res. New Ser. 23:244–247.

Khurana, M.P.S., N.S. Dhillon, and V.K. Nayyar. 1995. Critical level of sulphur deficiency and response of Indian mustard (*Brassica juncea*) to sulphur application in alluvial soil. Indian J. Agric. Sci. 65:528–530.

Khurana, M.P.S., V.K. Nayyar, B.S. Sidhu, and M.S. Gill. 2003. Response of raya and gobhi sarson to sulphur in Punjab. Fert. News 48(7):39–41.

Khurana, M.P.S., V.K. Nayyar, and R.L. Bansal. 1998. Direct and residual effects of sulfur and Zn on yield and their uptake in an Indian mustard (*Brassica Juncea* L.)-maize (*Zea mays* L.) cropping system. Acta Agron. Hung. 46:327–334.

Maity, S.K., and G. Giri. 2003. Influence of phosphorus and sulphur fertilization on productivity and oil yield of groundnut (*Arachis hypogaea*) and sunflower (*Helianthus annuus*) in intercropping with simultaneous and staggered planting. Indian J. Agron. 48:262–270.

Marok, A.S., and G. Dev. 1979. Response of sunflower (*Helianthus annuus*) to sulfur application and evaluation of the sulphur status of soils. J. Nucl. Agric. Biol. 8:100–102.

Pasricha, N.S., and M.S. Aulakh. 1986. Role of sulfur in the nutrition of groundnut. Fert. News 31(9):17–21.

Pasricha, N.S., and M.S. Aulakh. 1991. Twenty years of sulphur research and oilseed production in Punjab, India. Sulphur Agric. 15:17–23.

Pasricha, N.S., and M.S. Aulakh. 1997. Sulphur- an emerging deficient nutrient. p. 265–275. *In* J.S. Kanwar and J.C. Katyal (ed.) Plant nutrient needs, supply, efficiency and policy issues: 2000–2025. National Academy of Agricultural Sciences, New Delhi, India.

Pasricha, N.S., M.S. Aulakh, G.S. Bahl, and H.S. Baddesha. 1987. Nutritional requirements of oilseed and pulse crops in Punjab. Res. Bull. No. 15. Punjab Agricultural University, Ludhiana, India.

Pasricha, N.S., M.S. Aulakh, G.S. Bahl, and H.S. Baddesha. 1988. Fertilizer use research in oilseed crops. Fert. News 33(9):15-22.

Pasricha, N.S., and R.L. Fox. 1993. Plant nutrient sulphur in the tropics and subtropics. Adv. Agron. 50:209–269.

Pillai, P.B., and H.G. Singh. 1975. Effect of different sources of sulfur and Fe on flag leaf composition and grain yield of rice on calcareous soils. Indian J. Agric. Sci. 45:340–343.

Rana, K.S., and D.S. Rana. 2003. Response of mustard (*Brassica juncea*) to nitrogen and sulphur under dryland conditions. Indian J. Agron. 48:217–219.

Sakai, H. 1980. Some analytical results of sulphur deficient plants, soil and water. p. 35–39. *In* Proceedings of the Workshop on Sulphur Nutrition in Rice. Publication No. 41. Bangladesh Rice Research Institute, Dhaka, Bangladesh.

Sakal, R., A.P. Singh, B.C. Choudhary, and B. Shahi. 2001. Sulphur status of ustifluvents and response of crops to sulphur application. Fert. News 46(10):61–65.

Shinde, D.A. 1988. Agronomy of sulphur in Madhya Pradesh (India). Agric. Rev. 9:125–152.

Singh, A., and N.L. Meena. 2004. Effect of nitrogen and sulphur on growth, yield and attributes and seed yield of mustard (*Brassica juncea*) in eastern plains of Rajasthan. Indian J. Agron. 49:186–188.

Singh, D. 1987. Sulphur indexing of soils and crops in areas under maize-wheat cropping system. M.Sc. Thesis, Punjab Agricultural University, Ludhiana, India.

Singh, M., and R.C. Tiwari. 1985. Response of oilseed crops to fertilizers in dryland agriculture. p. 147–159. *In* Proceedings of FAI-NR Seminar, Varanasi. Fertilizer Association of India, New Delhi.

Singh, M.V. 1999. Sulphur management for oilseed and pulse crops. Bull. No. 3. Indian Institute of Soil Science, Bhopal, India.

Singh, M.V. 2001. Importance of sulphur in balanced fertilizer use in India. Fert. News 46(10):13–18, 21–28, 31–35.

Singh, N., B.V. Subbiah, and V.P. Gupta. 1970. Effect of sulphur fertilization on the chemical composition of groundnut and mustard. Indian J. Agron. 15:24–28.

Singh, S., and Ganga Saran. 1987. Effect of sulfur and N on growth, yield quality and nutrient uptake of Indian rape. Indian J. Agron. 32:474–475.

Sital, J.S., H.S. Aulakh, P. Sharma, G. Singh, and H.S. Sekhon. 2007. Effect of rhizobium inoculation and sulphur nutrition on growth parameters of mungbean (*Vigna radiata* L. Wilczek). Indian J. Ecol. 34:40–43.

Tandon, H.L.S. 1991. Sulphur– Research and Agricultural Production in India. The Sulphur Institute, Washington, DC.

Tandon, H.L.S. (ed.). 1995. Sulphur fertilizer for Indian agriculture–A guidebook. Fertilizer Development and Consultation Organization, New Delhi, India.

Tandon, H.L.S., and D.L. Messick. 2002. Practical sulphur guide. The Sulphur Institute, Washington, DC.

Tiwari, K.N. 1989. Sulphur research and agriculture production in U.P. C.S. Azad University of Agriculture and Technology, Kanpur, India.

Velayutham, M., D.K. Mandal, C. Mandal, and J. Sehgal. 1999. Agro-ecological sub-regions of India for planning and development. NBBS Pub. 35. National Bureau of Soil Survey and Land Use Planning, Nagpur, India.

3

Soil Sulfur Cycling in Temperate Agricultural Systems

Jørgen Eriksen
University of Aarhus, Tjele, Denmark

Abstract

To avoid sulfur deficiency in agricultural crops, sulfur must be available in the required forms and quantities and in synchrony with plant demand. Soil sulfur exists in numerous forms, and its dynamics play an important role in the sulfur application to plants. Soil organic sulfur has been separated into broad—mostly chemically defined—fractions, reflecting land use and fertilizer practice, but these are of limited value for predicting plant availability. There are several reasons for this: (i) soil organic sulfur consists of a continuum of fractions with different timescales for mineralization; (ii) association to soil particles provides physical protection of soil organic sulfur against decomposition; and (iii) mineralization only constitutes 0.5 to 3% per year of the soil organic sulfur pool. The addition of organic material can build up soil organic sulfur and may contribute significantly to plant sulfur supply, depending on the carbon/sulfur (C/S) ratio of the added material. In animal manure, some of the sulfur is plant-available in the application year, but there is no indication that residual sulfur mineralizes more readily than the bulk of soil organic sulfur. For crops with a short growing season, mineralization of residual and soil organic sulfur may not be in synchrony with the demand and may even lead to increased leaching losses of sulfate on freely draining soils with high winter rainfall; although, some catch crops have demonstrated an ability to reduce sulfate leaching and increase synchrony with crop demand. The transient nature of plant-available sulfur makes soil sulfur testing a difficult task and often sulfur balance considerations provide a better background for fertilizer sulfur recommendations, keeping in mind that availability and synchrony is ignored using this approach.

To avoid or overcome the problems of sulfur deficiency in agriculture, sulfur must be available when plants need it and in the required form and quantity. In soil, sulfur exists in a variety of forms and their fluxes and balances play an important role for the sulfur fertilization of plants. The objective of this chapter is to create an overview of sulfur cycling in soil and its dependency on agricultural system and management in relation to plant availability in the short and long term.

Sulfur in Soil

Soil sulfur exists in organic and inorganic forms. From a plant nutritional viewpoint, inorganic sulfate is the most important, since this is the form assimilated by plant roots. However, sulfate—which is the stable form of inorganic sulfur in aerobic soils—constitutes only a small part of total sulfur in soils. Generally, more than 95% of soil sulfur is organically bonded with several hundred kilograms of organic sulfur present in the upper horizons of most soils. Although not readily available, this large organic sulfur pool may potentially be an important source of sulfur to plants in deficiency situations.

Characterization of Organic Sulfur Pools in Soil

Organic sulfur in soils is a heterogeneous mixture of soil organisms and partly decomposed plant, animal, and microbial residues. Little is known about the identity of individual compounds. However, the sulfur containing amino acids cysteine and methionine, choline sulfate, sulfolipids, sulfonic acids, and sulfated polysaccharides have been found in soils (Freney, 1986). Several different approaches have been used to separate soil organic sulfur into broad fractions representing distinct forms and properties.

Separation According to Reactivity with Reducing Agents

The traditional way of separating organic sulfur is according to reactivity with reducing agents (Tabatabai, 1982). Two distinct groups of sulfur compounds are obtained: (i) organic sulfur not directly bonded to carbon, which can be reduced to H_2S by hydroiodic acid, and (ii) organic sulfur, which is directly bonded to carbon (C–S). The first group is composed primarily of sulfate esters (C–O–S), and the second includes sulfur containing amino acids, mercaptans, disulfides, sulfones, and sulfonic acids (Freney, 1986). Generally, total sulfur content decreases with depth in line with the organic carbon content, and the percentage of organic sulfur present as sulfate esters increases with depth (Tabatabai and Bremner, 1972; Eriksen, 1996).

Physical Separation into Mineral and Aggregate Size Fractions

Fractionation of soil and organic matter into primary particle-size separates has been used as a tool for studying soil organic matter distribution and dynamics, since a significant part of organic matter is closely associated with soil minerals (Christensen, 1992). Similar to what has been found for carbon and nitrogen, there was a considerable sulfur enrichment of the clay fractions in the soils studied. The ratios of C/S decreased dramatically with decreasing particle size and for some soils the nitrogen/sulfur (N/S) ratio also decreased, showing differences in the nature of the organic materials associated with different particle size fractions (Hinds and Lowe, 1980; Anderson et al., 1981).

The interaction with clay can protect some of the more easily decomposable organic matter from microbial breakdown (Ladd et al., 1993). This interaction seems even more important to sulfur than carbon and nitrogen. The increase in the percentage of hydroiodic acid-reducible sulfur (sulfate esters), which is believed to be the more labile form of organic sulfur (Biederbeck, 1978), with decreasing particle size suggests that organic sulfur–clay interactions are the major mechanisms protecting organic sulfur from mineralization.

Chemical Extraction Followed by Physical–Chemical Separations

Soil sulfur has been studied by conventional organic matter fractionation into humic acids, fulvic acids, and humin. By means of a sonification procedure in combination with the extraction, humic and fulvic acids are obtained, some of which are intimately associated with clay minerals and humin (Bettany et al., 1979; 1980). Although this type of chemical extraction has been widely used to study soil organic matter dynamics, it is questionable if it is useful for studying soil organic sulfur because of possible artifact formation caused by the strongly alkaline reagent (Freney, 1986).

Molecular Weight Fractionation

Sephadex gel-filtration was used in a number of studies to obtain fractions of organic-matter extracts with different molecular size and nominal molecular weight (MW) (e.g., Scott and Anderson, 1976; Keer et al., 1990). Generally, much of the organic sulfur was found in fractions with a very high MW (>100,000 Da), but a significant proportion also had MWs of less than 10,000 Da. Eriksen et al. (1995c) showed that in the short term, little sulfur cycling takes place in fractions >5000 Da. During 8 wk of incubation, carrier-free, radio-labeled sulfur (^{35}S) was initially incorporated into the <700 Da MW fraction and then recycled into the 700 to 5000 Da fraction. Keer et al. (1990) noted that high MW organic matter was enriched with compounds in the sulfate-ester form. More than 75% of total organic sulfur was in the form of sulfate esters in organic matter with MWs greater than 200,000 Da. This agrees with the finding that the percentage of sulfate esters increased with decreasing particle size, since high MW compounds are better adsorbed to clay particles because of the presumed aliphatic nature of high MW organic matter (Anderson et al., 1974).

Fractionation According to Physical Protection of Soil Organic Sulfur

Extraction of organic sulfur was performed with acetylacetone, which works by complexing with the metals that link organic matter and mineral particles together. The organic matter was then extracted by water (Giovannini and Sequi, 1976). Acetylacetone extraction of organic matter in conjunction with ultrasonic dispersion was used (e.g., Halstead et al., 1966; Scott and Anderson, 1976; Keer et al., 1990). Eriksen et al. (1995b) suggested that organic sulfur that could only be extracted by acetylacetone when subjected to ultrasonic dispersion was physically protected in soil aggregates, and that sulfur turnover in this fraction was slow. Thus, by means of the combined dispersion–extraction procedure by Eriksen et al. (1995b), organic matter was assumed to be divided into unprotected, protected, and insoluble organic sulfur, where protected means isolated from decomposers inside water-stable aggregates. It was demonstrated that physical protection of soil organic sulfur caused by soil aggregation plays an important role in the turnover of soil sulfur (Eriksen et al., 1995c). In an incubation experiment, it was found that much of the organic sulfur was physically protected inside aggregates, and that turnover in this protected fraction was slow. It has also been demonstrated that sulfur is initially immobilized into organic matter not protected by soil structure and then gradually into physically protected organic matter (Eriksen et al., 1997b).

XANES Spectroscopy

A nondestructive technique that has been widely used in speciation studies of environmental and geochemical samples is X-ray absorption fine-structure spectroscopy. A version of this, X-ray adsorption near-edge structure (XANES), provides specific information on the functional groups containing sulfur because of its sensitivity to electric structure, oxidation state, and geometry of neighboring atoms (Vairavamurthy et al., 1997). This technique has been used as a tool for understanding sulfur dynamics in soil organic matter. The advantage over wet chemical methods is that intermediate oxidation states can be identified.

Long-term agricultural management has been shown to have a significant effect on sulfur speciation (Solomon et al., 2003). Investigations by Zhao et al. (2006) indicate that sulfur species in the reduced and intermediate oxidation states are the main sources of organic sulfur for mineralization. One problem, however, is that XANES is applied to humic extracts from agricultural soils, which may not reflect the in situ speciation in soil. Soil organic sulfur, moreover, exists in many subcompartments in the soil with widely differing availability, a factor not reflected in the oxidation states. Still, XANES is a promising technique that needs to be used for more soils, and the relation to sulfur mineralization potential should be further investigated.

Microbial Biomass Sulfur

The microbial biomass plays a major role in soil sulfur cycling, and an understanding of the mechanisms and forms of sulfur involved is thus very important. Biomass sulfur was measured as the flush of extractable sulfur following chloroform fumigation, analogous to the way biomass carbon was determined (Saggar et al., 1981; Banerjee et al., 1993). Microbial biomass sulfur forms only a small part of soil organic sulfur, accounting for 0.9 to 2.6% of total organic sulfur in agricultural soils (Chapman, 1987) and 2.2 and 1.2% in a hardwood and a conifer forest, respectively (Strick and Nakas, 1984).

It has been suggested that fluctuations in microbial-biomass sulfur may be related to levels of inorganic sulfate in soils (Chapman, 1987). Biomass sulfur might become available to plants in a period of decreasing biomass, and sulfate may be immobilized during a period of biomass increase. However, results by Wu et al. (1993) showed that once sulfur was immobilized by the microbial biomass, it was directly transformed into soil organic sulfur, remaining unavailable to plants until remineralized. The activity of the microbial pool influences turnover rates rather than the size of microbial-biomass sulfur. In agricultural soils, a strong correlation was found between sulfur mineralization and microbial activity (Sparling and Searle, 1993). In forest soil, Autry and Fitzgerald (1993) similarly found a significant correlation between organosulfur formation and ATP content.

In experiments with ^{35}S-labeling of the sulfate pool, the labeling of microbial-biomass sulfur was shown to be relatively constant over time both under controlled and field conditions even though immobilization took place (Wu et al., 1995; Eriksen, 1997b). This indicated that ^{35}S immobilized by the microbial biomass was transformed directly into soil organic sulfur.

Mineralization of Soil Organic Sulfur

In soils, the processes responsible for sulfur transformations, such as mineralization, immobilization, oxidation, and reduction, are mainly microbially mediated. Therefore, factors that affect the microbial activity such as temperature, moisture, pH, and substrate availability also affect these processes. In aerobic agricultural soils, the main process of interest is the release of inorganic, plant-available sulfate from organic matter. Since mineralization and immobilization of sulfur happen concurrently (Maynard et al., 1983; Ghani et al., 1993, Eriksen,

1997a), the release or incorporation of inorganic sulfate is the net result of several processes.

Biological and Biochemical Mineralization

McGill and Cole (1981) proposed a conceptual model for the cycling of organic carbon, nitrogen, sulfur, and phosphorus through soil organic matter in which the mineralization of sulfur involved two different processes—biological and biochemical mineralization. Biological mineralization is believed to be driven by the microbial need for organic carbon to provide energy, and sulfur released as sulfate is a by-product of the oxidation of carbon to carbon dioxide. Biochemical mineralization is the release of sulfate from the sulfate-ester pool through enzymatic hydrolysis. Whereas, since mineralization of carbon-bonded sulfur is strictly dependent on microbial activity, the sulfate esters can be readily hydrolyzed by sulphatase enzymes in the soil, and therefore the biochemical mineralization is controlled by the supply of sulfur rather than the need for energy. In situations where microbial demands cannot be met by soil inorganic sulfate, sulphatase enzymes are used to hydrolyze sulfate esters, and, conversely, high levels of sulfate will inhibit biological mineralization.

Although the conceptual model can be criticized as an oversimplification of a much more complex system, it provides insight into the fundamental differences between the processes for different nutrients. Originally it was thought that because of the close relationship between sulfur and nitrogen in organic matter, the ratio between mineralized nitrogen and sulfur would be the same as in soil organic matter (Walker, 1957; White, 1959). This conflicts with many studies that show considerable deviations from this. Results range from a much wider ratio (Kowalenko and Lowe, 1975) to a narrower (Tabatabai and Al-Khafaji, 1980) N/S ratio in mineralization products than in soil organic matter. Considering the two-mineralization mechanisms in the McGill and Cole (1981) model for sulfur, these observations are not surprising. Whereas carbon-bonded sulfur and nitrogen are stabilized together and released through biological mineralization, the sulfate esters can be mineralized independently. Thus, net mineralization of sulfur depends on the rates of the two reactions and the N/S ratio in mineralized material will vary accordingly.

Sulphatase Enzymes

Because much of the soil organic sulfur exists as sulfate esters that can be mineralized through enzymatic hydrolysis, the responsible sulphatases have gained some interest. Bacteria and fungi are major sources in the soil, but also plant roots and possibly mammalian urine may contain these enzymes (Fitzgerald, 1976; Klose et al., 1999). Sulphatases are classified according to the sulfate esters they hydrolyze. The main groups are aryl-, alkyl-, steroid, gluco-, condro- and mycosulphatases, but only arylsulphatases have been measured in soils (Germida et al., 1992). Strong correlations have been found between arylsulphatase activity and soil organic carbon content (Elsgaard et al., 2002) as well as between sulfur mineralization and arylsulphatase activity (Lee and Speir, 1979; Castellano and Dick, 1991). However, this may be due to general factors that affect enzyme activity. Since arylsulphatase is only one of many enzymes involved in the mineralization process, it is unlikely that this alone can explain variations in mineralization of sulfur (Germida et al., 1992). Little is known about the substrate specificity of sulphatases, but the activity of arylsulphatases in the soil does not appear to constitute a rate-limiting factor in the hydrolysis of sulfate esters (Houghton and Rose, 1976; Ganeshamurthy and Nielsen, 1990).

Sulfur Mineralization Potential

The potential mineralization of soil organic sulfur has been estimated by kinetic equations for sulfur mineralization on the basis of the release of sulfur from incubated soils (Pirela and Tabatabai, 1988; Ghani et al., 1991). The validity of this approach may be questionable, since mineralization of sulfur strongly depends on the incubation technique used (Maynard et al., 1983; Valeur and Nilsson, 1993). It may be impossible to identify a kinetically homogeneous potential mineralizable pool, since the mineralization mechanisms consist of several substrates, biochemical pathways, and microbial communities (Ellert and Bettany, 1988). Besides, kinetic mechanisms change with temperature (Ellert and Bettany, 1992), which implies that a true measure of mineralizable sulfur in soils is best achieved when field conditions are closely simulated.

Under such conditions, Eriksen et al. (1995d) found that the contribution from mineralization to the supply of sulfur to plants was small (3–7 µg sulfur g soil^{-1} yr^{-1}), but differences between soils were very consistent. The amount of sulfur mineralized constituted 1.7 to 3.1% per year of the organic sulfur pool in the soil was in agreement with other findings on the mineralization of sulfur in the order of 0.5 to 3% of the soil organic sulfur pool (Freney, 1986; Keer et al., 1986). Mineralization following the addition of organic material to soils is discussed below.

Gross Sulfur Mineralization–Immobilization Turnover

The release or incorporation of inorganic nutrients is the net result of mineralization and immobilization processes. By means of isotopic-dilution techniques, quantification of gross nutrient fluxes has been used successfully to understand the fundamentals of these processes for both nitrogen (Murphy et al., 2003) and phosphorus (Di et al., 1994). The principle involved with this methodology is to label the mineral nutrient pool with an isotopic tracer and measure the change with time. The rates of influx to and outflux from the labeled pool are then calculated with equations based on tracer kinetics. Although these equations were developed as early as the 1950s (Kirkham and Bartholomew, 1954), this technique was not widely used until recently (Di et al., 2000; Murphy et al., 2003). A ^{35}S tracer was used to study sulfur transformations in soil without organic matter addition (e.g., Ghani et al., 1993; Eriksen, 1997a; Goh and Pamidi, 2003; Vong et al., 2003), following the addition of urinary or fecal sulfur (e.g., Blair et al., 1994; Nguyen and Goh, 1994b; Williams and Haynes, 2000) and plant residues (Wu et al., 1993; 1995). The use of the tracer–dilution method to determine gross sulfur transformation rates has rarely been attempted. Although the use of isotopic–dilution techniques has limitations because of the assumptions of the techniques, it has proven invaluable for the mechanistic modeling of gross nitrogen fluxes as well as understanding the fundamental processes of the soil internal nitrogen cycle and individual microbial pathways (Murphy et al., 2003). Some recent studies indicate that the isotopic–dilution technique has a similar potential to increase our knowledge of the soil sulfur cycle (Eriksen, 2005; Nziguheba et al., 2005).

Soil Inorganic Sulfur

Sulfur can have any oxidation number from −2 (sulfide) to +6 (sulfate). In agricultural soils where conditions are mostly aerobic, the dominant and stable form of inorganic sulfur is sulfate, and only negligible quantities of lower-oxidation-state compounds are present (Bohn et al., 1986). Consequently, sulfate is often referred to as inorganic sulfur in the literature. Concentrations of sulfate in soils fluctuate throughout the year because of changes in the balance between atmospheric inputs, decomposition of plants, fertilizer addition, leaching, plant uptake, and microbial activity. Usually low levels of sulfate are observed over winter and spring because of leaching, plant uptake, and low mineralization rates associated with low temperatures (Ghani et al., 1990; Castellano and Dick, 1990). Sulfate exists as water-soluble salt and as sulfate adsorbed to soil inorganic components. The soluble sulfate plus most of the adsorbed sulfate is generally believed to be plant available.

The retention of sulfate in soils is dependent on the nature of the colloidal system, the pH, the sulfate concentration, and the concentration of other ions in the solution (Harward and Reisenauer, 1966). Sulfate is adsorbed by hydrous oxides of iron and aluminum and by edges of clay particles (Parfitt, 1978). It has been proposed that sulfate is adsorbed by purely electrostatic mechanisms (Marsh et al., 1987), but also chemisorption occurs (Parfitt and Smart, 1978). The amount of sulfate adsorbed depends on the surface area of the clay and the surface charge, and therefore the higher the aluminum content, the greater the anion adsorption (Bohn et al., 1986).

The effect of pH on sulfate adsorption is related to net charge of the iron and aluminum oxides. If pH in the soil is lower than zero point of charge (ZPC), it will lead to a positive surface because of hydration of the metal oxides, and sulfate will be adsorbed in the soil. Thus, acid subsoils (for example under forest vegetation) and peat soils typically have large storage capacities for adsorbed sulfate. In agricultural soils, however, the pH is often higher than the ZPC, leading to a low retention of sulfate in the soils. Thus, Curtin and Syers (1990) found that virtually all sulfate in soils with pH >6 was in solution. Liming, moreover, has been demonstrated to increase sulfur leaching (Chao et al., 1962b; Bolan et al., 1988) because of desorption of sulfate and increased mineralization. Even for soils with a marked capacity to retain sulfate, the strength of the retention seemed weak, and Chao et al. (1962a) found that repeated extraction with water removed adsorbed sulfate.

Sulfate adsorption is influenced by the presence of other anions. The order of adsorption strength of anions in soils is: hydroxyl > phosphate > sulfate > nitrate = chloride (Tisdale et al., 1984). The stronger adsorption of phosphate than sulfate is the basis for extraction of adsorbed sulfate (Tabatabai, 1982), and addition of phosphate to soils has been shown to increase sulfur leaching (Chao et al., 1962b; Bolan et al., 1988).

Conceptual Model for Sulfur Cycling in Temperate Agricultural Soils

Because of the complex nature of soil organic matter, any procedure attempting to divide organic sulfur into a few biologically meaningful fractions is a

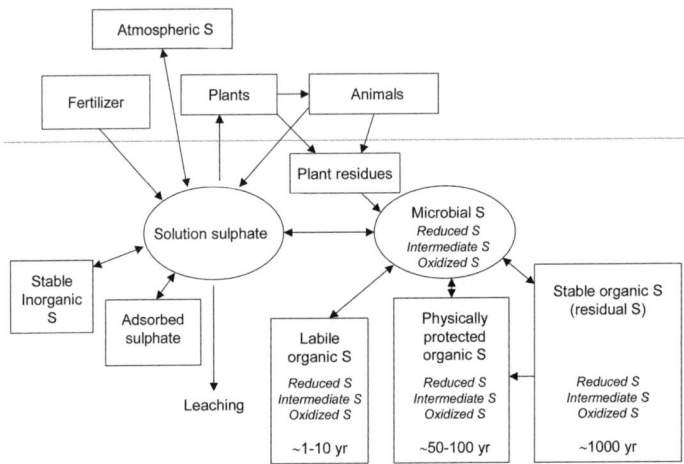

Fig. 3–1. Conceptual diagram of the sulfur cycle.

rough simplification. On the other hand, there is some truth to the various fractions identified above. Clearly, the carbon-bonded sulfur and the sulfate esters have chemically distinct properties important in sulfur cycling. In some stable and clay-protected organic sulfur pools, however, these differences are overruled by the physical properties of the soils, as illustrated in Fig. 3–1, where the organic sulfur compounds of different oxidation states, such as sulfate-esters and carbon-bonded sulfur fractions, are integrated parts of other mainly physically determined pools. It is, therefore, not surprising that carbon-bonded sulfur and sulfate ester determination on whole soils offers little information regarding plant availability of the organic sulfur. In a review, Scherer (2001) concluded that the complicated dynamics of organic sulfur compounds in soil make it difficult to estimate the sulfur delivery to plants. However, because short-term sulfur cycling probably involves only a small fraction of total organic sulfur, it is essential to identify the more active parts to obtain a better understanding of factors determining plant availability. A key point is physical protection of soil organic sulfur.

Sulfur Amendments to Soil

Inorganic Sulfur Fertilizer

The use of sulfur fertilizers has been reviewed elsewhere (e.g., Messick et al., 2002; Blair, 2002), but, generally, there are two types of sulfur fertilizers: (i) those where sulfur is in the sulfate form and readily available to plants and (ii) those where sulfur needs to be oxidized to sulfate to become available to plants.

The advantage of sulfate-containing fertilizers, in addition to their being in a plant-available form, is that they are easily incorporated in multinutrient fertilizers, which is a cost-efficient way of fertilizer application. However, in some cases their use may give an unbalanced nutrient supply. One example is the widely used ammonium sulfate, where a direct application as a nitrogen source applies much more sulfur than is typically required. Since sulfate is readily leached from the soil, there is no point in attempting to raise soil sulfur levels by excessive fertilization (Eriksen, 1996; Knights et al., 2000).

Studies have shown the importance of the availability of sulfur during grain filling (Eriksen et al., 2001; Fitzgerald et al., 1999; Monaghan et al., 1999), where shortage is caused by the limited redistribution of sulfur from the vegetative tissue to grain (Eriksen et al., 2001; Zhao et al., 1999). The application of sulfate-containing fertilizer in the spring carries the risk of sulfur intended for the grain-filling period to either be leached or immobilized in soil–microbial biomass or in vegetative plant parts. The use of elemental sulfur fertilizer offers several advantages including slow release over time as it is gradually converted to sulfate. The effectiveness of elemental sulfur is governed by oxidation performed principally by bacteria that depends on temperature, moisture, and particle size. Blair et al. (1993) developed a model for matching the oxidation rate of sulfur from elemental sulfur to plant requirements. Furthermore, elemental sulfur is the most concentrated form of sulfur that cuts transport and application costs, offers reserve availability because of the oxidation process, and may be incorporated in compound fertilizers (Messick et al., 2002).

Organic Sulfur Sources

Animal Manure

On a global scale, total sulfur excretion from domestic animals is estimated at around 8×10^6 Mg per year, corresponding to about 13% of world sulfur production or 80% of the world consumption for mineral fertilizer manufacture (Eriksen, 2002). However, it is difficult to utilize this potential source of fertilizer sulfur.

Of this total, developed countries account for the 2.6×10^6 Mg and developing countries contribute 5.5×10^6 Mg. In developed countries, 54% of the excretion occurs during grazing, whereas in the developing countries this figure is 65%. The sulfur excreted during grazing is difficult to utilize. This sulfur may be lost from the main grazing area to unproductive sites by excremental transfer (Nguyen and Goh, 1994a), and even when deposited within the productive area, it may not be efficiently recycled because of nonuniform distribution and losses (Nguyen and Goh, 1994b; Till et al., 1994; Williams and Haynes, 1992).

Much of the sulfur excreted within a housing system may be collected and applied to agricultural land as a fertilizer, provided the technical facilities are sufficient. Especially in developed countries the sulfur content of manure collected from cattle and pigs (1×10^6 Mg per year) has potential as a sulfur fertilizer, since regulations already stipulate on the utilization of manure nitrogen and/or phosphorus that animal wastes are used as fertilizers. In countries with a large animal production, the potential sulfur contribution can be significant; it is important to establish to what extent the sulfur content in animal manure is available to plants.

Sulfur Content in Farm Animal Manure

Animal manure varies in substance given that its composition is a product of many factors, e.g., animal species, feed composition, production system and time, and conditions of storage. Because stored manure may be a mixture of excreta differing in age from animals fed different diets, and possibly even from different animal species, the sulfur content can fluctuate considerably. The concentrations of total sulfur in slurry typically vary between 0.15 and 0.7 kg sulfur m^{-3} of slurry (Eriksen et al., 1995a), and the mean value in farmyard manure is around 1 kg S Mg^{-1}.

The question is how much does slurry sulfur contribute to the sulfur supply of plants? Average values of around 0.35 kg total sulfur m^{-3} are commonly found in cattle slurry (Eriksen et al., 1995a; Lloyd, 1994; Watson and Stevens, 1986). On the basis of this value, a dressing of 50 m^3 cattle slurry would provide 17.5 kg sulfur ha^{-1}, corresponding to the sulfur requirement of e.g., cereals. However, the total sulfur content is of limited value in predicting fertilizer value.

Plant Availability of Manure Sulfur in the Short Term

The composition of sulfur in slurry and the content of plant-available sulfate may vary depending on feeding and storage. As a consequence, different slurries may be expected to have different levels of plant availability of sulfur. Unfortunately, there is a lack of information on feeding and slurry storage in the literature covering plant availability of manure sulfur, which makes it difficult to generalize results.

Lloyd (1994) found an effectiveness of sulfur in cattle slurry of 55% compared with sulfur in gypsum when applied to grass for silage. This was much higher than the effectiveness of 5% found by Eriksen et al. (1995a), when applying slurry to spring oilseed rape in a pot experiment. The low plant availability of sulfur in slurry was most likely due to transformations in the slurry during storage, where plant-available inorganic sulfate was incorporated into organic sulfur or reduced to sulfide. Sulfide was expected to be readily oxidized to plant-available sulfate when the slurry was applied to soil. However, the low plant uptake of slurry sulfur suggested this did not happen. The possibilities are that sulfide was either emitted from the slurry as H_2S or immobilized in the soil as metallic sulfides or by sorption to soil particles (Bremner and Steele, 1978).

The differences in effectiveness between studies may be explained by differences in feeding and storage. Under Danish conditions the sulfur content of feed is normally tailored to animal requirements and the storage time is usually many months. This combination minimizes the content of inorganic plant-available sulfur. The results from the pot experiment are supported by experiments by Pedersen et al. (1998). In eight Danish field trials, they found a response to a mineral fertilizer application of 40 kg sulfur ha^{-1} to winter oilseed rape despite applications of organic manure.

Plant Availability of Manure Sulfur in the Long Term

A low efficiency of sulfur in animal slurry is mainly attributed to slurry sulfur being in organic forms not available to plants. This suggests that soils with an annual application of slurry or other organic manures will release more plant-available inorganic sulfate than unmanured soils. To get an idea of the residual effect of sulfur added in animal manure, some plots in the Askov long-term field experiments in Denmark were examined for their content of organic carbon and sulfur and inorganic sulfate (Eriksen and Mortensen, 1999). The experiments, started in 1894 on both sandy and loamy soils, had the objective of comparing the effect of animal manure with equal dressings of nitrogen, phosphorus, and potassium (NPK) in mineral fertilizers and unmanured treatments (Christensen, 1996). Fertilization with animal manure or NPK fertilizer increased the content of soil organic carbon compared with unfertilized plots (Fig. 3–2). On the sandy soil, the build-up of organic carbon was followed by a similar build-up of organic sulfur. Thus, in the mineral-fertilized plots, organic carbon and sulfur were, on average, increased by 26 and 19%, respectively, compared with the unfertilized plots. In the organic-manured plots, organic carbon and sulfur were increased

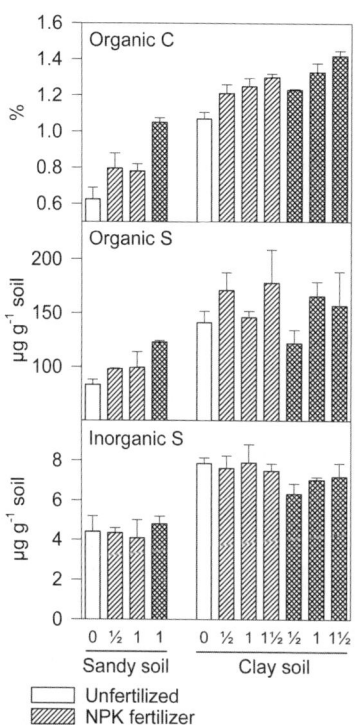

Fig. 3–2. The effect of fertilizer history on the soil content of organic carbon, organic, and inorganic sulfur in selected plots in the Askov long-term field experiment. Error bars denote SE.

by 51 and 56%, respectively. On the clay soil, the build-up of organic carbon, on average 17% in the NPK plots and 24% in organic-manured plots, was followed by insignificant increases in organic sulfur content of 24 and 5%, respectively.

The increased organic sulfur content did not significantly affect soil inorganic sulfate levels in the spring. This could be due to leaching losses during the winter, especially in the sandy soil, or alternatively, it could indicate that organic sulfur in the soil originating from increased fertilization was not more readily mineralized than the bulk of soil organic sulfur. Similarly, Knights et al. (2001) found that 153 yr of manure application in the Broadbalk experiment in England increased soil organic carbon and sulfur contents, but the long-term application of inorganic sulfur-containing fertilizers had little effect. They found similarly low inorganic sulfate levels in all arable plots. However, they did find increased mineralization rates when incubating soil from the farmyard manure plot compared with the inorganic fertilizer treatments, indicating that organic manure application indeed does have a potential long-term effect on the sulfur-supplying capacity to crops. Reddy et al. (2001) found increased sulfur mineralization when incubating soil subjected to 27 yr of manure application, but also mineral sulfur applications showed increased mineralization levels.

Annual applications of organic manure increase the soil organic sulfur content and thus the sulfur mineralization rate. The extent of this increase depends on soil type, cropping system, and management. Therefore, a residual sulfur effect of long-term organic manure application must be expected, although there is no indication that the sulfur from manure will mineralize more readily than the bulk of soil organic sulfur. The ability of a cropping system to use mineralized sulfur depends on the length of the growing season of the crops, but mineralization is unlikely to fully meet the sulfur demand of a crop.

Other Organic Materials

Other organic materials such as sewage sludge, green manure, and compost may be applied to agricultural land. The plant availability of organic-bonded sulfur in these materials depends on the mineralization rates. It is generally believed that sulfate is released from organic material when C/S ratios are less than 200 and is immobilized if C/S is above 400, whereas a C/S between 200 and 400 can cause both mineralization and immobilization (Barrow, 1960). This rule seems to apply across different organic materials such as sludge, animal manure, and plant material. Figure 3–3 gives data from three studies involving incorporation of different types of organic material in pot or incubation experiments. The regression line explains 65% of the variation in mineralization, which is a rather good correlation considering the comprehensiveness of soils and organic materials used in the experiment. Straw or other strongly immobilizing materials were not included because they do not exhibit the same relationship; therefore, it was advisable to use supplemental sulfur fertilizers during the incorporation of such materials (Chowdhury et al., 2000). For straw, it was estimated that a C/S ratio of 340 was the critical level for immobilization (Chapman, 1997). Sewage sludge was a good source of sulfur in this example (Fig. 3–3) as sulfur mineralization was high because of low C/S ratios. The C/S ratios of solid animal manure ranged from 22 for poultry manure to 297 for horse manure. These differences—probably largely due to differences in straw content—resulted in much decreased sulfur mineralization levels in soils amended with a high C/S ratio material.

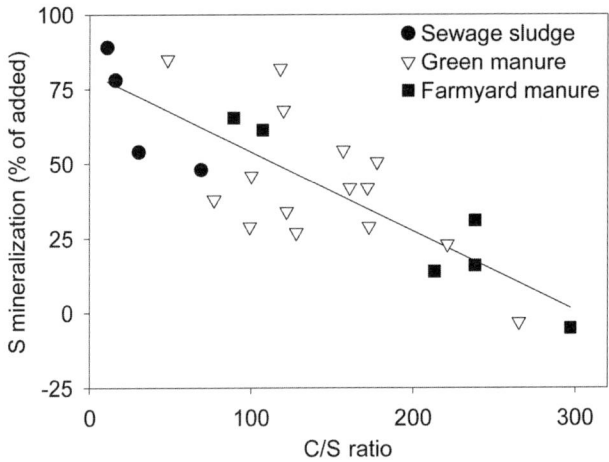

Fig. 3–3. Mineralization of sulfur after soil incorporation of different types of organic material. Data are for the incorporation of sewage sludge (Tabatabai and Chae, 1991), green manure (Eriksen and Thorup-Kristensen, 2002; Reddy et al., 2002; Tabatabai and Chae, 1991), and farmyard manure (Reddy et al., 2002; Tabatabai and Chae, 1991) as a function of total C/S ratio.

Green Manure–Catch Crop

A green manure is a crop grown primarily for the purpose of being plowed in to add nutrients and organic matter to the soil. It has been demonstrated that a catch crop succeeding the main crop can absorb sulfate from the root zone during autumn and winter and thereby reduce nitrate leaching (Eriksen and Thorup-Kristensen, 2002). The catch crop–green manure is incorporated before the sowing of the next crop and ideally the nutrient content of the residues should quickly become plant available. To maintain sufficient soil sulfate levels, it is important not only to capture the sulfate leached in the autumn, but also to make sure that it is released in time for the following crop to use it efficiently. If the mineralization of catch-crop sulfur is not synchronized with plant sulfur uptake, there is a risk of leaching losses before or after the growing season of the following crop. Most nonlegume species release both nitrogen and sulfur soon after incorporation, especially the crucifers having low C/N and C/S ratios (Eriksen et al., 2004). Legumes are generally not as effective because they have high C/S ratios, which may retard sulfur mineralization considerably or even lead to immobilization of sulfur.

In sulfur-deficient crop rotations, the release of sulfur from incorporated catch crops will not be able to fulfill the needs of sulfur-demanding crops such as oilseed rape, grass ley, kale, or onion. Those crops will more than likely need supplemental sulfur fertilizer. However, the mineralization of catch-crop sulfur will be able to contribute considerably to the sulfur nutrition of cereals, especially from cruciferous catch crops (Eriksen and Thorup-Kristensen, 2002).

Soil Sulfur Accumulation and Losses in Farming Systems

Sulfur Mass Balances

Mass balances may give indications of the status of a particular crop or farming system at a local, national or global level. This can be illustrated by an example from Danish agriculture, where reduced inputs from atmospheric deposition and fertilizer have resulted in the development of a negative sulfur balance (Table 3–1; Eriksen, 1997c). Because of reduced SO_2 emissions in Denmark and neighboring countries, atmospheric deposition was reduced from about 38 kg sulfur ha^{-1} yr^{-1} in the late 1960s to 10 to 15 kg sulfur in the 1990s. The use of low sulfur-containing fertilizers, especially the change from single super phosphate

Table 3–1. Sulfur balance in Danish agricultural soils.

	1967	1970–1975	1989	1994
		kg S ha^{-1} yr^{-1}		
Atmosphere	38	30	17	12
Fertilizer	25	21	8	12
Animal manure	15	13	8	8
Irrigation	–	–	2	2
Crop removal	10–30	15	16	14
Volatilization	–	–	1	1
Leaching	30–40	28	34	30
Total	8–38	21	–16	–11

(12% sulfur) to triple super phosphate (1.3% sulfur) also increased the deficit. As a consequence of this development, sulfur deficiency in crops started to appear from the late 1980s onward, first in sulfur-demanding crops such as oilseed rape and later in cereals. Since the mid-1990s, sulfur fertilization was recommended for all crops (Pedersen et al., 1998).

Generally, the advantages of the balances are their quantitative nature and their value as a management tool, but they also have some serious shortcomings. The main shortcoming is the inability to estimate internal flows, which is a major drawback when applying nutrient balances to estimate potential losses, and the fact that they only address nutrient amounts and ignore availability (Öborn et al., 2003). This also applies to sulfur balances and especially when used at farm or field level. One pitfall is the inclusion in animal manure of sulfur that is partly in organic form and therefore not immediately plant available. The underlying assumption is that annual applications of organic sulfur in animal manure will eventually increase sulfur mineralization. However, the organic fertilizer history may not always be well described, so it is virtually impossible to know if equilibrium will ever be reached between organic sulfur input and mineralization for the individual soil. Besides, as mentioned above, the mineralization of organic sulfur, whether in manure or in crop residues, may not be synchronized with plant demand. Much of the mineralization may take place during summer when soil temperatures are highest, which, in the case of cereals, may be too late to prevent sulfur deficiency (Haneklaus et al., 1995). The additional consequence of this situation is that substantial sulfate leaching can take place in the autumn and winter following a sulfur-responsive spring-sown cereal (Eriksen et al., 2002).

In the same way that a positive sulfur balance may not always result in crop sulfur sufficiency, a negative balance may not always mean sulfur deficiency. When inputs such as atmospheric deposition are declining, sulfate leaching will for a while—depending on soil type and climatic conditions—reflect previous inputs more than the actual inputs. Obviously, when basing sulfur balances partly on historical data or literature values, special care should be taken that sulfate leaching reflects current sulfur inputs.

Long-Term Experiments

The soil sulfur status caused by a particular land use or fertilizer practice may also be evaluated from the changes in the total soil pool. Since most management-related changes have relatively little effect on the large soil sulfur pool in the short term, long-term experiments offer the best possibility to investigate these changes.

Generally, the long-term experiments point in the same direction: a build-up of the soil organic sulfur pool only occurs if there is a concomitant input of carbon to the soil, be it in the form of manure or plant material. Thus, long-term applications (35–153 yr) of inorganic sulfur have, in some studies, hardly had any effect on soil organic sulfur contents (Kirchmann et al., 1996; Knights et al., 2001), probably because the additions were in excess of crop uptake and therefore lost through leaching. In a situation with sulfur deficiency in pasture, the inorganic sulfur application was shown to increase the amount of sulfur in the soil because of increased crop production and hence greater organic carbon input to the soil (Bhupinderpal-Singh et al., 2004).

Changes in land use also affect soil sulfur stocks. When comparing permanent pasture with medium- to long-term cultivation (11 and 30 yr, respectively), soil sulfur was reduced, and it was significantly related to the decrease in soil carbon, a well-known consequence of pasture cultivation (Bhupinderpal-Singh et al., 2004). In the Broadbalk experiment, organic sulfur accumulation in permanent grassland and woodland was found to be much higher than in arable cropping systems, but annual applications of FYM, equivalent to 7 Mg dry mass ha^{-1}, to the arable system was able to maintain soil total sulfur content at the same level as permanent pasture and woodland, which in turn was 2.7 times higher than the sulfur content of mineral-fertilized arable land (Knights et al., 2001). In the Askov long-term experiment on animal manure and mineral fertilizer (initiated in 1894), soil carbon and sulfur contents were found to be correlated, but their isotopic signatures were not (Bol et al., 2005). This indicates that although sulfur in the longer term is stabilized in association with carbon in soil organic matter, the short-term cycles are not necessarily so closely linked.

Conclusions

Organic sulfur in soil forms a highly complex system of compounds the mineralization of which occurs along a continuum of different timescales. This makes prediction of potentially plant-available soil organic sulfur difficult, and methods based on identifications of compounds or groups of compounds with distinct chemical properties have failed, although they often reflect the long-term agricultural management. This is at least partly due to association with soil particles and aggregates that provide extensive physical protection of soil organic sulfur against decomposition. As an alternative, balance considerations will often be able to provide a better background for fertilizer recommendations, keeping in mind that availability and synchrony are ignored in this approach.

The efficiency of sulfur in organic manures, especially animal manure, seems under explored considering the focus on animal-manure management in many parts of the world. Although part of the sulfur in animal manure is plant available to the crop in the year of application, to what extent depends on feeding and storage conditions—most of the residual sulfur will probably not mineralize any more readily than the bulk of soil organic sulfur. Furthermore, as a source of sulfur for plants, mineralization of residual and soil organic sulfur may not be in synchrony with the demand for crops with a short growing season and may even lead to increased leaching losses of sulfate.

In some cases catch crops, especially crucifers, have been shown to serve the double purpose of reducing sulfate leaching and increasing synchrony with crop demand. However, when planning catch crop strategies, priority is usually given to maximizing nitrogen efficiency, and it is in this context that the effect of catch crops on sulfur nutrition should be seen.

Acknowledgments

Figures 3–2 and 3–3 are used with permission of The International Fertiliser Society.

References

Anderson, D.W., E.A. Paul, and R.J.S.T. Arnaud. 1974. Extraction and characterization of humus with reference to clay-associated humus. Can. J. Soil Sci. 54:317–323.

Anderson, D.W., S. Saggar, J.R. Bettany, and J.W.B. Stewart. 1981. Particle size fractions and their use in studies of soil organic matter: I. The nature and distribution of forms of carbon, nitrogen and sulfur. Soil Sci. Soc. Am. J. 45:767–772.

Autry, A.R., and J.W. Fitzgerald. 1993. Relationship between microbial activity, biomass and organosulfur formation in forest soil. Soil Biol. Biochem. 25:33–39.

Banerjee, M.R., S.J. Chapman, and K. Killham. 1993. Factors influencing the determination of microbial biomass sulfur in soil. Commun. Soil Sci. Plant Anal. 24:939–950.

Barrow, N.J. 1960. A comparison of the mineralization of nitrogen and of sulfur from decomposing organic materials. Aust. J. Agric. Res. 11:960–969.

Bettany, J.R., S. Saggar, and J.W.B. Stewart. 1980. Comparison of the amount and forms of sulfur in soil organic matter fractions after 65 years of cultivation. Soil Sci. Soc. Am. J. 44:70–75.

Bettany, J.R., J.W.B. Stewart, and S. Saggar. 1979. The nature and forms of sulfur in organic matter fractions of soils selected along an environmental gradient. Soil Sci. Soc. Am. J. 43:981–985.

Bhupinderpal-Singh, M.J. Hedley, S. Saggar, and G.S. Francis. 2004. Chemical fractionation to characterize changes in sulfur and carbon in soil caused by management. Eur. J. Soil Sci. 55:79–90.

Biederbeck, V.O. 1978. Soil organic sulfur and fertility. p. 273–310. *In* M. Schnitzer and S.U. Khan (ed.) Soil organic matter. Elsevier, Amsterdam.

Blair, G.J. 2002. Sulfur fertilisers: A global perspective. Proceedings No 498. International Fertiliser Society, York, UK.

Blair, G.J., R.B. Lefroy, M. Dana, and G.C. Anderson. 1993. Modelling of sulfur oxidation from elemental sulfur. Plant Soil 155/156:379–382.

Blair, G.J., A.R. Till, and C. Boswell. 1994. Rate of recycling of sulfur from urine, faeces and litter applied to the soil surface. Aust. J. Soil Res. 32:543–554.

Bohn, H.L., N.J. Barrow, S.S.S. Rajan, and R.L. Parfitt. 1986. Reactions of inorganic sulfur in soils. p. 233–249. *In* M.A. Tabatabai (ed.) Sulfur in agriculture. ASA, CSSA, and SSSA, Madison, WI.

Bol, R., J. Eriksen, P. Smith, M.H. Garnett, K. Coleman, and B.T. Christensen. 2005. The natural abundance of ^{13}C, ^{15}N, ^{34}S and ^{14}C in archived (1923–2000) plant and soil samples from the Askov long-term experiments on animal manure and mineral fertilizer. Rapid Commun. Mass Spec. 19:3216–3226.

Bolan, N.S., J.K. Syers, R.W. Tillman, and D.R. Scotter. 1988. Effect of liming and phosphate additions on sulfate leaching in soils. J. Soil Sci. 39:493–504.

Bremner, J.M., and C.G. Steele. 1978. Role of microorganisms in the atmospheric sulfur cycle. p. 273–310. *In* M. Alexander (ed.) Advances in microbial ecology. Plenum Press, London.

Castellano, S.D., and R.P. Dick. 1990. Cropping and sulfur fertilization influence on sulfur transformations in soil. Soil Sci. Soc. Am. J. 54:114–121.

Castellano, S.D., and R.P. Dick. 1991. Modified calibration procedure for the measurement of microbial sulfur in soil. Soil Sci. Soc. Am. J. 55:283–285.

Chao, T.T., M.E. Harward, and S.C. Fang. 1962a. Movement of 35S tagged sulfate through soil columns. Soil Sci. Soc. Am. Proc. 26:27–31.

Chao, T.T., M.E. Harward, and S.C. Fang. 1962b. Adsorption and desorption phenomena of sulfate ions in soils. Soil Sci. Soc. Am. Proc. 26:234–237.

Chapman, S.J. 1987. Microbial sulfur in some Scottish soils. Soil Biol. Biochem. 19:301–305.

Chapman, S.J. 1997. Barley straw decomposition and sulfur immobilisation. Soil Biol. Biochem. 29:109–114.

Chowdhury, M.A.K., K. Kouno, T. Ando, and T. Nagaoka. 2000. Microbial biomass, sulfur mineralisation and sulfur uptake by African millet from soil amended with various composts. Soil Biol. Biochem. 32:845–852.

Christensen, B.T. 1992. Physical fractionation of soil and organic matter in primary particles and density separates. Adv. Soil Sci. 20:1–87.

Christensen, B.T. 1996. The Askov long-term experiments on animal manure and mineral fertilizers. p. 301–312. In D.S. Powlson et al. (ed.) Evaluation of soil organic matter models. NATO ASI Series, Vol. I 38, Springer-Verlag, Berlin.

Curtin, D., and J.K. Syers. 1990. Extractability and adsorption of sulfate in soils. J. Soil Sci. 41:295–304.

Di, H.J., K.C. Cameron, and R.G. McLaren. 2000. Isotopic dilution methods to determine the gross transformation rates of nitrogen, phosphorus, and sulfur in soil: A review of the theory, methodologies, and limitations. Aust. J. Soil Res. 38:213–230.

Di, H.J., R. Harrison, and A.S. Campbell. 1994. Assessment of methods for studying the dissolution of phosphate fertilizers of differing solubility in soil. I. An isotopic method. Fert. Res. 38:1–9.

Ellert, B.H., and J.R. Bettany. 1988. Comparison of kinetic models describing net sulfur and nitrogen mineralization. Soil Sci. Soc. Am. J. 52:1692–1702.

Ellert, B.H., and J.R. Bettany. 1992. Temperature dependence of net sulfur and nitrogen mineralisation. Soil Sci. Soc. Am. J. 56:1133–1141.

Elsgaard, L., G. Hastrup Andersen, and J. Eriksen. 2002. Measurement of arylsulphatase activity in agricultural soils using a simplified assay. Soil Biol. Biochem. 34:79–82.

Eriksen, J. 1996. Incorporation of sulfur into soil organic matter in the field as determined by the natural abundance of stable sulfur isotopes. Biol. Fertil. Soils 22:149–155.

Eriksen, J. 1997a. Sulfur cycling in Danish agricultural soils: Inorganic sulfate dynamics and plant uptake. Soil Biol. Biochem. 29:1379–1385.

Eriksen, J. 1997b. Sulfur cycling in Danish agricultural soils. Turnover in organic sulfur fractions. Soil Biol. Biochem. 29:1371–1377.

Eriksen, J. 1997c. Animal manure as sulfur fertilizer. Sulfur Agric. 20:27–30.

Eriksen, J. 2002. Organic manures as sources of fertiliser sulfur. Proceedings No. 505. International Fertiliser Society, York, UK.

Eriksen, J. 2005. Gross sulfur mineralisation-immobilisation turnover in soil amended with plant residues. Soil Biol. Biochem. 37:2216–2224.

Eriksen, J., R.D.B. Lefroy, and G.J. Blair. 1995b. Physical protection of soil organic sulfur studied using acetylacetone at various intensities of ultrasonic dispersion. Soil Biol. Biochem. 27:1005–1010.

Eriksen, J., R.D.B. Lefroy, and G.J. Blair. 1995c. Physical protection of soil organic sulfur studied by extraction and fractionation of soil organic matter. Soil Biol. Biochem. 27:1011–1016.

Eriksen, J., J.V. Mortensen, J. Dissing Nielsen, and N.E. Nielsen. 1995d. Sulfur mineralisation in five Danish soils as measured by plant uptake in a pot experiment. Agric. Ecosyst. Environ. 56:43–51.

Eriksen, J., and J. Mortensen. 1999. Soil sulfur status following long-term annual application of animal manure and mineral fertilizers. Biol. Fertil. Soils 28:416–421.

Eriksen, J., J. Mortensen, V.K. Kjellerup, and O. Kristjansen. 1995a. Forms and plant-availability of sulfur in cattle and pig slurry. Z. Pflanzenernähr. Bodenk. 158:113–116.

Eriksen, J., M. Nielsen, J.V. Mortensen, and J.K. Schjørring. 2001. Redistribution of sulfur during generative growth of barley with different sulfur and nitrogen status. Plant Soil 230:239–246.

Eriksen, J., J.E. Olesen, and M. Askegaard. 2002. Sulfate leaching and sulfur balances of an organic crop rotation on three Danish soils. Eur. J. Agron. 17:1–9.

Eriksen, J., and K. Thorup-Kristensen. 2002. The effect of catch crops on sulfate leaching and availability of sulfur in the succeeding crop on sandy loam soil in Denmark. Agric. Ecosyst. Environ. 90:247–254.

Eriksen, J., K. Thorup-Kristensen, and M. Askegaard. 2004. Plant-availability of catch crop sulfur following spring incorporation. J. Plant Nutr. Soil Sci. 167:609–615.

Fitzgerald, J.W. 1976. Sulfate ester formation and hydrolysis: A potentially important yet often ignored aspect of the sulfur cycle of aerobic soils. Bacteriol. Rev. 40:698–721.

Fitzgerald, M.A., D.T. Ugalde, and J.W. Anderson. 1999. Sulfur nutrition changes the sources of sulfur in vegetative tissue of wheat during generative growth. J. Exp. Bot. 50:499–508.

Freney, J.R. 1986. Forms and reactions of organic sulfur compounds in soils. p. 207–232. In M.A. Tabatabai (ed.) Sulfur in agriculture. ASA, CSSA, and SSSA, Madison, WI.

Ganeshamurthy, A.N., and N.E. Nielsen. 1990. Arylsuphatase and the biochemical mineralisation of soil organic sulfur. Soil Biol. Biochem. 22:1163–1165.

Germida, J.J., M. Wainwright, and V.V.S.R. Gupta. 1992. Biochemistry of sulfur cycling in soil. p. 1–53. In G. Stotzky and J.-M. Bollag (ed.) Soil Biochemistry 7. Marcel Dekker, New York.

Ghani, A., R.G. McLaren, and R.S. Swift. 1990. Seasonal fluctuations of sulfur and soil microbial biomass-S in the surface of a Wakanui soil. N.Z. J. Agric. Res. 33:467–472.

Ghani, A., R.G. McLaren, and R.S. Swift. 1991. Sulfur mineralisation in some New Zealand soils. Biol. Fertil. Soils 11:68–74.

Ghani, A., R.G. McLaren, and R.S. Swift. 1993. Mobilization of recently-formed soil organic sulfur. Soil Biol. Biochem. 25:1739–1744.

Giovannini, G., and P. Sequi. 1976. Iron and aluminium as cementing substances of soil aggregates. I. Acetylacetone in benzene as an extractant of fractions of soil iron and aluminium. J. Soil Sci. 27:140–147.

Goh, K.M., and J. Pamidi. 2003. Plant uptake of sulfur as related to changes in the HI-reducible and total sulfur fractions in soil. Plant Soil 250:1–13.

Halstead, R.L., G. Anderson, and N.M. Scott. 1966. Extraction of organic matter from soils by means of ultrasonic dispersion in aqueous acetylacetone. Nature 211:1430–1431.

Haneklaus, S., D.P.L. Murphy, G. Nowak, and E. Schnug. 1995. Effects of the timing of sulfur application on grain yield and yield components of wheat. Z. Pflanzenernähr. Bodenk. 158:83–85.

Harward, M.E., and H.M. Reisenauer. 1966. Reactions and movement of inorganic soil sulfur. Soil Sci. 101:326–335.

Hinds, A.A., and L.E. Lowe. 1980. Distribution of carbon, nitrogen, sulfur and phosphorus in particle-size separates from gleysolic soils. Can. J. Soil Sci. 60:783–786.

Houghton, C., and F.A. Rose. 1976. Liberation of sulfate from sulfate esters by soils. Appl. Environ. Microbiol. 31:969–976.

Kirkham, D., and W.V. Bartholomew. 1954. Equations for following nutrient transformations in soil, utilizing tracer data. Soil Sci. Soc. Am. Proc. 18:33–34.

Keer, J.I., R.G. McLaren, and R.S. Swift. 1986. The sulfur status of intensive grassland sites in southern Scotland. Grass Forage Sci. 41:183–190.

Keer, J.I., R.G. McLaren, and R.S. Swift. 1990. Acetylacetone extraction of soil organic sulfur and fractionation using gel chromatography. Soil Biol. Biochem. 22:97–104.

Kirchmann, H., F. Pichlmayer, and M.H. Gerzabek. 1996. Sulfur balances and sulfur-34 abundance in a long-term fertilizer experiment. Soil Sci. Soc. Am. J. 59:174–178.

Klose, S., J.M. Moore, and M.A. Tabatabai. 1999. Arylsulfatase activity of microbial biomass in soils as affected by cropping system. Biol. Fertil. Soils 29:46–54.

Knights, J.S., F.J. Zhao, B. Spiro, and S.P. McGrath. 2000. Long-term effects of land use and fertiliser treatments on sulfur cycling. J. Environ. Qual. 29:1867–1874.

Knights, J.S., F.J. Zhao, S.P. McGrath, and N. Magan. 2001. Long-term effects of land use and fertiliser treatments on sulfur transformations in soils from the Broadbalk experiment. Soil Biol. Biochem. 33:1797–1804.

Kowalenko, C.G., and L.E. Lowe. 1975. Mineralisation of sulfur from four soils and its relationship to soil carbon, nitrogen and phosphorus. Can. J. Soil Sci. 55:9–14.

Ladd, J.N., R.C. Foster, and J.O. Skjemstad. 1993. Soil structure: Carbon and nitrogen metabolism. Geoderma 56:401–434.

Lee, R., and T.W. Speir. 1979. Sulfur uptake by ryegrass and its relationship to inorganic and organic sulfur levels and sulphatase activity in soil. Plant Soil 53:407–425.

Lloyd, A. 1994. Effectiveness of cattle slurry as a sulfur source for grass cut for silage. Grass Forage Sci. 49:203–208.

Marsh, K.B., R.W. Tillman, and J.K. Syers. 1987. Charge relationship of sulfate sorption by soils. Soil Soc. Am. J. 51:318–323.

Maynard, D.G., J.W.B. Stewart, and J.R. Bettany. 1983. Sulfur and nitrogen mineralization in soils compared using two incubation techniques. Soil Biol. Biochem. 17:127–134.

McGill, W.B., and C.V. Cole. 1981. Comparative aspects of cycling or organic C, N, sulfur and P through soil organic matter. Geoderma 26:267–286.

Messick, D.L., C. de Brey, and M.X. Fan. 2002. Sources of sulfur, their processing and use in fertiliser manufacture. Proceedings No 502. International Fertiliser Society, York, UK.

Monaghan, J.M., C. Scrimgeour, F.J. Zhao, and E.J. Evans. 1999. Sulfur accumulation and redistribution in wheat (*Triticum aestivum*): A study using stable sulfur isotope ratios as a tracer system. Plant Cell Environ. 22:831–840.

Murphy, D.V., S. Recous, E.A. Stockdale, I.R.P. Fillery, L.S. Jensen, D.J. Hatch, and K.W.T. Goulding. 2003. Gross nitrogen fluxes in soil: Theory, measurement and application of ^{15}N pool dilution techniques. Adv. Agron. 79:69–118.

Nguyen, M.L., and K.M. Goh. 1994a. Sulfur cycling and its implications on sulfur fertilizer requirements of grazed grassland ecosystems. Agric. Ecosyst. Environ. 49:173–206.

Nguyen, M.L., and K.M. Goh. 1994b. Distribution, transformations and recovery of urinary sulfur and sources of plant-available soil sulfur in irrigated pasture soil-plant systems treated with ^{35}sulfur-labelled urine. J. Agric. Sci. (Cambridge) 122:91–105.

Nziguheba, G., E. Smolders, and R. Merckx. 2005. Sulfur immobilization and availability in soils assessed using isotope dilution. Soil Biol. Biochem. 37:635–644.

Öborn, I., A.C. Edwards, E. Witter, O. Oenema, K. Ivarsson, P.J.A. Withers, S.I. Nilsson, and A. Richert Stinzing. 2003. Element balances as a tool for sustainable nutrient management: A critical appraisal of their merits and limitations within an agronomic and environmental context. Eur. J. Agron. 20:211–225.

Parfitt, R.L. 1978. Anion adsorption by soils and soil materials. Adv. Agron. 30:1–50.

Parfitt, R.L., and R.S.T.C. Smart. 1978. The mechanism of sulfate adsorption on iron oxides. Soil Sci. Soc. Am. J. 42:48–50.

Pedersen, C.A., L. Knudsen, and F. Schnug. 1998. Sulfur fertilization. p. 115–134. *In* E. Schnug (ed.) Sulfur in agroecosystems. Kluwer Academic Publishers, Dordrecht, the Netherlands.

Pirela, H.J., and M.A. Tabatabai. 1988. Sulfur mineralisation rates and potentials of soils. Biol. Fertil. Soils 6:26–32.

Reddy, K.S., M. Singh, A. Swarup, A.S. Rao, and K.N. Sing. 2002. Sulfur mineralisation in two soils amended with organic manures, crop residues, and green manures. J. Plant Nutr. Soil Sci. 165:167–171.

Reddy, K.S., M. Singh, A.K. Tripathi, A. Swarup, and A.K. Dwivedi. 2001. Changes in organic and inorganic sulfur fractions and sulfur mineralisation in a Typic Haplustert after long-term cropping with different fertiliser and organic manure inputs. Aust. J. Soil Res. 39:737–748.

Saggar, S., J.R. Bettany, and J.W.B. Stewart. 1981. Sulfur transformations in relation to carbon and nitrogen in incubated soils. Soil Biol. Biochem. 13:499–511.

Scott, N.M., and G. Anderson. 1976. Sulfur, carbon, and nitrogen contents of organic fractions from acetylacetone extracts of soils. J. Soil Sci. 27:324–330.

Scherer, H.W. 2001. Sulfur in crop production—Invited paper. Eur. J. Agron. 14:81–111.

Solomon, D., J. Lehmann, and C.A. Martinez. 2003. Sulfur K-edge XANES spectroscopy as a tool for understanding sulfur dynamics in soil organic matter. Soil Sci. Soc. Am. J. 67:1721–1731.

Sparling, G.P., and P.L. Searle. 1993. Dimethyl sulphoxide reduction as a sensitive indicator of microbial activity in soil: The relationship with microbial biomass and mineralization of nitrogen and sulfur. Soil Biol. Biochem. 25:251–256.

Strick, J.E., and J.P. Nakas. 1984. Calibration of a microbial sulfur technique for use in forest soils. Soil Biol. Biochem. 16:289–291.

Tabatabai, M.A. 1982. Sulfur. p. 501–538. *In* A.L. Page (ed.) Methods of soil analysis, Part 2. SSSA, Madison, WI.

Tabatabai, M.A., and A.A. Al-Khafaji. 1980. Comparison of nitrogen and sulfur mineralisation in soils. Soil Sci. Soc. Am. J. 44:1000–1006.

Tabatabai, M.A., and J.M. Bremner. 1972. Forms of sulfur, and carbon, nitrogen and sulfur relationships in Iowa soils. Soil Sci. 114:380–386.

Tabatabai, M.A., and Y.M. Chae. 1991. Mineralization of sulfur in soils amended with organic wastes. J. Environ. Qual. 20:684–690.

Till, A.R., G.J. Blair, and C.C. Boswell. 1994. Sulfur leaching from soil columns treated with 35S-labelled urine. Aust. J. Soil Res. 32:535–542.

Tisdale, S.L., W.L. Nelson, and J.D. Beaton. 1984. Soil fertility and fertilizers. Macmillan Publishing Company, New York.

Vairavamurthy, M.A., D. Maletic, S. Wang, B. Manowitz, T. Eglinton, and T. Lyons. 1997. Characterization of sulfur-containing functional groups in sedimentary humic substances by X-ray absorption near-edge structure spectroscopy. Energ. Fuel 11:546–553.

Valeur, I., and I. Nilsson. 1993. Effects of lime and two incubation techniques on sulfur mineralisation in a forest soil. Soil Biol. Biochem. 25:1343–1350.

Vong, P.-C., O. Dedourge, F. Lasserre-Joulin, and A. Guckert. 2003. Immobilized-S, microbial biomass-S and soil arylsulfatase activity in the rhizosphere soil of rape and barley as affected by labile substrate carbon and nitrogen additions. Soil Biol. Biochem. 35:1651–1661.

Walker, T.W. 1957. The sulfur cycle in grassland soils. J. Brit. Grassl. Soc. 12:10–18.

Watson, C.J., and R.J. Stevens. 1986. The sulfur content of slurries and fertilizers. Records Agric. Res. 34:5–7.

White, J.G. 1959. Mineralisation of nitrogen and sulfur in sulfur-deficient soils. N.Z.J. Agric. Res. 2:255–259.

Williams, P.H., and R.J. Haynes. 1992. Transformations and plant uptake of urine-sulfate in urine-affected areas of pasture soil. Plant Soil 145:167–175.

Williams, P.H., and R.J. Haynes. 2000. Transformations and plant uptake of urine nitrogen and sulfur in long- and short-term pastures. Nutr. Cycl. Agroecosyst. 56:109–116.

Wu, J., A.G. O'Donnell, and J.K. Syers. 1993. Microbial growth and sulfur immobilization following the incorporation of plant residues into soil. Soil Biol. Biochem. 25:1567–1573.

Wu, J., G. O'Donnell, and J.K. Syers. 1995. Influences of glucose, nitrogen and plant residues on the immobilization of sulfate-S in soil. Soil Biol. Biochem. 27:1363–1370.

Zhao, F.J., M.J. Hawkesford, and S.P. McGrath. 1999. Sulfur assimilation and effects on yield and quality of wheat. J. Cereal Sci. 30:1–17.

Zhao, F.J., J. Lehmann, D. Solomon, M.A. Fox, and S.P. McGrath. 2006. Sulfur speciation and turnover in soils: Evidence from sulfur K-edge XANES spectroscopy and isotope dilution studies. Soil Biol. Biochem. 38:1000–1007.

4

History of Sulfur Deficiency in Crops

Silvia Haneklaus, Elke Bloem, and Ewald Schnug
Julius-Kühn Institute, Braunschweig, Germany

Abstract

Sulfur is unique in that its reputation has changed from undesired pollutant to a major limiting factor in plant production within just a few years in Northern Europe. The reduction in natural availability of this nutrient coincided with the change of major targets of agricultural production from maximizing crop production in the 1980s to supplying high-quality produce in the new millennium. This required a much more holistic approach to all inputs, including sulfur to develop ecologically sound standards that both will maintain sustainable production as well as not interact unfavorably with neighboring nonagricultural ecosystems. It is the objective of this chapter to deliver a concise chronicle about sulfur in the soil–plant environment that includes a brief presentation of sulfur deficiency symptoms from plant to paddock for oilseed rape (*Brassica napus* L.), cereals, and sugar beet (*Beta vulgaris* L.).

Sulfur—Yellow Poison or Jilted Plant Nutrient?

Era of Industrialization

In the preindustrial period, anthropogenic sulfur emissions were negligible, and global sulfur depositions were about 43 Tg sulfur yr^{-1} (Andreae, 1986). Since then, annual atmospheric sulfur depositions have nearly doubled, and the combustion of fossil fuels accounts for 80 to 85% of the total anthropogenic sulfur emissions (Whelpdale, 1992). Industrialization has been accompanied by increased levels of sulfite and sulfur dioxide, which contributed to the acid rain phenomenon. Acid rain was a serious problem in Europe in the 1970s (Ulrich, 1980), and research on sulfur as a plant nutrient was at that time limited and mostly restricted to the sulfur deficient areas in the world (Coleman, 1966; Freney and Spencer, 1967; Saalbach, 1973). An issue of prime relevance in climate change is the significance of particulate sulfur in the global sulfur cycle. In East Asia, where under current legislation restrictions of sulfur dioxide (SO_2) emissions are expected to increase further by 34% until 2030 (Ichikawa et al., 2001), aspects of sulfur pollution remain a major concern.

Clean Air Acts: Technological Impact Assessment

At the start of the 1980s, clean air acts came into force and atmospheric sulfur depositions were reduced rapidly in Western Europe (Anonymous, 1983). They declined further in the 1990s after the political transition of Eastern European

---

Copyright © 2008. American Society of Agronomy, Crop Science Society of America, Soil Science Society of America, 677 S. Segoe Rd., Madison, WI 53711, USA. *Sulfur: A Missing Link between Soils, Crops, and Nutrition.* Agronomy Monograph 50.

countries (Downing et al., 1993). On production fields, the appearance of significant sulfur deficiency can be retraced to the early 1980s (Schnug and Pissarek, 1982), and the nutritional disorder spread with decreasing atmospheric sulfur inputs. Widespread symptoms of macroscopic sulfur deficiency can be observed consistently over time in *Brassica* species such as oilseed rape since the introduction of so-called double-low varieties in 1986, in cereals since 1992, and in sugar beet since 1995 (Haneklaus et al., 2006c). The chronological appearance of visual symptoms reflects the order of sulfur demand of these crops in decreasing sequence. Since then, severe sulfur deficiency has become the main nutrient disorder in agricultural crops.

Sulfur-response experiments under field conditions were initiated in Northern Europe, targeted first at maximizing crop productivity (Schnug, 1988; Walker and Booth, 1992) and later at improving crop quality (Hu and Sparks, 1992; Lencioni et al., 1992; Schnug, 1990; Sexton et al., 1998). Scientific research in the field of agronomic aspects of sulfur nutrition was, however, not rapidly implemented on farms because the perception of sulfur as a pollutant; furthermore, the availability of suitable fertilizer products was restricted. Noteworthy in this context is an incident that occurred in Denmark in the early 1990s. Here, one draconian penalty was reported where advisors had to pay farmers $10,500 to compensate for yield losses as a result of having been advised to fertilize winter oilseed rape with sulfur (Pedersen, personal communication). Meanwhile, a major shortcoming of agricultural production in developed countries is the lack of investments in official advisory systems (Maene, 2007), whereby the implementation of advances in fertilizer management relies on the farmer himself.

By now, in sulfur-deficient areas, sulfur fertilization is regularly applied to satisfy the nutrient demand of the crop. However, recommendations for crop-specific fertilizer rates are quite variable (Haneklaus et al., 2006a, 2006b). Sulfur is commonly regarded as biocompatible at excessive rates. An evaluation of various studies in this field indicate that crop plants may react disproportionately to sulfur inputs with losses in productivity and quality of the produce (Haneklaus et al., 2006a).

The impact of excessive sulfur input in temperate regions is dealt with only sporadically. An exception is when atmospheric sulfur pollution affects plant growth (see above). In comparison, in desert agriculture, extremely high sulfur rates are applied regularly for the amelioration of salinity and alkalinity and in the course of cultivating post-mining land (Haneklaus et al., 2006a).

Generally, excess sulfur may cause a premature leaf fall (Motavalli et al., 2006). It seems possible that a uniform application rate of 134 kg ha^{-1} sulfur causes site-specific yield increases and depressions, as was shown in forage grass (Kowalenko 2000). Forage yield was reduced by about 5% at stem extension when 224 kg ha^{-1} sulfur had been applied, while the corresponding value for grain yield was as high as 11% (Girma et al., 2005). Cruciferous crops appear to be able to utilize higher sulfur rates than noncruciferous crops, but results are contradictory. Sulfur applications up to 670 kg ha^{-1} sulfur proved to be useful for broccoli, *Brassica oleracea* var. *botrytis* L. (Sanderson, 2003). In contrast, 45 to 90 kg ha^{-1} sulfur in cabbage (*Brassica oleracea* var. *capitata* L.) resulted in a significant reduction of head size (Rhoads and Olson, 2001).

More notably than the detrimental effects of an over-rated sulfur supply on crop parameters is the possible deleterious effect on animal health. Prominent

■ History of Sulfur Deficiency in Crops

Fig. 4–1. Macroscopic sulfur deficiency in winter oilseed rape before winter.

Fig. 4–2. Chlorosis, together with spoon-like deformations of younger, fully developed leaves of oilseed rape induced by severe sulfur deficiency at start of stem elongation.

Fig. 4–3. White flowering oilseed rape (top) and morphological changes in petals (bottom) under conditions of severe sulfur deficiency.

Fig. 4–4. Deformation of pods together with reduction of number of seeds per pod and enrichment of anthocyanins in oilseed rape under conditions of severe sulfur deficiency during ripening.

Fig. 4–5. Bird's eye view of the small-scale spatial variability of sulfur deficiency in oilseed rape at flowering and winter wheat at stem elongation.

Fig. 4–6. Macroscopic sulfur deficiency in winter wheat at tillering (top) and oats at panicle development (bottom).

■ History of Sulfur Deficiency in Crops 5

Fig. 4–7. Reduced number of kernels per head in winter wheat induced by severe sulfur deficiency.

Fig. 4–8. Macroscopic symptoms of sulfur deficiency in sugar beet in back light at row closing (top) and on field scale (bottom).

Fig. 4–9. Macroscopic sulfur deficiency in leaves of sugar beet at row closing with characteristic distortion of the center rib at the leaf tip.

examples of the adverse effects of high sulfur intake on ruminants are polioencephalomalacia, a neurological disorder, and hemolytic anemia (Stoewsand, 1995; Gould et al., 2002). The risk of polioencephalomalacia exists when grass ingested by the animals contains more than 0.38% sulfur (Gould et al., 2002).

A Major Nutrient Disorder: Sulfur

In 1986, winter oilseed rape varieties with a significantly reduced seed glucosinolate content were first grown on production fields. The corresponding summer varieties all come under the well-known brand name canola. These winter cultivars are traded as double low (00) varieties, whereby the first 0 indicates that seeds are free of erucic acid and the second 0 that the variety has only a low glucosinolate content.

Research during this time period concentrated on isolation and identification of individual glucosinolates (Sorensen, 1985), development of analytical methods for the determination of total and individual glucosinolates (Schnug and Haneklaus, 1990; Wathelet et al., 1995), as well as investigations on glucosinolate synthesis and degradation (Schnug, 1988; Underhill, 1980). The focus of the first international sulfur workshop (Rennenberg et al., 1990) was based on the regulatory aspects of sulfur uptake, metabolism of organic sulfur compounds, and ecological aspects of sulfur metabolism.

Though it was known from greenhouse studies that sulfur deficiency impairs the baking quality of bread-making wheat, *Triticum aestivum* L. (Byers et al., 1987; Randall et al., 1981; Yoshino and McCalla, 1966), it was not until the early 1990s that the detrimental effect of sulfur deficiency on wheat quality was verified in field experimentation (Haneklaus et al., 1992; Haneklaus and Schnug, 1992). Wheat samples from a variety of trials in England and Germany revealed that the baking quality was diminished before crop productivity was reduced. In these samples, the sulfur content in the flour was directly related to the baking quality with each 0.1% of sulfur equaling 40- to 50-mL loaf volume. Noteworthy also is that a lack of protein or sulfur could partly be compensated by increased concentrations of either compound (Haneklaus et al., 1992; Haneklaus and Schnug, 1992).

Recent studies by Muttucumaru et al. (2006) revealed that sulfur deficiency strongly favors the accumulation of acrylamide during the baking process, presumably because of an enrichment of asparagine, which is a precursor of acrylamide.

Severe sulfur deficiency not only impairs crop productivity and quality but also interferes with environmental quality. Under conditions of sulfur deficiency, nitrogen utilization efficiency is reduced so that nitrogen losses from agricultural soils through volatilization and leaching may increase drastically (Schnug 1991). On average, each kilogram of sulfur shortfall causes 15 kg of nitrogen to be lost to the environment. Such nitrogen inputs strongly endanger the stability of natural communities, such as, the growth of algae in water bodies (Wild, 1993). Correcting sulfur deficiency by fertilization is environmentally safe as sulfate is, compared with nitrogen, relatively abundant.

Another ecologically relevant impact of sulfur deficiency is that on honeybees. In relation to the duration of sulfur deficiency, scent and color are affected as well as the size and shape of oilseed rape petals (Schnug and Haneklaus, 2005; Brauer, 2007). These three features of a flowering plant trigger their attraction and number of visiting honeybees. Breakdowns of the sulfur supply for a short time are the reason for white petals that are normally shaped petals. This phenomenon is characteristic on sites where sulfur deficiency is due to decreasing environmental sulfur inputs at the beginning of development. In regions with an established low-sulfur input, as in Northern European growing areas, sulfur-deficient white rapeseed flowers are significantly smaller with more oval-shaped petals (Schnug and Haneklaus, 1994). Even more pivotal, presumably, is the fact that crops visited by insects showed earlier petal fall (Williams, 1985; Sabbahi et al., 2005), the yellow petal color vanishes, and the petals shrink quickly before falling to the ground. Thus, the optical characteristics of a nonpollinated, sulfur-deficient rapeseed flower resemble that of a pollinated, sufficiently sulfur supplied, fading rapeseed flower. Nectar is the bee's source of carbohydrate. The hovering bee is one of the most energy-expensive forms of flight and requires reliable acquisition of food that is based not only on scent, morphology, and color but also on the reflective pattern of flowers.

Historically, plant analysis proved to be a reliable tool to evaluate the nutritional status of plants. Balanced nutrient ratios are vital for crop productivity, quality, and plant health. With PIPPA (Professional Interpretation Program for Plant Analysis), software became available that not only evaluates the status of individual, essential plant nutrients but also appraises results from multiple elemental analyses (Schnug and Haneklaus, 1992). Since then, the database has been updated annually and extended from oilseed rape and cereals only to sugar beet (Schnug and Haneklaus, 2008). In comparison with plant analysis, because of the high spatial and temporal variability of sulfate in soils soil, analytical methods have proven to be unreliable in assessing the status of an agricultural site (Schnug and Haneklaus, 1998; Bloem et al., 2001).

Studies performed in the 1990s revealed that landscape architecture and spatiotemporal variability are main factors governing the sulfur supply on agricultural production fields. Sulfur is often in excess in natural systems, even if the atmospheric sulfur input is low, because the turnover of sulfur is much lower in these systems than in agricultural systems. Another factor is sulfate adsorption at pH values <5. Consequently, natural vegetation, land without plant production, and forests show a positive sulfur balance even when atmospheric depositions are low (Eriksen et al., 1998). Thus, under humid conditions, forests can be sulfur accumulating areas, while agricultural farmland is sulfur consuming. Agricultural crops may benefit from the sulfur pool of adjacent ecosystems if groundwater reservoirs of both ecosystems are connected. As a result, in forest-rich landscapes, the risk of sulfur deficiency in agricultural crops is reduced.

At the field level, sources and sinks commonly included in the sulfur balance are inputs by depositions from atmosphere, fertilizers, plant residues, and mineralization, and outputs include sulfur removed by crop harvest and sulfur losses due to leaching. A frequent problem when establishing such simple sulfur balance is that the budget does not correspond with the actual sulfur supply. The reason is that under temperate conditions, it is the spatiotemporal variation of hydrological soil properties that controls the plant available sulfate-sulfur content. It was not until the end of the 1990s that the significance of plant-available soil water as a source and storage for sulfur was fully acknowledged. As a matter of fact, under humid growing conditions, plant-available soil water proved to be the largest contributor to the sulfur balance (Bloem, 1998). There are three ways groundwater contributes to the sulfur nutrition of plants: (i) if a direct sulfur input that is normally sufficient to satisfy the sulfur demand of a high-yielding crop exists when the groundwater level is 1 to 2 m below surface; (ii) if groundwater that is used for irrigation can supply up to 100 kg ha^{-1} sulfur to the crop (Preuschoff, 1995; Schlichting, 1996; Bloem, 1998; Pedersen et al., 1998), when applied at the start of the main growth period of the crop; and (iii) if capillary rise of groundwater under conditions of water saturation deficit in the upper soil layers leads to a sulfur input.

In general, the sulfur supply of a crop increases with the amount of plant-available water or shallow groundwater. The higher the water storage capacity of a soil, the less likely there will be a loss of water and sulfate sulfur by leaching. The greater the pool of porous-soil water, the more it is likely that there will be an enrichment of sulfate sulfur from subsequent evaporation. Thus, heavier soils have a higher charging capacity for sulfate sulfur than lighter ones.

Since 1995, severe sulfur deficiency has been observed in crops such as sugar beet that have a low sulfur demand. The sugar beet root contains low levels of sulfur in the form of proteins or sulfate so that sulfur removal is low with an average of 0.08 kg sulfur Mg^{-1} of beet roots (Haneklaus et al., 1998). However, a sufficient sulfur supply is required for growth of the canopy. The correlation between total sulfur in leaf tissue and α-amino-nitrogen content in the root of the beet was positive (Haneklaus and Schnug, 1998). A negative correlation as reported by Koch (1996) seems possible in the range of severe sulfur deficiency because an excess of nitrogen promotes the synthesis of amides (Schnug, 1997). In the range of macroscopic sulfur deficiency, increasing sulfur content will therefore decrease the α-amino-nitrogen content because of a more favorable nitrogen/sulfur (N/S) ratio.

Fertilizer research in the 1990s focused on the question of timing of sulfur fertilization, crop-specific sulfur rates, and favorable sulfur form and application mode. The morphogenetic development of cereals and oilseed rape is completely different. Experimentation revealed that a sufficient sulfur supply of cereal crops is required in early growth stages for the development of tillers and for limiting reduction processes that otherwise cannot be fully compensated by sulfur fertilization during later growth (Haneklaus et al., 1995). In contrast, oilseed rape may fully compensate for sulfur starvation at the beginning of flowering. The recommended crop-specific levels of sulfur rates by official advisory services at that time do not always reflect the status of scientific results (Walker, 2002). Foliar sprays with Epsom salt were used because of the lack of sulfur-containing fertilizer products. However, with a dose of about 1.5 kg sulfur ha^{-1} that are commonly applied by this source, the rate was far too low to satisfy the sulfur demand of crops. Additionally, although foliar applications of Epsom salts do favor a rapid uptake of sulfate, Schnug et al. (1995b) showed that most of it is metabolically inactive because it is translocated into the vacuoles. In contrast, elemental sulfur proved to contribute continuously to the sulfur nutrition of the crop because of its continuous release after oxidation.

Elemental sulfur has been widely used in agricultural production since the end of the 19th century because of its fungicidal effect (Hoy, 1987). Recently, it was shown that repeated, foliar applications of elemental sulfur significantly reduced the infection rate of Fusarium head blight (caused by *Fusarium* spp.) by 30% after artificial inoculation under field conditions (Haneklaus et al., 2007a). The clear fungicidal action of elemental sulfur has to be strictly distinguished from the induced effect of sulfur previously taken up by the plant. In Scotland, the infection of oilseed rape plants by fungal pathogens such as *Pyrenopeziza brassicae* Sutton & Rawlinson and *Leptosphaeria maculans* (Desm.) Ces & de Not. increased during the 1980s (Brokenshire et al., 1984), when atmospheric sulfur depositions declined drastically in this region. Sulfur was found to play a key role in the defense system of plants and soil-applied sulfur in the form of sulfate that increased the resistance to various fungal diseases in different crops under greenhouse (Luong et al., 1993; Wang et al., 2003) and field conditions (Dubuis et al., 2005; Klikocka et al., 2005; Schnug et al., 1995a). In greenhouse and field experimentation, soil-applied sulfur fertilization in the form of sulfate reduced the disease index for various host–pathogen relationships. In 1997, Schnug coined the term *Sulfur Induced Resistance* (SIR) to describe the complex biological phenomenon behind these observations. This term denotes the reinforcement of the natural resistance of plants to fungal pathogens through triggering the stimulation of metabolic processes involving sulfur by targeted sulfate-based and soil-applied fertilizer strategies (Haneklaus et al., 2009). It is important to note that SIR is one constituent of the complex phenomenon of *Induced Resistance* (IR) (Haneklaus et al., 2009). Research in this field has strengthened since then and advances made are discussed comprehensively by Bloem et al. (2007) and Haneklaus et al. (2007b, 2009).

In the late 1990s, modeling and computation of plant availability of soil sulfate and evaluation of the sulfur nutritional status, respectively, were a major practical outcome of agronomic research. Analytical methods need to be validated on production fields if critical nutrient values are based on interpretation by regression analysis of individual field and greenhouse experiments, since optima may be different because of other yield-limiting nutrients or growth factors (Bergmann, 1983). In comparison, the BOundary LIne DEvelopment System (BOLIDES) evaluates the relationship between individual growth factors and yield and determines optimum values and ranges of the soil and plant nutrient status, not only of field-grown crops, but also of experimental data (Haneklaus and Schnug, 1998). The principle of this approach and critical nutrient values derived for different crops are given by (Haneklaus and Schnug, 1998; Haneklaus et al., 2006b). Schnug and Haneklaus (2008) provide an overview and evaluation of mathematical approaches for assessing the sulfur nutritional status of a crop and supply verified algorithms for interpretation and comparative weighting of plant analytical data.

Diagnostic tools used for the prognosis of the sulfur nutritional status need to be rapid, precise, and cost-efficient. Neither soil, nor plant analysis, meet all of these requirements. Here, MOPS (Model for the Prognosis of Sulfur-deficiency) provided a solution as a site-specific sulfur balance, which takes into account soil hydrological and physical parameters, and proved to be well suited to predict the sulfur requirement of different crops (Bloem, 1998).

Though the sulfur requirement of agricultural crops grown on organic farms is usually lower than from conventional production because of the lower yield level, an impaired baking quality of wheat is still a serious problem (Hagel, 2000). The problem was aggravated by the fact that conventional varieties that are used in organic farming require a higher nitrogen and sulfur input to give the best results in yield and quality.

The Future

During the last decade, several reference books, which dealt with various aspects of sulfur in the soil–plant system, have been published and are recommended for further reading (Schnug, 1998; Abrol and Ahmad, 2003; Barker and Pilbeam, 2006; Hawkesford and De Kok, 2006; Datnoff et al., 2007). In the new millennium, one key future direction of sulfur research in the agronomic sector has been the involvement of genetic approaches, as the presentations at the Sixth International Sulfur Workshop clearly revealed (Saito et al., 2005). Employing these technologies may accelerate the substitution of crude chemicals for biological expertise, as targeted for instance in SIR (see above). Transdisciplinary cooperation is required to unravel the interrelationships of plants and their environment. Here, not only aspects of sulfur metabolism but interrelationships between different nutrients, beneficial elements, and their metabolic pathways need to be elucidated to develop sophisticated fertilizer management systems for agricultural production. The question of why sulfur deficiency symptoms are more pronounced at high nitrogen levels is still unanswered. Another application of genetic approaches could be the comparative analysis of seed material from organic and conventional breeding programs to verify possible genetic and metabolic differences. Limited information also exists about the relationship of sulfur nutrition, sulfur-betaines (3-dimethylsulfopropionate, DMSP), and osmoprotection (Schnug, 1997).

Other research focuses on natural methods to enhance the quality of phytopharmaceuticals and on the substitution of antibiotics by alternative feed additives. A promising example is the use of nasturtium (*Tropaeolum majus* L.), an herb that meets this demand (Bloem et al., 2008). For this plant, guidelines for cultivation and conditioning of the harvested plant material have been elaborated; this has substantially contributed to a defined and elevated level of glucotropaeolin, the bioactive constituent in the supplement. The parameters for sulfur fertilization, harvesting techniques, drying procedures, and climatic conditions may cause a change in the glucotropaeolin content is about ±50 to 80% (Bloem et al., 2008). In addition, a first-feeding experiment with weaning piglets has been performed (Bloem et al., 2008). Over a period of 5 wk, direct and graded supplementation of *T. majus* was supplemented at an upper dosage of 1 g/kg with the feed, equaling 22.6 mg/kg glucotropaeolin. On an average, 4.5 to 9.2% of the glucotropaeolin taken up by the animals was excreted as bioactive benzyl-isothiocyanate. In principle, it can be expected that the concentration of benzyl-isothiocyanate in the urine was sufficiently high to yield an antimicrobial effect against bacterial infections. Supplementation with *T. majus* had no effect on feed intake and growth performance of piglets (Bloem et al., 2008).

Sulfur nutrition in agricultural crops is one key factor stimulating the nutritional–physiological, sensory, and technical–physical quality of plants and food quality. To overcome possible shortages brought about by an unbalanced diet, the food industry offers nutraceuticals. Dietary supplements and nutraceuticals should, however, be critically evaluated because their regular intake may support, or even encourage, malnutrition with, as yet, unknown consequences for health. Thus, sulfur fertilization together with a balanced nutrient supply are the best guarantees for producing a health-promoting foodstuff.

From an ecological point of view, interactions between sulfur status and biodiversity of plant communities, visiting insects, and soil biota have not been explored sufficiently. The investigation of allelopathic effects on pest and diseases (Auger et al., 1993; Dugravot et al., 2004) as well as sulfur compounds detoxifying allelopathic substances (Li et al., 1993), also deserves further attention.

Summing up almost four decades of plant sulfur research, it can be stated that sulfur, as one of the three philosophic essentials, was of prominent importance in alchemy (Junius, 1979) and is still in the 21st millennium.

Visual Diagnosis of Severe Sulfur Deficiency

Colored images of macroscopic symptoms of sulfur deficiency in oilseed rape, cereals, and sugar beet have been published in Schnug and Haneklaus (1994), Schnug and Haneklaus (2005), and Haneklaus et al. (2006b). Below is a short description of characteristic symptoms given together with selected images.

Oilseed Rape

A detailed description of the symptomatology of sulfur deficiency in *Brassica* crops from seed to harvest including colored images and physiological background information is provided by Schnug (1988), Schnug and Haneklaus (1994), Schnug and Haneklaus (1998), Schnug and Haneklaus (2005), and Haneklaus et al. (2006b).

Severe sulfur deficiency symptoms are very specific in *Brassica* species such as oilseed rape and can be observed throughout the vegetation period and on all plant parts (leaves, flowers, pods) (Fig. 4–1; see color section in this chapter for all illustrations). Symptoms typically appear in the youngest leaves because sulfur and is fairly immobile within the plant. The symptomatological value of 3.5 mg g^{-1} sulfur in younger leaves of oilseed rape at the start of stem elongation reflects the threshold sulfur content when macroscopic deficiency symptoms will appear if the sulfur concentration falls below this value (Haneklaus et al., 2006b).

Leaves that are sulfur deficient begin to develop chlorosis that starts at the leaf's edge, spreading over intercostal areas, but the zones along the veins remain green (Schnug, 1988) (Fig. 4–2). Chlorosis caused by sulfur deficiency never becomes necrotic (Schnug, 1988), as it does with nitrogen and magnesium deficiency; this is an important criterion for differential diagnosis. Even under extreme sulfur deficiency, when an oilseed rape plant shows severe disorders, it will not wither. An increasing nitrogen supply promotes the intensity of sulfur-deficiency symptoms on leaves.

During a long period of severe sulfur deficiency, the chlorotic parts of *Brassica* leaves turn reddish purple because of an enrichment of anthocyanins. Under field conditions, the formation of anthocyanins starts 4 to7 d after chlorosis appears. Many other nutrient deficiencies are also accompanied by the formation of anthocyanins, a less specific indicator for sulfur deficiency. Leaves that are not fully expanded produce regular, spoon-like deformations under conditions of sulfur deficiency. A reduced cell growth rate in the chlorotic areas together with a normal cell growth in the green areas along the veins cause this characteristic. This deformation will be more prominent the less expanded the leaf is at the time of sulfur deficiency occurs.

During the flowering stage, sulfur deficiency causes one of the most impressive symptoms of nutrient deficiency: the "white blooming" of oilseed rape flowers (Fig. 4–3). This symptom is more apparent during periods of high photosynthetic activity, which is the case with anthocyanin accumulation as well (Fig. 4–4). In addition to the remarkable change in color, size, and shape of oilseed rape, the petals are modified (Schnug and Haneklaus, 1994; Brauer, 2007; see above). The petals of sulfur-deficient oilseed rape flowers are smaller and oval shaped compared with the larger and rounder shape of plants without sulfur-deficiency symptoms. Pollen and nectar content of flowers that are sulfur deficient are not affected (Brauer, 2007). Flowers of severely sulfur-deficient oilseed rape plants develop no pods or pods with significantly lower numbers of seeds than normal. The leaves as well as the branches and pods of sulfur-deficient plants are often red or purple as a result of the accumulation of anthocyanins.

The spatial variation of soil hydrological and physical parameters yields a characteristic site-specific patchwork of sulfur deficiency symptoms, which differ in their strength (see above). The reason is the underlying spatiotemporal variability of plant-available sulfate contents in the soil. Sulfur deficiency occurs first on light sandy summits and compacted headland where rooting is hampered, whereas in more clay soil, visual symptoms are seen particularly in conjunction with capillary ascending water or groundwater (Fig. 4–5). Such spatial patterns of sulfur deficiency are consistent over time in the crop rotation as physical soil properties are stable and long term (Schnug and Haneklaus, 1994, 1998; Haneklaus and Schnug, 2006). In principal, sulfur deficiency in oilseed rape can be balanced by adequate fertilizer rates until the start of flowering if weather conditions favor a rapid availability of the product.

Cereals

Since the early 1990s, macroscopic sulfur deficiency symptoms have been regularly observed in cereals. A detailed description of symptomatology of sulfur deficiency in cereals including colored images is given by Haneklaus and Schnug (1998) and Haneklaus et al. (2006b).

Sulfur deficiency symptoms in gramineous crops such as cereals and maize are less specific than in dicotyledonous crops. The symptomatological value 1.2 mg g^{-1} sulfur in the whole above-ground biomass of the crop reflects the threshold sulfur content for the appearance of macroscopic sulfur deficiency symptoms (Haneklaus et al., 2006b).

In early-growth stages of wheat, plants remain smaller, their appearance is stunted, and color is lighter than in plants without symptoms (Fig. 4–6). Chlorosis is often accompanied by light green stripes around along the veins (Fig. 4–6).

Neither of deformations that are seen in rape has been observed; furthermore, cereal plants do not accumulate anthocyanines regularly. Because symptoms are nonspecific, they can easily be misinterpreted as symptoms of nitrogen deficiency. However, specific patterns in fields, as described previously for oilseed rape, provide sufficient evidence for sulfur deficiency. Because of an early reduction of fertile flowers per head, sulfur deficient cereals have a reduced number of kernels per head (Fig. 4–7). Thus, the sulfur demand of cereals needs to be fully met from seeding to harvest. This may require sulfur fertilization of winter crops in autumn.

Sugar Beet

Since 1995, severe sulfur deficiency has affected crops with a low sulfur demand such as sugar beet. A detailed description of symptomatology of sulfur deficiency in sugar beet including colored images is given by Haneklaus and Schnug (2000) and Haneklaus et al. (2006b). The symptomatological value, which reflects the threshold sulfur content for the appearance of macroscopic sulfur deficiency symptoms is 1.7 mg g^{-1} sulfur in younger, fully developed leaves of sugar beet at row closing (Haneklaus et al., 2006b). In particular, back-light sugar beet leaves are bright yellow (Fig. 4–8). Growth is reduced, and the whole paddock exhibits chlorosis. Plants with severe symptoms have leaves that are in an upright position. In comparison, leaves that are sufficiently supplied are fan-shaped.

Younger, fully-developed leaves are longer (ratio of basal width to length of leaf blade = 0.5–0.65) than well-supplied plants, which show a ratio of 0.75 to 0.76 (Fig. 4–9). The absence of lobuli at the basis of the leaf blade (Fig. 4–9) and torsion-like deformations at the acropetal part of the middle ridge are characteristic. This effect is more pronounced in bigger leaves. Leaves show a characteristic convex curving because of reduced cell growth at the leaf edge when compared with leaf areas along the veins. This symptom is less pronounced in oilseed rape, as noted, where leaves develop spoon-like deformations under conditions of severe sulfur deficiency. Another discriminative feature is that oilseed rape plants show dark green areas around the veins, while the sugar beet crop shows a diffuse yellowing in the corresponding plant parts. The sugar beet root contains low levels of sulfur in the form of proteins or sulfate so that sulfur removal is on average 0.08 kg sulfur Mg^{-1} of beet roots (Haneklaus et al., 1998). However, a sufficient sulfur supply is required for canopy growth.

References

Abrol, Y.P., and A. Ahmad. 2003. Sulphur in plants. Kluwer Academic Publishers, Dordrecht, the Netherlands.

Andreae M.O. 1986. The ocean as a source of atmospheric sulfur compounds. p. 331–362. *In* P. Buat-Menard (ed.) The role of air-sea exchange in geochemical cycling. Reidel, Dordrecht, the Netherlands.

Anonymous. 1983. Verordnung über Großfeuerungsanlagen. Dreizehnte Verordnung zur Durchfuehrung des Bundes-Immissionsschutzgesetzes (13 BimschV) vom 22. June 1983. BGBl. IS.719.

Auger, J., C. Lecomte, and E. Thibout. 1993. Allium-sulfur compounds—Biological activities in insects and biosynthesis. Acta Bot. Gallica 140:157–168.

Barker, A.V., and D.J. Pilbeam. 2006. Handbook of plant nutrition. CRC Press, Boca Raton, FL.

Bergmann, H. 1983. Ernährungsstoerungen bei Kulturpflanzen. Gustav Fischer Verlag, Stuttgart.

Bloem, E. 1998. Schwefel-Bilanz von Agraroekosystemen unter besonderer Beruecksichtigung hydrologischer und bodenphysikalischer Standorteigenschaften. Dissertation TU Braunschweig. Landbauforschung Voelkenrode Sonderheft 192, 1–156.

Bloem, E., A. Berk, S. Haneklaus, and E. Schnug. 2008. Influence of *Tropaeolum majus* supplements on growth and antimicrobialcapacity of glucotropaeolin in piglets. FAL– Agric. Res. (in press).

Bloem, E., S. Haneklaus, I. Salac, P. Wickenhäuser, and E. Schnug. 2007. Facts and fiction about sulphur metabolism in relation to plant-pathogen interactions. J. Plant Biol. 9:596–607.

Bloem, E., S. Haneklaus, G. Sparovek, and E. Schnug. 2001. Spatial and temporal variability of sulphate concentration in soils. Commun. Soil Sci. Plant Anal. 32:1391–1403.

Brauer, A. 2007. Einfluss der Schwefelversorgung auf morphologogische und physiologische Parameter von Rapsblüten (*Brassica napus* L.) und deren Wirkung auf das Verhalten von Honigbienen. PhD thesis, Der Andere Verlag, ISBN 978-3-89959-643-4.

Brokenshire, T., A.G. Channon, and S. Wale. 1984. Recognising oilseed rape disease. Publ. 135, The Scottish Agricultural Colleges, Edinburgh.

Byers, M., J. Franklin, and S.J. Smith. 1987. The nitrogen and sulfur nutrition of wheat and its effect on the composition and baking quality of the grain. Aspects Appl. Biol. 15:327-344.

Coleman, R. 1966. The importance of sulfur as a plant nutrient in world crop production. Soil Sci. 101:230-239.

Datnoff, L., W. Elmer, and D. Huber. 2007. Mineral nutrition and plant diseases. APS Press, St. Paul, MN.

Downing, R.J., J.-P. Hettelingh, and P.A.M. Sment. 1993. Calculation and mapping of critical loads in Europe: Status report 1993. RIVM report No. 259101003.

Dubuis, P.H., C. Marazzi, E. Städler, and F. Mauch. 2005. Sulphur deficiency causes a reduction in antimicrobial potential and leads to increased disease susceptibility of oilseed rape. J. Phytopathol. 153:27-36.

Dugravot, S., E. Thibout, A. Abo-Ghalia, and J. Huignard. 2004. How a specialist and a nonspecialist insect cope with dimethyl disulfide produced by *Allium porrum*. Entomol. Exp. Appl. 13:173-179.

Eriksen, J., M.D. Murphy, and E. Schnug. 1998. The soil sulphur cycle. p. 39-74. *In* E. Schnug (ed.) Sulphur in agroecosystems. Kluwer Academic Publishers, Dordrecht, the Netherlands.

Freney, J.R., and K. Spencer. 1967. Diagnosis of sulphur deficiency in plants. J. Aust. Inst. Agric. Sci. 33:284-288.

Girma, K., J. Mosali, K.W. Freeman, W.R. Raun, K.L. Martin, and W.E. Thomason. 2005. Forage and grain yield response to applied sulfur in winter wheat as influenced by source and rate. J. Plant Nutr. 28:1541-1553.

Gould, D.H., D.A. Dargatz, F.B. Garry, D.H. Hamar, and P.F. Ross. 2002. Potentially hazardous sulfur conditions on beef cattle ranches in the United States. JAVMA 221:673-677.

Hagel, I. 2000. Differenzierung und Charakterisierung von Weizen verschiedener Anbausysteme und Sorten durch Proteinfraktionierung. FAL- Agric. Res. 208 (special issue).

Haneklaus, S., E. Bloem, U. Funder, and E. Schnug. 2007a. Effect of foliar-applied elemental sulphur on Fusarium infections in barley. FAL Agric. Res. 57:213-217.

Haneklaus, S., E. Bloem, and E. Schnug. 1998. Sulphur balance of a sugar beet rotation Aspects Appl. Biol. 52:121-126.

Haneklaus, S., E. Bloem, and E. Schnug. 2006a. Sulphur interactions in crop ecosystems. p. 17-58. *In* M.J. Hawkesford and L.J. DeKok (ed.) Sulfur in plants—An ecological perspective. Springer, Dordrecht, the Netherlands.

Haneklaus, S., E. Bloem, and E. Schnug. 2007b. Sulfur and plant disease. p. 101-118. *In* L. Datnoff et al. (ed.) Mineral nutrition and plant diseases. APS Press, St. Paul, MN.

Haneklaus, S., E. Bloem, and E. Schnug. 2009. Plant disease control by nutrient management: Sulphur. *In* D. Walters (ed.) Disease control in crops: Biological and environmentally-friendly approaches. Wiley-Blackwell Publishers, Chichester, UK (in press).

Haneklaus, S., E. Bloem, E. Schnug, L. De Kok, and I. Stulen. 2006b. Sulphur. p. 183-238. *In* A.V. Barker and D.J. Pilbeam (ed.) Handbook of plant nutrition. CRC Press, Boca Raton, FL. Haneklaus, S., J. Fleckenstein, and E. Schnug. 1995. Comparative studies of plant and soil analysis for the evaluation of the sulphur status of oilseed rape and wheat Z. Pflanzenernaehrung Bodenkunde 158:109-111.

Haneklaus, S., E. Evans, and E. Schnug. 1992. Baking quality and sulphur content of wheat. I. Relations between sulphur and protein content and loaf volume. Sulphur Agric. 16:31-34.

Haneklaus, S., and E. Schnug. 1992. Baking quality and sulphur content of wheat. II. Evaluation of the relative importance of genetics and environment including sulphur fertilisation. Sulphur Agric.16:335-338.

Haneklaus, S., and E. Schnug. 1998. Evaluation of critical values for soil and plant analysis of sugar beets by means of boundary lines applied to field survey data. Aspects Appl. Biol. 52:87–94.

Haneklaus, S., and E. Schnug. 2006. Site specific nutrient management—Objectives, current status and future research needs. p. 91–151. *In* A Srinivasan (ed.) Precision farming – A global perspective., Marcel Dekker, New York.

Haneklaus, S., K.C. Walker, and E. Schnug. 2006c. A chronicle of sulfur research in agriculture. p. 249–256. *In* K. Saito et al. (ed.) Sulfur transport and assimilation in plants in the post genomic era. 2005. Backhys Publ., Leiden, the Netherlands.

Hawkesford, M.J., and L.J. DeKok. 2006. Sulfur in plants—An ecological perspective. Springer, Dordrecht, the Netherlands.

Hoy, M.A. 1987. Sulfur as a control agent for pest mites in agriculture. p. 1:51–61. *In* Proceedings of the International Symposium on Elemental Sulphur in Agriculture, Nice.

Hu, H.N., and D. Sparks. 1992. Nitrogen and sulfur interaction influences net photosynthesis and vegetative growth of pecan. J. Am. Soc. Hortic. Sci. 117:59–64.

Ichikawa, Y., H. Hayami, T. Sugiyama, M. Amann, and W. Schoepp. 2001. Forecast of sulfur deposition in Japan for various energy supply and emission control scenarios. Water Air Soil Pollut. 130:301–306.

Junius, M.M. 1979. Alchimia verde–Spagyrica vegetale. Rome.

Klikocka, H., S. Haneklaus, E. Bloem, and E. Schnug. 2005. Influence of sulphur fertilization on the infection of potato tubers (*Solanum tuberosum*) with *Rhizoctonia solani* and *Strepomyces scabies*. J. Plant Nutr. 28:819–833.

Koch, H.-J. 1996. Schwefelduengung in Zukunft auch zu Zuckerrueben? Landpost vom 17. Feb. 1996, 52–54.

Kowalenko, C.G. 2000. Response of forage grass to sulphur applications on coastal British Columbia soils. Can. J. Soil Sci. 84:227–236.

Lencioni, L., G. Lotti, and A. Ranieri. 1992. Effects of sulfur on rapeseed growth and rapeseed-oil composition. Agrochimica 36:185–192.

Li, H.H., H. Nishimura, K. Hasegawa, and J. Mizutani. 1993. Some physiological effects and the possible mechanism of action of juglone in plants. Weed Res. 38:214–222.

Luong, H., E.J. Booth, and K.C. Walker. 1993. Utilisation of sulphur nutrition to induce the natural defence mechanisms of oilseed rape. Agriculture Group Symposium; Novel aspects of crop nutrition J. Sci. Food Agric. 63:119–120.

Maene, L. 2007. Plant nutrition and human well-being: An industry perspective. p. 3–13. *In* Proceedings Int. Symposium of CIEC Mineral versus Organic Fertilization–Conflict or Synergism? Gent, Belgium.

Motavalli, P., T. Marler, F. Cruz, and J. McConnell. 2006. Essential plant nutrients. ; see http://www.cartage.org.lb/en/themes/Sciences/BotanicalSciences/PlantHormones/EssentialPlant/EssentialPlant.htm; verified 28 March 2008.

Muttucumaru, N., N.G. Halford, J.S. Elmore, A.T. Dodson, M. Parry, P.R. Shewry, nad D.S. Mottram. 2006. Formation of high levels of acrylamide during the processing of flour derived from sulfate-deprived wheat. J. Agric. Food Chem. 54: 8951–8955.

Pedersen, C.A., L. Knudsen, and E. Schnug. 1998. Sulphur fertilisation. p. 115–134. *In* E. Schnug (ed.) Sulphur in agroecosystems. Kluwer Academic Publishers, Dordrecht, the Netherlands.

Preuschoff, M. 1995. Untersuchungen zur Schwefelversorgung von Weißkohl an zwei Loeßstandorten. PhD, University Hanover, Verlag Ulrich E. Grauer, Stuttgart.

Randall, P.J., K. Spencer, and J.R. Freney. 1981. Sulphur and nitrogen fertiliser effects on wheat. I. Concentrations of sulphur and nitrogen and the nitrogen to sulphur ratio in grain in relation to the yield response. Aust. J. Agric. Res. 32:203–212.

Rennenberg, H., C. Brunold, L.J. De Kok, and I. Stulen. 1990. Sulfur nutrition and sulfur assimilation in higher plants. SPB Academic Publ., The Hague.

Rhoads, F.M., and S.M. Olson. 2001. Cabbage response to sulfur and nitrogen rate. Soil Crop Sci. Soc. Florida Proc. 60:37–40.

Saalbach, E. 1973. The effect of sulphur, magnesium and sodium on yield and quality of agricultural crops. Pontificiae Academiae Scientiarum scripta Varia 38: 1–49.

Sabbahi, R., D. De Oliveira, and J. Marceau. 2005. Influence of honey bee (Hymenoptera: Apidae) density on the production of canola (Crucifera: Brassicacae). J. Econ. Entomol. 98:367–372.

Saito, K., L.J. De Kok, S. Stulen, M.J. Hawkesford, E. Schnug, A. Sirko, and H. Rennenberg. 2005. Sulfur transport and assimilation in plants in the post genomic era. Backhys Publ., Leiden, the Netherlands.

Sanderson, K.R. 2003. Broccoli and cauliflower response to supplemental soil sulphur and calcium. Acta Hortic. 627:171–179.

Schlichting, M. 1996. Der Sulfatgehalt des Grundwassers in Abhaengigkeit von bodennutzungsspezifischen Stoffeintraegen und dessen Bedeutung fuer die Nutzung des Grundwassers zur Feldberegnung– Beispiel Fuhrberger Feld. Master thesis, University of Hanover.

Schnug, E., and H.-P. Pissarek. 1982. Kalium und Schwefel, Minimumfaktoren des schleswig-holsteinischen Rapsanbaus. Kali-Briefe (Büntehof) 16:77–84.

Schnug, E. 1988. Quantitative und qualitative Aspekte der Diagnose und Therapie der Schwefelversorgung von Raps (*Brassica napus* L.) unter besonderer Berücksichtigung glucosinolatarmer Sorten. Habilitationsschrift (Dsc thesis), Kiel, Germany (published under www.pb.fal.de; verified 15 March 2008).

Schnug, E. 1990. Sulphur nutrition and quality of vegetables. Sulphur Agric. 14:3–7.

Schnug, E., and S. Haneklaus. 1990. Quantitative glucosinolate analysis in Brassica seeds by X-ray fluorescence spectroscopy. Phytochem. Anal. 1:40–43.

Schnug, E. 1991. Sulphur nutritional status of European crops and consequences for agriculture. Sulphur Agric. 15:7–12.

Schnug, E. 1997. Significance of sulphur for the nutritional and technological quality of domesticated plants. p. 109–130. *In* W.J. Cram et al (ed.) Sulfur nutrition and sulfur assimilation in higher plants II. Backhuys Publ., Leiden, the Netherlands.

Schnug, E. 1998. Sulfur in agroecosystems. Kluwer Academic Publishers, Dordrecht, the Netherlands.

Schnug, E., E. Bloem, and S. Haneklaus. 2000. Schwefelmangel—jetzt auch in Rüben? Top Agrar 3:122–125.

Schnug, E., E. Booth, S. Haneklaus, and K.C. Walker. 1995a. Sulphur supply and stress resistance in oilseed rape. p. 229–231. *In* Proc. 9th Int. Rapeseed Congress, Cambridge, UK.

Schnug, E., and S. Haneklaus. 1992. PIPPA: Un programme d'interprétation des analyses de plantes pour le colza et les céréales. Perspec.Agricol. Supp.71:30–32.

Schnug, E., and S. Haneklaus. 1994. Sulphur deficiency in Brassica napus- biochemistry, symptomatology, morphogenesis. Landbauforschung Völkenrode, FAL-Braunschweig, Sonderheft 144.

Schnug, E., and S. Haneklaus. 1998. Diagnosis of sulphur nutrition. p. 1–38. *In* E. Schnug (ed.) Sulphur in agroecosystems. Kluwer Academic Publishers, Dordrecht, the Netherlands.

Schnug, E., and S. Haneklaus. 2005. Sulphur deficiency symptoms in oilseed rape (*Brassica napus* L.)—The aesthetics of starvation. Phyton 45: 79–95.

Schnug, E., and S. Haneklaus. 2008. Evaluation of the relative significance of sulfur and other essential mineral elements in oilseed rape, cereals, and sugar beet production. p. 219–234. *In* J. Jez (ed.) Sulfur: A missing link between soils, crops, and nutrition. Agron. Monogr. 50. ASA, CSSA, SSSA, Madison, WI.

Schnug, E., H.-M. Paulsen, H. Untiedt, and S. Haneklaus. 1995b. Fate and physiology of foliar applied sulphur compounds in *Brassica napus*. p. 91–100. Proc. IAOPN Symposium, Cairo.

Sexton, P.J., N.C. Paek, and R. Shibles. 1998. Soybean sulfur and nitrogen balance under varying levels of available sulfur. Crop Sci. 4:975–982.

Sorensen, H. 1985. Limitations and possibilities of different methods suitable to quantitative analysis of glucosinolates occurring in double low rapeseed and products hereof. *In* Advances in the production and utilization of cruciferous crops, p. 73–84.

Stoewsand, G.S. 1995. Bioactive organosulfur phytochemicals in *Brassica oleracea* vegetables—A review. Food Chem. Toxicol. 33:537–543.

Ulrich, B. 1980. Die Waelder in Mitteleuropa: Messergebnisse ihrer Umweltbelastung, Theorie ihrer Gefaehrdung, Prognose ihrer Entwicklung. Allgemeine Forstzeitschrift 44 (special issue).

Underhill, E.W. 1980. Glucosinolates. Encyclopedia Plant Physiol. 8:493–511.

Walker, K.C., and E.J. Booth. 1992. Sulphur research on oilseed rape in Scotland. Sulphur Agric.16:15–19.

Walker, K.C. 2002. Sulphur fertiliser recommendations in Europe. Proc. Int. Fert. Soc. 506: 0–20.

Wathelet, J.P., M. Marlier, M. Severin, A. Boenke, and P.J. Wagstaffe. 1995. Measurement of glucosinolates in rapeseeds. Nat. Toxins 3:299–304.

Wang, J., J. Zhang, Y. Ma, L. Wang, L. Yang, S. Shi, L. Liu, and E. Schnug. 2003. Crop resistances to diseases as influenced by sulphur application rates. Proc. 12th World Fertiliser Congress, Beijing, China, p. 1285–1296.

Whelpdale, D.M. 1992. An overview of the atmospheric sulphur cycle. p. 5–26. *In* R.W. Howarth et al. (ed.) Sulphur cycling on the continents: Wetlands, terrestrial ecosystems and associated water bodies, SCOPE 48. John Wiley & Sons, Chichester, UK.

Wild, A. 1993. Umweltorientierte Bodenkunde. Spektrum Akademischer Verlag: Heidelberg.

Williams, I.H. 1985. The pollination of Swede rape (*Brassica napus* L). Bee World 66(1):16–22.

Yoshino, D., and A.G. McCalla. 1966. The effects of sulfur content on the properties of wheat gluten. Can. J. Biochem. 44:339–346.

5

Availability of Sulfur to Crops from Soil and Other Sources

Warren A. Dick, David Kost, and Liming Chen
The Ohio State University, Wooster, Ohio

Abstract

Sulfur deficiencies in crops have increased worldwide because of decreases in sulfur inputs to the soil system and increases in sulfur outputs. This has made many crops vulnerable to yield reductions. The goal of soil and crop nutrient management systems is to make sure that there is an adequate, but not an excessive, amount of nutrients available to the growing crop. Excessive application of nutrients is not only inefficient but can negatively affect the environment. It is important to understand what limits sulfur availability to crops and the resources available to remove sulfur availability limitations. This chapter reviews sulfur inputs to soil and sulfur outputs from soil and how they affect sulfur availability to crops. General approaches and tests developed to measure the level of sulfur in soils that is available to crops and potential crop responses to inputs of fertilizer sulfur are also reviewed. Research needs to improve our ability to estimate soil-sulfur availability, and crop responses to sulfur fertilizer inputs conclude the chapter.

Sulfur as a Major Plant Nutrient

Sulfur is an element essential for plant growth. It is a macronutrient and—like nitrogen, phosphorus, potassium, calcium, and magnesium—must be available in relatively large amounts for good crop growth. Total sulfur concentrations in plants normally vary from 0.1 to 0.3% but under some conditions may range from a low of 0.05% to a high of 0.9% (Blanchar, 1986). Sulfur is a constituent of the amino acids cysteine, cysteine, and methionine and hence of protein. When sulfur is deficient, the sulfur-containing amino acid content in plants decreases, and the synthesis of proteins is inhibited (Marschner, 1995). Sulfur in plants is also a structural constituent of many coenzymes and secondary plant products or acts as a functional group directly involved in metabolic reactions. Sulfur requirements for different crops vary considerably.

Sulfur as an Overlooked Plant Nutrient

Sulfur sources for crop production are, in general, the soil and the atmosphere above the soil. Sulfur is similar to nitrogen in that gaseous compounds

---

Copyright © 2008. American Society of Agronomy, Crop Science Society of America, Soil Science Society of America, 677 S. Segoe Rd., Madison, WI 53711, USA. *Sulfur: A Missing Link between Soils, Crops, and Nutrition.* Agronomy Monograph 50.

play an important role in cycling but differs from nitrogen in that there is no large reservoir of sulfur compounds in the atmosphere.

In recent years, deficiencies of sulfur in crops have increased worldwide. This is attributed to decreases in sulfur inputs to the soil system and increases in sulfur outputs. Decreased sulfur inputs are primarily due to use of highly concentrated fertilizers containing little or no sulfur (Scherer, 2001) and less sulfur deposition from the atmosphere (National Atmospheric Deposition Program, 2007). Increased sulfur outputs include intensive cropping systems that increase crop yields and, thus, more removal of sulfur from the field (Ohio Department of Agriculture, 2006).

Crop sulfur deficiency was recently reported to be widespread in the Madhya Pradesh state of India because of use of concentrated fertilizers and intensive cropping (Chibber, 2007). In Ohio, average yields of hay (mostly alfalfa, *Medicago sativa* L.) increased from 5.5 Mg ha^{-1} in 1977 to 1979 to 6.5 Mg ha^{-1} in 2003 to 2005. Corn (*Zea mays* L.) yields increased from 6.8 to 9.6 Mg ha^{-1}, soybean [*Glycine max* (L.) Merr.] yields from 2.3 to 2.9 Mg ha^{-1}, and wheat (*Triticum aestivum* L.) yields from 3.0 to 4.5 Mg ha^{-1} during the same period (Ohio Department of Agriculture, 2006). This means that approximately 18 to 50% more sulfur is removed by crops from the soils each year compared to 25 yr ago. Also in Ohio, annual sulfur deposition gradually decreased by 37% from 34.8 kg ha^{-1} in 1979 to 22.0 kg ha^{-1} in 2005 (National Atmospheric Deposition Program, 2007). As the sulfur dioxide (SO_2) produced during fuel burning is removed from flue gases via some type of scrubbing technology to meet clean air regulations, less sulfur is deposited on the soil for subsequent crop uptake. Because of these trends in sulfur inputs and outputs, crop response to sulfur application on agricultural soils will probably occur with greater frequency in the future.

Crop response to sulfur on any particular soil will vary depending on the soil type and sulfur requirement of the crop, which is high for alfalfa and relatively low for corn and soybean. Soils that exhibit sulfur deficiencies are generally those low in organic matter, coarse-textured, well drained, and subject to leaching. Alfalfa yields were increased by sulfur application on sandy loam soils but not on silt loam soils in Minnesota (O'Leary and Rehm, 1989; Sloan et al., 1999). Alfalfa yields were not affected by sulfur application in central Maryland in the USA (Vough et al., 1986) and on fine sandy loam soils on Prince Edward Island in Canada (Gupta and MacLeod, 1984). Alfalfa yields were increased by sulfur application on a silt loam soil in Ohio (Chen et al., 2005). In substantial amounts of farmland in western Canada, the yields of corn and soybean were significantly increased by sulfur fertilizer treatments (Beaton and Soper, 1986). Soybean and corn also responded to sulfur application on some experimental sites in Ohio (Chen et al., 2005). When experimental sites were located on soils containing a higher concentration of organic matter and receiving greater amounts of sulfur from precipitation and air pollution, soybean yields were not increased (Chen et al., 2005).

Sources of Available Sulfur

Mineralization of Soil Organic Matter

Organic matter and plant residues are major sources of plant-available sulfur in the soil. Any practice that removes crop residues, in addition to harvested grain, will eventually have a negative impact on available sulfur. Organic sulfur levels

Table 5–1. Ratios of organic carbon, nitrogen, phosphorus, and sulfur in soil.

Soils sampled	Organic ratios	Organic ratio values
Soil humus†	C/N/P/S	140:10:1.3:1.3
	C/S	108:1
	N/S	7.7:1
Brazilian soils‡	C/N/P/S	194:10:1.2:1.6
	C/S	121:1
	N/S	6.3:1
Iowa soils‡	C/N/P/S	110:10:1.4:1.3
	C/S	85:1
	N/S	7.7:1

† Stevenson and Cole (2000).
‡ Tabatabai (2005).

may be maintained or increased under intensive cropping conditions if adequate masses of crop residues are returned to the soil. Wheat straw and corn stover generally contain about 0.10 to 0.22% sulfur (Miller, 1993). A silty loam soil that was cropped to corn, and well fertilized with nitrogen, also received annual additions of either alfalfa residue or cornstalks (Larson et al., 1972). After 11 yr, organic sulfur increased by 16% when 2 Mg alfalfa residue ha^{-1} yr^{-1} and by 49% when 16 Mg alfalfa residue ha^{-1} yr^{-1} were applied to the soil. If cornstalks were applied at similar rates, the corresponding increases in organic sulfur were 9% (2 Mg ha^{-1} yr^{-1}) and 40% (16 Mg ha^{-1} yr^{-1}). Amendment of soil with biosolids, manures, and composts will generally increase available sulfur. Biosolids may contain 0.3 to 1.2% sulfur (Miller, 1993). Most animal manures contain 0.25 to 0.30% sulfur except for sheep manure and poultry manure, which average approximately 0.35% sulfur and 0.50% sulfur, respectively (Banwart and Bremner, 1975).

The carbon/sulfur (C/S) ratio of organic materials determines whether there is a net release of mineral sulfate-sulfur from organic materials or a net immobilization into microbial biomass. Generally, a C/S ratio <200 results in a net release of plant available mineral sulfur and a ratio >400 results in a net immobilization. A C/S ratio of an organic material between 200 and 400 results in either no change in mineral soil-sulfur concentrations or a small net increase or decrease (Freney, 1967). Once the organic materials applied to the soil have undergone their mineralization or immobilization reactions, the C/S and nitrogen/sulfur (N/S) ratios of the soil organic matter tend to stabilize near values shown in Table 5–1.

The amount of plant-available sulfur released by mineralization of organic matter each year depends on the amount of organic sulfur in the soil and its turnover rate. Organic sulfur can be measured directly but is often estimated on the basis of its presumed ratio to organic carbon. The ratio of organic carbon to organic sulfur varies widely among soils, but a value of 100:1 is often used (Tabatabai, 2005). The annual turnover rate of soil organic sulfur is not precisely known but has been estimated at 2% on the basis of the rate of organic matter turnover (McGrath and Zhao, 1995). On the basis of these values, a soil that contains 1% organic carbon and a bulk density of 1325 kg m^{-3} to 20-cm depth would release 5.3 kg sulfur ha^{-1} each year by mineralization of organic sulfur, as follows:

$$\frac{1\,\text{kg org C}}{100\,\text{kg soil}} \times \frac{10^4\,\text{m}^2}{\text{ha}} \times 0.2\,\text{m depth} \times \frac{1325\,\text{kg soil}}{1\,\text{m}^3\,\text{soil}} \times \frac{1\,\text{kg org S}}{100\,\text{kg org C}} \times \frac{2\,\text{kg S min}}{100\,\text{kg org S}} = \frac{5.3\,\text{kg S min}}{\text{ha}}$$

Soils that contain more than 1% organic carbon would release proportionately more available sulfur each year. For example, a soil that contains 3% organic carbon and has a turnover rate of 2% would release 15.9 kg sulfur ha^{-1} each year. The amount of sulfur mineralized is also quite sensitive to the turnover rate of organic sulfur. If the turnover rate of organic sulfur were 3% instead of 2% (a 50% increase), a soil with 3% organic carbon would release 23.8 kg sulfur ha^{-1} each year. An increase of 7.9 kg sulfur ha^{-1} each year, compared with the increase of sulfur in soil that will occur when there is a 2% turnover rate value, could be the difference between sulfur sufficiency and deficiency of a crop.

Inorganic Transformations

Inorganic sulfur is generally much less abundant in most agricultural soils from humid and semihumid regions than is organically bound sulfur, with the inorganic sulfur accounting for <5% of the total sulfur. The form of inorganic sulfur found in soils is mostly sulfate (SO_4^{2-}). However, in strongly acid soils or in reduced soils, small amounts of reduced sulfur compounds may be found. These compounds include sulfides (S^{2-}), elemental sulfur (S^0), thiosulphate ($S_2O_3^{2-}$), tetrathionate ($S_4O_6^{2-}$), and sulfite (SO_3^{2-}) (Konopka et al., 1986). Sulfur minerals in drier regions of the world include the moderately soluble gypsum ($CaSO_4 \cdot 2H_2O$) and the more soluble epsomite ($MgSO_4 \cdot 7H_2O$). In humid regions, these minerals are leached from the soil and are rarely found.

Sulfate in soils may be found in the soil solution, adsorbed onto soil colloids, and/or exist in insoluble mineral forms (Tabatabai, 1987). Surface horizons of most well-drained temperate soils contain only small amounts of water-soluble sulfate, and this sulfate in the soil solution is in equilibrium with the solid phase forms. Soils vary widely in their capacity to adsorb sulfate, and adsorption usually increases with the clay content of soils, with hydrous oxides of Al and Fe especially showing marked tendencies to retain sulfate. Adsorption of sulfate in soil systems is also favored by strongly acid conditions. Acid subsoils with high clay content can contain sizeable reserves of available sulfate. Phosphate may displace or reduce the adsorption of sulfate. In arid regions, sulfates of calcium, magnesium, potassium, and sodium are the predominant inorganic sulfur forms. Gypsum ($CaSO_4 \cdot 2H_2O$) is widely employed as a soil amendment to improve physical and chemical properties and to provide a soluble source of calcium and sulfur for plant nutrition (Shainberg et al., 1989).

Because plants take up sulfur from soil in the form of sulfate, reduced sulfur compounds must first be oxidized to sulfate to be available for uptake by crops. Oxidation of those reduced compounds to sulfate under aerobic conditions can be chemical, microbiological, or a combination of both (Konopka et al., 1986). Large and diverse groups of microorganisms have been shown to be capable of oxidizing one or more reduced sulfur compounds in the biosphere. Autotrophic sulfur oxidizers such as the genus *Thiobacillus* were once believed to dominate these processes, but the involvement of heterotrophs is now given more importance. Heterotrophs were the most abundant elemental sulfur oxidizers in Canadian (Saskatchewan) soils (Schoenau and Germida, 1992). The sulfur oxidation pathway, starting with S^{2-}, is shown below:

$$S^{2-} \to S^0 \to [S_2O_3^{2-} \to S_4O_6^{2-}] \to SO_3^{2-} \to SO_4^{2-}$$

Acid sulfate soils are formed when sulfide-bearing soil materials are oxidized, when minerals are weathered by the sulfuric acid produced, and when new mineral phases are formed from the dissolution products (Fanning et al., 2002). Acid sulfate soils can often be found along coastal areas of continents. Modern disturbance of the earth's surface because of activities such as surface mining and highway construction have exposed other areas that result in the formation of acid sulfate soils. Such soils can cause severe agricultural and environmental problems, but if properly managed can also be quite productive.

Available sulfur can be changed into unavailable sulfide-sulfur forms, e.g., pyrite, in water-logged soils. Most of the sulfides that occur in soils have formed under ambient conditions of temperature and pressure in aquatic environments. The basic reactions are as follows:

$H_2SO_4 + 2\ CH_2O \rightarrow H_2S + 2\ H_2CO_3$

$4\ Fe_2O_3 + 9\ H_2S \rightarrow 8\ FeS + H_2SO_4 + 2\ HCO_3^-$

$Fe^{2+} + S^{2-} \rightarrow FeS$ (formation of pyrite under strictly anaerobic conditions)

$FeS + S^0 \rightarrow FeS_2$ (formation of pyrite from elemental sulfur)

Wet and Dry Deposition Inputs of Sulfur via Precipitation

In the global sulfur cycle there are exchanges between the atmosphere and the soils and waters of the earth's surface. Sulfur enters the atmosphere by a variety of processes including combustion of fuels and other sulfur-containing materials, roasting of sulfur-containing ores, volcanic and soil emissions, and as gases and aerosols from the oceans. Global emissions of sulfur to the atmosphere (185 million Mg yr^{-1}) were partitioned into approximately 35% anthropogenic (fuel combustion, ore roasting), 34.5% from soils, 24.0% from oceans, and 6.5% from volcanoes (Noggle et al., 1986).

The total amount of sulfur deposition will vary across the geographic landscape (Fig. 5–1). Industrialized areas will have greater amounts because of fuel combustion and other processes and coastal areas because of oceanic aerosols. An extreme example of a localized effect would be the city of Gary, IN, where 141 kg sulfur ha^{-1} was deposited by rainfall in 1947 compared with an average of 30 kg sulfur ha^{-1} in 10 other Indiana cities (Bertramson et al., 1950).

In the atmosphere, sulfur is generally converted to sulfuric acid and then is washed out by rainfall or falls as dry deposition. Approximately 81% of total atmospheric sulfur falls as wet deposition in a pristine area of Wyoming and 19% as dry deposition (Zeller et al., 2000). However, the ratio of wet to dry deposition is greatly affected by local conditions.

In industrialized countries, sulfur emissions to the atmosphere from combustion and industrial processes are decreasing because of increasingly stricter emission regulations. In the USA, the first Clean Air Act in 1963 was enacted with the purpose of reducing air pollution from stationary sources such as power plants and steel mills. Major revisions of the Act were enacted in 1970 and then updated in 1977 and 1990. Title IV of the 1990 Clean Air Act imposed a 15 million Mg national cap on major point source emissions of SO_2. This is expected to eventually reduce SO_2 emissions about 50% from 1980 levels (Gbondo-Tugbawa and Driscoll, 2002). The decrease in sulfur emissions to the atmosphere has resulted in a decrease in atmospheric sulfur deposition to soils. For example, in Wooster,

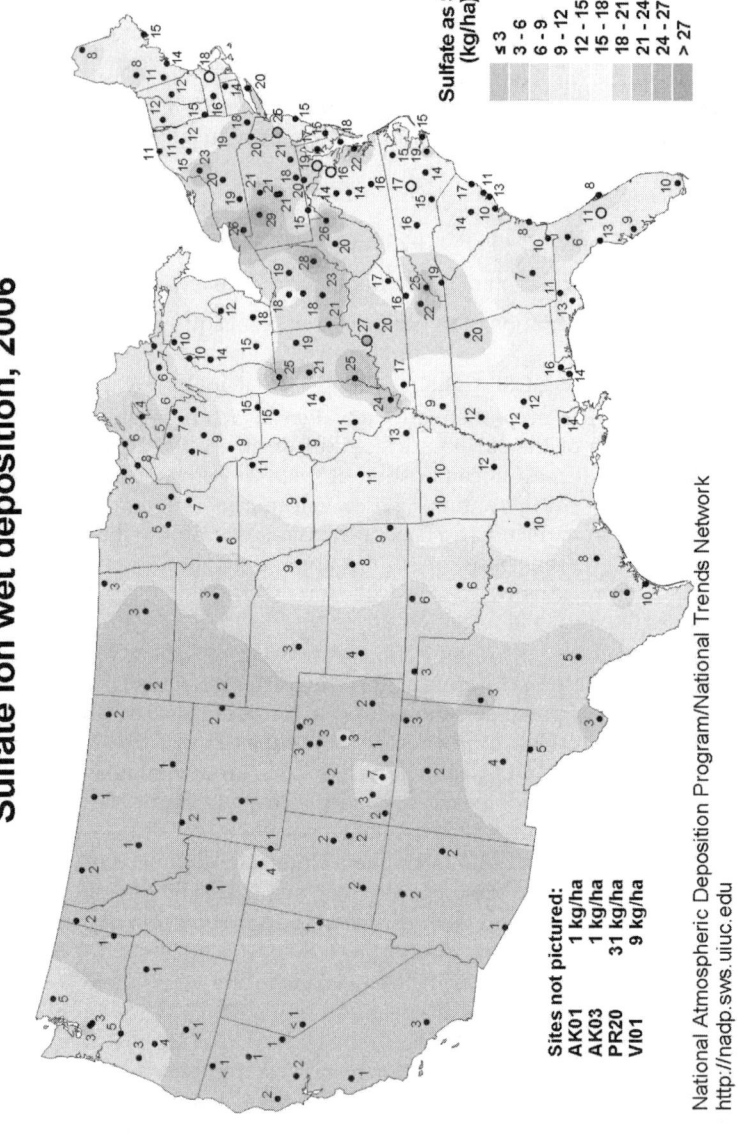

Fig. 5–1. Wet deposition of sulfur (as sulfate) in the contiguous United States during the year 2006.

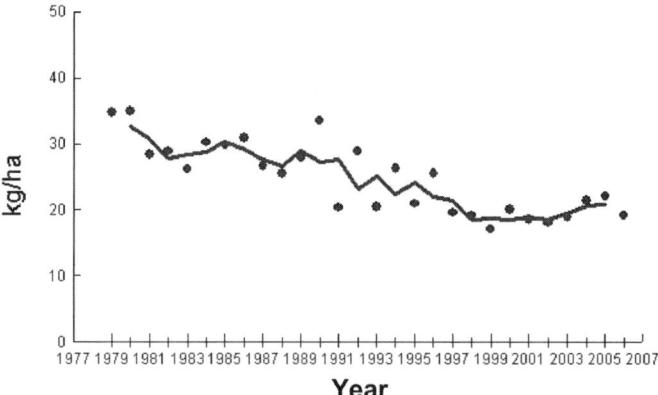

Fig. 5–2. Decline of total sulfur deposition in north-central Ohio (Wooster, OH) from 1978 to 2006 (National Atmospheric Deposition Program, 2007).

OH, annual wet sulfate deposition decreased from 34.8 kg ha^{-1} in 1979 to 22.0 kg ha^{-1} in 2005 (Fig. 5–2).

The source of gypsum for many gypsic soils is eolian dust from weathering of gypsum-containing rocks (Dixon, 1994; Schaetzl and Anderson, 2005). Thus dust may be an important source of sulfur for soils in dry climates. In wetter climates, the importance of dust as a sulfur source for soils is less evident. Simonson (1995) made little mention of sulfur in his review of dust as an input to soils.

Direct Absorption of Sulfur Gases by Soil and Plants

Sulfur gases, e.g., SO_2, are potentially phytotoxic. However, if root uptake of sulfur is not sufficient, the sulfur gases may also be taken up and metabolized as beneficial sulfur sources (Van der Kooij et al., 1997; Westerman et al., 2000). Atmospheric sulfur levels of 60 µg L^{-1} or higher appear to be sufficient to cover the total sulfur requirement of plants. The foliar uptake of SO_2 is generally dependent on the degree of opening of the stomates, because the internal resistance to this gas is low. Excessive S, after conversion to sulfate, is transferred into the vacuole and enhanced foliar sulfate levels are characteristic of plants exposed to SO_2 (DeKok, 1990; DeKok and Tausz, 2001). The foliar uptake of another sulfur gas, hydrogen sulfide, appears to be directly dependent on the rate of its metabolism into cysteine and subsequently into other sulfur compounds. Sulfur that is taken up directly by the plant as a gas is eventually deposited in the soil as plant litter.

Sulfur can also enter the soil via direct sorption of sulfur gases by soil components, including the soil solution (Bohn, 1972). Many sulfur gases are water-soluble and an equilibrium develops between concentrations in the atmosphere, soil solution, and the solid soil matrix. The soil uptake of sulfur gases is due to a combination of factors including diffusion into the soil, solubility in the soil solution, and sorption by sinks in the litter and soil material.

Moisture content of the soil affects the processes of diffusion, chemical–physical sorption, and biological activity. As more sulfur is sorbed out of the soil solution, more can be dissolved from the atmosphere into the soil solution. Sorption of sulfur gases in soil will tend to saturate as sites are filled. However, sites for sorption can be renewed or newly created by erosion, leaching, plant uptake or formation of soil minerals.

Soils absorb SO_2 rapidly and convert it to sulfuric acid at intermediate water contents (Gumerman and Carlson, 1966) that tend to support optimum microbial activity. The solubility of SO_2 is also pH dependent and only small amounts of SO_2 are absorbed by acid soils and the amount is related to the content of exchangeable alkali and alkaline earth cations (Faller and Herwig, 1969). Soils can also absorb hydrogen sulfide and the absorption rate is increased with higher soil pH and in the presence of transition metal cations, particularly Cu^{2+}. The absorbed hydrogen sulfide is then oxidized to eventually form sulfate and the absorption sites for hydrogen sulfide are renewed. Soil can also absorb mercaptans at rates that depend on the soil microbial population (Carlson et al., 1970; Gumerman and Carlson, 1966). Of a large number of sulfur gases tested, soils had a much larger capacity to absorb H_2S, SO_2, and CH_3SH than CH_3SCH_3, CH_3SSCH_3, COS, or CS_2 (Bremner and Banwart, 1976). Moist soils absorbed more of these gases than did air-dry soils and biological activity was partially responsible.

Fertilizer Inputs of Sulfur

Total sulfur fertilizer production in the USA in the past few years has remained almost stagnant (Table 5–2). Although the demand is expected to increase in the future, much of this demand will probably be met by inexpensive, but high quality, byproduct gypsum.

Fertilizer sources of sulfur may be categorized as organic or inorganic in nature (Table 5–3). They may also be separated into those that contain sulfur as the primary nutrient and those in which some other element is of primary interest but sulfur is also present. Elemental sulfur and sulfuric acid only provide sulfur and are used both to supply sulfur nutrient to enhance plant growth and to lower the pH of alkaline soils. Another major source of fertilizer sulfur is gypsum, which contains 23.3% calcium and 18.6% sulfur. It is also used as a calcium source for crops with large calcium requirements (i.e., peanuts, *Arachis hypogaea* L.) and as a source of exchangeable calcium to ameliorate sodic soils or acid sub-

Table 5–2. Sulfur fertilizer production in the USA (1998–2007). (United States Census Bureau, 2008).

Year	Ammonium sulfate	Sulfuric acid
	Thousands of Mg	
2007†	2040	27,200
2006	2550	35,900
2005	2640	37,100
2004	2730	38,000
2003	2600	37,300
2002	2670	36,100
2001	2350	36,300
2000	2550	39,600
1999	2610	40,600
1998	2530	44,000

† Three quarters of the year only reported for 2007.

Table 5–3. Inorganic and organic sources of sulfur fertilizers.

Fertilizer sources	Nutrient concentrations	
	N–P–K	S
	%	
Inorganic sources		
Elemental S	0–0–0	88–98
Gypsum	0–0–0	18
Ammonium sulfate	21–0–0	24
Ammonium thiosulfate	12–0–0	26
Ordinary superphosphate	0–9–0	11–12
Magnesium sulfate	0–0–0	14
Potassium magnesium sulfate	0–0–18.2	22
Potassium sulfate	0–0–41.5	18
Organic sources		
Biosolids	–†	0.3–1.2
Manures	–	
Most animals	–	0.25–0.3
Sheep	–	0.35
Poultry	–	0.5
Composts	–	
Biosolids	–	0.44
Dairy manure	–	0.22
Crop residues	–	0.10–0.22

† Nitrogen, phosphorus, and potassium levels in organic fertilizer sources are widely variable and only the typical ranges in sulfur concentrations for these materials are provided.

soils. In addition to mined gypsum and gypsum recovered from flue gases when coal is burned, gypsum is also produced from a variety of processes involving neutralization of sulfuric acid such as what occurs during production of titanium dioxide, lactic acid, citric acid, and phosphorus fertilizers.

Thiosulfates, especially ammonium thiosulfate, can be used as sulfur fertilizer materials. Ammonium thiosulfate has the benefit of supplying both sulfur (26%) and nitrogen (12%). It can be applied in irrigation water and is compatible with many fertilizer solutions such as aqua ammonia, nitrogen solutions containing ammonium nitrate, urea solutions, and most nitrogen, nitrogen–phosphorus, or complete fertilizer solutions.

There are a number of fertilizers in which sulfur is of secondary interest, but these fertilizers still contain substantial amounts of sulfate. Ordinary superphosphate is a mixture of monocalcium phosphate [$Ca(H_2PO_4)_2$] and gypsum. It is primarily applied as a phosphorus source (7.0–9.5% phosphorus) but also contains 11 to 12% sulfur. Ammonium sulfate [$(NH_4)_2SO_4$] is a nitrogen source (21.2% nitrogen) but also contains 24.2% sulfur. Potassium sulfate (K_2SO_4) is 41.5% potassium and 18.4% sulfur. Potassium magnesium sulfate ($K_2SO_4 \cdot 2MgSO_4$) is 18% potassium, 11% magnesium, and 22% sulfur. The latter is only produced from commercially valuable deposits of the mineral langbeinite located east of Carlsbad, NM.

The trend to use fertilizers that have a high percentage (high analysis) of nitrogen, phosphorus, or potassium has resulted in decreased fertilizer inputs of sulfur. A prime example is the use of triple superphosphate (19–23% phosphorus) as a phosphorus source instead of ordinary superphosphate. At required rates for phosphorus fertilization, ordinary superphosphate provides substantial sulfur for crop growth but triple superphosphate does not because it is applied at a lower overall rate and contains much less sulfur (i.e., only 0–3% sulfur) than ordinary superphosphate. Another example is the use of ammonium nitrate (NH_4NO_3) as a more concentrated source of nitrogen (34% nitrogen) instead of ammonium sulfate. Ammonium nitrate provides no sulfur and its long-term use may lead to sulfur deficiencies.

The sulfur content of animal manures is often ignored when manure management systems are being considered. Animal manures, as well as other organic amendments such as composts or biosolids, can be economical and effective sources of sulfur for enhancing crop production. Land application of organic amendments must, however, follow best management practices. Factors to consider are the manure's total nutrient content, the best application times and methods, and the percentage of total sulfur in the manure that becomes available during the initial and subsequent growing seasons. An excellent table showing the sulfur content of a variety of animal manures is posted on-line (Camberato et al., 1996).

Industrial by-products may also serve as important sources of sulfur fertilizers. Usually the sulfur is a potential air pollutant that is recovered to reduce environmental effects. Ammonium sulfate is produced from ammonia recovered from coke oven gases during the production of metallurgical coke. Ammonium sulfate is also produced from flue gases by using ammonia as the sorbent to react with the sulfur oxides in the gas. A fertilizer grade ammonium sulfate is produced as a commercial product at a coal gasification plant in North Dakota (Wallach, 1997). However, the process is not widely used because of the expense of the ammonia sorbent.

Another large source of sulfur is from combustion of fossil fuels. This produces sulfur dioxide (SO_2) at a rate proportional to the sulfur concentration in the fuel. Industrialized nations have adopted flue gas desulfurization (FGD) technologies to reduce sulfur dioxide emissions to the environment. Increasingly large volumes of calcium sulfite ($CaSO_3$) and calcium sulfate ($CaSO_4$) are produced from these FGD processes with calcium-based sorbents that react with the sulfur oxides. Sorbents used in the processes include calcitic ($CaCO_3$) and dolomitic [$CaMg(CO_3)_2$] limestone and more reactive materials such as hydrated lime [$Ca(OH)_2$] that will affect the performance of the final material created. For example, the dolomitic limestone will produce $MgSO_4 \cdot 7H_2O$ (epsomite), which is much more soluble in soil than gypsum.

Flue gas desulfurization processes are classified as either wet or dry (Srivastava and Jozewicz, 2001). Wet FGD processes produce a wet slurry waste or product, and dry processes produce a dry waste or product. The "wet scrubber sludge" product is primarily calcium sulfite but also contains some gypsum. This material may cause sulfite toxicity to plants if applied to soil (Ritchey et al., 1995b), but the potential for toxicity is probably short-lived because the calcium sulfite is rapidly oxidized to gypsum after soil application (Lee et al., 2007; Chen et al., 2007). In the process called wet limestone forced oxidation, compressed air is blown through a solution of calcium sulfite and gypsum to oxidize calcium sulfite. This yields relatively pure gypsum that is suitable as a soil amendment. Approximately 12.1 million Mg of FGD gypsum were produced in 2006 (American Coal Ash Association, 2008). This represents a vast potential source of material that can be used as a sulfur fertilizer.

Elemental sulfur is a common sulfur fertilizer. Before sulfur can be absorbed by plants, it must first be oxidized to sulfate. This oxidation process is mostly biological and is dependent on the size and activity of the microbial biomass, sulfur particle size, soil temperature and moisture, and fertilizer formulation. Although most agricultural soils contain very few *Thiobacillus* species, numbers increase when elemental sulfur fertilizer is added to soils. Soils treated with elemental sulfur produce $S_2O_3^{2-}$ and $S_4O_6^{2-}$. The oxidation of elemental sulfur increases as the particle size of the sulfur decreases (Donald and Chapman, 1998). Soil temperature also affects sulfur oxidation because of normal temperature influences on microbial activity. Sulfur oxidation in soils generally is optimum at soil moisture tensions near field capacity. Some commercial formulations of sulfur fertilizer inhibit or slow the sulfur oxidation process by *Thiobacillus* species (Lindemann et al., 1991).

Removal of Available Sulfur from Soil

Crop Uptake and Removal

The major exports of plant available sulfur from the soil system result from crop removal and leaching. Crop removal of sulfur is a function of yield and sulfur concentration in the harvested grain or biomass (Table 5–4).

Crops differ in their sulfur requirements from relatively low (15 kg sulfur ha^{-1}) as in corn and wheat to relatively high (30 kg sulfur ha^{-1}) as in alfalfa. There are also variations in the amount of sulfur removed with different parts of a crop. Wheat straw removes 17 kg sulfur ha^{-1} and wheat grain removes 8 kg sulfur ha^{-1}. Alfalfa, a high-sulfur requirement crop, removes almost 40 kg sulfur ha^{-1} from the soil each year when yield is 15 Mg ha^{-1} (Troeh and Thompson, 1993). Corn, a low-sulfur requirement crop, removes 11 kg sulfur ha^{-1} from the soil each year when grain yield is 9.4 Mg ha^{-1} (Hoeft and Fox, 1986). Similarly, cruciferous forages, alfalfa, and rapeseed (*Brassica napus* L.) are known to accumulate, and thus remove from soil, large amounts of sulfur or approximately 70 kg sulfur ha^{-1} (Spencer, 1975). Sugarcane (*Saccharum officinarum* L.), coffee (*Coffea arabica* L.), and coconut (*Cocos nucifera* L.) accumulate moderate amounts of sulfur (50 kg sulfur ha^{-1}) (Spencer, 1975) and field crops and grass forages accumulate between 20 and 30 kg sulfur ha^{-1} (Kamprath and Jones, 1986; Hoeft and Fox, 1986).

As previously mentioned, increased crop yields lead to increased outputs of all elements including sulfur. Crop yield increases of 18 to 50% would also result in sulfur removal increases of 18 to 50% if there were no changes in crop sulfur concentrations.

Leaching

Leaching is the removal of soluble materials from soil by percolating water. The main factors influencing leaching loss of available sulfur from the soil are well known, but the amount of leaching is difficult to quantify. Any factor that

Table 5–4. Sulfur removal by various crops at specific yields (from Lamond, 1997).

Crop	Portion of crop	Yield	S removed
		Mg ha^{-1}	kg ha^{-1}
Corn	grain	13	17
	stover	–†	20
Grain sorghum	grain	9.4	25
	stover	–	18
Wheat	grain	5.4	8
	straw	–	17
Canola	grain	2.2	13
	straw	–	10
Soybean	grain	4.0	13
	stover	–	15
Sunflower	seed	3.9	7
	stover	–	11
Alfalfa	biomass	13	34
Cool-season grass	biomass	9.0	18
Cotton	lint	1.7	45
Peanut	tuber	4.5	24
Rice	grain	7.8	13
Sugar beet	tuber	67	50
Orange	fruit	60	31
Tomato	fruit	67	46
Potato	tuber	56	25

† Removal of sulfur in stover and straw are estimates based on typical values of a harvest index (i.e., the ratio of harvested grain to stover or straw). In many instances the stover or straw is not harvested and the sulfur would thus not be removed from the field.

promotes water movement through the soil will increase the rate of leaching. Amount of precipitation is the most important factor controlling sulfur leaching. Sulfate is readily leached in wet climates but in dry climates accumulates as gypsum in the subsoil as a gypsic or petrogypsic horizon.

Soil texture influences sulfate leaching by influencing pore water velocity. Coarse-textured (sandy) soils have high pore water velocities in which there is relatively rapid movement of water and dissolved solutes through the soil. There is a high probability that precipitation will drain internally through the soil rather than externally as surface runoff. Fine-textured (clayey) soils have smaller pore water velocities. There is a greater chance that precipitation rate will exceed the rate of water movement through the soil and result in a greater percentage of the precipitation draining externally rather than internally.

Sulfate adsorption is the chemical or physical binding of sulfate ions to solid surfaces in the soil. Sulfate adsorption reduces sulfate leaching because the sulfate ions are not free to move with the soil solution. Adsorption is affected by soil pH, presence of competing anions such as phosphate, presence of iron and aluminum oxides, clay content, and clay mineralogy. Sulfate adsorption decreases with increasing pH from 4 to 7 and is essentially nil above pH 7 (Bohn et al., 1986). Phosphate adsorbs more strongly than sulfate and at high concentrations can cause complete desorption of adsorbed sulfate. Organic anions such as citrate, oxalate, and fulvate most likely absorb more so than sulfate. Iron and aluminum oxides facilitate sulfate adsorption. Adsorption also increases with clay content, and is greater for kaolinitic than montmorillonitic clay (Tisdale et al., 1986).

Many soils have an accumulation of adsorbed sulfate in the subsoil because conditions (low pH, high clay) are favorable for adsorption. Alfisols and ultisols are soils that developed under forest cover and have an accumulation of clay in the subsoil called an argillic horizon (Buol et al., 2003). In the southeastern USA, concentrations of adsorbed sulfate-sulfur in the subsoil of ultisols varied from 26 to 234 mg kg^{-1} (Kamprath and Jones, 1986). In ultisols, the range of water soluble sulfate-sulfur was 0.09 to 0.16 mM in surface horizons and 0.20 to 0.36 mM in argillic horizons (Camberato and Kamprath, 1986). This sulfate may make a substantial contribution to crop sulfur nutrition if it can be tapped by the plants. Utilization of adsorbed sulfate in subsoil is affected by inherent crop rooting depth, the depth to a clay horizon that contains adsorbed sulfate, and presence of impediments to deep rooting by the crop. Sulfur fertilization sometimes increased tobacco (*Nicotiana tabacum* L.) yields on ultisols when depth to the argillic horizon was greater than 0.45 m but not when less than 0.30 m (Smith et al., 1987).

Crop response to sulfur fertilization may occur where subsoils are rich in sulfate if roots are prevented from reaching the subsoil by tillage pans or toxic exchangeable aluminum levels. On an ultisol with high sulfate in the subsoil, wheat yield was increased by sulfur fertilization where a tillage pan was intact but not where the pan was broken to allow rooting into the subsoil (Oates and Kamprath, 1985). On another ultisol (Kline et al., 1989), corn yield responded to sulfur fertilization where rooting into the subsoil was possibly limited by physical impedance and chemical impedance (aluminum saturation levels greater than 50%). Gypsum amendment not only provides sulfur but also ameliorates aluminum toxicity in the subsoil to promote rooting into the subsoil.

Immobilization of Available Sulfur and Mineral Transformations

Available sulfur in soil can be converted to unavailable forms via the processes of immobilization of mineral sulfate into organic sulfur forms and/or conversion into stable soil minerals. As previously noted, the C/S ratio of an organic materials in soil where immobilization will occur is 400:1. Since most sulfur in soil is organic in nature, the cycling of sulfur minerals between available and unavailable forms of sulfur is less important than that of mineralization–immobilization of organic sulfur forms. However, it is possible for conditions in soil to develop, such as when water logging occurs for an extended period of time, where sulfate is reduced to sulfides, which are more easily precipitated by solubilized metals and thus removed from the sulfur availability pool.

Gaseous Losses from Soil

Rates and composition of sulfur gases emitted from soils vary greatly by soil type and among sites having the same soil type (Noggle et al., 1986). Gaseous emissions of sulfur from nonwetland soils are usually less than 1 kg sulfur ha yr^{-1}. Average emissions rates (kg sulfur ha yr^{-1}) in the USA varied from 0.13 for ultisols and spodosols to 1.1 for mollisols. For specific sites, emission rates varied from 0.0 to 1.3 for seven alfisols and from 0.1 to 1.8 for four mollisols. The dominant gas emitted by most soil types is H_2S. Other gases released in general order of importance are COS (carbonyl sulfide), CS_2 (carbon disulfide), $(CH_3)_2S$ (dimethyl sulfide), and CH_3SH (methyl mercaptan).

Predicting Crop Responses to Available Soil Sulfur

Sulfur deficiencies are most commonly observed when crops are grown on sandy soils where there is a high level of leaching, on soils with low organic matter content, and on soils where atmospheric inputs are low. Thus crop responses to sulfur fertilizer additions are more apt to be observed in limed soils in humid regions, in regions with low atmospheric inputs, and in soils with low sulfur status of the parent material. As the need to add supplemental sulfur to soil to obtain optimum crop growth increases, greater attention is being devoted to developing tests that will accurately predict responses to the added sulfur (Table 5–5).

Sulfur availability tests must meet several important criteria. They must predict crop uptake of sulfur and crop yield responses to a specified amount of

Table 5–5. Summary of tests used to assess availability of soil sulfur to crops.

Test method	Comments
Soil extractants	Many different extractants have been used to estimate sulfur availability to crops. The most common are water, $Ca(H_2PO_4)_2$, Mehlich III, and KCl. There are strengths and weaknesses associated with each extractant. All such extractants provide only a one-time estimate of available sulfur concentrations in soil. These concentrations can change rapidly. Water and weak salt solutions, such as KCl, will only extract that which is immediately soluble and cannot predict what may become available in the future. Weak alkaline extractants such as $Ca(H_2PO_4)_2$ will also extract a portion of the soil organic sulfur and Mehlich III is slightly acid and will generally extract some additional sulfur from different sulfur minerals in soil. Critical levels for most extractants ranges from 5 to 10 mg sulfur kg^{-1} soil.
Soil fraction	Since the majority of sulfur in soil is in the organic form, tests to predict availability of sulfur from soil to plants have included measuring the total organic sulfur concentration, or some fraction of the total soil organic S. This includes measurements of the ester sulfate fraction, the amino acid sulfur fraction, or some other active fraction of the soil organic matter. This approach has promise provided the size of the fractions measured is placed into a proper soil sulfur mineralization model that includes the soil, weather and plant components of the system.
Soil mineralization	Actual measurements of sulfur mineralization can be made in the laboratory and then equations developed that relate plant uptake with the amount of sulfur mineralized. These tests are time consuming and expensive. Results are often highly variable, even in the laboratory, because of a lack of understanding of all that affects the sulfur mineralization process. Also, laboratory results often correlate poorly with field mineralization results because field results are highly dependent on weather and crop management conditions.
Plant diagnostics	Sampling of plants or plant parts early in the plant's development has been tested as a means of determining if sulfur availability may be limiting crop production and yield. Variables that affect such tests include the crop sampled, the growth phase of the crop, and the plant part to be sampled. Critical concentrations values of sulfur in crop plants must be established that relate these concentrations to some reasonable chance that addition of fertilizer sulfur will result in an economic growth response. An N/S ratio may also be useful to assess the sulfur nutritional status of crops.

sulfur fertilizer added to the soil. The tests must also be relatively inexpensive to perform and be reproducible.

Soil tests for measuring sulfur and predicting crop response to available sulfur in soil have generally not been very effective, even though many academic and commercial laboratories offer such tests. This is because sulfur, like nitrogen, has a gaseous biogeochemical cycle as opposed to the sedimentary cycle of other nutrients such as phosphorus, potassium, and calcium. A test for sulfur does not, or cannot, measure the atmospheric sulfur that is continuously supplied to plants and also has difficulty allowing for inorganic sulfur that is mineralized from organic sulfur. Other complications include our inability to accurately predict weather. Weather greatly influences organic matter mineralization, sulfate leaching, and atmospheric inputs. Atmospheric inputs vary depending on the time of year, amount of rainfall, and relative location from an atmospheric sulfur source. Fields located downwind from the prevailing winds of a gaseous sulfur source, such as a coal-fired electricity-generating power station, will receive greater sulfur inputs than a field in the same general area on the same soil type upwind of the power station. Also, uptake of gaseous sulfur compounds directly into the plant leaf through the leaf stomata between rainfall events requires knowledge of sulfur concentrations in the air.

Another complicating factor is that sometimes addition of sulfur fertilizers can result in increased crop yields, but the reason has nothing to do with sulfur nutrition of the crop. Instead soils with acid subsoil, if treated with sulfur fertilizers such as gypsum, can be ameliorated by the calcium in the gypsum that increases rooting depth and exploitation of the soil volume to enhance overall water and nutrient uptake (Radcliffe et al., 1986; Ritchey et al., 1995a).

To devise the perfect soil test, we would need to know the factors that control the available sulfur pool in soil, climate, and plant physiology, etc. Several sulfur tests have been developed and are discussed in the sections that follow. These tests are based on (i) model results, (ii) soil properties, and (iii) plant nutrient information.

Equations of Inputs and Outputs Defining Plant Available Sulfur in Soil

Plant available sulfur is a function of atmospheric deposition and mineralization of soil organic sulfur output due to leaching and crop removal. An equation describing the sources and sinks for plant available sulfur in the soil can focus attention on those components that are most important in controlling crop sulfur nutrition. The starting point for a material balance on a component within a system is the equation (Felder and Rousseau, 2000):

Input + Generation = Output + Accumulation

Using this equation, we can develop an annual balance of *plant available sulfur* and not on *total sulfur* in the soil. We assume that the total sulfur concentrations in the soil remain unchanged (i.e., they are at steady state). The *input* term includes atmospheric deposition of sulfur, generally as sulfate in acid rain, and sulfur applications in fertilizers. The *generation* term represents plant available sulfur produced by mineralization of organic matter and by weathering of sulfur-containing minerals. As the soil organic matter is decomposed (or mineralized), sulfur will be released and will become available for plant uptake. For the genera-

Table 5–6. Number and percentage of Ohio soils classified in each soil sulfur availability category for crops requiring either 15 or 30 kg sulfur ha^{-1} yr^{-1}.

Crop S requirement	Soils†	Soils identified in each sulfur status category				
		Highly deficient	Moderately deficient	Variably deficient	Adequate	Highly sufficient
15 kg S ha^{-1} yr^{-1}	No.	2	17	1010	416	28
	Pct.	0.14	1.2	68.6	28.2	1.9
30 kg S ha^{-1} yr^{-1}	No.	77	655	636	96	9
	Pct.	5.2	44.5	43.2	6.5	0.6

† No. = number and Pct. = percentage of soils in each soil sulfur availability classification.

tion term to remain constant, sufficient organic matter containing organic sulfur must be returned to the soil each year to maintain the pool of mineralizable sulfur at a constant level. The primary *outputs* or losses from the soil are leaching of sulfate and crop removal of sulfur as harvested grain or forage. The *accumulation* term represents a change in plant available sulfur in the soil and could be positive, negative, or zero. Substituting these terms into the general equation yields:

$$S_{deposited} + S_{fertilizers} + S_{mineralized} + S_{weathering} = S_{leached} + S_{removed\ by\ crop} + S_{accumulated}$$

Rearranging:

$$S_{deposited} + S_{fertilizers} + S_{mineralized} + S_{weathering} - S_{leached} = S_{removed\ by\ crop} + S_{accumulated}$$

If the amount of sulfur needed for good crop growth is greater than that made available in the soil by atmospheric deposition, fertilizer application, organic matter mineralization, and mineral weathering, the deficit in sulfur could be satisfied, at least in the short term, by a negative sulfur accumulation. This would represent a decrease in plant available sulfur in the soil. We usually lack information on the accumulation of plant available sulfur in the soil on an annual basis, so we assume that it is at steady state and therefore set the value for sulfur accumulation in soil to zero. This leads to the following equation that may be used to evaluate soils in terms of their abilities to supply sufficient sulfur for good crop growth:

$$S_{deposited} + S_{fertilizers} + S_{mineralized} + S_{weathering} - S_{leached} = S_{removed\ by\ crop}$$

If the value of the left side of the equation is less than that required by the growing crop, the crop will suffer some degree of sulfur deficiency. In this equation, the terms that are most difficult to estimate for magnitude are weathering and leaching. For most soils, sulfur input due to weathering of inorganic minerals is probably small and inconsequential. For soils in wet climates, sulfur lost by leaching is great and thus important.

This model approach to predicting crop responses to S fertilizer additions has been developed by McGrath and Zhao (1995) for conditions in the UK and for production of corn and alfalfa in Ohio, USA (Table 5–6). For an alfalfa crop requiring 30 kg S ha^{-1} yr^{-1}, 43.2% of soils in Ohio were classified as variably deficient, but 49.7% were classified as moderately or highly deficient. These latter two categories imply that a response will generally occur for alfalfa when S fertilizer is added to soil.

Soil Availability Tests

A soil sulfur fertility tests begins with collecting of samples. Several variables will influence the available sulfur status in soil and these variables must be standardized to obtain a valid soil test. The depth from which a sample is taken is especially important when sampling a field that has been tilled versus a field under no-tillage. This is because organic matter plays such an important role in providing sulfur to a growing crop. For a crop that has deep rooting characteristics, it may be necessary to take a sample from a larger soil layer since such a plant has the ability to explore a large soil volume and extract sulfur.

Other important variables include whether soils have been manured just before sampling and other factors that affect soil organic matter concentrations and the time of year a sample is taken. The amount of sulfur mineralized annually is approximately 0.5 to 3.1% (Eriksen et al., 1998). This may seem like a small amount but, if the soil contains enough organic matter, it is sufficient to meet most or all of the sulfur requirements of low-sulfur requiring crops (i.e., 10–30 kg sulfur ha^{-1}). For crops with higher sulfur demands, this may not be the case. The soil sulfate status is often highest in the spring because of mineralization processes occurring during the months when little plant uptake is occurring. However, spring in some parts of the world is also when soils tend to be water saturated and sulfate may be lost via leaching.

Finally, how samples are composited so as to be considered representative of a field or an area of a field and whether the sample is processed in its field-moist state or after drying before applying the soil test can also be important variables that affect the soil sulfur test. Despite the need for a soil sulfur test, only limited research has been conducted to predict plant sulfur available concentrations in soil. This contrasts with the hundreds of laboratory and field studies that have been conducted to develop widely approved soil tests for potassium and phosphorus.

Under most conditions, it is the spatial-temporal variation of soil properties that control the plant available sulfur content in the soil. A soil test only represents the conditions at the time and for the places from which the sample was obtained. The assumption is that the results obtained for a soil test represent the conditions of the soil through much or all of the growing season of the crop. Thus, if the soil test predicts sulfur deficiencies, it is assumed this deficiency will remain and will not naturally be easily corrected. If the soil test predicts sulfur availability in soil is more than sufficient to meet the demands of the growing crop, it is also assumed that this condition will remain long enough as to not inhibit final crop growth and yield.

Soil tests for sulfur measure soluble sulfate only, soluble sulfate plus some fraction of adsorbed sulfate, or soluble and adsorbed sulfate plus some fraction of organic sulfur. Total sulfur content or total organic sulfur content in the soil is a poor measure of correlation with crop uptake and plant growth response. The reason is that in most temperate soils, organic sulfur makes up about 90% or more of the total sulfur in the soil.

More than 20 extractants have been studied for predicting sulfur response; these include water, calcium or potassium phosphate, sodium bicarbonate, and potassium chloride. Sulfur that is extracted with water or weak salt solutions of acid or base include sulfates that are found in soil pores, on anion exchange surfaces, in soluble organic forms, or that which is solubilized from soil minerals during extraction. Other extractants to predict the sulfur fertility remove a fraction of the organic sulfur that can be easily mineralized.

Because many sulfates in soil are soluble, water extraction will provide a reasonable estimate of the sulfur that is immediately available. However, this is a very transient value. $Ca(H_2PO_4)_2$ may also be used as an extractant because the phosphate ions displaces absorbed sulfate. The calcium helps flocculate the soil and can make the subsequent analytical steps easier to perform, especially if older methods such as turbidity are used to estimate sulfate concentrations. Mehlich III is commonly used to extract multiple elements from soil and so its use precludes a separate extraction just for sulfur determination. Because of its composition, it will often extract greater amounts of sulfur from the soil than water or $Ca(H_2PO_4)_2$.

Approximately 30% of the sulfur extracted with KCl in New Zealand was in organic form and the remainder was inorganic sulfate (Tan et al., 1994). Organic sulfur extracted with potassium phosphate was a better indicator predictor of maximum crop yield than initial sulfate, sulfate mineralized during a short-term incubation, organic sulfur extracted by calcium phosphate, or organic sulfur extracted by sodium bicarbonate (Watkinson and Kear, 1996). Extractants that are slightly alkaline and can partially solubilize organic matter will extract a greater proportion of sulfur as organic sulfur than a neutral or acidic salt.

Sulfur testing is only moderately reliable and research is greatly needed to develop more robust and accurate tests. However, in combination with information such as the crop being grown, soil type, atmospheric inputs, leaching potential and inputs from other sources such as manures, irrigation waters and phosphate fertilizers, a reasonable estimation of whether a soil will respond to S fertilizer inputs can be made. In fact, this approach has been used to develop a soil S availability model for Ohio to help guide S fertility decisions (http://www.oardc.ohio-state.edu/sulfurdef/; verified 28 March 2008).

The concentrations of sulfur in the soil solution need not be too great to satisfy crop needs. However, the sulfur must be continually replenished in the soil solution so that as the water flows to the crop root, it will also take sulfur with it. This is because sulfur is very soluble and moves through the soil primarily as water moves. The critical soil sulfur concentration value for crop growth in nutrient solution is 0.23 mM SO_4–sulfur for soybean and 0.01 mM for wheat. Critical levels for three extractants (KCl, calcium phosphate, and sodium bicarbonate) in soils from New South Wales, Australia, ranged from 6.5 to 8.4 mg kg^{-1} (Blair et al., 1991). In Ohio, the critical SO_4–sulfur level by water extraction was predicted as being approximately 3.8 mg kg^{-1} on a silt loam soil with 3.0% organic matter (Chen et al., 2008).

Plant Tests

Plants obtain sulfur primarily by absorbing sulfate from the soil or sulfur dioxide (SO_2) from the air. Sulfur deficiencies decrease crop yields and can lower the quality of the crop and nutritional value. This is because sulfur is part of several amino acids, i.e., cysteine–cysteine and methionine, required for production of proteins. Other compounds in plants that have a sulfur component include the coenzymes thiamine, biotin, coenzyme A, and ferrodoxin. Ferrodoxin is a nonheme iron protein involved in photosynthesis and other electron transfer processes. Sulfur fertilizers are often noted to cause a greening of the crop, forage, or

Table 5–7. Critical sulfur concentrations in crop plants (from Oenema and Postma, 2003; Schultz and Kelling, 1992).

Crop	Part sampled†	Time of sampling	Critical concentrations at various uptake levels			
			Deficient	Low	Sufficient	High
					%	
Alfalfa	top 15 cm	early bud	<0.20	0.20–0.25	0.26–0.50	>0.50
Corn	ear leaf	silking	<0.10	0.10–0.20	0.21–0.50	>0.50
Oats	top leaves	boot stage	<0.15	0.15–0.20	0.21–0.40	>0.40
Soybean	first trifoliate	early flower	<0.15	0.15–0.20	0.21–0.40	>0.40
Barley	YEB	mid-late tillering			0.15–0.40	
Canola/rape	YMB	before flowering			0.35–0.47	
Cotton	YMB	early flowering			0.20–0.25	
Ryegrass	young herbage	active growth			0.10–0.25	
Peanut	YML	pre-flowering			0.20–0.35	
Sugar cane	top visible dewlap	active growth			0.12–0.13	
White clover	young herbage	active growth			0.18–0.30	
Wheat	YEB/YMB	mid-late tillering			0.15–0.40	
Rice	whole top	maximum tillering			0.14	
Rice	whole top	active tillering			0.23	

† YEB, youngest emerged leaf blade; YMB, youngest mature leaf blade; and YML, youngest mature leaf.

turf because of increased chlorophyll production. However, the effect on actual yield may not follow the visual response noted when sulfur fertilizer inputs are applied.

Sulfur deficiencies in plants exhibit symptoms similar to that of nitrogen deficiencies, i.e., a yellowing of leaves because of lack of protein and chlorophyll synthesis. Legumes are especially affected by lack of sulfur because of their high protein content and the need for sulfur to aid in nitrogen fixation. Yields of alfalfa can be doubled and tripled by additions of sulfur to sulfur-deficient soil. The critical biomass value for alfalfa is 0.15 to 0.20% sulfur (Westermann, 1975).

The stage of growth and the plant part sampled are the two most important variables in the development of a plant sulfur test. Sulfur can be translocated from one part of the plant to another and thus the older leaves may first show symptoms of sulfur deficiency. However, sulfur may also be taken up above and beyond the needs of the plant and accumulate in the plant as sulfates. This also makes a plant diagnostic test difficult. Thomas et al. (1950) found that the organic sulfur content of nearly 1000 leaf samples all fell within a rather narrow range of 0.2 to 0.4%. Thus a critical value of less that 0.2% is implied for most crops. A range of critical values for several tropical crops was reported as being 0.07 for forage grasses to 0.25% for plantain (Fox and Blair, 1986). Critical values for other crops and plant parts are shown in Table 5 7.

Sulfur response is not always noted or corrected because of its link to nitrogen nutrition. If nitrogen availability is limiting crop growth, the lack of sulfur will not be noted. If nitrogen is not limiting, then sulfur deficiencies are more apt to be observed. Although few studies have documented the interaction of sulfur and nitrogen fertility, it has long been recognized that nitrogen uptake efficiency can be enhanced by making sure the sulfur levels in the soil are also adequate

(Chen et al., 2008). In fact, the use of an N/S ratio of plant tissue has been proposed as a way of estimating sulfur needs (Epstein, 1972).

Research Needs

Renewed research attention must be paid to the sulfur status of soils and crops. Although sulfur is a major plant nutrient, it has long been overlooked as an important soil fertility variable compared with the other macronutrients of nitrogen, phosphorus, and potassium. Advances are needed to better predict which soils and crops will respond to sulfur fertilizer additions. These responses are becoming increasingly evident because of increased crop yields and removal of sulfur from the soil as harvested grain and biomass, reduced atmospheric inputs, and use of concentrated fertilizers that contain little or no sulfur as a by-product.

Some specific research needs are listed below, but it is to be recognized that this list is not exhaustive. It is easy to overlook important research topics because of rapid changes in agricultural production systems.

1. Development of soil sulfur tests that accurately predict the sulfur fertility status of a soil and crop response to added sulfur fertilizers and that are crop, soil, and climate specific. This includes specific recommendations on sampling and sample processing as well as analysis of the soil sample.
2. Development of plant sulfur tests, in place of soil tests that are crop, soil, and climate specific. This includes specific recommendations on sampling and sample processing as well as analysis of the plant sample.
3. Evaluation of various sources of sulfur as fertilizer inputs, including the many industrial by-products that are available.
4. Evaluation of the best time and the best methods of applying sulfur fertilizer to soil.
5. Development of better models to predict sulfur release from soil organic matter to the soil solution and the growing crop. This is especially important in areas of the world where the organic sulfur fraction comprises 90% or more of the total sulfur in the soil.
6. Obtaining a better understanding of the weathering of sulfur-containing minerals in soil and better knowledge of leaching losses of sulfur from soil.
7. Evaluations of interactions of sulfur with other macronutrients (e.g., nitrogen) and micronutrients.
8. The role of the plant in both uptake and release of sulfur gases.

References

American Coal Ash Association. 2008. CCP survey. Available at http://acaa.affiniscape.com/associations/8003/files/2006_CCP_Survey_(Final-8-24-07).pdf. (verified 28 Mar. 2008).

Banwart, W.L., and J.M. Bremner. 1975. Identification of sulfur gases evolved from animal manures. J. Environ. Qual. 4:363–366.

Beaton, J.D., and R.J. Soper. 1986. Plant response to sulfur in western Canada. p. 375–403. *In* M.A. Tabatabai (ed.) Sulfur in agriculture. Agron. Monogr. 27. ASA, CSSA, SSSA, Madison, WI.

Bertramson, B.R., M. Fried, and S.L. Tisdale. 1950. Sulfur studies of Indiana soils and crops. Soil Sci. 70:27–42.

Blair, G.J., N. Chinoim, R.D.B. Lefroy, G.C. Anderson, and G.J. Crocker. 1991. A soil sulfur test for pastures and crops. Aust. J. Soil Res. 29:619–626.

Blanchar, R.W. 1986. Measurement of sulfur in soils and plants. p. 455–490. *In* M.A. Tabatabai (ed.) Sulfur in agriculture. Agron. Monogr. 27. ASA, CSSA, SSSA, Madison, WI.

Bohn, H.L. 1972. Soil absorption of air pollutants. J. Environ. Qual. 1:372–377.

Bohn, H.L., N.J. Barrow, S.S.S. Rajan, and R.L. Parfitt. 1986. Reactions of inorganic sulfur in soils. p. 233–249. *In* M.A. Tabatabai (ed.) Sulfur in agriculture. Agron. Monogr. 27. ASA, CSSA, SSSA, Madison, WI.

Bremner, J.M., and W.L. Banwart. 1976. Sorption of sulfur gases by soils. Soil Biol. Biochem. 8:79–83.

Buol, S.W., R.J. Southard, R.C. Graham, and P.A. McDaniel. 2003. Soil Genesis and Classification, Fifth ed. Iowa State Press, Ames, IA.

Camberato, J.J., and E.J. Kamprath. 1986. Solubility of adsorbed sulfate in Coastal Plain soils. Soil Sci. 142:211–213.

Camberato, J., B. Lippert, J. Chastain, and O. Plank. 1996. Land application of animal manures. http://hubcap.clemson.edu/~blpprt/manure.html#nutrient_content (verified 23 Mar. 2008).

Carlson, D.A., C.P. Leiser, and R. Gumerman. 1970. The soil filter: A treatment process for removal of odorous gases. Fed. Water Pollut. Contr. Assoc. Rep. WP 00883–03.

Chen, L., W.A. Dick, and D. Kost. 2008. Flue gas desulfurization products as sulfur sources for corn. Soil Sci. Soc. Am. J. 72:1464–1470.

Chen, L., W.A. Dick, and S. Nelson, Jr. 2005. Flue gas desulfurization products as sulfur sources for alfalfa and soybean. Agron. J. 97:265–271.

Chen, L., C. Ramsier, J. Bigham, B. Slater, D. Kost, Y.B. Lee, and W.A. Dick. 2007. Oxidation of FGD-$CaSO_3$ and effect on soil properties when applied to the soil surface. *In* Proceedings of the 2007 World of Coal Ash Conference. University of Kentucky, Lexington, KY.

Chibber, N. 2007. Sulphur deficiency in Madhya Pradesh soil leads to poor harvest. Down to Earth (15 Aug. 2007). Available at http://www.downtoearth.org.in/full6.asp?foldername=20070815&filename=news&sec_id=4&sid=3 (verified 23 Mar. 2008).

DeKok, L.J. 1990. Sulfur metabolism in plants exposed to atmospheric sulfur. p. 111–130. *In* H. Rennenberg et al. (ed.) Sulfur nutrition and sulfur assimilation in higher plants. SPB Academic Publishing, The Hague.

DeKok, L.J., and M. Tausz. 2001. The role of glutathione in plant reaction and adaptation to air pollutants. p. 185–201. *In* D. Grill et al. (ed.) Significance of glutathione to plant adaptation to the environment. Kluwer Academic Publ., Dordrecht, the Netherlands.

Dixon, J.C. 1994. Aridic soils, patterned ground, and desert pavements. p. 64–81. *In* A.D. Abrahams and A.J. Parsons (ed.) Geomorphology of desert environments. Chapman & Hall, London.

Donald, D., and S.J. Chapman. 1998. Use of powdered elemental sulphur as a sulphur source for grass and clover. Commun. Soil Sci. Plant Anal. 29:1315–1328.

Epstein, E. 1972. Mineral nutrition of plants: Principles and perspectives. John Wiley & Sons, New York.

Eriksen, J., M.D. Murphy, and E. Schnug. 1998. The soil sulphur cycle. p. 39–74. *In* E. Schnug (ed.) Sulphur in agroecosystems. Kluwer Academic Publ., Dordrecht, the Netherlands.

Faller, N., and W. Herwig. 1969. Untersuchungen uber die SO_2-oxydation in verschiedenen Boden. Geoderma 3:45–54.

Fanning, D.S., M.C. Rabenhorst, S.N. Burch, K.R. Islam, and S.A. Tangren. 2002. Sulfides and sulfates. p. 229–260. *In* J.B. Dixon and D.G. Schulze (ed.) Soil mineralogy with environmental applications. SSSA Book Ser. 7. SSSA, Madison, WI.

Felder, R.M., and R.W. Rousseau. 2000. Elementary principles of chemical processes: 3rd ed. John Wiley & Sons, New York.

Fox, R.L., and G.J. Blair. 1986. Plant responses to sulfur in tropical soils. p. 405–434. *In* M.A. Tabatabai (ed.) Sulfur in agriculture. Agron. Monogr. 27. ASA, CSSA, SSSA, Madison, WI.

Freney, J.R. 1967. Sulfur-containing organics. p. 229–259. *In* A.D. McLaren and G.H. Peterson (ed.) Soil biochemistry. Marcel Dekker, New York.

Gbondo-Tugbawa, S.S., and C.T. Driscoll. 2002. Evaluation of the effects of future controls on sulfur dioxide and nitrogen oxide emissions on the acid–base status of a northern forest ecosystem. Atmos. Environ. 36:1631–1643.

Gumerman, R., and D.A. Carlson. 1966. Hydrogen sulfide and methyl mercaptan removal with soil columns. p. 172–191. *In* Proc. 21st Industrial Waste Conference, Purdue University, West Lafayette, IN.

Gupta, U.C., and J.A. MacLeod. 1984. Effect of various sources of sulfur on yield and sulfur concentration of cereals and forages. Can J. Soil Sci. 64:403–409.

Hoeft, R.G., and R.H. Fox. 1986. Plant response to sulfur in the Midwest and Northeastern United States. p. 345–356. *In* M.A. Tabatabai (ed.) Sulfur in agriculture. Agron. Monogr. 27. ASA, CSSA, SSSA, Madison, WI.

Kamprath, E.J., and U.S. Jones. 1986. Plant response to sulfur in the southeastern United States. p. 323–343. *In* M.A. Tabatabai (ed.) Sulfur in agriculture. Agron. Monogr. 27. ASA, CSSA, SSSA, Madison, WI.

Kline, J.S., J.T. Sims, and K.L. Schilke-Gartley. 1989. Response of irrigated corn to sulfur fertilization in the Atlantic Coastal Plain. Soil Sci. Soc. Am. J. 53:1101–1108.

Konopka, A.E., R.H. Miller, and L.E. Sommers. 1986. Microbiology of the sulfur cycle. p. 23–55. *In* M.A. Tabatabai (ed.) Sulfur in agriculture. Agron. Monogr. 27. ASA, CSSA, SSSA, Madison, WI.

Lamond, R.E. 1997. Sulphur in Kansas: Plant, soil and fertilizer considerations (MF-2264). Kansas State University Agricultural Experiment Station and Cooperative Extension Service. Available at http://www.oznet.ksu.edu/library/CRPSL2/mf2264.pdf (verified 23 Mar. 2008).

Larson, W.E., C.E. Clapp, W.H. Pierre, and Y.B. Morachan. 1972. Effects of increasing amounts of organic residues on continuous corn: II. Organic carbon, nitrogen, phosphorus, and sulfur. Agron. J. 64:204–208.

Lee, Y.B., J.M. Bigham, W.A. Dick, F.S. Jones, and C. Ramsier. 2007. Influence of soil pH and application rate on the oxidation of calcium sulfite derived from flue gas desulfurization. J. Environ. Qual. 36:298–304.

Lindemann, W.C., J.J. Aburto, W.M. Haffner, and A.A. Bona. 1991. Effect of sulfur source on sulfur oxidation. Soil Sci. Soc. Am. J. 55:85–90.

Marschner, H. 1995. Mineral nutrition of higher plants. Academic Press, London.

McGrath, S.P., and F.J. Zhao. 1995. A risk assessment of sulphur deficiency in cereals using soil and atmospheric deposition data. Soil Use Manage. 11:110–114.

Miller, F.C. 1993. Minimizing odor generation. p. 219–241. *In* H.A.J. Hoitink and H.M. Keener (ed.) Science and engineering of composting: Design, environmental, microbiological, and utilization aspects. Renaissance Publications, Worthington, OH.

National Atmospheric Deposition Program. 2007. Annual data summary for site OH 71. Available at http://nadp.sws.uiuc.edu/sites/ntnmap.asp (verified 23 Mar. 2008).

Noggle, J.C., J.F. Meagher, and U.S. Jones. 1986. Sulfur in the atmosphere and its effects on plant growth. p. 251–278. *In* M.A. Tabatabai (ed.) Sulfur in agriculture. Agron. Monogr. 27. ASA, CSSA, SSSA, Madison, WI.

Oates, K.M., and E.J. Kamprath. 1985. Sulfur fertilization of winter wheat grown on deep sandy soils. Soil Sci. Soc. Am. J. 49:925–927.

Oenema, O., and R. Postma. 2003. Managing sulphur in agroecosystems. p. 45–70. *In* Y. P. Abrol and A. Ahmad (ed.) Sulphur in plants. Kluwer Academic Publishers, Dordrecht, the Netherands.

O'Leary, M.J., and G.W. Rehm. 1989. Effect of sulfur on forage yield and quality of alfalfa. J. Fert. Issues 6:6–11.

Ohio Department of Agriculture. 2006. Ohio agricultural statistics annual report 2005. Columbus, OH.

Radcliffe, D.E., R.L. Clark, and M.E. Sumner. 1986. Effect of gypsum and deep-rooting perennials on subsoil mechanical impedance. Soil Sci. Soc. Am. J. 50:1566–1570.

Ritchey, K.D., C.M. Feldhake, R.B. Clark, and D.M.G. Souza. 1995a. Improved water and nutrient uptake from subsurface layers of gypsum-amended soils. p. 157–181. *In* D.L. Karlen et al. (ed.) Agriculture utilization of urban and industrial by-products. ASA Spec. Publ. 58. ASA, Madison, WI.

Ritchey, K.D., T.B. Kinraide, and R.R. Wendell. 1995b. Interaction of calcium sulfite with soils and plants. Plant Soil 173:329–335.

Schaetzl, R., and S. Anderson. 2005. Soils: Genesis and geomorphology. Cambridge Univ. Press, New York.

Scherer, H.W. 2001. Sulphur in crop production– invited paper. Eur. J. Agron. 14:81–111.

Schoenau, J.J., and J.J. Germida. 1992. Sulphur cycling in upland agricultural systems. p. 261–277. *In* R.W. Howarth et al. (ed.) Sulphur cycling on the continents: Wetlands, terrestrial ecosystems and associated water bodies. SCOPE 48. John Wiley & Sons Ltd., Chichester, UK.

Schultz, E.E., and K.A. Kelling. 1992. Soil and applied sulfur. Wisconsin Cooperative Extension Publication A2525. University of Wisconsin, Madison, WI.

Shainberg, I., M.E. Sumner, W.P. Miller, M.P.W. Farina, M.A. Pavan, and M.V. Fey. 1989. Use of gypsum on soils: A review. Adv. Soil Sci. 9:1–111.

Simonson, R.W. 1995. Airborne dust and its significance to soils. Geoderma 65:1–43.

Sloan, J.J., R.H. Dowdy, M.S. Dolan, and G.W. Rehm. 1999. Plant and soil responses to field-applied flue gas desulfurization residue. Fuel 78:169–174.

Smith, W.D., G.F. Peedin, W.K. Collins, M.R. Tucker, G.S. Miner, and E.J. Kamprath. 1987. Tobacco response to sulfur on soils differing in depth to the argillic horizon. Tob. Sci. 31:36–39.

Spencer, K. 1975. Sulphur requirements of plants. p. 98–116. *In* K. D. McLachlan (ed.) Sulphur in Australasian agriculture. Sydney Univ. Press, Sydney, Australia.

Srivastava, R.K., and W. Jozewicz. 2001. Flue gas desulfurization: The state of the art. J. Air Waste Manage. Assoc. 51:1676–1688.

Stevenson, F.J., and M. Cole. 2000. Cycles of soil: Carbon, nitrogen, phosphorus, sulfur, micronutrients. 2nd ed. John Wiley & Sons, New York.

Tabatabai, M.A. 1987. Physicochemical fate of sulfate in soils. J. Air Pollut. Control Assoc. 37:34–38.

Tabatabai, M.A. 2005. Chemistry of sulfur in soils. p. 193–226. *In* M.A. Tabatabai and D.L. Sparks (ed.) Chemical Processes in Soils. SSSA Book Ser. 8. SSSA, Madison, WI.

Tan, Z., R.G. McLaren, and K.C. Cameron. 1994. Forms of sulfur extracted from soils after different methods of sample preparation. Aust. J. Soil Res. 32:823–834.

Thomas, M.D., R.H. Hendricks, and G.R. Hill. 1950. Sulfur content of vegetation. Soil Sci. 70:9–18.

Tisdale, S.L., R.B. Reneau, Jr., and J.S. Platou. 1986. Atlas of sulfur deficiencies. p. 295–322. *In* M.A. Tabatabai (ed.) Sulfur in agriculture. Agron. Monogr. 27. ASA, CSSA, SSSA, Madison, WI.

Troeh, F.R., and L.M. Thompson. 1993. Soils and soil fertility. 5th ed. Oxford Univ. Press, New York.

United States Census Bureau. 2008. Current industrial reports (CIR). MQ325B-Fertilizers and Related Chemicals. Available at http://www.census.gov/cir/www/325/mq325b.html (verified 23 Mar. 2008).

Van der Kooij, T.A.W., L.J. De Kok, S. Haneklaus, and E. Schnug. 1997. Uptake and metabolism of sulphur dioxide by *Arabidopsis thaliana*. New Phytol. 135:101–107.

Vough, L.R., R.R. Weil, and A.M. Decker. 1986. Fertilizing alfalfa for higher yields. p. 230–234. *In* Proceedings of the Forage Grassland Conference. University of Kentucky, Lexington, KY.

Wallach, D.L. 1997. Ammonium sulfate fertilizer as by-product in flue gas desulfurization: The Dakota Gasification Company Experience. p. 240–254. *In* J.E. Rechcigl and H.C. MacKinnon (ed.) Agricultural uses of by-products and wastes. ACS Symposium Series 668. American Chemical Society, Washington, DC.

Watkinson, J.H., and M.J. Kear. 1996. Sulfate and mineralisable organic sulfur in pastoral soils of New Zealand. II. A soil test for mineralisable organic sulfur. Aust. J. Soil Res. 34:405–412.

Westermann, D.T. 1975. Indexes of sulfur deficiency in alfalfa. II. Plant analyses. Agron. J. 67:265–268.

Westerman, S., L.J. De Kok, C. Stuiver, E. Elisabeth E., and I. Stulen. 2000. Interaction between metabolism of atmospheric H2S in the shoot and sulfate uptake by the roots of curly kale (*Brassica oleracea*). Physiologia Plantarum 109:443–449.

Zeller, K., D. Harrington, A. Riebau, and E. Donev. 2000. Annual wet and dry deposition of sulfur and nitrogen in the Snowy Range, Wyoming. Atmos. Environ. 34:1703–1711.

6 S

Sulfur and Cysteine Metabolism

Rainer Hoefgen and **Holger Hesse**
Max-Planck-Institut für Molekulare Pflanzenphysiologie, Potsdam, Germany

Abstract

The influence of the plant macronutrient sulfur on plant metabolism and on crop productivity has long been underestimated. Sulfur enters cellular metabolism usually as sulfate ion, and a suite of sulfate transporters is responsible for its uptake via roots and for its distribution within the plant or between subcellular compartments. The inert sulfate ion needs to be activated by binding to ATP to be assimilated and to enter cellular metabolism, essentially being subject to two fates then, both essential for plant life. First, the ATP-bound sulfate is used for sulfatation reactions of biomolecules without further reduction steps. Second, sulfate is reduced in a series of reduction steps to sulfide, which is then fixed into the proteinogenic thiol amino acid cysteine. Cysteine is the common precursor of all organic molecules containing a reduced sulfur moiety such as glutathione, which is involved in redox control and stress response, the essential amino acid methionine, the versatile one-carbon donor S-adenosylmethionine, hormones, vitamins, and cofactors such as acetyl-CoA. Uptake and assimilation of sulfate and its integration with other plant metabolic pathways such as nitrogen and carbon levels is tightly regulated through control of gene expression or enzyme activity. The most important control points are the reduction step to sulfite by the enzyme APS reductase and the cysteine synthase complex, which uses interaction of the enzymes serine acetyltransferase and O-acetylserine-(thiol)-lyase to control flux through the pathway.

Crop productivity is influenced by a complex mixture of environmental, biotic, and abiotic factors (Hesse and Hoefgen, 2001; Hawkesford and de Kok, 2006; Hoefgen and Hesse, 2007). Availability of the indispensable inorganic macronutrients such as nitrate, phosphate, potassium, and sulfate is a key determinant for plant growth, crop yield, and, eventually, product quality. Plants react to environmental conditions through adaptation processes at the biochemical, molecular, and physiological level to maintain homeostasis and to allow growth and reproduction. Nitrogen is the main nutrient required for plant growth, and the need for sulfate is about 10% that of nitrogen. Natural sources of sulfate are soil minerals, bacterial degradation of deteriorating plant material, and gaseous compounds from volcanoes or marine algae being deposited by rainfalls as sulfite or sulfate, resulting in an ocean-cloud-land-ocean sulfur cycle (Stefels, 2007). Sulfate is an integral part of plant metabolism. An insufficient supply negatively affects numerous metabolic pathways. Thus, sulfate determines plant productivity and vitality and, hence, crop quality and crop yield.

Copyright © 2008. American Society of Agronomy, Crop Science Society of America, Soil Science Society of America, 677 S. Segoe Rd., Madison, WI 53711, USA. *Sulfur: A Missing Link between Soils, Crops, and Nutrition.* Agronomy Monograph 50.

Plants possess the capacity to assimilate sulfate while animals have to take up reduced sulfur compounds, mainly the amino acids cysteine and methionine, with their diet. These proteinogenic amino acids are essential for humans and livestock such as monogastric animals and birds. While methionine is an essential amino acid, cysteine is viewed as semi-essential amino acid because animals are able to convert methionine to cysteine. However, the need for methionine is increased in low cysteine diets. Both sulfur-containing amino acids are low in most crops, especially in staple crops, such as cereals and potatoes (*Solanum tuberosum* L.). Thus, sulfate availability is directly linked to crop nutritional quality. This nutritional dependency indicates the importance of sulfur metabolism for crop plants and, hence, human and animal nutrition (Hesse and Hoefgen, 2001; Hell et al., 2002; Galili and Hoefgen, 2002; Galili et al., 2005; Hawkesford and de Kok, 2006; Hawkesford et al., 2006; Hesse and Hoefgen, 2006; Hoefgen and Hesse, 2007).

In agricultural production systems, even situations of only moderate sulfate starvation will lead to negative effects on yield and plant performance and a reduced ability of plants to cope with additional abiotic or biotic stresses. Even before visible symptoms of sulfur-starvation are apparent, plant performance is affected, while severe insufficiencies result in acute growth and yield depressions. The symptoms of sulfate starvation such as leaf yellowing or growth reduction are often misinterpreted as nitrogen deficiencies, leading to an over-application of nitrogen by farmers, which when sulfate is low, cannot be taken up and negatively affects the environment (Blake-Kalff et al., 2000; Hesse and Hoefgen, 2001; Haneklaus et al., 2003). Management of sulfate fertilization in field conditions has, however, gained more attention in the past three decades after the unexpected observation that reduced air pollution by gaseous sulfite (SO_2) leads to widespread sulfate deficiencies in soils. Sulfite is mainly derived from fossil fuels or coal being converted to sulfuric acid (acid rain) when dissolved in water. Though excess pollution results in environmental problems, especially when soils are not well buffered, a certain input is necessary to sustain optimal plant growth.

These agronomical problems together with the growing interest in plant sulfur metabolism over recent years and the progress in plant biochemistry, molecular biology, and physiology dramatically increased our respective knowledge base (Anderson, 1990; Hell, 1997; Hell and Rennenberg, 1998; Leustek and Saito, 1999; Saito, 2000; Leustek et al., 2000; Hell et al., 2002; Saito, 2004; Wirtz and Droux, 2005; Kopriva, 2006; Hesse and Hoefgen, 2006; Hoefgen and Hesse, 2007). The basic principles have been worked out in bacterial systems though plant-specific differences were identified later (Bryan, 1980, 1990; Kredich, 1996). Understanding the links between soil chemistry, agricultural practices, and plant sulfate metabolism and its integration into plant metabolism will contribute to improving agricultural performance, yield, plant vigor, and product quality.

Sulfur Assimilation and Reduction

Uptake and assimilation processes of all nutrient anions (sulfate, phosphate, and nitrate) share similar strategies. They are taken up by specialized transporters and are transported in the transpiration stream. Nitrate and sulfate are either reduced in successive reaction steps and bound to organic molecules or stored in vacuoles (Kopriva and Rennenberg, 2004; Hesse et al., 2004a). After uptake from the environment and delivery to cells, sulfate is either deposited and stored in the

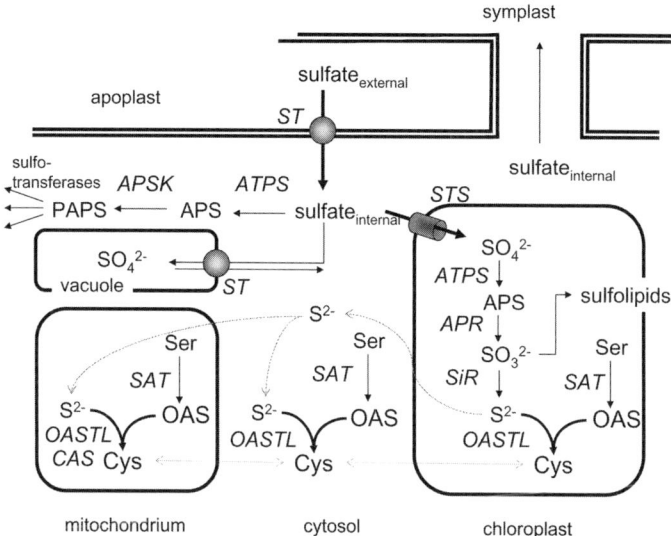

Fig. 6–1. Subcellular organization of the sulfate assimilation pathway in plants. Sulfate transporters (ST) facilitate uptake into the cell and storage in and retrieval from the vacuole. No ST has been identified for the plastids and a sulfate translocation system (STS) corresponding to permeases can be assumed. In the cytosolic sulfatation pathway ATP sulfurylase (ATPS) activates the inert sulfat ion through binding to ATP generating adenosine 5¢-phosphosulfate (APS) that is further converted to 3¢-phosphoadenosine 5¢-phosphosulfate (PAPS) by APS kinase (APSK). PAPS serves as substrate for a huge family of sulfotransferases producing numerous sulfated metabolites. Within the plastidic reduction pathway a gluthathione dependent APS reductase (APR) reduces sulfate to sulfite (2e−) that is then reduced to sulfide (6e−) by a ferredoxin dependent sulfite reductase (SiR). Cysteine synthesis itself takes place in all compartments, cytosol, plastids and mitochondria by the cysteine synthase complex composed of serine acetyltransferase (SAT) generating O-acetylserine (OAS) using serine and acetyl-CoA as precursors and O-acetylserine thiol lyase (OAS-TL) synthesizing cysteine from OAS and sulfide. In the chloroplast sulfide derives from the reduction activity of SiR, for the other compartments a diffusion of sulfide out of the plastids has to be assumed, additionally to local degradation processes. In the mitochondria of several plant species a mitochondrial cyanoalanine synthase (CAS) functions as cysteine synthesizing enzyme, while Arabidopsis contains an OASTL isoform. Cysteine as well has to be assumed to cross the compartment membranes. In a side reaction sulfite is utilized to synthesize sulfolipids for the photosynthetic chloroplast membranes.

vacuole or enters the sulfatation or the reductive assimilation pathway to become an essential part of plant metabolism (Hawkesford and De Kok, 2006) (Fig. 6–1).

Sulfate Uptake

Most crop plants take up sulfur in the form of sulfate from the soil. Plants can utilize gaseous sulfur compounds such as SO_2 or hydrogen sulfide (H_2S). Under agricultural conditions, this contributes only a minor fraction (de Kok et al., 2007). H_2S is immediately incorporated into cysteine, while SO_2 is most likely retrieved through catabolic processes utilizing a peroxisomal sulfite oxidase (EC 1.8.3.1) to oxidize sulfite, generated when SO_2 is dissolved in water, to sulfate (Hänsch and Mendel, 2005; Durenkamp and de Kok, 2004; Hänsch et al., 2005, 2007; Lang et al., 2007; de Kok et al., 2007). The nature of import and export processes over the peroxisomal membrane is so far unknown.

Sulfate supply in the agricultural environment is neither stable nor uniform in time or space. Free sulfate might move with the water table, vary according to the amount of rain or irrigation or nutrient regimes under agricultural practice. Additionally, sulfate is slowly released from organic soil matter through degrading activities of bacteria. It is not clear whether plants are able to actively stimulate beneficial bacterial or fungal associations by providing carbohydrates or organic acids (Kertesz and Mirleau, 2004). Thus, plants that devise strategies to cope with this situation and maintain homeostasis are able to actively enrich sulfate from micromolar concentrations in soil against a concentration gradient to millimolar concentrations in the cell.

This is achieved by a family of sulfate transporters that facilitate uptake from the environment and distribution to the cell and subcellular compartments (Saito, 2000; Hawkesford, 2003; Hawkesford et al., 2003; Maruyama-Nakashita et al., 2004; Buchner et al., 2004a, 2004b; Kataoka et al., 2004a, 2004b; Hawkesford, 2007). While bacteria, fungi, and seawater organisms such as diatoms use sulfate permeases, plants developed highly specific sulfate transporters that can be grouped into high (K_m ~1–10 µM) and low (K_m ~0.1–1 µM) affinity proton–symporters (Anderson, 1990; Hawkesford, 2000; Saito, 2000; Grossman and Takahashi, 2001). Uptake and transport of sulfate within the plant conceivably takes the same combined apoplastic–symplastic route as nitrate and phosphate (Kopriva and Rennenberg, 2004; Hesse et al., 2004a). Expression analysis of sulfate transporters demonstrated that they are present in root hairs and epidermis for sulfate acquisition and in vascular bundles of roots and leaves for the interorgan allocation of sulfate. Root membrane localized transporters (Group 1) take up sulfate into the symplast from the environment either directly from the soil capillary system or the root apoplast.

Sulfate is then symplastically transported into root cells or into the transpiration stream to supply the rest of the plant (Group 2). Sulfate is assimilated in root and leaf tissues. Excess cellular sulfate is stored in the vacuole. The sulfate ion appears to be less mobile after deposition in vacuoles of the sink tissues than other ions; though, it might be retrieved under insufficient supply situations from the soil or during grain filling (Hawkesford, 2003).

Sulfate transporters have been cloned and functionally characterized from several species. In *Arabidopsis thaliana* (L.) Heynh., 14 sulfate transporters assigned to five groups based on sequence homologies have been identified and partially functionally assigned to various tissues or subcellular localizations (Hawkesford, 2003; Hawkesford et al., 2003; Maruyama-Nakashita et al., 2004). Group 1 contains the high affinity transporters (Sultr1;1, Sultr1;2, and Sultr1;3) that are induced in root tissues on sulfate deprivation (Shibagaki et al., 2002; Yoshimoto et al., 2002, 2003). Group 2 members comprise low affinity transporters and may be involved in long-distance transport (Sultr2;1 and Sultr2;2) and seed loading (Sultr2;1); these are expressed in vascular tissue and seed nourishing tissues (Kataoka et al., 2004a; Awazuhara et al., 2005). Group 3 is not fully characterized. For Sultr3;5, a function in transport of sulfate from root to shoot has been suggested; however, Sultr3;5 seems to be active only when complexed with Sultr2;1 (Kataoka et al., 2004b). Group 4 transporters are also not fully characterized, but Sultr4;1 and Sultr4;2 have been shown to act as efflux transporters from the vacuole, specifically in the tonoplast membrane, which disproves their prior functional assignment as chloroplast transporters (Kataoka et al., 2004b). Group 5 sulfate transporters (Sultr5;1 and Sultr5;2) are distantly related by sequence to the other groups. Recently they have been characterized as molybdate rather than sulfate transporters (Tomatsu et al., 2007; Tejada-Jimenez et al., 2007; Baxter et al., 2008). The localization of sulfate transporters to distinct tissues or subcellular locations remains to be completed.

Individual sulfate transporters may function at various locations, depending on their expressional control or on protein–protein interactions (Rouached et al., 2005; Hawkesford, 2003; Hawkesford and de Kok, 2006). For example, the C-terminal STAS domain of Sultr1;2 may be involved in protein–protein interactions. Surprisingly, no chloroplast sulfate transporters have been identified so far; even though sulfate reduction exclusively occurs in this organelle. Tentatively, ABC-type transporters and sulfate permeases, such as those identified in *Chlamydomonas*, may serve as a chloroplastic sulfate transport system (Fig. 6-1) instead of sulfate transporters family members (Melis and Chen, 2005).

Crop-specific differences need to be considered for breeding and agriculture. The high sulfate need of rape (*Brassica napus* L.) seed may be a consequence of insufficient release from internal stores and plant-specific features of the respective sulfate transporters rather than from high needs for reduced sulfur compounds, such as glucosinolates (Koralewska et al., 2007). The regulation of these sulfate uptake processes, as well as the signals controlling sulfate metabolism at the biochemical level, are not fully understood (Hawkesford, 2003; Hawkesford et al., 2003; Hawkesford and de Kok, 2006; Kopriva, 2006). Several sulfate transporter genes (but not all) are induced within hours of sulfate deficiency and are rapidly repressed on resupply of sulfate. These individual responses indicate regulatory differences within the functionally distinct genes of the sulfate transporter family. For example, the low affinity transporter Sultr2;1 of *Arabidopsis thaliana* reaches its maximum of expression on sulfate deprivation later than the high affinity sulfate transporter Sultr1;1 (Saito, 2000; Hawkesford, 2003; Maruyama-Nakashita et al., 2005; Hawkesford and de Kok, 2006; Yoshimoto et al., 2007).

Sulfate could be assumed to be reduced in situ when allocated to source or sink tissues, as with seeds. However, supply of sink tissues with sulfur does not only occur through supply of sulfate but also through reduced sulfur compounds via the phloem. Glutathione (GSH) and S-methylmethionine (SMM) appear to have a role in transport (Bourgis et al., 1999; Ranocha et al., 2001; Hawkesford, 2003; Kopriva, 2006). In plants such as wheat (*Triticum aestivum* L.), substantial amounts of reduced sulfur are transported as SMM from source leaves to sink tissues.

Although the interplay of complex regulatory and signaling mechanisms remain largely unknown, these processes imply that crop quality is dependent on the capacity of local sulfate reduction systems to produce cysteine and derived products for protein biosynthesis and other complex biochemical activities as a supply of ready-made compounds. Furthermore, plant-specific differences have to be taken into account; the effectiveness of overexpression of sulfur-rich sink proteins in grain crops which in effect direct the flux of methionine and cysteine has yielded diverse results depending on the plant species. Overexpression and successful accumulation of the 2S SSA in rice (*Oryza sativa* L.) did not result in a net increase of reduced sulfur compounds in the seed protein fraction but rather evoked signs of sulfur starvation, indicating either an inability of rice seed to effectively convert sulfate into reduced sulfur compounds in the seeds or an insufficient supply of reduced compounds from source leaves to the seed (Hagan et al., 2003). In contrast, a similar approach in lupines resulted in a substantial net increase of reduced sulfur compounds in the seed protein fraction and, thus, a substantial improvement of grain protein quality (Molvig et al., 2003; Chiaiese et al., 2004). Vegetative storage tissues might differ completely from the situation observed in grain crops. It could be shown that most of the cysteine and methionine biosynthetic pathway is inactive in tubers, a vegetative storage and reproductive organ, making them entirely dependent on an external supply of reduced sulfur compounds (Harms et al., 2000; Maimann et al., 2000; Zeh et al., 2001, 2002; Kreft et al., 2003; Hesse et al., 2004b; Hawkesford et al., 2006).

Biochemical Activation of Sulfate

Sulfate (SO_4^{2-}) is a relatively inert bivalent anion with low reactivity that needs to be activated before entering cellular metabolism by the enzyme ATP-sulfurylase (ATPS; EC 2.7.7.4). ATPS couples sulfate as an anhydride to ATP to form adenosine 5'-phosphosulfate (APS) with the release of pyrophosphate (Anderson, 1990; Hell, 1997; Hell and Rennenberg 1998; Leustek and Saito, 1999; Saito, 2000; Leustek et al., 2000; Hell et al., 2002; Saito, 2004; Wirtz and Droux, 2005; Kopriva and Koprivov, 2004; Kopriva, 2006; Hesse and Hoefgen, 2006; Hoefgen and Hesse, 2007). APS is the starting point of sulfate assimilation in plants. The reaction is highly reversible and under cellular conditions is driven by removal of inorganic phosphate by pyrophosphatases and downstream enzymes that utilize APS.

In plants, ATPS isoforms and activity are mainly located in plastids to feed sulfate reduction, but a minor activity is found in the cytosol to initialize the cytosolic sulfatation pathway. The cDNAs for these isoforms were first isolated from potato (Klonus et al., 1994). In *Arabidopsis*, at least three plastidic and one putatively cytosolic ATPS isoforms were identified on genomic basis. Isolated chloroplasts can reduce exogenously supplied sulfate and convert it to cysteine. Thus, a fully functional reduction pathway is found in chloroplasts.

APS serves as a substrate for two pathways: the sulfatation pathway located in the cytosol and the reduction pathway located in the plastids. In the first case, APS is phosphorylated to 3'-phosphoadenosine-5'-phosphosulfate (PAPS) through the activity of tadenosine-5'-phosphosulfate kinase or APS-kinase (APSK; EC 2.7.1.25) (Schwenn, 1994; Lillig et al., 2001). The sulfate anhydrid moiety is not further reduced. Three genes are present in the *Arabidopsis* genome and at least one is localized in the chloroplast. PAPS is the substrate for a diverse family of sulfotransferases that produce a variety of O-sulfated metabolites, including flavanols, choline, betaines, extracellular polysaccharides, and glucosinolates (Klein and Papenbrock, 2004). The precise physiological roles of these compounds, especially of the flavonol derived secondary metabolites, are unknown. Glucosinolates act as defense compounds against herbivores releasing toxic cyanide (Halkier and Gershenzon, 2006). Increasing evidence indicates a regulatory role for enzyme or substrate activity by sulfatation processes. For example, phytosulfokines, a family of pentapeptides processed from precursorpeptides are assumed to regulate cell proliferation and are activated through O-phosphorylation of a tyrosine residue (Lorbiecke et al., 2005).

The Reductive Pathway to Cysteine

In the reductive assimilatory pathway, the ATP-bound sulfate in the APS molecule is stepwise reduced to yield sulfide, which is incorporated into serine to form cysteine, the precursor molecule of all downstream molecules containing fully reduced sulfur (Fig. 6-1). This process is not energetically efficient for the plant and requires ATP and three electrons for sulfate import into the cell (which might occur repetitively as sulfate is moved between tissues and organelles) and eight electrons for full reduction. Thus, there is a need to tightly control the assimilation process. This dependency on reductants and ATP produced by photosynthesis is substantiated by the chloroplastic localization of most of the enzymes and the presence of the full assimilation pathway in the chloroplast. However, parts of the pathway are present in the cytosol and mitochondria as cysteine is also produced in these organelles. A need for control arises also because toxic, highly reactive intermediates are produced during this process, i.e., sulfite and sulfide. Cysteine levels are tightly controlled in plants (Anderson, 1990; Hell, 1997; Hell and Rennenberg, 1998; Leustek and Saito, 1999; Saito, 2000; Leustek et al., 2000; Hell et al., 2002; Saito, 2004; Wirtz and Droux, 2005; Kopriva and Koprivov, 2004; Kopriva, 2006; Hesse and Hoefgen, 2006; Hoefgen and Hesse, 2007).

Step One: Sulfate Reduction to Sulfite

In the reductive pathway, APS synthesized by the plastidial ATPS isoform is reduced to sulfite by the plastid-localized APS reductase (APR; EC 1.8.4.9) (Leustek and Saito, 1999, Leustek et al., 2000). It is assumed that glutathione acts as the reductant in this reaction and provides the necessary electrons via a carboxy terminal domain of the APR resembling a thioredoxin. Because this domain of the reductase enzyme is glutathione-dependent, it is termed a glutaredoxin and probably functions in avoiding thioredoxin competition for sulfate reduction in the chloroplast (Setya et al., 1996).

Next, APR produces free instead of bound sulfite when reducing the APS-bound sulfate (Suter et al., 2000; Kopriva and Koprivova, 2004; Kopriva, 2006). PAPS reductase, an enzyme present in bacteria and mosses lacks an integral redox factor and is dependent on thioredoxin when reducing PAPS-bound sulfate to sulfite (Kredich, 1996; Kopriva and Koprivova, 2004). The N-terminal domains of APR and PAPS share high homologies. The enzyme provides an alternative pathway linking the reductive and the sulfatation pathway. However, in higher plants the APR pathway is the exclusive entry of sulfate to the reductive pathway to cysteine. Speculatively, this allows higher plants to establish APR as a tight control point of sulfate. In a side reaction sulfite is utilized to synthesize sulfolipids for the photosynthetic chloroplast membranes (Benning, 1998; Sanda et al., 2001; Frentzen, 2004)

APR is highly regulated at the molecular and biochemical level through numerous stimuli such as thiols, sulfate availability, cadmium exposure, oxidative stress, hormones, nitrogen, and salt and, thus, integrates various signals from other anabolic pathways (Kopriva et al., 2002; Hesse et al., 2003; Kopriva, 2006; Durenkamp et al., 2007). This agrees with the characteristics of a pathway control point fitting to its position at a branch point. Tentatively, the first enzyme of the sulfate assimilation pathway, ATPS, has been speculated to be a flux controlling key enzyme; however, overexpression in tobacco cells provided no evidence for substantial effects on metabolism (Hatzfeld et al., 1998), while overexpression in *Brassica juncea* L. (Indian mustard) (Pilon-Smits et al., 1999) resulted in a higher resistance to a natural toxic sulfate analog, selenate. In contrast to this, manipulating APR expression in plants resulted in drastic effects corroborating APR as a control point in the sulfate reductive pathway (Tsakraklides et al., 2002;

Loudet et al., 2007). As ATPS favors the back reaction, APR, together with phosphatases, plays a crucial role in removing the reaction products of ATPS, i.e., APS and pyrophosphate from the equilibrium. Further, APR activity increases and its transcript accumulates during sulfur starvation, suggesting it as a key step for controlling sulfate reduction. This is further supported by the fact that other sulfate assimilation genes and enzymes are less subject to external regulation, such as ATPS, or are constitutively expressed, such as sulfite reductase (SiR).

Step Two: Sulfite Reduction to Sulfide

Sulfite is reduced to sulfide in a ferredoxin-dependent reaction by a plastidial sulfite reductase (SiR; EC 1.8.7.1) (Bork et al., 1998). Currently, only a plastid-localized isoform has been identified, suggesting that a complete reduction pathway for sulfate only occurs in plastids. SiR contains an iron–sulfur cluster [4Fe-4S] and a siroheme as prosthetic groups for the redox centers with ferredoxin acting as an electron donor (Kessler and Papenbrock, 2005). It has been reported that the active enzyme forms a homo-oligomer (Schwenn, 1994). SiR and nitrite reductase share substantial homologies at the sequence and protein level of the catalytic domains. Both use siroheme–iron–sulfur clusters as prosthetic groups. In addition, both employ ferredoxin as an electron donor and each enzyme's substrate is chemically similar. Whether this results in competition or whether this contributes to the regulation of the crosstalk between nitrogen and sulfur assimilation pathways is still an open question.

Free sulfite is toxic to cells and needs to be effectively removed. A peroxisomal molybdenum enzyme, sulfite oxidase (SOX), converts sulfite to sulfate (Lang et al., 2007; Mendel, 2007). Although free sulfite may be produced when plants are exposed to SO_2, internal sources must exist to justify this costly detoxification mechanism (Hänsch et al., 2007; Hänsch and Mendel, 2005). Free cellular sulfite originating from SO_2 thus needs to be transported to the peroxisomes where it is oxidized to sulfate rather than channeled into the reductive pathway. As the free reduction pathway seems to be generally accepted, free sulfite should be able to enter the reduction pathway as a substrate for SiR. As this is not the case, this suggests that possible substrate channeling of SO_3^{2-} from APR to SiR and probably even to the cysteine synthase complex, may occur. Substrate channeling is also suggested by the observation that free sulfite is hardly detectable in plant extracts, and an excess of gaseous H_2S supplied to plants leads to cysteine formation instead of the reverse production of sulfite and/or sulfate. However, enzyme complexes that allow substrate channeling and exclusion of free sulfite from the reduction pathway have not yet been demonstrated. Alternatively, SO_2 conversion to sulfite may predominantly occur in the cytosol and with import into the chloroplast might be insufficient. But this raises the question of why should there be a functional peroxisomal transport system leading to a costly derouting of sulfite to sulfate and back via sulfite to sulfide/cysteine?

Unexpectedly, SiR has been shown to bind chloroplast DNA and is assumed to be involved in the regulation of the transcriptional activity of plastid nucleoids, which are structures of compacted chloroplast DNA and protein complexes resembling that of chromatin (Chi-Ham et al., 2002; Sekine et al., 2002, 2007). SiR, when dephosphorylated, contributes to condensation of the nucleoids, thereby reducing transcriptional activity of chloroplast encoded genes. This putative bifunctional role of SiR and the rationale behind these completely unrelated functions cannot be satisfactorily explained yet. It is also unclear whether the nucleoid bound SiR or the free form is active within the assimilatory reductive sulfate pathway.

Cysteine Synthesis

Cysteine synthesis is a complex process resulting from an interplay between two enzymes, serine acetyltransferase (SAT) that catalyses the formation of O-acetylserine (OAS) and O-acetylserine(thiol) lyase (OASTL) that generates cysteine from OAS and sulfide. This enzyme complex is termed the cysteine synthase complex (CSC). Both enzymes are encoded by small multigene families and are located in the cytosol, chloroplasts, and mitochondria.

Cysteine is the first organic molecule bound to reduced sulfide and the common precursor of all following metabolic steps requiring reduced sulfur. Cysteine is an integral part of proteins determining structure and function. The free sulfhydryl group allows the formation of tertiary and even quaternary structures via disulfide bridges and it protects proteins against oxidation or even regulates enzyme activities. Cysteine is part of active centers of enzymes and is involved in redox reactions through its capacity of reversible disulfide bonding. Fatty acid biosynthesis would be impossible without binding of the growing fatty acid chain to a thiol group of the acyl carrier protein and the repetitive delivery of acetyCoA. Further, enzyme activities depend on Fe/S clusters as prosthetic groups and on sulfur-containing vitamin cofactors such as biotin and thiamine. Cysteine, eventually, is also the source of a wide range of sulfur-containing metabolites, predominant among them glutathione (GSH) and S-adenosylmethionine (SAM) and numerous derived compounds such as phytochelatins, thioredoxins, glutaredoxins, CoA, S-methylmethionine, S-methylcysteine, S-alkylcysteine, ethylene, and polyamines which all dispose of essential functions in plant metabolism. Also, secondary metabolites such as protective cyanogenic glucosinolates or even fragrances and tastes are often determined through sulfur containing compounds or their breakdown products. (Anderson, 1990; Azevedo et al., 1997; Hell, 1997; Hell and Rennenberg, 1998; Matthews, 1999; Leustek and Saito, 1999; Saito, 2000; Leustek et al., 2000; Hell et al., 2002; Saito, 2004; Wirtz and Droux, 2005; Kopriva, 2006; Hesse and Hoefgen, 2006; Wirtz and Hell, 2006; Hesse et al., 2004a; Hesse and Hoefgen 2003, 2007).

The control of cysteine and methionine biosynthesis has been the target of numerous studies at the biochemical and molecular level. Various systems-oriented approaches were used to describe and interpret the global response of sulfur starvation. As sulfate is an integral part of plant metabolism, the response to sulfate starvation and resupply is complex and affecting not only sulfate assimilation but in a pleiotropic manner numerous other pathways and responses (Hoefgen et al., 2001; Hell et al., 2002; Galili and Hoefgen 2002; Maruyama-Nakashita et al., 2003; Hesse et al., 2004b; Riemenschneider et al., 2005; Nikiforova et al., 2002, 2003, 2004, 2005a, 2005b; Hirai et al., 2003, 2005; Hirai and Saito, 2004; Hesse and Hoefgen, 2006; Hawkesford and de Kok, 2006; Yoshimoto et al., 2007; Saito, 2008).

Providing the Carbon Backbone for Cysteine Biosynthesis

The carbon backbone for cysteine biosynthesis is derived from serine, which is either generated by the glycolate pathway associated with photorespiration or from 3-phosphoglycerate (Ho and Saito, 2001; Rébeillé et al., 2006). Cysteine is essentially formed by exchanging the hydroxyl group of serine against a sulfhydryl group forming a thiol (Anderson, 1990; Hell, 1997; Hell and Rennenberg, 1998; Leustek and Saito, 1999; Saito, 2000, 2004; Leustek et al., 2000; Hell et al., 2002; Wirtz and Droux, 2005; Kopriva, 2006; Hesse and Hoefgen, 2006; Hesse and Hoefgen, 2007). Before sulfhydrylation, performed by OASTL, serine is activated by acetylation of its hydroxyl group.

This reaction is performed by the activity of serine acetyltransferase (SAT, EC 2.3.1.30) employing serine and acetyl-CoA as substrates to synthesize *O*-acetyl-L-serine (OAS). The five SAT isoforms of *Arabidopsis* are localized in all three organelles capable of synthesizing proteins autonomously—the plastids, cytosol, and mitochondria. The plastidial isoforms function as part of the reductive sulfate assimilation pathway. For the isoforms of the other plant compartments, a discussion is still ongoing regarding their respective functions since, in the cytosol and mitochondria, the preceding reactions of the reduction pathway are missing. A reasonable explanation could be the detoxification of the reactive sulfide groups or a necessity for local cysteine production in all compartments, but also a regulatory function for the cytosolic isoforms is discussed. Further, the presence of these truncated pathways requires a substantial amount of metabolite shuffling between the compartments of at least sulfide.

Current evidence suggests that SAT consists of a hexameric structure composed of two associated homotrimers. However, OAS formation from serine and acetyl-CoA only occurs when the SAT complex is associated with two homodimers of OASTL in the CSC and when OASTL is in a 400-fold excess over SAT—essentially pushing all free SAT hexamers into the CSC. Monomer sizes of SAT in different species range between 29 and 34 kDa, whereas OASTL monomer sizes range between 68 and 75 kDa, resulting in CSC quaternary sizes of about 320 kDa (Droux et al., 1998; Droux, 2004; Wirtz et al., 2004; Wirtz and Hell, 2006, 2007). SAT activity limits the rate of cysteine formation and the cytosolic isoforms are feedback-sensitive to micromolar concentrations of cysteine (Kredich et al., 1969; Saito, 2004). Cysteine inactivates SAT through dissociation from the SAT-OASTL complex. However, it should be noted that plant-specific differences have been identified. For example, in *Arabidopsis*, the plastidial SAT is inhibited by cysteine; conversely, in pea (*Pisum sativum* L.) the cytosolic SAT is inhibited. This should be taken into account when deducing models or devising engineering strategies for genetically improving crop plants. The K_m values for serine have been determined with 2.4 and 0.35 µM for acetylcoenzyme A. In a second study of free and OASTL bound SAT, respectively, K_m values for serine shifted from 2.8 to 1.0 µM for serine and from 0.9 to 0.34 µM for acetyl CoA corroborating results obtained for SAT from *Salmonella typhimurium*. Cytosolic SAT from soybean [*Glycine max* (L.) Merr.] is further subject to phosphorylation that inactivates the enzyme in dependence of cysteine levels (Droux, 2004).

Under conditions of sulfate deprivation, the mRNA levels of the plastidic SAT increase nearly 2-fold. Otherwise the genes appear to be constitutively expressed. The ectopic expression and overproduction of feed-back insensitive or also sensitive SAT isoforms, such as the *Arabidopsis thaliana* SAT-A or an *E. coli* cysE derivative, resulted in transgenic tobacco and potato plants displaying increased OAS levels accompanied by up to 6- and 3-fold increases in cysteine and GSH levels, respectively (Blaszczyk et al., 1999; Harms et al., 2000; Stiller et al., 2007). The accumulation of OAS poses questions as to whether the reduction capacity is insufficient to cope with increased SAT levels that might be due to thiol-based downregulation of ATPS and APR activity, or whether the excess amounts of the OASTL indeed function in cysteine biosynthesis.

Similar to nitrate assimilation, sulfate or downstream metabolites may trigger changes in the mRNA levels of the sulfate transporter and APR genes. OAS, a marker for sulfate starvation (Nikiforova et al., 2005a) is discussed as a signal for sulfate assimilation. An activating effect on uptake and APR activity has been demonstrated for OAS as well as an increase of gene expression of the sulfate assimilatory genes. However, as the responses of OAS treatment and sulfate starvation do not entirely overlap regulation of sulfate, metabolism may be achieved through further components (summarized in Kopriva, 2006). An investigation of changes in OAS levels may suggest a link between sulfur, nitrogen, and carbon metabolism, since external supplies of these macronutrients mutually affect at least single steps within the assimilatory activities of each and photosynthesis (Fig. 6–1).

Synthesizing Cysteine

O-Acetylserine (OAS) synthesized by SAT when bound to OASTL is released from the complex and serves, together with sulfide provided by SiR, as a substrate for the free O-acetylserine (thiol) lyase (OASTL; EC 4.2.99.8) (Wirtz and Hell, 2007). Free OASTL with a molecular weight between 57 and 72 kDa is in excess of up to 400-fold over SAT. OASTL activity is reduced when present in the complex. OASTL also catalyzes a backward reaction with cysteine and cyanide as substrates to produce β-cyanoalanine and sulfide (Burandt et al., 2002). This corresponds to the reaction catalyzed by β-cyanoalanine synthases (CAS; EC 4.4.1.9) that detoxifies toxic cyanides that can inhibit the mitochondrial respiration chain. The CAS and OASTL enzymes are evolutionarily related members of the pyridoxal-phosphate enzyme family, though they differ in their biochemical parameters. Thus, CAS is also able to act as cysteine synthase. In several plant species, CAS enzymes synthesize cysteine in mitochondria. Only in *Arabidopsis* has a mitochondrially targeted OASTL has been identified (Hesse et al., 1999). The catalytic properties of OASTL with affinities for OAS in a range from 1 to 7 μM and from 20 to 60 μM for sulfide require concentrations of the substrate usually not detected in plants; thus, either the biosynthesis proceeds far below its maximum rate in cells or an effective substrate channeling has to be assumed to provide a suitable microenvironment.

In *Arabidopsis*, nine genes encode for OASTL-like proteins. The historically derived nomenclature of SAT and OASTL proteins, cDNAs, and genes was summarized recently (http://www.Arabidopsis.org; verified 6 April 2008). Expression analyses in *Arabidopsis* revealed that four genes—OASTL A1 (At4 g14880), A2 (At3 g22460), OAS-TL B (At2 g43750), and OAS-TL C (At3 g59760)—are probably responsible for most of the OASTL activity. The A isoforms are localized in the cytosol, the B isoforms in the plastids, and the C isoforms in the mitochondria and the plastids (Hesse et al., 1999). In plant families other than the Brassicaceae, the mitochondrial cysteine synthase activity may be performed by CAS enzymes, since OASTL appears to be missing. Expression of OASTL is increased up to 3-fold by factors including sulfate deficiency and development of the plant that seem to have little or no impact on mRNA contents of OAS-like genes. However, cadmium and salt stress substantially elevate OASTL A1 expression and activity (Dominguez-Solis et al., 2001). Moderate increases of mRNA in response to reduction of carbon, nitrogen, or sulfate supply also indicate some function in integrating nutrient signals. Abiotic stresses, such as salt and heavy metals, increase cytosolic levels of OASTL 7-fold and reduce the strong expression of the same isoforms in leaf trichomes (Hell et al., 2002).

For functional elucidation of different isoforms, genetic approaches have been followed. OASTL overexpression results in increased tolerance to abiotic stresses due to cysteine and GSH overproduction (Youssefian et al., 1993, 2001; Saito et al., 1994). Alternatively, OASTL antisense inhibition of cytosolic or plastidial OASTL genes resulted unexpectedly in increased levels of thiols, which led to the assumption that the free OASTL acts additionally to biosynthesis as a desulfhydrase in regulating the homeostasis of the free cysteine pool (Hopkins et al., 2005; Riemenschneider et al., 2005; Papenbrock et al., 2007). As a result of that same study, an intensive exchange of sulfur-related metabolites has to be assumed between cytosol and plastids, actually, then counter arguing substrate channeling of at least the sulfide substrate, given the catalytic properties of the OASTL. Together with the diverse, species specific differences in cysteine feedback sensitivity of SAT (Droux, 2003, 2004) a species specific, respectively, and an organelle specific adaptation of the model for cysteine biosynthesis needs to be considered.

The Cysteine Synthase Complex

The understanding of the cellular cysteine biosynthesis in plants has been increased by our knowledge of the cysteine synthase complex (CSC) composed of SAT and OASTL (Fig. 6–2). A widely accepted model (Kredich et al., 1969; Saito, 2000; Wirtz and Hell, 2006) provides the basis for future research or applied approaches to manipulate the pathway flux. Generally, when SAT is associated with OASTL in the CSC, it is able to synthesize OAS. The free form of OASTL appears to be responsible for cysteine synthesis. However, certain aspects of regulation, subcellular distribution, and metabolic control need to be investigated in detail with regard to species-specific differences The functional role of the free OASTL isoform versus substrate channeling might need revisiting (Riemenschneider et al., 2005). The CSC is formed in all three compartments where autonomous protein synthesis does occur. In *Arabidopsis*, the subcellular localization of the proteins of the CSC have been experimentally confirmed for all SAT isoforms by means of GFP fusion proteins and transient expression in plants. OASTL A1 is lacking any transit peptide and is apparently localized in the cytosol; OASTL B and C are attributed to the chloroplasts and mitochondria (Hesse et al., 1999; Wirtz and Hell, 2006).

In vitro sulfide stabilizes the CSC, thus promoting SAT activity and cysteine formation, whereas OAS destabilizes the CSC (Kredich et al., 1969). SAT displays two domains involved in protein–protein interactions. The N-terminal extension and tandem hexapeptide repeats have been shown to be important for interaction between SAT subunits, while the C terminus, which contains the catalytically active site, mediates SAT-OASTL interaction (Francois et al., 2006; Kumaran and Jez, 2007). OAS and sulfide driven dissociation and association kinetics have recently been established for the CSC by means of BIAcore technology. Substrate concentrations, at which the complex was dissociated or associated to 50% have

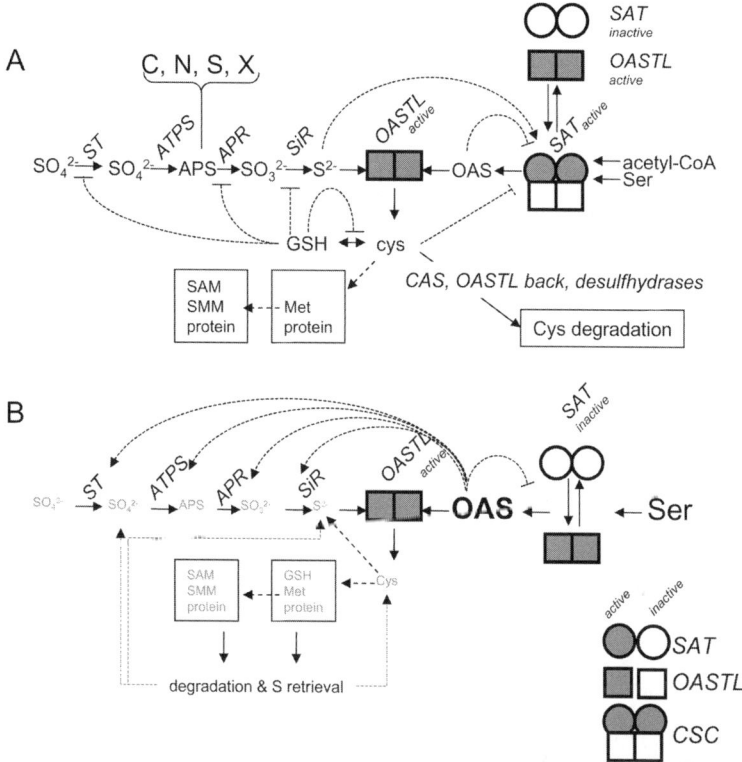

Fig. 6–2. Model for metabolic regulation of the cysteine synthase complex. Sulfate reduction is a tightly regulated process, both at the enzymatic and molecular level. APR integrates various cellular signals such as carbon supply (C), nitrogen status (N), sulfate status (S), and hormonal signals (X) to regulate the flux by changes in expression. The cysteine synthase complex (CSC) composed of OASTL and SAT serves as a control point to fine-tune cysteine biosynthesis. Under sufficient sulfate supply conditions (A) the CSC bound SAT provides OAS that is converted by free OASTL to cysteine using sulfide. When OAS accumulates, the complex dissociates preventing further OAS synthesis, while increasing sulfide concentrations favor association of the complex. Cysteine downregulates SAT in a feedback manner to reduce precursor synthesis when cysteine accumulates to adjust biosynthesis to downstream biosynthetic needs. GSH, a tripeptide synthesized from cysteine, downregulates the expression of several assimilation genes, i.e., ST, ATPS, APR, and SiR. Cysteine concentrations are additionally controlled by various degrading activities of beta-cyanoalanine synthases (CAS), desulfhydrases, and the desulfhydrase like backreaction of OASTL. Under sulfate deplete conditions (B) sulfide concentrations and, eventually, cysteine and GSH concentrations drop, while OAS accumulates. The free OAS then induces the expression of upstream genes, i.e., ST, ATPS, APR, and SiR. GSH repression of these very genes is alleviated as GSH concentrations are reduced. When sulfate is resupplied, the system is set to quickly take up and reduce sulfate, which results in an overshoot of cysteine and downstream metabolites until equilibrium (A) is reached again. In a parallel pleiotropic response degradation of metabolites and proteins occurs to retrieve sulfur and to refuel cysteine synthesis.

been determined to be 77 µM and approximately 50 µM for OAS and sulfide, respectively (Wirtz and Hell, 2006).

This interplay of metabolite effects and reversible changes in enzyme activities and associations would result in the following model of a regulatory circuit (Fig. 6–2). Binding of SAT to the complex enhances SAT activity. Conversely, free SAT is rendered more inactive. Vice versa, OASTL while inactive in the complex is active when released from the complex. However, because of the excess of OASTL, this is of no relevance for the model. As OASTL is in excess over SAT, it can be assumed that under normal growth conditions, SAT is to its majority complexed to OASTL readily providing OAS that together with sulfide is converted to cysteine by the free OASTL. This explains the usually low cellular concentrations of both substrates. Accumulation of cysteine downregulates SAT activity and reduces flux. Local OAS concentrations above 77 µM dissociate the complex, and sulfide concentrations above 50 µM promote association of the CSC. Thus, under conditions of sulfate deprivation, the accumulation of OAS as well as low sulfide concentrations favor disruption of the complex, reducing further OAS production. Reduction of cellular cysteine concentrations quickly accompany situations of sulfate starvation (Hirai et al., 2003; Nikiforova et al., 2003) and alleviate SAT inhibition. Sulfate resupply leads to sulfide provision and reduction of OAS, favoring complex association and resumption of OAS synthesis. Thus, the CSC senses sulfide levels and controls cysteine synthesis. However, this model applies for the cytosolic CSC of *Arabidopsis* not to the plastidial CSC and in the case of pea vice versa, which indicates that the *Arabidopsis* model is not generally applicable to other species, especially when taking metabolite shuffling between compartments into account (Droux, 2003, 2004; Riemenschneider et al., 2005). Furthermore, the accumulation of OAS, when sulfide supply is low, such as under conditions of sulfate starvation or increased need for thiols under heavy metal or oxidative stress conditions, might act as one of the signals controlling sulfate assimilation in plants, e.g., positively regulating sulfate uptake. Transcriptome analysis in response to OAS treatment in *Arabidopsis thaliana* revealed changes in mRNA transcript abundance (Saito, 2004; Hesse et al., 2004a; Galili et al., 2005; Kopriva 2006; Hirai et al., 2003). These partially resemble those observed during sulfur starvation, indicating that OAS accumulation on sulfur deficiency controls gene expression (Fig. 6–2). The transcription of genes encoding sulfate transporters and enzymes of the sulfate reduction pathway is enhanced in an attempt to provide sufficient sulfide for retaining homeostasis. OAS acts positively on the transcript and activity levels of sulfate transporters, ATP sulfurylase, APR, SiR, plastidial OASTL, and cystosolic SAT, as shown in OAS-feeding experiments (summarized in Kopriva, 2006).

Because cysteine has multiple cellular functions, it is an interesting target for genetic manipulation to increase its content in plants. Cysteine biosynthesis can be regulated at several steps (Hesse et al., 2004a). Manipulation of cellular pools of thiols, and their derivatives can be accomplished through overexpression of homologous or heterologous pathway genes or pathway genes mutated in properties such as feedback inhibition. The introduction of entire new pathways from other organisms might also be an option (Matityahu et al., 2006). Interestingly, by pushing sulfate reduction through overexpression of APR, a greater flux through the pathway could be achieved, yielding increased levels of thiols and methionine. When sulfite and thiosulfate are increased more than the thiols cysteine and glutathione, it is assumed that the provision of the carbon backbone precursor OAS by SAT limits cysteine biosynthesis more than the reduction capacity (Tsakraklides et al., 2002). Hence, from a biotechnological point of view, shifting the ratio of SAT to OASTL in favor of SAT is the method of choice, resulting in an increased accumulation of up to 6-fold of the downstream products cysteine, GSH, and methionine. This could also be achieved by overexpression of feedback-insensitive or -sensitive SAT isoforms either in plastids or in the cytosol by respective antisense inhibition of OASTL (Galili et al., 2005; Blaszczyk et al., 1999; Harms et al., 2000; Riemenschneider et al., 2005; Wirtz and Hell, 2006). A positive side effect is an increased stress tolerance because of increased GSH levels (Youssefian et al., 2001). The current working model for the CSC does not entirely explain all the experimental observations. Why should overproduction of OASTL, which increases the excess pool of the enzyme and, thus, further decreases the SAT to OASTL ratio, result in elevated cysteine and GSH levels? Further regulatory aspects have thus to be assumed.

In summary, the association–dissociation kinetics of the CSC controls upstream processes of cysteine synthesis by sensing the plant sulfide status, adjusting gene expression and enzyme activity through steering CSC architecture and generating a potential signal molecule, OAS. The CSC also senses downstream products as cysteine accumulation regulates precursor synthesis in a feedback fashion in some isoforms.

Summary

Plants and plant associations are usually adapted to certain soil types and environmental conditions (Hoefgen and Hesse, 2007). As environmental conditions, such as mineral nutrient availability, vary under natural conditions, plants have developed mechanisms to adapt within certain limits to these alterations. Crop plants have been bred for specific agricultural production schemes and climates. Modern elite varieties are usually adapted to optimal- and high-nutrient inputs through mineral fertilizers and to produce maximal yield under these conditions. In this context, sulfate metabolism of crop plants has long been neglected and the effects of insufficient supplies underestimated as industry-based pollutions provided sufficient sulfur input. Less than optimal provision of sulfate, though not having immediate effects, impairs plant performance and plant health as well as nutritional or processing quality of the crop.

The most common source of sulfur for plants is soil-borne sulfate. Sulfate is actively transported into the cell and distributed throughout the plant. It is then subject to different fates as it is stored in the vacuole or activated through binding to ATP and used either for the sulfatation pathway or reduced to sulfide and incorporated into cysteine, the exclusive entry point for organic bound reduced sulfur. Cysteine thus constitutes the intersection between sulfate assimilation and downstream metabolization in plants. Therefore, it is mandatory that its biosynthesis be tightly controlled during various stages of sulfate assimilation. The main regulatory steps are the integration of general metabolic signals, such as nitrogen and carbon status, by APR and the fine-tuning of thiol biosynthesis by the CSC. Cysteine serves as precursor for numerous plant metabolites involved in numerous and often essential cellular functions. Thus, sulfate metabolism is as important as nitrogen assimilation into glutamine for plant growth and human consumption.

Optimal crop yield and nutritional quality can only be obtained under agricultural conditions of optimal sulfate provision together with all other nutrients. Situations of sulfate starvation during sensitive phases of crop production must be avoided. This can be achieved through the application of mineral fertilizer provisions adapted to the respective plant species and cultivar. Additionally, plant breeding should aim at improving sulfate use efficiency in crop plants. In conclusion, it is obvious that crop quality depends on a combination of parameters reaching from soil-sulfate availability over effective transport and remobilization processes of sulfate within the cell facilitated by a family of specialized transporters to perfectly regulated biosynthetic activities to provide the reduced sulfur compounds necessary for plant growth and vigor. The detailed knowledge of sulfur assimilation obtained in model and in crop plants in the past years helps to devise strategies to produce optimized crop varieties with improved features of sulfate uptake and use, with improved nutritional qualities such as protein content and composition, essential amino acid content, and vitamin contents. Such novel cultivars would fit into strategies of increasing sustainability of agriculture and protection of the environment.

References

Anderson, J.W. 1990. Sulfur metabolism in plants. p. 327–381. *In* B.J. Miflin and P.J. Lea (ed.) The biochemistry of plants. Vol. 16. Academic Press, New York.

Awazuhara, M., T. Fujiwara, H. Hayashi, A. Watanabe-Takahashi, H. Takahashi, and K. Saito. 2005. The function of SULTR2;1 sulfate transporter during seed development in *Arabidopsis thaliana*. Physiol. Plant. 125:95–105.

Azevedo, R.A., P. Arruda, W.L. Turner, and P.J. Lea. 1997. The biosynthesis and metabolism of the aspartate derived amino acids in higher plants. Phytochemistry 46:395–419.

Baxter, I., B. Muthukumar, H.C. Park, P. Buchner, B. Lahner, J. Danku, K. Zhao, J. Lee, M.J. Hawkesford, M.L. Guerinot, and D.E. Salt. 2008. Variation in molybdenum content across broadly distributed populations of *Arabidopsis thaliana* is controlled by a mitochondrial molybdenum transporter (MOT1). PLoS Genet. 4:e1000004, doi:10.1371/journal.pgen.1000004

Benning, C. 1998. Biosynthesis and function of the sulfolipid sulfoquinovosyl diacylglycerol. Annu. Rev. Plant Physiol. Plant Mol. Biol. 49:53–75.

Blake-Kalff, M.M.A. 2000. Diagnosing sulfur deficiency in field-grown oilseed rape (*Brassica napus*). Plant Soil 225:95–107.

Blaszczyk, A., R. Brodzik, and A. Sirko. 1999. Increased resistance to oxidative stress in transgenic tobacco plants overexpressing bacterial serine acetyltransferase. Plant J. 20:237–243.

Bork, C., J.D. Schwenn, and R. Hell. 1998. Isolation and characterization of a gene for assimilatory sulfite reductase from *Arabidopsis thaliana*. Gene 212:147–153.

Bourgis, F., S. Roje, M.L. Nuccio, D.B. Fisher, M.C. Tarczynski, C.J. Li, C. Herschbach, H. Rennenberg, M.J. Pimenta, T.L. Shen, D.A. Gage, and A.D. Hanson. 1999. S-methylmethionine plays a major role in phloem sulfur transport and is synthesized by a novel type of methyltransferase. Plant Cell 11:1485–1497.

Bryan, J.K. 1990. Advances in the biochemistry of amino acid biosynthesis. p. 161–196. *In* B.J. Miflin (ed.) The biochemistry of plants. Vol. 5. Academic Press, New York.

Bryan, J.K. 1980. Synthesis of the aspartate family and branched-chain amino acids. p. 403–452. *In* B.J. Miflin (ed.) The biochemistry of plants. Vol. 5. Academic Press, New York.

Buchner, P., C.E.E. Stuiver, S. Westerman, M. Wirtz, R. Hell, M.J. Hawkesford, and L.J. de Kok. 2004a. Regulation of sulfate uptake and expression of sulfate transporter genes in *Brassica oleracea* as affected by atmospheric H_2S and pedospheric sulfate nutrition. Plant Physiol. 136:3396–3408.

Buchner, P., H. Takahashi, and M.J. Hawkesford. 2004b. Plant sulfate transporters: Co-ordination of uptake, intracellular and long distance transport. J. Exp. Bot. 55:1765–1773.

Burandt, P., A. Schmidt, and J. Papenbrock. 2002. Three O-acetyl-L-serine(thiol)lyase isoenzymes from Arabidopsis catalyse cysteine synthesis and cysteine desulfuration at different pH values. J. Plant Physiol. 159:111–119.

Chiaiese, P., N. Ohkama-Ohtsu, L. Molvig, R. Godfree, H. Dove, C. Hocart, T. Fujiwara, T.J.V. Higgins, and L.M. Tabe. 2004. Sulphur and nitrogen nutrition influence the response of chickpea seeds to an added, transgenic sink for organic sulphur. J. Exp. Bot. 55:1889–1901.

Chi-Ham, C.L., M.A. Keaton, G.C. Cannon, and S. Heinhorst. 2002. The DNA-compacting protein DCP68 from soybean chloroplasts is ferredoxin: Sulfite reductase and co-localizes with the organellar nucleoid. Plant Mol. Biol. 49:621–631.

de Kok, L.J., M. Durenkamp, L. Yang, and I. Stulen. 2007. Atmospheric sulfur. p. 91–1006. *In* M.J. Hawkesford and L.J. de Kok (ed.) Sulfur in plants: An ecological perspective. Springer, Dordrecht, the Netherlands.

Dominguez-Solis, J., G. Gutierrez-Alcala, L. Romero, and C. Gotor C. 2001. The cytosolic O-acetylserine(thiol)lyase gene is regulated by heavy metals and can function in cadmium tolerance. J. Biol. Chem. 276:9297–9302.

Droux, M. 2003. Plant serine acetyltransferase: New insights for regulation of sulphur metabolism in plant cells. Plant Physiol. Biochem. 41:619–627.

Droux, M. 2004. Sulfur assimilation and the role of sulfur in plant metabolism: A survey. Photosynth. Res. 79:331–348.

Droux, M., M.-L. Ruffet, R. Douce, and D. Job. 1998. Interactions between serine acetyltransferase and O-acetylserine (thiol) lyase in higher plants. Structural and kinetic properties of the free and bound enzymes. Eur. J. Biochem. 255:235–245.

Durenkamp, M., and L.J. de Kok. 2004. Impact of pedospheric and atmospheric sulphur nutrition on sulphur metabolism of *Allium cepa* L., a species with a potential sink capacity for secondary sulphur compounds. J. Exp. Bot. 55:1821–1830.

Durenkamp, M., L.J. de Kok, and S. Kopriva S. 2007. Adenosine 5'-phosphosulphate reductase is regulated differently in *Allium cepa* L. and *Brassica oleracea* L. upon exposure to H2S. J. Exp. Bot.. 58:1571–1579.

Francois, J.A., S. Kumaran, and J.M. Jez. 2006. Structural basis for interaction of O-acetylserine sulfhydrylase and serine acetyltransferase in the Arabidopsis cysteine synthase complex. Plant Cell 18:3647–3655.

Frentzen, M. 2004. Phosphatidylglycerol and sulfoquinovosyldiacylglycerol: Anionic membrane lipids and phosphate regulation. Curr. Opin. Plant Biol. 7:270–276.

Galili, G., R. Amir, R. Hoefgen, and H. Hess. 2005. Improving the levels of essential amino acids and sulfur metabolites in plants. Biol. Chem. 386:817–831.

Galili, G., and R. Hoefgen. 2002. Metabolic engineering of amino acids and storage proteins in plants. Metab. Eng. 4:3–11.

Grossman, A., and H. Takahashi. 2001. Macronutrient utilization by photosynthetic eukaryotes and the fabric of interactions. Annu. Rev. Plant. Phys. Plant. Mol. Biol. 52:163–210.

Hagan, N.D., N. Upadhyaya, L.M. Tabe, and T.J.V. Higgins. 2003. The redistribution of protein sulfur in transgenic rice expressing a gene for a foreign, sulfur-rich protein. Plant J. 34:1–11.

Halkier, B.A., and J. Gershenzon. 2006. Biology and biochemistry of glucosinolates. Annu. Rev. Plant Biol. 57:303–333.

Haneklaus, S., E. Bloem, and E. Schnug. 2003. The global sulfur cycle and its links to plant environment. p. 1–28. *In* Y.P. Abrol and A. Ahmad (ed.) Sulfur in plants. Kluwer Academic Publishers, Dordrecht, the Netherlands.

Hänsch, R., C. Lang, H. Rennenberg, and R.R. Mendel. 2007. Significance of plant sulfite oxidase. Plant Biol. 9:589–595.

Hänsch, R., and R.R. Mendel. 2005. Sulfite oxidation in plant peroxisomes. Photosynth. Res. 86:337–343.

Harms, K., P. von Ballmoos, C. Brunold, R. Hoefgen, and H. Hesse. 2000. Expression of a bacterial serine acetyltransferase in transgenic potato plants leads to increased levels of cysteine and glutathione. Plant J. 22:335–343.

Hatzfeld, Y., N. Cathala, C. Grignon, and J.C. Davidian. 1998. Effect of ATP sulfurylase overexpression in Bright Yellow 2 tobacco cells–regulation of ATP sulfurylase and SO_4^{2-} transport activities. Plant Physiol. 116:1307–1313.

Hawkesford, M.J. 2000. Plant responses to sulphur deficiency and the genetic manipulation of sulphate transporters to improve S-utilization efficiency. J. Exp. Bot. 51:131–138.

Hawkesford, M.J. 2003. Transporter gene families in plants: The sulfate transporter gene family—Redundancy or specialization? Physiol. Plant. 117:155–163.

Hawkesford, M.J. 2007. Sulfur and Plant ecology: A central role of sulfate transportersin response to sulfur availability. p. 61–79. In M.J. Hawkesford and L.J. de Kok (ed.) Sulfur in plants: An ecological perspective. Springer, Dordrecht, the Netherlands.

Hawkesford, M.J., P. Buchner, L. Hopkins, and J.R. Howarth. 2003. The plant sulfate transporter family: Specialized functions and integration with whole plant nutrition. p. 1–10. In J.C. Davidian et al. (ed.) Sulfur transport and assimilation in plants—Regulation, interaction and signaling. Backhuys Publishers, Leiden, the Netherlands.

Hawkesford, M.J., and L.J. de Kok. 2006. Managing sulphur metabolism in plants. Plant Cell Environ. 29:382–395.

Hawkesford, M., R. Hoefgen, G. Galili, R. Amir, G. Angenon, H. Hesse, D. Rentsch, J. Schaller, I. Van der Meer, J. Rouster, Z. Banfalvi, P. Zsolt, L. Szabados, J. Szopa, and A. Sirko. 2006. Optimising nutritional quality of crops. p. 85–116. In P.K. Jaiwal (ed.) Plant genetic engineering. Vol. 7, Plant metabolic engineering and molecular farming. Studium Press LLC, Houston, TX.

Hell, R. 1997. Molecular physiology of plant sulfur metabolism. Planta 202:138–148.

Hell, R., R. Jost, O. Berkowitz, and M. Wirtz. 2002. Molecular and biochemical analysis of the enzymes of cysteine biosynthesis in the plant *Arabidopsis thaliana*. Amino Acids 22:245–257.

Hell, R., and H. Rennenberg. 1998. The plant sulfur cycle. Sulfur Agroecosys. 2:135–173.

Hesse, H., and R. Hoefgen. 2006. On the way to understand biological complexity in plants: S-nutrition as a case study for systems biology. Cell. Mol. Biol. Lett. 11:37–55.

Hesse, H., and R. Hoefgen. 2001. Application of genomics in agriculture. p. 61–79. In M.J. Hawkesford and P. Buchner (ed.) Molecular analysis of plant adaptation to the environment. Kluwer AP, Dordrecht, the Netherlands.

Hesse, H., and R. Hoefgen. 2003. Molecular aspects of methionine biosynthesis in Arabidopsis and potato. Trends Plant Sci. 8:259–262.

Hesse, H., O. Kreft, S. Maimann, M. Zeh, and R. Hoefgen. 2004b. Current understanding of the regulation of methionine biosynthesis in plants. J. Exp. Bot. 55:1799–1808.

Hesse, H., J. Lipke, T. Altmann, and R. Hoefgen. 1999. Molecular cloning and expression analysis of mitochondrial and plastidic isoforms of cysteine synthase (O-acetylserine (thiol)lyase) from *Arabidopsis thaliana*. Amino Acids 16:113–131.

Hesse, H., V. Nikiforova, B. Gakière, and R. Hoefgen. 2004a. Molecular analysis and control of cysteine biosynthesis: Integration of nitrogen and sulfur metabolism. J. Exp. Bot. 55:1283–1292.

Hesse, H., N. Trachsel, M. Suter, S. Kopriva, P. von Ballmoos, H. Rennenberg, and C. Brunold. 2003. Effect of glucose on assimilatory sulphate reduction in *Arabidopsis thaliana* roots. J. Exp. Bot. 54:1701–1709.

Hirai, M.Y., T. Fujiwara, M. Awazuhara, T. Kimura, N. Masaaki, and K. Saito. 2003. Global expression profiling of sulfur-starved Arabidopsis by DNA macroarray reveals the role of O-acetyl-L-serine as a general regulator of gene expression in response to sulfur nutrition. Plant J. 33:651–663.

Hirai, M.Y., M. Klein, Y. Fujikawa, M. Yano, D.B. Goodenowe, Y. Yamazaki, S. Kanaya, Y. Nakamura, M. Kitayama, H. Suzuki, N. Sakurai, D. Shibata, J. Tokuhisa, M. Reichelt, J. Gershenzon, J. Papenbrock, and K. Saito. 2005. Elucidation of gene-to-gene and metabolite-to-gene networks in Arabidopsis by integration of metabolomics and transcriptomics. J. Exp. Biol. Chem. 280:25590–25595.

Hirai, M.Y., and K. Saito. 2004. Post-genomics approaches for the eleucidation of plant adaptive mechanisms to sulphur deficiency. J. Exp. Bot. 55:1871–1879.

Ho, C.L., and K. Saito. 2001. Molecular biology of the plastidic phosphorylated serine biosynthetic pathway in *Arabidopsis thaliana*. Amino Acids 20:243–259.

Hoefgen, R., and H. Hesse. 2007. Sulfur in plants as part of a metabolic network. p. 107–142. In M.J Hawkesford and L.J. de Kok (ed.) Sulfur in plants: An ecological perspective. Springer, Dordrecht, the Netherlands.

Hoefgen, R., O. Kreft, L. Willmitzer, and H. Hesse. 2001. Manipulation of thiol contents in plants. Amino Acids 20:291–299.

Hopkins, L., S. Parmar, A. Blaszczyk, H. Hesse, R. Hoefgen, and M.J. Hawkesford. 2005. O-acetylserine and the regulation of expression of genes encoding components for sulfate uptake and assimilation in potato. Plant Physiol. 138:433–440.

Kataoka, T., N. Hayashi, T. Yamaya, and H. Takahashi. 2004a. Root-to-shoot transport of sulfate in Arabidopsis. Evidence for the role of SULTR3;5 as a component of low-affinity sulfate transport system in the root vasculature. Plant Physiol. 136:4198–4204.

Kataoka, T., A. Watanabe-Takahashi, N. Hayashi, M. Ohnishi, T. Mimura, P. Buchner, M.J. Hawkesford, T. Yamaya, and H. Takahashi. 2004b. Vacuolar sulfate transporters are essential determinants controlling internal distribution of sulfate in *Arabidopsis*. Plant Cell 16:2693–2704.

Kertesz, M.A., and P. Mirleau. 2004. The role of microbes in plant sulphur nutrition. J. Exp. Bot. 55:1939–1945.

Kessler, D., and J. Papenbrock. 2005. Iron-sulfur cluster biosynthesis in photosynthetic organisms. Photosynth. Res. 86:391–407.

Klein, M., and J. Papenbrock. 2004. The multi-protein family of Arabidopsis sulphotransferases and their relatives in other plant species. J. Exp. Bot. 55:1809–1820.

Klonus, D., R. Hoefgen, L. Willmitzer, and J.W. Riesmeier. 1994. Isolation and characterization of two cDNA clones encoding ATP-sulfurylases from potato by complementation of a yeast mutant. Plant J. 6:105–112.

Kopriva, S. 2006. Regulation of sulfate assimilation in Arabidopsis and beyond. Ann. Bot. (Lond.) 97:479–495.

Kopriva, S., and A. Koprivova. 2004. Plant adenosine 5'-phosphosulphate reductase: The past, the present, and the future. J. Exp. Bot. 55:1775–1783.

Kopriva, S., and H. Rennenberg. 2004. Control of sulphate assimilation and glutathione synthesis: Interaction with N and C metabolism. J. Exp. Bot. 55:1831–1842.

Kopriva, S., M. Suter, P. von Ballmoos, H. Hesse, U. Krahenbuhl, H. Rennenberg, and C. Brunold. 2002. Interaction of sulfate assimilation with carbon and nitrogen metabolism in *Lemna minor*. Plant Physiol. 130:1406–1413.

Koralewska, A., F.S. Posthumus, C.E.E. Stuiver, P. Buchner, M.J. Hawkesford, and L.J. De Kok. 2007. The characteristic high sulfate content in *Brassica oleracea* is controlled by the expression and activity of sulfate transporters. Plant Biol. 9:654–661.

Kredich, N.M. 1996. Biosynthesis of cysteine in *Escherichia coli* and *Salmonella typhimurium*. p. 514–527. In F.C. Neidhardt et al. (ed.) Cellular and molecular biology. ASM Press, Washington, DC.

Kredich, N.M., M.A. Becker, and G.M. Tomkin. 1969. Purification and characterization of cysteine synthetase, a bifunctional protein complex, from *Salmonella typhimurium*. J. Biol. Chem. 244: 2428–2439.

Kreft, O., R. Hoefgen, and H. Hesse. 2003. Functional Analysis of cystathionine gamma-synthase in genetically engineered potato plants. Plant Physiol. 131:1843–1854.

Kumaran, S., and J.M. Jez. 2007. Thermodynamics of the interaction between O-acetylserine sulfhydrylase and the C-terminus of serine acetyltransferase. Biochemistry 46:5586–5594.

Lang, C., Popko, M. Wirtz, R. Hell, C. Herschbach, J. Kreuzwieser, H. Rennenberg, R.R. Mendel, and R. Hänsch. 2007. Sulphite oxidase as key enzyme for protecting plants against sulphur dioxide. Plant Cell Environ. 30:447–455.

Leustek, T., M.N. Martin, J.A. Bick, and J.P. Davies. 2000. Pathways and regulation of sulphur metabolism revealed through molecular and genetic studies. Annu. Rev. Plant Physiol. Plant Mol. Biol. 51:141–165.

Leustek, T., and K. Saito. 1999. Sulfate transport and assimilation in plants. Plant Physiol. 120:637–643.

Lillig, C.H., S. Schiffmann, C. Berndt, A. Berken, R. Tischka, and J.D. Schwenn. 2001. Molecular and catalytic properties of Arabidopsis thaliana adenylyl sulfate (APS)-kinase. Arch. Biochem. Biophys. 392:303–310.

Lorbiecke, R., M. Steffens, J.M. Tomm, S. Scholten, P. von Wiegen, E. Kranz, U. Wienand, and M. Sauter. 2005. Phytosulphokine gene regulation during maize (*Zea mays* L.) reproduction. J. Exp. Bot. 56:1805–1819.

Loudet, O., V. Saliba-Colombani, C. Camilleri, F. Calenge, V. Gaudon, A. Koprivova, A.K. North, S. Kopriva, and F. Daniel-Vedele. 2007. Natural variation for sulfate content in *Arabidopsis thaliana* is highly controlled by APR2. Nat. Genet. 39:896–900.

Maimann, S., C. Wagner, O. Kreft, M. Zeh, L. Willmitzer L, R. Hoefgen R, and H. Hesse. 2000. Transgenic potato plants reveal the indispensable role of cystathionine β-lyase in plant growth and development. Plant J. 23:747–758.

Maruyama-Nakashita, A., E. Inoue, A. Watanabe-Takahashi, T. Yarnaya, and H. Takahashi. 2003. Transcriptome profiling of sulfur-responsive genes in Arabidopsis reveals global effects of sulfur nutrition on multiple metabolic pathways. Plant Physiol. 132:597–605.

Maruyama-Nakashita, A., Y. Nakamura, T. Yamaya, and H. Takahashi. 2004. Regulation of high-affinity sulfate transporters in plants: Towards systematic analysis of sulfur signalling and regulation. J. Exp. Bot. 55:1843–1849.

Maruyama-Nakashita, A., A. Watanabe-Takahashi, E. Inoue, T. Yarnaya, and H. Takahashi. 2005. Identification of a novel cis-acting element conferring sulfur deficiency response in Arabidopsis roots. Plant J. 42:305–314.

Matityahu, I., L. Kachan, I.B. Ilan, and R. Amir. 2006. Transgenic tobacco plants overexpressing the Met25 gene of *Saccharomyces cerevisiae* exhibit enhanced levels of cysteine and glutathione and increased tolerance to oxidative stress. Amino Acids 30:185–194.

Matthews, B. 1999. Lysine, threonine and methionine biosynthesis. p. 205–225. *In* B.K. Singh (ed.) Plant amino acids. Marcel Dekker, New York.

Melis, A., and H.C. Chen. 2005. Chloroplast sulfate transport in green algae- genes, proteins and effects. Photosynth. Res. 86:299–307.

Mendel, R.R. 2007. Biology of the molybdenum cofactor. J. Exp. Bot. 58:2289–2296.

Molvig, L., L.M. Tabe, J. Hamblin, V. Ravindran, W.L. Bryden, C.L. White, and T.J.V. Higgins. 2003. Nutritional improvement of lupin seed protein using gene technology. Appl. Gen. Legumin. Biotech. 10:213–219.

Nikiforova, V.J., C.O. Daub, H. Hesse, L. Willmitzer, and R. Hoefgen. 2005b. Integrative gene-metabolite network with implemented causality deciphers informational fluxes of sulfur stress response. J. Exp. Bot. 56:1887–1896.

Nikiforova, V., J. Freitag, S. Kempa, M. Adamik, H. Hesse, and R. Hoefgen. 2003. Transcriptome analysis of sulfur depletion in *Arabidopsis thaliana*: Interlacing of biosynthetic pathways provides response specificity. Plant J. 33:633–650.

Nikiforova, V.J., B. Gakière B, S. Kempa, M. Adamik, L. Willmitzer L, H. Hesse, and R. Hoefgen. 2004. Towards dissecting nutrient metabolism in plants: A systems biology case study on sulfur metabolism. J. Exp. Bot. 55:1861–1870.

Nikiforova, V., S. Kempa, M. Zeh, S. Maimann, O. Kreft, A.P. Casazza, K. Riedel, E. Tauberger, R. Hoefgen, and H. Hesse. 2002. Engineering of cysteine and methionine biosynthesis in potato. Amino Acids 22:259–278.

Nikiforova, V.J., J. Kopka, V. Tolstikov, O. Fiehn, L. Hopkins, M.J. Hawkesford, H. Hesse, and R. Hoefgen. 2005a. Systems rebalancing of metabolism in response to sulfur deprivation, as revealed by metabolome analysis of Arabidopsis plants. Plant Physiol. 138:304–318.

Papenbrock, J., A. Riemenschneider, A. Kamp, H.N. Schulz-Vogt, and A. Schmidt. 2007. Characterization of cysteine-degrading and H_2S-releasing enzymes of higher plants- From the field to the test tube and back. Plant Biol. 9:582–588.

Pilon-Smits, E.A.H., S.B. Hwang, C.M. Lytle, Y.L. Zhu, J.C. Tai, R.C. Bravo, Y.C. Chen, T. Leustek, and N. Terry. 1999. Overexpression of ATP sulfurylase in Indian mustard leads to increased selenate uptake, reduction, and tolerance. Plant Physiol. 119:123–132.

Ranocha, P., S.D. McNeil, M.J. Ziemak, C.J. Li, M.C. Tarczynski, and A.D. Hanson. 2001. The S-methylmethionine cycle in angiosperms: Ubiquity, antiquity and activity. Plant J. 25:575–584.

Rébeillé, F., S. Ravanel, S. Jabrin, R. Douce, S. Storozhenko, and D. Van Der Straeten. 2006. Folates in plants: Biosynthesis, distribution, and enhancement. Physiol. Plant. 126:330–342.

Riemenschneider, A., K. Riedel, R. Hoefgen, J. Papenbrock, and H. Hesse. 2005. Impact of reduced O-acetylserine(thiol)lyase isoform contents on potato plant metabolism. Plant Physiol. 137:892–900.

Rouached, H., P. Berthomieu, E. El Kassis, N. Cathala, V. Catherinot, G. Labesse, J.C. Davidian, and P. Fourcroy. 2005. Structural and functional analysis of the C-terminal STAS (sulfate transporter and anti-sigma antagonist) domain of the *Arabidopsis thaliana* sulfate transporter SULTR1.2. J. Biol. Chem. 280:15976–15983.

Saito, K. 2000. Regulation of sulfate transport and synthesis of sulfur-containing amino acids. Curr. Opin. Plant Biol. 3:188–195.

Saito, K. 2004. Sulfur assimilatory metabolism. The long and smelling road. Plant 136:2443–2450.

Saito, K. 2008. Decoding genes with coexpression networks and metabolomics– 'majority report by precogs'. Trends Plant Sci. 13:36–43.

Saito, K., M. Kurosawa, K. Tatsuguchi, Y. Takagi, and I. Murakoshi. 1994. Modulation of cysteine biosynthesis in chloroplasts of transgenic tobacco overexpressing cysteine synthase [O-acetylserine(thiol)-lyase]. Plant Physiol. 106:887–895.

Sanda, S., T. Leustek, M.J. Theisen, R.M. Garavito, and C. Benning. 2001. Recombinant Arabidopsis SQD1 converts UDP-glucose and sulfite to the sulfolipid head group precursor UDP-sulfoquinovose in vitro. J. Biol. Chem. 276:3941–3946.

Schwenn, J.D. 1994. Photosynthetic sulfate reduction. Z. Natforsch. C-A. J. Biosci. 49:531–539.

Sekine, K., M. Fujiwara, M. Nakayama, T. Takao, T. Hase, and N. Sato. 2007. DNA binding and partial nucleoid localization of the chloroplast stromal enzyme ferredoxin: Sulfite reductase. FEBS J. 274:2054–2069.

Sekine, K., T. Hase, and N. Sato. 2002. Reversible DNA compaction by sulfite reductase regulates transcriptional activity of chloroplast nucleoids. J. Biol. Chem. 277:24399–24404.

Setya, A., M. Murillo, and T. Leustek. 1996. Sulfate reduction in higher plants– molecular evidence for a novel 5'-adenylylsulfate reductase. Proc. Natl. Acad. Sci. USA 93:13383–13388.

Shibagaki, N., A. Rose, J. Mcdermott, T. Fujiwara, H. Hayashi, T. Yoneyama, and J. Davies. 2002. Selenate-resistant mutants of *Arabidopsis thaliana* identify Sultr1;2, a sulfate transporter required for efficient transport of sulfate into roots. Plant J. 29:475–486.

Stefels, J. 2007. Sulfur in the marine environment. p. 77–90. *In* M.J Hawkesford and L.J. de Kok (ed.) Sulfur in plants: An ecological perspective. Springer, Dordrecht, the Netherlands.

Stiller, I., G. Dancs, H. Hesse, R. Hoefgen, and Z. Banfalvi. 2007. Improving the nutritive value of tubers: Elevation of cysteine and glutathione contents in the potato cultivar White Lady by marker-free transformation. J. Biotechnol. 128:335–343.

Suter, M., P. von Ballmoos, S. Kopriva, R.O. den Camp, J. Schaller, C. Kuhlemeier, P. Schurmann, and C. Brunold. 2000. Adenosine 5'-phosphosulfate sulfotransferase and adenosine 5¢-phosphosulfate reductase are identical enzymes. J. Biol. Chem. 275:930–936.

Tejada-Jimenez, M., A. Llamas, E. Sanz-Luque, A. Galvan, and E. Fernandez. 2007. A high-affinity molybdate transporter in eukaryotes. Proc. Natl. Acad. Sci. USA 104:20126–20130.

Tomatsu, H., J. Takano, H. Takahashi, A. Watanabe-Takahashi, N. Shibagaki, and T. Fujiwara. 2007. An *Arabidopsis thaliana* high-affinity molybdate transporter required for efficient uptake of molybdate from soil. Proc. Natl. Acad. Sci. USA 104:18807–18812.

Tsakraklides, G., M. Martin, R. Chalam, M.C. Tarczynski, A. Schmidt, and T. Leustek. 2002. Sulfate reduction is increased in transgenic Arabidopsis thaliana expressing 5'-adenylylsulfate reductase from *Pseudomonas aeruginosa*. Plant J. 32:879–889.

Wirtz, M., and M. Droux. 2005. Synthesis of the sulfur amino acids: Cysteine and methionine. Photosynth. Res. 86:345–362.

Wirtz, M., M. Droux, and R. Hell. 2004. O-acetylserine(thiol)lyase: An enigmatic enzyme of plant cysteine biosynthesis revisited in *Arabidopsis thaliana*. J. Exp. Bot. 55:1785–1798.

Wirtz, M., and R. Hell. 2006. Functional analysis of the cysteine synthase protein complex from plants: Structural, biochemical and regulatory properties. Plant Physiol. 163:273–286.

Wirtz, M., and R. Hell. 2007. Dominant-negative modification reveals the regulatory function of the multimeric cysteine synthase protein complex in transgenic tobacco. Plant Cell 19:625–639.

Yoshimoto, N., E. Inoue, K. Saito K, T. Yamaya, and H. Takahashi. 2003. Phloem-localizing sulfate transporter, Sultr1;3, mediates re-distribution of sulfur from source to sink organs in Arabidopsis. Plant Physiol. 131:1511–1517.

Yoshimoto, N., E. Inoue, A. Watanabe-Takahashi, K. Saito, and H. Takahashi. 2007. Posttranscriptional regulation of high-affinity sulfate transporters in Arabidopsis by sulfur nutrition. Plant Physiol. 145:378–388.

Yoshimoto, N., H. Takahashi, F.W. Smith, T. Yamaya, and K. Saito. 2002. Two distinct high-affinity sulfate transporters with different inducibilities mediate uptake of sulfate in Arabidopsis roots. Plant J. 29:465–473.

Youssefian, S., M. Nakamura, E. Orudgev, and N. Kondo. 2001. Increased cysteine biosynthesis capacity of transgenic tobacco overexpressing an O-acetylserine(thiol) lyase modifies plant responses to oxidative stress. Plant Physiol. 126:1001–1011.

Youssefian, S., M. Nakamura, and H. Sano. 1993. Tobacco plants transformed with the O-acetylserine(thiol) lyase gene of wheat are resistant to toxic levels of hydrogen sulphide gas. Plant J. 4:759–769.

Zeh, M., A.P. Casazza, O. Kreft, U. Roessner, K. Bieberich, L. Willmitzer, R. Hoefgen, and H. Hesse. 2001. Antisense inhibition of threonine synthase leads to high methionine content in transgenic potato plants. Plant Physiol. 127:792–802.

Zeh, M., G. Leggewie, R. Hoefgen, and H. Hesse. 2002. Cloning and characterization of a cDNA encoding a cobalamin-independent methionine synthase from potato (*Solanum tuberosum* L.). Plant Mol. Biol. 48:255–265.

7

Sulfur Response Based on Crop, Source, and Landscape Position

Dave Franzen
North Dakota State University, Fargo

Cynthia A. Grant
Agriculture and Agri-Food Canada, Brandon, Manitoba, Canada

Abstract

Sulfur responses can be seen in almost any crop in the northern Plains of North America if soil sulfur availability is low. Canola (*Brassica napus* L.) is a crop grown in this region with special requirements for sulfur. Crops growing on soils with low organic matter that are coarse-textured are more susceptible to sulfur deficiency. Within fields, eroded hilltops and slopes, especially when consisting of coarse-textures, respond more to sulfur application than higher organic matter soils in depressional areas. In this region, sulfur source is important. Elemental sulfur, even if formulated with bentonite, does not alleviate sulfur deficiency as well as sulfate-containing amendments.

In 1860, sulfur was recognized as an essential nutrient, required by all plants for growth (Woodard, 1922). It is frequently referred to as the "fourth major nutrient" following nitrogen, phosphorus, and potassium and is required by plants at levels comparable to phosphorus. The amount of sulfur in crops was initially grossly underestimated by the use of ash in chemical analysis (Crocker, 1923). Following more nondestructive means of sulfur analysis, it became generally recognized that plant content of sulfur was much higher than previously believed. This led to renewed interest in pursuing sulfur fertility work in crops.

Sulfur Deficiency and Response on the Northern Great Plains

Sulfur deficiency and response to sulfur application has been observed in many crops around the world (Tisdale et al., 1986). The likelihood of a deficiency occurring within a region is related to the balance between sulfur supply, crop removal, and losses. These factors include differences in sulfur content in soils and its release, source of fertilizers used by individual growers, atmospheric deposition of sulfur, presence of sulfur or absence in irrigation water, and sulfur content in pesticides, balanced by losses through erosion, leaching, and crop

---

Copyright © 2008. American Society of Agronomy, Crop Science Society of America, Soil Science Society of America, 677 S. Segoe Rd., Madison, WI 53711, USA. *Sulfur: A Missing Link between Soils, Crops, and Nutrition.* Agronomy Monograph 50.

removal. From the mid-1960s until the mid-1980s, sulfur deficiencies increased in western Canada (Beaton, 1987). Reasons given for the increase in deficiencies were declining organic matter levels because of erosion, higher crop removal, and a shift in production to crops with a high sulfur requirement, including canola and alfalfa (*Medicago sativa* L.), use of fertilizer low in sulfur, lower atmospheric contributions, more awareness of possible sulfur deficiency among farmers, and better laboratory methods for sulfur deficiency prediction and identification. Differences in climate, soil, and choice of crops are instrumental in affecting the frequency and severity of sulfur deficiencies.

In the 1920s, there was already concern in the USA regarding the loss of organic matter and, thus, sulfur in soils because of cropping and erosion (Crocker, 1923). In addition, there was an understanding that sulfur dissolved in rainfall could be an important source of sulfur for the growing crop. Crocker reports that 40 to 45 kg ha^{-1} sulfur was found in annual rainfall at Urbana, IL, in the 1920s. In contrast, using National Atmospheric Deposition Program data (http://nadp.sws.uiuc.edu; verified 8 April 2008), the amount of sulfur deposited in 2002 at Urbana was approximately 15 kg ha^{-1}. In the northern Plains, including Montana, North Dakota, South Dakota, and northwest Minnesota, the annual sulfate ion deposition was 2 to 6 kg ha^{-1} yr^{-1} in 2002, with 2 to 4 kg ha^{-1} yr^{-1} being most common in the western part of the region (Fig. 7–1; see the section of color figures within this chapter). The relatively low aerial deposition of sulfate, related to the relatively low industrialization, is a major contributing factor to the high occurrence of sulfur deficiency in the northern Great Plains (Fig. 7–1).

The occurrence of sulfur deficiency on the northern Great Plains is strongly affected by soil characteristics. Sulfur deficiency is most common on soils that are low in organic matter and thus will not release significant amounts of sulfur over the growing season through mineralization, and in soils where the sulfur has leached out from the rooting zone over time. The northern Plains are blanketed with a combination of exposed glacial and residual soils with very little loess covering compared with the central USA. Wind erosion during the past 100 yr has resulted in the loss of organic matter especially at the higher landscape elevations. Many of the soils at hilltop and shoulder landscape positions are coarse in texture, consisting of loamy sands or sandy loam soils because of sorting due to glacial melt. Sandy textures may also be present in these and other landscape positions as a result of soil formation in glacial outwash and residual materials from the weathering of sandstones (Bluemle, 1991). These sandy soils are susceptible to sulfur deficiency both because they tend to be low in organic matter and because sulfate-sulfur will readily leach out of the rooting zone, if precipitation is sufficient. In Canada, Gray Luvisolic soils are particularly susceptible to sulfur deficiency, because of the high leaching and relatively low organic matter content of the soils (Nuttal et al., 1987). The combination of soils that are both coarse textured and low in organic matter make the northern Plains susceptible to sulfur deficiency. However, there are also significant areas of soils in the northern Great Plains that contain a reservoir of calcium sulfate and magnesium sulfate within the rooting zone, reducing the risk of sulfur deficiency.

Sulfur deficiency is most commonly observed in the U.S. northern Plains in coarse-textured soils after periods of greater than normal precipitation that moves the sulfate below the rooting zone. Sulfur deficiency in wheat is expressed as yellowing in the newly emerged leaves (Fig. 7–2).

Table 7–1. Yield of canola with sulfur rate, source, and landscape position, Rock Lake, ND, no-till system. Data are from Deibert et al. (1996).

S rate	S source	Yield from different soil series/landscape position		
		Buse/hilltop	Barnes/slope	Svea/footslope
kg ha^{-1}		kg ha^{-1}		
0		35	267	1639
22	AS	1852	1876	1928
44	AS	2026	2094	2434
44	ES	697	1190	1827

Sulfur availability in the northern Great Plains of the USA is often associated with landscape position. Lower landscape positions tend to have higher total sulfur and sulfates than upper landscape positions (Roberts and Bettany, 1985). Part of the availability difference between upper- and lower-slope positions is due to decreased organic matter and associated organic sulfur of the upper landscape surfaces. Higher water tables in lower landscapes result in higher subsoil sulfate. Sulfur deficiencies are most often observed on hilltop or ridge positions, especially on eroded, coarse textured soils. However, an exception can occur where gypsum occurs close to the surface on eroded knolls, providing there is a readily available source of sulfate for crop growth. Sulfur deficiency is less common on foot slope and depressional positions with medium to heavy textured soils high in organic matter. It is not unusual to find extremely high sulfates and sulfur deficiencies in the same field. A 12.5-ha field near Valley City, ND, was sampled in a 33-m grid. A range of sulfate-sulfur at the 0- to 60-cm depth was from 5 to 1000 kg ha^{-1} (Fig. 7–3). Low sulfate was associated with sandy textured hilltops and ridges. High sulfate was associated with depressional positions. The high variability in sulfur concentration within a field poses challenges for soil testing. If soil samples are blended, an extremely high sample can elevate the results of the soil test and may lead to the conclusion that the field is well-supplied with sulfur when the majority of the field is sulfur deficient.

As sulfur distribution in the field varies across a field, the response of canola to sulfur is also often strongly related to landscape position. Differential yield response to landscape position was documented by Deibert et al. (1996) (Table 7–1). The response to sulfur was over 6000% in the Buse-hilltop soils, characterized by organic matter levels less than 2% because of a history of past erosion. Responses of over 800% were seen in the Barnes-slope positions, and responses were also measured in the Svea, higher organic matter footslope soil.

Crop Demand for Sulfur

Sulfur is a component of two major amino acids, cysteine and methionine, and thus plays a crucial role in plant metabolism (Bennett, 1993; Droux, 2004). Cysteine and methionine function not only as structural components of proteins but also serve as precursors for the synthesis of compounds including glutathione and a wide range of enzymes, vitamins, cofactors, and sulfur compounds involved in growth and development of plant cells. Sulfur plays a role in basic plant functions such as photosynthesis, carbon and nitrogen metabolism, protein synthesis, synthesis of oils in oilseed crops, and detoxification mechanisms (Droux, 2004).

Visual symptoms of sulfur deficiency will vary with crop type and the degree of deficiency (Havlin et al., 1999). If deficiencies are severe, the symptoms may be clear and characteristic. Symptoms include yellowing of the younger leaves, with lighter colored veins. The yellowing from sulfur deficiency may be differentiated from yellowing from nitrogen deficiency in that sulfur-related symptoms tend to occur first in the younger leaves, as sulfur is nonmobile in the plant. In contrast, nitrogen can be remobilized within the plant so yellowing occurs first in the older leaves. Sulfur-deficient plants are smaller, less robust, grow slowly, and maturity may be delayed. In crops that flower, such as canola and alfalfa, seed set may be reduced and canola flowers may be paler yellow in color. With severe deficiencies, canola leaves show a characteristic cupping and purpling (Fig. 7–4). However, with mild deficiency, the symptoms may be less definitive. Therefore, visual symptoms may not always be reliable indicators of a mild to moderate sulfur deficiency.

Crops Differ in Sulfur Requirements and Responsiveness

Crops differ substantially in the amount of sulfur they require and in their responsiveness to sulfur supply. As with any nutrient, sulfur demand and removal will depend on crop yield and portion of the crop harvest. Forage crops tend to remove large quantities of sulfur from the soil, since the majority of the biomass is removed at harvest. In contrast, when only the seed of the crop is harvested, a large proportion of the sulfur accumulated by the crop that is recycled when the crop residues are returned to the soil. Since sulfur in an important component of protein, as a constituent of the sulfur-containing amino acids cysteine and methionine, high protein crops such as legumes tend to have a higher sulfur demand relative to their yield potential than do cereal crops (Mills and Jones, 1996).

Of the crops commonly grown for seed on the northern Great Plains, canola, corn (*Zea mays* L.) and oat (*Avena sativa* L.) tend to remove the greatest amount of sulfur (Canadian Fertilizer Institute, 1998) with canola as a high protein oilseed crop having particularly high sulfur requirements relative to its yield potential (Jackson, 2000). Alfalfa, as a forage legume crop, removes even greater amounts of sulfur. Wheat (*Triticum aestivum* L.), barley (*Hordeum vulgare* L.), sunflower (*Helianthus annuus* L.), and pea (*Pisum sativa* L.) tend to be intermediate in sulfur requirements, while flax (*Linum usitatissinum* L.) and buckwheat (*Fagopyrum esculentum* Moench) remove only a small amount of sulfur (Canadian Fertilizer Institute, 1998). Crop sulfur demand will be affected by crop yield potential and by nitrogen supply. Increasing the crop yield potential and protein content through application of nitrogen fertilizer may accentuate sulfur deficiencies. This condition is observed in Fig. 7–5a, where banded ammonia fertilizer application has provided adequate nitrogen availability in the band but not in the area between the bands. Sulfur deficiency is seen over the fertilizer band but not between the bands where nitrogen is lower.

Crop response to sulfur in western Canada, including the Prairie Provinces, was reviewed through about 1984 by Beaton and Soper (1986). Yield increases from 9 to 125% were observed in oat, 8 to 345% in wheat, and 11 to 145% in barley (Beaton et al., 1966). Sulfur responses in corn were observed at five of six locations in Minnesota, including a silt loam textured soil (Table 7–2). The magnitude

Table 7–2. Response of corn to S fertilization at six locations in Minnesota to sulfur. Data are from Rehm (2005).

Site	Texture	Yield	
		0 kg S ha⁻¹	6.7 kg S ha⁻¹
		Mg ha⁻¹	
1	loamy fine sand	10.55	11.06*
2	silty clay loam	11.74	11.73
3	loamy fine sand	6.33	6.89*
4	loam	9.57	10.23*
5	sandy loam	8.90	9.79*
6	silt loam	9.45	10.17*

* Response is significant at $P > 0.05$.

of responses in the Minnesota study ranged from 5 to 11%. Responses were seen at both high- and low-yielding sites. Sulfur application has also increased yield in field pea (Tables 7–3 and 7–4) when soil levels were inadequate (McKay, 1996; Haderlein and Dowbenko, 2002).

Yield increases to sulfur have also occurred in barley and wheat (Table 7–5). In addition to its effects on yield, sulfur may have important effects on wheat

Table 7–3. Response of field pea to sulfur on a low organic matter, coarse-textured Soil near Garrison, ND. Table adapted from McKay (1996).

Source	Rate	Yield
	kg ha⁻¹	
None	0	1325
Ammonium thiosulfate (ATS)	22	1744
Elemental S (ES) †	22	1738
ES	44	1268
50/50 blend ATS and ES	22	1591

† A bentonite clay formulated elemental S.

Table 7–4. Response of field pea to sulfur application. Table adapted from Haderlein and Dowbenko (2002).

Source	Rate	Yield
	kg ha⁻¹	
None	0	1871
Ammonium sulfate (AS)	22	2602
50/50 blend AS and ES	22	2503
Elemental S (ES)†	22	2366

† A bentonite clay formulated elemental S.

Table 7–5. Wheat and barley yield response to S fertilizer, as a mean of three site-years.

Crop	Reference	Yield		
		0 kg S ha⁻¹	20 kg S ha⁻¹	Increase
		kg ha⁻¹		%
Spring wheat	McKay (1996)	1915	2280	19
Barley	Nyborg et al. (1974)	1100	3540	221

quality because of its important role in protein formation (Zhao et al., 1999; Wooding et al., 2000). The production of a tall, fine-textured loaf of bread requires elastic dough that can trap the gases produced as the yeast metabolizes sugars and expands, causing the bread to rise. The elastic structure is formed by cross-linkages between the sulfhydryl groups of proteins. Hence, an adequate sulfur supply is essential for good bread quality. In the UK, grain protein increased with sulfur application as grain sulfur increased (Zhao et al., 1999). With these increases, loaf volume and gel protein also increased. Elastic modulus of gel protein decreased. Dough resistance decreased, while dough extensibility increased. The beneficial effect of sulfur fertilization on bread making is associated with the composition of glutenin polymers (Zhao et al., 1999). In a New Zealand study (Wooding et al., 2000), sulfur fertilization also increased grain protein and grain sulfur. As a result, protein composition increased in gliadins. Sulfur fertilization therefore resulted in improved bread making parameters. Similar results were found by Unger (2002) and Unger et al. (2002), in Manitoba, where sulfur fertilizer led to improved loaf quality but had no significant effect in increasing wheat grain yield.

Of the crops commonly grown in the northern Great Plains, canola may be the crop most susceptible to sulfur deficiency. Large responses have been documented when sulfur was applied to canola (Janzen and Bettany, 1984a; Nuttal et al., 1987; Grant and Bailey, 1993; Endres, 1998; Jackson, 2000; Grant et al., 2003a, 2003b, 2004; Karamanos et al., 2005, 2007; Malhi et al., 2007). Adequate sulfur can mean the difference between no crop and a normal yield in canola. Deficiency symptoms can be expressed early or later in the season. Early season deficiency is expressed as mottling of leaves and plants lingering in the rosette stage longer than with adequate sulfur supply (Fig. 7–4). Branching is limited, flowering is sparse, and flowers are paler than normal. Midseason sulfur deficiency in canola is expressed as purpling along the leaf edges, appearing as bronze areas from the field edge (Fig. 7–5b). When nitrogen is applied under sulfur deficiency, yields may be reduced. It appears that with excess nitrogen, the nitrogen to sulfur (N/S) ratio becomes excessive and sulfur deficiency is enhanced because protein production and the demand for sulfur-containing amino acids increases. More nitrogen and sulfur are immobilized in sulfur-containing compounds in older tissues and remobilized from old to newer tissue (Beaton, 1987). An adequate supply of both nitrogen and sulfur are required for optimum canola production. A fertilizer application ratio of 7N/1S was suggested by Janzen and Bettany (1984a, 1984b) for canola.

Fertilizer Management of Sulfur

Where sulfur deficiencies are limiting to crop yield or quality, effective utilization of sulfur fertilizer is necessary to optimize crop production. Woodard (1922) reviewed the early history of sulfur use and refers to the documented use of gypsum in Germany as early as 1768. Benjamin Franklin brought knowledge of gypsum use by French and British farmers to his own farm outside of Philadelphia.

In the northern Great Plains, sulfur is normally applied either as elemental sulfur or as a sulfate source. Plant can only utilize sulfur as sulfate, so before elemental sulfur can be used by the plant, it must be converted to sulfate (Janzen and Bettany 1984a, 1984b). If the oxidation of elemental is not rapid enough to

■ Sulfur Response Based on Crop, Source, and Landscape Position

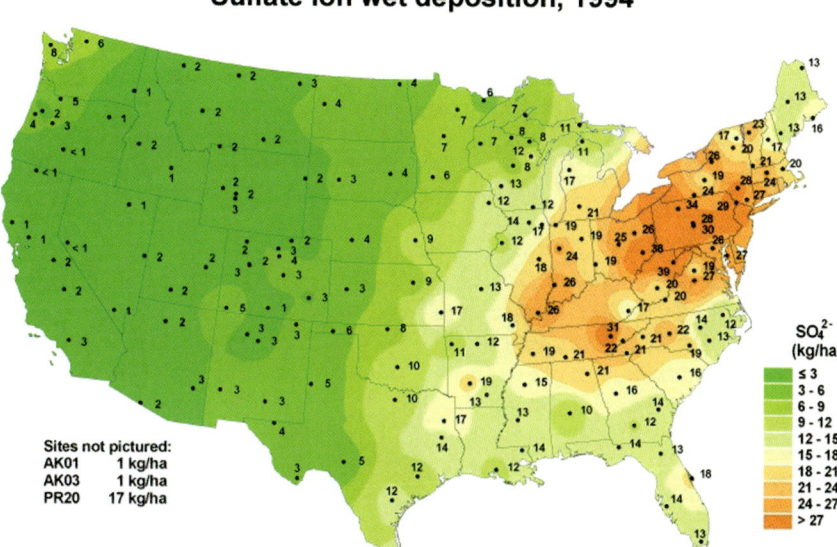

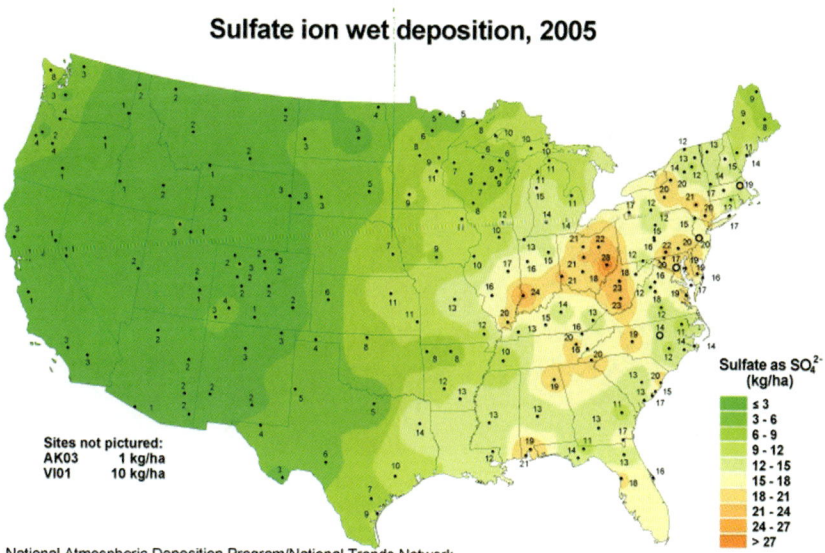

Fig. 7–1. Comparison of sulfate ion wet deposition, 1994–2005. National Atmospheric Deposition Program/National Trends Network (http://nadp.sws.uiuc.edu; verified 8 April 2008).

Fig. 7–2. Sulfur deficiency in spring wheat, Valley City, ND. Observed in sandy loam textured soil, organic matter less than 2%, after above normal spring precipitation.

■ Sulfur Response Based on Crop, Source, and Landscape Position

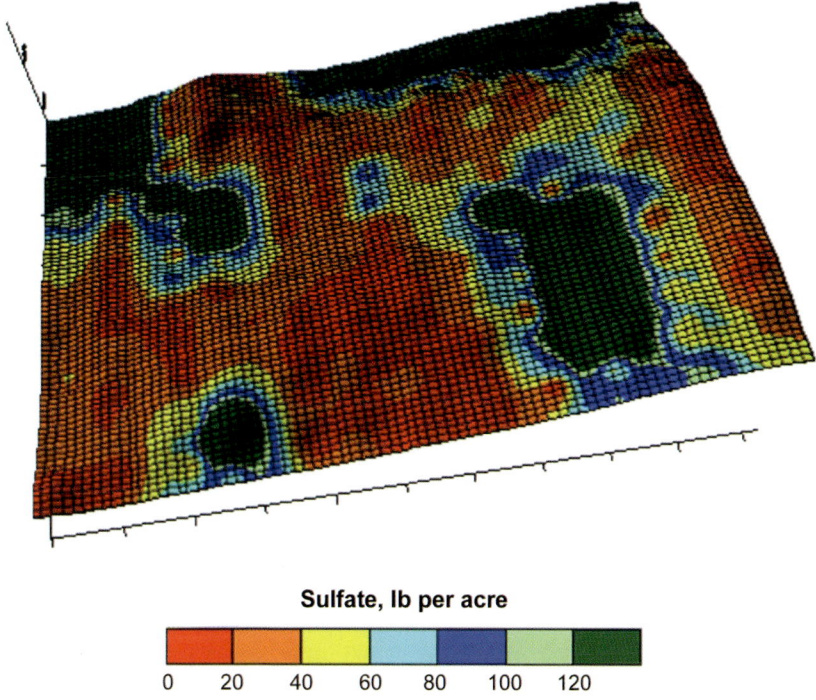

Fig. 7–3. Sulfate-sulfur levels, Valley City, ND, 12.5-ha field. Low sulfur areas associated with coarse-textured ridges. High sulfur areas associated with loam-textured, higher organic matter local depressions.

Fig. 7–4. Early season sulfur deficiency in canola.

Fig. 7–5. Sulfur deficiency symptoms. (A) In the field, yellowing above the nitrogen fertilizer bands, but not between the bands. (B) Midseason sulfur deficiency in canola, showing reddish leaves, especially leaf edges.

supply sufficient sulfate to meet crop demand, crop response and ultimately crop yield may be restricted.

Under the environmental conditions in the northern Great Plains, crop yield is generally maximized when soluble sulfate forms of sulfur, such as ammonium sulfate, ammonium thiosulfate, or even gypsum are applied rather than elemental sources (Karamanos and Janzen, 1991; Deibert et al., 1996; Johnston et al., 1998; Lukach and Deibert, 2000; Deibert and Lukach, 2000; Haderlein and Dowbenko, 2002; Grant et al., 2003a, 2003b, 2004; Solberg et al., 2006). Forms of elemental sulfur, even dispersible forms, have not been as effective at equivalent rates as soluble sulfate forms. Elemental sulfur availability is maximized with very small particle size and adequate dispersion of particles (Janzen, 1990). Dispersion can be enhanced through soil mixing (Burns, 1967; Germida and Janzen, 1993). Some elemental sulfur products are formulated with bentonite clay. Bentonite clay expands and aids in forming small particle sizes as well as in facilitating the dispersion process.

Elemental sulfur must be oxidized by soil organisms—autotrophic bacteria (*Thiobacillus*) heterotrophic bacteria, fungi, and actinomycetes (Burns, 1967; Germida and Janzen, 1993). The activity of these organisms is affected by moisture content, temperature, and soil pH. Saskatchewan studies showed that most sulfur oxidation in the soils of the northern Plains was controlled by heterotrophic bacteria (Lawrence and Germida, 1988; Lawrence et al., 1988). Autotrophic bacteria, which are the most effective sulfur oxidizers, were not detected in high numbers. Oxidation was generally stimulated initially by sulfur addition but not always. Because most of the oxidation was heterotrophic, decomposition of residues was linked with activity, since these organisms use carbon-based compounds as their primary energy source (Lawrence et al., 1988).

Most studies comparing sulfate-sulfur sources with elemental sulfur formulations, including bentonite formulations, conclude that in the initial year of application, sulfate-sulfur sources are more effective in increasing yield (Solberg et al., 2006). Some studies show that residual sulfur from elemental sulfur fertilizers may become available over time and effectively increase yields in subsequent crops (Janzen and Bettany, 1987; Wen et al., 2003; Solberg et al., 2006), while others find little value in residual sulfur (Karamanos and Janzen, 1991). Application of elemental sulfur is sometimes used in the hope that its slow conversion to available and leachable sulfate forms would assist availability after periods of high rainfall. However, elemental sulfur will be subject to leaching losses between its conversion to plant-available sulfate and its uptake by the crop. Grant et al. (2004) found that the residual benefits from elemental sulfur forms and ammonium sulfate were similar 3 yr after fertilizer application. Lack of sulfur bacteria in the soil biosphere may contribute to the poor performance of elemental sulfur. Most sulfur is transformed to sulfate through heterotrophic bacteria and fungi (Lawrence et al., 1988), not *Thiobacillus*. Cool conditions, dry soils and the relatively short growing season that occurs on the northern Great Plains may also restrict the oxidation of elemental sulfur sources.

The demand for sulfur by canola is greatest during flowering and seed set; therefore, it may be possible for a crop to recover to a great extent from early-season sulfur deficiency if sulfur becomes available by the rosette to early-bolting stage (Malhi 1999). Crops displaying sulfur deficiency symptom can recover from their deficiency when the roots contact a subsurface layer of gypsum that serves

Table 7–6. Canola yield with sulfur applied as ammonium thiosulfate at bolting. Data are for Rock Lake, ND (Lukach and Deibert, 2000).

S rate	Canola yield
kg ha^{-1}	
0	325
11	690
22	970

Table 7–7. Relative effectiveness of Sulfate (30 kg ha^{-1}) fertilizer applied at different growth stages in increasing seed yield of canola near Star City, 1998. Table adapted from Malhi (1999).

Stage of growth and method of application	Yield increase
	kg ha^{-1}
Preseeding incorporated	620
Side-band at seeding	567
Seed-row at seeding	558
Topdress at bolting	457
Foliar at bolting	511
Topdress at flowering	205
Foliar at flowering	310

as a sulfur source. In a greenhouse study, Janzen and Bettany (1984b) found that canola yield was similar whether sulfur was applied from seeding through to the rosette stage. Lukach and Deibert (2000) and Deibert and Lukach (2000) also found yield increases even with sulfur applications during the bolting stage (Table 7–6). Although late treatments of sulfate or thiosulfate have been found effective in increasing canola yield, greatest yields are possible when sulfate sulfur is applied at planting to provide an adequate sulfur supply throughout the growing season (Table 7–7, Malhi, 1999). In addition, for the crop to respond to late applications of sulfur, the fertilizer must be in an available form that is accessible by the crop. Dry conditions may leave fertilizer stranded on the soil surface, restricting the crop's ability to utilize the applied sulfur.

Determining a Need for Sulfur

Soil tests have been widely used for many years to predict crop requirements for sulfur. Many of the methods measure extractable sulfate as an indication of plant-available sulfur (Kowalenko and Grimmett, 2008). Various extractants have been used, including acetates, carbonates, chlorides, phosphates, citrates, and oxalates (Kowalenko and Grimmett, 2008). Soil solution sulfate is frequently measured by use of water, or more commonly a weak solution of calcium chloride or lithium chloride. The monocalcium phosphate extractant soil test for sulfur is widely used in the northern Plains and elsewhere to predict sulfur response (Combs et al., 1998). There are many shortcomings to the procedure and its predictive ability that were identified early in its use (Hoeft et al., 1973); however, there have not been any substituting laboratory analyses that have proven any more effective. Although the analysis is offered in many states, most recommendations come with cautions against under- or overfertilization that are based on the test results alone. Some possible reasons for the low confidence in the test are

(i) lack of consideration of soil texture or organic matter in the recommendation, (ii) lack of information regarding large sulfate mineral crystals in the soil pulverized during soil test preparation that makes them more available in the test than the solubility would make them in situ, and (iii) leaching of sulfate following sampling under high rainfall. The large degree of field variability in sulfate distribution can also create problems in soil testing and recommendation.

Some U.S. states include organic matter in their charts for determining whether or not to apply sulfur and what rate to apply (Ferguson et al., 2000). Other states do not include organic matter explicitly in recommendation tables but imply that sulfur responses are most often seen in low organic matter soils (Franzen, 2007; Franzen and Lukach, 2007), along with coarse-textured soils. Some states do not include organic matter in their recommendations but include coarse-textured soils as having a higher susceptibility to sulfur deficiency (Rehm et al., 2006; Gerwing and Gelderman, 2005).

The lack of a good correlation between soil tests and crop response has led to the consideration of N/S ratios in soils as an indication of sulfur supply (Janzen and Bettany, 1984a; Bailey, 1985). Total sulfur in a selected group of Canadian soils was highly correlated with organic carbon and total nitrogen (Bailey, 1985). Total sulfur (0–15 cm) in these soils ranged from 120 to 1110 mg kg^{-1}. Sulfate-sulfur extracted with $CaCl_2$ was 1.2 to 6.0% of total sulfur. The N/S ratio for these soils averaged 8.3. It was suggested that soils with a high N/S ratio could be prone to sulfur deficiency.

Plant-tissue testing for sulfur can also be used as an indication of sulfur status of the crop (Mills and Jones, 1996). For canola, a tissue concentration of less than 0.2% total sulfur at flowering is considered low, 0.2 to 0.25% marginal, and greater than 1.0% excessive (Jones 1986). Plant N/S ratio has also been suggested as an indication of sulfur deficiency. Bailey (1986) suggested that for maximum yield, canola should have a total N/S ratio of 12 in the tissue at flowering while barley required a ratio of 16 in the tissue at shotblade. However, Zhao et al. (1997) did not find seed N/S ratios useful in diagnosis of sulfur deficiency and reported that sulfur-deficient rapeseed had similar N/S ratios as those with adequate sulfur. One of the problems with relying on plant analysis to diagnose sulfur problems is that if the problem is found, application of sulfur may come too late to be of greatest benefit during that growing season (Mahli et al., 2005).

The nature of landscape, soil texture, and organic matter roles in both the response of crops to sulfur and the likelihood of deficiency lend themselves to site-specific sampling of sulfur and variable-sulfur application. Using the field in Fig. 7–3 as an example, one could easily imagine delineating sampling zones on the basis of organic matter, soil texture, and landscape position. The resulting sampling would result in zones with very high residual sulfate that would not benefit from sulfur application and also low organic matter, coarse-textured hilltops that probably would benefit from sulfur application. The sulfur response shown in Fig. 7–3 was a similar site with these potentially sulfur-deficient characteristics.

Summary

Sulfur deficiencies are found around the world and affect many different crops. When sulfur is deficient, most crop yields may increase from detectable increments to over 300% with sulfur fertilization. Canola, however, has a special requirement for sulfur that appears to be unique. Yield increases over 6000% have been recorded, and inadequate sulfur can result in a total crop failure. Sulfur sources greatly affect efficacy. Although elemental sulfur–bentonite clay formulations are more useful in the northern Plains and Canada than pure elemental forms, they are still not as effective in the initial and often subsequent years than soluble sulfate products, such as ammonium sulfate.

Landscape and topography play an important role within fields in sulfur availability from the soil and may lend themselves to site-specific management of sulfur nutrition. The response of crops to sulfur fertilization is often related to landscape position and the organic matter and minerals related to the position. The soil test for sulfur has been inconsistent in its prediction of sulfur status. The soil test is often coupled with statements regarding organic matter levels and soil texture to improve prediction of possible response. Plant analysis more effectively relates to sulfur status of crops, particularly when the N/S ratio is considered, but identification of sulfur deficiency may be too late to remedy.

References

Bailey, L.D. 1985. The sulphur status of eastern Canadian prairie soils: The relationship of sulphur, nitrogen and organic carbon. Can. J. Soil Sci. 65:179–185.

Bailey, L.D. 1986. The sulphur status of eastern Canadian prairie soils: Sulphur response and requirements of alfalfa (*Medicago sativa* L.), rape (*Brassica napus* L.) and barley (*Hordeum vulgare* L.). Can. J. Soil Sci. 66:209–216.

Beaton, J.D. 1987. Recent developments in sulphur use in western Canada. Manitoba Agri-Forum, 15–17 Dec., Winnipeg, MB, Canada.

Beaton, J.D., J.T. Harapiak, and S.L. Tisdale. 1966. Crop responses to sulphur in western Canada. Sulphur Inst. J. 2:9–16.

Beaton, J.D., and R.J. Soper. 1986. p. 375–403. *In* M.A. Tabatabai (ed.) Sulfur in agriculture. Agron. Monogr. 27. ASA, CSSA, SSSA, Madison, WI.

Bennett, W.F. 1993. Plant nutrient utilization and diagnostic plant symptoms. p. 1–7. *In* W. F. Bennett (ed.) Nutrient deficiencies and toxicities in crop plants. APS Press, St. Paul, MN.

Bluemle, J. 1991. The Face of North Dakota. revised ed. North Dakota Geological Survey, Bismarck, ND.

Burns, G.R. 1967. Oxidation of sulphur in soils. Technical Bulletin Number 13. The Sulfur Institute, Washington, DC.

Canadian Fertilizer Institute. 1998. Nutrient uptake and removal by field crops– Western Canada. 1998. Can. Fert. Inst. Ottawa, ON.

Combs, S.M., J.L. Denning, and K.D. Frank. 1998. p. 35–40. *In* J.R. Brown (ed.) Recommended chemical soil test procedures. North Central Reg. Res. Pub. No. 221 (Revised). Missouri Agri. Exp. Sta. SB 1001. Columbia, MO.

Crocker, W. 1923. The necessity of sulfur carriers in artificial fertilizers. J. Am. Soc. Agron. 15:129–141.

Deibert, E.J., S. Halley, R.A. Utter, and J. Lukach. 1996. Canola response to sulfur fertilizer applications under different tillage and landscape positions. 1996 Annual Report to USDA/CSREES/Special Programs Northern Region Canola and North Dakota Oilseed Council.

Deibert, E.J., and J.R. Lukach. 2000. Notill canola (Brassica napus, L.) response to different rates and sources of sulfur fertilizer. p. 215–221. *In* A.J. Schlegel (ed.) 2000 Great Plains Soil Fertility Proceedings.7–8 March 2000. Denver, CO. Kansas State University, Tribune, KS.

Droux, M. 2004. Sulfur assimilation and the role of sulfur in plant metabolism: A survey. Photosynth. Res. 79: 331–348.

Endres, G. 1998.Canola response to sulfur and nitrogen fertilization. NDSU Carrington Research Extension Center Report. Carrington, ND.

Ferguson, R.B., G.W. Hergert, and E.J. Penas. 2000. Corn. *In* R.B. Ferguson and K.M. DeGroot (ed.) Nutrient management for agronomc crops. Nebraska. Univ. of Nebraska Extension Publication EC155. Lincoln, NE.

Franzen, D.W. 2007. North Dakota Fertilizer Recommendation Tables and Equations. North Dakota St. Univ. Ext. Circ. SF-882 (Revised). Fargo, ND.

Franzen, D.W., and J. Lukach. 2007. Fertilizing canola and mustard. North Dakota St. Univ. Ext. Circ. SF-1122 (Revised).

Germida, J.J., and H.H. Janzen. 1993. Factors affecting the oxidation of elemental sulfur in soils. Fert. Res. 35:101–114.

Gerwing, J., and R. Gelderman. 2005. Fertilizer Recommendation Guide. South Dakota St. Univ. EC750. Brookings, SD.

Grant, C.A., and L.D. Bailey. 1993. Fertility management in canola production. Can. J. Plant Sci. 73:651–670.

Grant, C.A., G.W. Clayton, and A.M. Johnston. 2003a. Sulphur fertilizer and tillage effects on canola seed quality in the Black soil zone of western Canada. Can. J. Plant Sci. 83:745–758.

Grant, C.A., A.M. Johnston, and G.W. Clayton. 2003b. Sulphur fertilizer and tillage effects on early season sulphur availability and N:S ratio in canola in western Canada. Can. J. Soil Sci. 83:451–463.

Grant, C.A., A.M. Johnston, and G.W. Clayton. 2004. Sulphur fertilizer and tillage management of canola and wheat in western Canada. Can. J. Plant Sci. 84:453–462.

Haderlein, L.K., and R.E. Dowbenko. 2002. Effect of sulfur source on grain yield and S recovery in crop rotations at three sites in western Canada. p. 199–204. *In* A. Schlegel (ed.) Proceedings of the 2002 Great Plains Soil Fertility Conference. 5–6 March 2002. Denver, CO. Potash & Phosphate Inst., Brookings, SD.

Havlin, J.L., J.D. Beaton, S.L. Tisdale, and W.L. Nelson. 1999. Soil fertility and fertilizers (sixth ed.) Prentice Hall, Upper Saddle River, NJ.

Hoeft, R.G., L.M. Walsh, and D.R. Keeney. 1973. Evaluation of various extractants for available soil sulfur. Soil Sci. Soc. Am. Proc. 37:401–404.

Jackson, G.D. 2000. Effects of nitrogen and sulfur on canola yield and nutrient uptake. Agron. J. 92:644–649.

Janzen, H.H. 1990. Elemental sulfur oxidation as influenced by plant growth and degree of dispersion within soil. Can. J. Soil Sci. 70:499–502.

Janzen, H.H., and J.R. Bettany. 1984a. Sulfur nutrition of rapeseed: I. Influence of fertilizer nitrogen and sulfur rates. Soil Sci. Soc. Am. J. 48:100–107.

Janzen, H.H., and J.R. Bettany. 1984b. Sulfur nutrition of rapeseed: II. Effect of time of sulfur application. Soil Sci. Soc. Am. J. 48:107–112.

Janzen, H.H., and J.R. Bettany. 1987. Oxidation of elemental sulfur under field conditions in central Saskatchewan. Can. J. Soil Sci. 67:609–618.

Johnston, A.M., C.A. Grant, and G.W. Clayton. 1998. Sulphur fertilizer management for conventional and no-till canola. p. 190–195. *In* A.J. Schlegel (ed.) 1998 Great Plains Soil Fertility Proceedings. 3–4 March 2000. Denver, CO. Kansas State University, Tribune, KS.

Jones, M.B. 1986. Sulfur availability indexes. p. 549–566. *In* M.A. Tabatabai (ed.) Sulfur in agriculture. Agron. Monogr. 27. ASA, CSSA, SSSA, Madison WI.

Karamanos, R.E., T.B. Goh, and D.P. Poisson. 2005. Nitrogen, phosphorus, and sulfur fertility of hybrid canola. J. Plant Nutr. 28:1145–1161.

Karamanos, R.E., T.B. Goh, and D.N. Flaten. 2007. Nitrogen and sulphur fertilizer management for growing canola on sulphur sufficient soils. Can. J. Plant Sci. 87:201–210.

Karamanos, R.E., and H.H. Janzen. 1991. Crop response to elemental sulfur fertilizers in central Alberta. Can. J. Soil Sci. 71:213–225.

Kowalenko, C.G., and M. Grimmett. 2008. Chemical Characterization of Soil Sulfur. p. 251–263. *In* M.R. Carter and E.G. Gregorich (ed.) Soil sampling and methods of analysis. Second ed. CRC Press, Boca Raton, FL.

Lawrence, J.R., and J.J. Germida. 1988. Relationship between microbial biomass and elemental sulfur oxidation in agricultural soils. Soil Sci. Soc. Am. J. 52:672–677.

Lawrence, J.R., V.V.S.R. Gupta, and J.J. Germida. 1988. Impact of elemental sulfur fertilization on agricultural soils. II. Effects on sulfur-oxidizing populations and oxidation rates. Can. J. Soil Sci. 68:475–483.

Lukach, J.R., and E.J. Deibert. 2000. Canola (Brassica napus, L.) response to source, rate and timing of sulfur fertilizer. p. 209–214. *In* A.J. Schlegel (ed.) 2000 Great Plains Soil Fertility Proceedings. 7–8 March 2000. Denver, CO. Kansas State University, Tribune, KS.

Malhi, S.S. 1999. Restoring canola yield by applying sulphur fertilizer during the growing season. p. 51–55. *In* D.W. Lee (ed.) Sulfur Fertility and Fertizers, Proceedings of the Agrium Symposia. Dec. 10, 1998, Calgary, AB, Canada and Jan. 7, 1999, Spokane, WA. Agrium New Products, R&D, Redwater, AB, Canada.

Malhi, S.S., Y. Gan, and J.P. Raney. 2007. Yield, seed quality, and sulfur uptake of *Brassica* oilseed crops in response to sulfur fertilization. Agron. J. 99:570–577.

Mahli, S.S., J.J. Schoenau, and C.A. Grant. 2005. A review of sulphur fertilizer management for optimum yield and quality of canola in the Canadian Great Plains. Can. J. Plant Sci. 85:297–307.

McKay, K. 1996. Fertility Study, McClean, Co. North Central Research Experiment Station Report. Minot, ND.

Mills, H.A., and J.B. Jones, Jr. 1996. Plant analysis handbook II. MicroMacro Publishing, Inc., Jefferson City, MO. 422 pp.

Nuttal, W.F., H. Ukainetz, J.W.B. Stewart, and D.T. Spurr. 1987. The effect of nitrogen, sulphur and boron on yield and quality of rapeseed (Brassica napus L. and B. campestris, L.). Can. J. Soil Sci. 67:545–559.

Nyborg, M., C.F. Bentley, and P.B. Hoyt. 1974. Effect of sulphur deficiency on seed yield of turnip rape. Sulphur Inst. J. 10:14–15.

Rehm, G. 2005. Sulfur management for corn growth with conservation tillage. Soil Sci. Soc. Am. J. 69:709–717.

Rehm, G., G. Randall, J. Lamb, and R. Eliason. 2006. Fertilizing corn in Minnesota. University of Minnesota Extension F-O-03790. St. Paul, MN.

Roberts, T.L., and J.R. Bettany. 1985. The influence of topography on the nature and distribution of soil sulfur across a narrow environmental gradient. Can. J. Soil Sci. 65:419–434.

Solberg, E.D., S.S. Malhi, M. Nyborg, B. Henriquez, and K.S. Gill. 2006. Crop response to elemental S and sulfate-S sources on S-deficient soils in the Parkland region of Alberta and Saskatchewan. J. Plant Nutr. 30:321–333.

Tisdale, S.L., R.B. Reneau, Jr., and J.S. Platou. 1986. Atlas of sulfur deficiencies. p. 295–322. *In* M.A. Tabatabai (ed.) Sulfur in agriculture. Agron. Monogr. 27. ASA, CSSA, SSSA, Madison, WI.

Unger, C.J.H. 2002. The impact of sulphur on the breadmaking quality of Canadian western red spring wheat in western Canada. M. Sc. Thesis. University of Manitoba.

Unger, C., D. Flaten, C. Grant, and O. Lukow. 2002. Impact of sulphur fertilizer on spring wheat breadmaking quality. Proceedings of the Great Plains Soil Fertility Conference. Vol. 9. Denver, Colorado. 5–6 March 2002. www.ppi-ppic.org; verified 8 April 2008.

Wen, G., J.J. Schoenau, S.P. Mooleki, S. Inanaga, T. Yamamoto, K. Hamamura, M. Inoue, and P. An. 2003. Effectiveness of an elemental sulfur fertilizer in an oilseed-cereal-legume rotation on the Canadian prairies. J. Plant Nutr. Soil Sci. 166:54–60.

Woodard, J. 1922. Sulphur as a factor in soil fertility. Bot. Gaz. 73:81–109.

Wooding, A.R., S. Kavale, F. MacRitchie, F.L. Stoddard, and A. Wallace. 2000. Effects of nitrogen and sulfur fertilizer on protein composition, mixing requirements, and dough strength of four wheat cultivars. Cereal Chem. 77:798–807.

Zhao, F.J., P.E. Bilsborrow, E.J. Evans, and S.P. McGrath. 1997. Nitrogen to sulphur ratio in rapeseed and in rapeseed protein and its use in diagnosing sulphur deficiency. J. Plant Nutr. 20:549–558.

Zhao, F.J., S.E. Salmon, P.J.A. Withers, E.J. Evans, J.M. Monaghan, P.R. Shewry, and S.P. McGrath. 1999. Responses of breadmaking quality to sulphur in three wheat varieties. J. Sci. Food Agric. 79:1865–1874.

8

Sulfur Management for Soybean Production

Kiyoko Hitsuda
Japan Development Service Co. Ltd., Tokyo, Japan

Kazunobu Toriyama, Guntur V. Subbarao, and Osamu Ito
Japan International Research Center for Agricultural Sciences, Ibaraki, Japan

Abstract

Sulfur is an essential nutrient for the growth and productivity of high-quality soybean [*Glycine max* (L.) Merr.]. While genetic approaches have been proposed to improve the nutritional value of soybean through the increase of the contents of sulfur-containing amino acids in seed, the negative relationships between yield level and quality traits have remained a problem. Although the distribution and assimilation of sulfur in soybean tissues have been studied to improve the effective use of remobilized sulfur from the vegetative plant parts to the seeds, measures for practical use have yet to be developed. Field management requires the assessment of the soil sulfur fertility status. However, because of the heterogeneity of sulfur concentrations within the soil horizons, this remains a challenging option. The sulfur concentration of the uppermost mature leaves of soybean at the flowering stage has been commonly used for diagnosing the sulfur status in plant. This approach is limited because latent sulfur deficiency and seed quality cannot be evaluated adequately. Recently, analysis of sulfur concentration in seed has been proposed, as it became possible to determine latent sulfur deficiency in relation to the seed quality, reflected by the concentration of sulfur-containing amino acids in seed proteins. To determine the need for sulfur fertilization in preparation for the subsequent soybean cropping, we suggest that sulfur levels in seed is a good indicator of the sulfur fertility status in the field.

World Soybean Production and Sulfur-Deficient Areas

World soybean production has been increasing because of the improvement of the yield potential and expansion of the cultivation area. Soybean production in 2005 reached 213 million megagrams (Mg), about eight times that in 1961, and the cultivation area covered 95 million hectares, about four times that in 1961 (FAOSTAT, 2007). Consequently, soybean yield doubled from 1.1 to 2.2 Mg ha^{-1} during the last 45 yr, indicating the acceleration of soil nutrient mining to support these higher soybean seed yields. Soybean requires 7 to 8 kg sulfur to produce 1 Mg of seed yield (FAO Sulphur Network, 1992; Lantmann and Castro, 2004).

Copyright © 2008. American Society of Agronomy, Crop Science Society of America, Soil Science Society of America, 677 S. Segoe Rd., Madison, WI 53711, USA. *Sulfur: A Missing Link between Soils, Crops, and Nutrition*. Agronomy Monograph 50.

Accordingly, 2.2 Mg ha^{-1} of seed yield requires 15 to 18 kg sulfur ha^{-1}, and about 5.7 kg of this amount will be taken from the field as seeds, assuming that seed sulfur concentrations will reach about 3.0 g kg^{-1} on a dry-weight basis. The amount of sulfur removed corresponds to 24 kg ha^{-1} of ammonium sulfate, 48 kg ha^{-1} of single superphosphate, or 32 kg ha^{-1} of potassium sulfate. While these amounts are small, compared with the amount of fertilization for general crops, the world consumption of sulfur-containing fertilizers has not increased as much as nitrogenous fertilizer use. The consumption of ammonium sulfate has been constant for more than 40 yr, and the consumption rate of all the nitrogenous fertilizers has decreased from 22% (1961) to 2.5% (2002) as a percentage of all nitrogenous fertilizers. The consumption of single superphosphate accounted for 15 to 20% of the total phosphate fertilizer used for these two decades, and that of potassium sulfate accounted for about 3% of the total potash fertilizer used since the 1990s (FAOSTAT, 2007). These data indicate that attention should be given to the nutritional balance of sulfur in fields.

The USA, Brazil, Argentina, China, and India are the world's major soybean producers, together accounting for more than 90% of the world soybean cultivation area and production. Because of the increasing awareness that soybean provides important nutrients for human and animal diet, cultivation of soybean has expanded to many other countries, which were not soybean producers previously. In 1961, soybean production was significant in only 48 countries, compared with nearly 85 countries in 2005 (FAOSTAT, 2007), and sulfur deficiency of crops has been reported in 60% of the total-producing countries (Blair, 1979; Tisdale et al., 1986; FAO Sulphur Network, 1992; Jansson, 1995; Beaton and White, 1997). Latent sulfur-deficient areas are expected to be observed at more sites, since sulfur deficiency became widespread along with the expansion of cultivation areas with low fertility, reduced use of sulfur-containing fertilizers, intensive agriculture, increase of yield, reduced atmospheric inputs, soil degradation caused by erosion or leaching, or decreased use of compost (Craswell and Karjalainen, 1990; Ceccotti, 1996; Knights et al., 2000). Field sulfur management for soybean production is, therefore, a worldwide concern.

Sulfur Nutrition of Soybean

Uptake, Distribution, and Mobilization of Sulfur in the Soybean Plant

Shoot-dry weight of soybean reaches a maximum value at the full-seed stage, decreases as leaves and petioles fall off, and the seed weight continues to increase until harvest (Fig. 8–1) (Hanway et al., 1984). Overall, the total amount of sulfur assimilated by the plant varies in parallel with the whole-plant dry weight (Naeve and Shibles, 2005), while the sulfur concentration in leaves decreases with the progression of the growth stages (Fig. 8–2) (Fontanive et al., 1996). When the soybean seed yield reached 7.3 Mg ha^{-1} (experimental maxima in the field), the leaves sampled at the early flowering stage contained 2.4 g sulfur kg^{-1}. When soybean yielded 6.8 Mg ha^{-1} of seed, 220 kg of seeds were produced per kilogram of absorbed sulfur (Munson and Nelson, 1990).

Most of the sulfate entering soybean primary leaves is exported through the phloem to sinks found elsewhere (first trifoliate leaf, apex, petiole, or stem) in the

Sulfur Management for Soybean Production

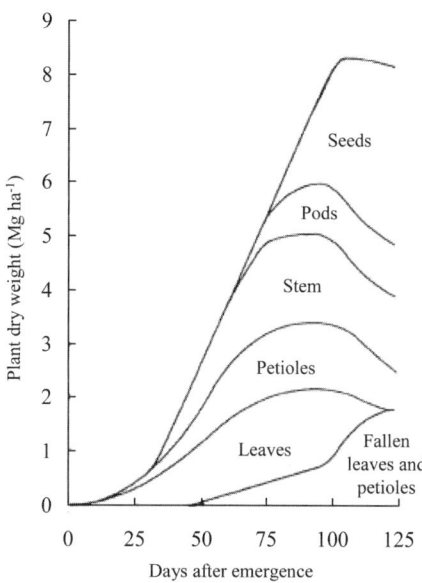

Fig. 8–1. Dry matter accumulation in soybean plants during the growth season. Beginning seed and full maturity stages: 72 and 124 d after emergence, respectively. Source: Hanway et al. (1984).

plant (Smith and Lang, 1988). A study on sulfur redistribution in soybean plants using labeled ^{35}S showed that the largest newly expanded soybean leaf plays an important role as an intermediary in the transport of sulfur from the roots to the youngest expanding leaves, nearly 25% of this mobilized sulfur being recycled through the roots as sulfate (Sunarpi and Anderson, 1998). The sulfur content of an expanding leaf increases until the expansion reaches about 70%, sulfur being imported as sulfate mostly from other leaves, and then the content decreases by almost one-half, as sulfate is exported to new expanding leaves (Sunarpi and Anderson, 1996). The translocation of newly acquired sulfate to expanding leaves has been confirmed in other studies (Sunarpi and Anderson, 1997b; Smith and Lang, 1988; Anderson and Fitzgerald, 2001). As a result, soybean leaves at

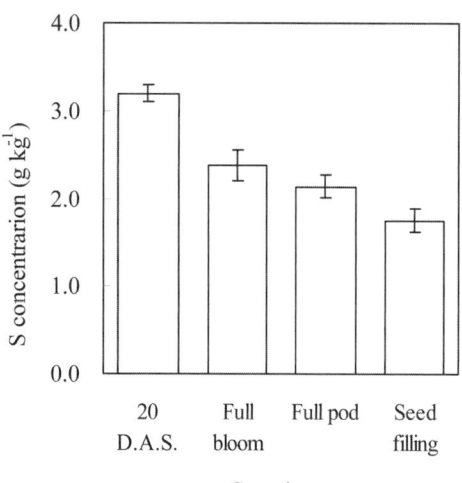

Fig. 8–2. Changes in average sulfur concentration of six varieties in uppermost mature leaves with growth stages. Bars indicate standard errors among six varieties. Data source Fontanive et al. (1996).

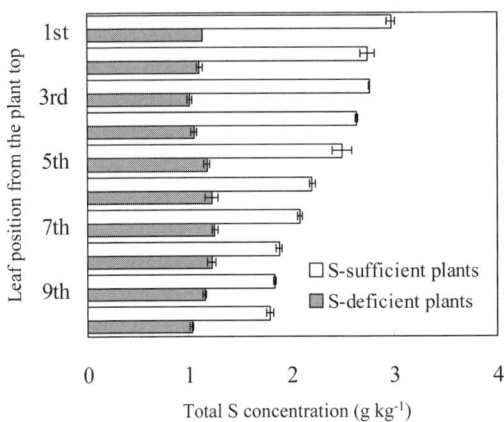

Fig. 8–3. Total sulfur concentration of leaf at each leaf position in the sulfur-deficient and sulfur-sufficient plants at flowering time. The sulfur-deficient plants were sampled from the 2.5 mg sulfur kg^{-1} soil treatment, and the sulfur-sufficient plants from the 40 mg sulfur kg^{-1} treatment. Bars indicate standard errors (n = 6). Source: Hitsuda et al. (2004).

a higher position tended to exhibit higher sulfur concentrations than those at a lower position under sufficient sulfur supply (Fig. 8–3) (Hitsuda et al., 2004). Leaves can accumulate nearly 60% of the total plant-sulfur by the end of vegetative development.

Likewise, pods accumulate sulfate from roots, stems, or leaves until they expand to half of their full length (Naeve and Shibles, 2005). Then, as seed growth starts, the amount of sulfate in pods decreases, and conversely, the contents of homoglutathione (hGSH), cysteine, and methionine increase. These assimilated substances are considered to be mobilized into grains and to be transformed into seed proteins (Sunarpi and Anderson, 1997a). Anderson and Fitzgerald (2001) also indicated that developing seeds contain hGSH but negligible amounts of sulfate. However, Naeve and Shibles (2005) insisted that seeds are capable of receiving and accumulating sulfate from pods because seeds contain significant quantities of sulfate throughout development under sufficient sulfur supply in the growth medium. Most likely, sulfate is metabolized to hGSH in the pods, although the sulfate uptake and assimilation in seeds cannot to be ruled out.

Seed-protein production requires a large mobilization of stored sulfur, as well as nitrogen, from the vegetative tissues of soybean plant. Leaves supply sulfur to developing seeds for approximately 20% of the total sulfur required, and pods contribute almost 10%. The amount of sulfur mobilized from the vegetative plant parts (leaves, pods, stems, and roots) into seeds reached a total of 36 to 40% of the seed-sulfur by the time of late seed development under sulfur-sufficient growth conditions (Naeve and Shibles, 2005) but to only 13% under low-sulfur growth conditions (Sunarpi and Anderson, 1997a). Absorbed sulfur during the reproductive stage is mainly transformed into seed protein until the late-seed filling stage (Sexton et al., 1998c). Sulfur acquired by leaves later in seed development is presumably exported to pods and seeds without entering the leaf protein because of the much faster seed growth than that of vegetative parts (Naeve and Shibles, 2005).

Sulfur Effects on Seed Quality

Soybean-seed quality is evaluated on the basis of oil and protein contents. Sulfur is one of the components of amino acids in seeds, such as methionine and cysteine. Since methionine, which can be metabolized to cysteine, is an essential

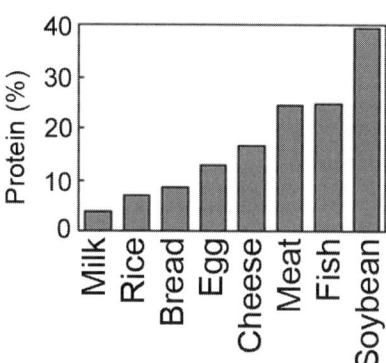

Fig. 8–4. Protein content per fresh weight of commonly consumed food sources. Source: Krishnan (2005).

amino acid for nonruminant animals, its content determines the nutritive value of foodstuffs. While the soybean protein content (35–50%) per fresh weight is the highest among ordinary human foods (Fig. 8–4) (Krishnan, 2005), the methionine content is not sufficient to meet the dietary needs of nonruminants. Compared with egg protein (i.e., the nutritionally complete FAO reference protein), soybean methionine and cysteine value on a protein proportion basis is only as high as 0.56 (Clarke and Wiseman, 2000).

Soybean-storage proteins are predominantly salt-soluble globulins designated as 2S, 7S, 11S, or 15S, depending on the sedimentation coefficients of proteins in sucrose density centrifugation. The 7S globulin, which is designated as β-conglycinin, and 11S globulin (or glycinin) are groups of proteins assembled from various subunits that are responsible for the nutritive value of soybean. Namely, β-conglycinin and glycinin contribute to nearly 70% of the amount of total seed proteins (Fig. 8–5) (Koshiyama, 1983; Nielsen et al., 1989; Sexton et al., 1998a). The sulfur-containing amino acids are mostly components of glycinin. The amino acid residues of glycinin are a family of five different proteins, and cysteine and methionine amount to a total of 24 to 45 g kg^{-1} of the residues (Nielsen et al.,

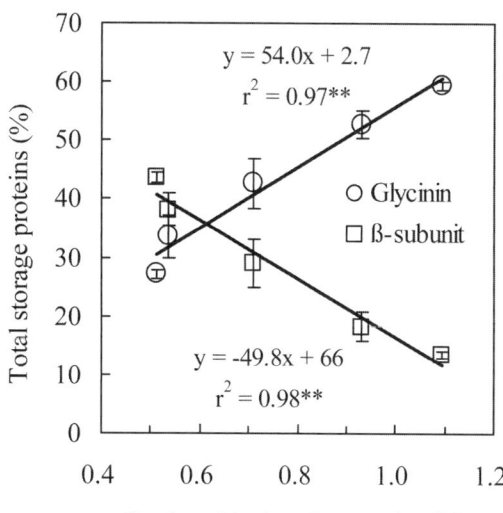

Fig. 8–5. Relative abundance of β-subunit of β-conglycinin and of glycinin storage proteins versus the amount of methionine plus cysteine in soybean seed. Source: Sexton et al. (1998a).

Table 8–1. Sulfur-containing amino acid contents in β-conglycinin (7S) and glycinin (11S) storage proteins of soybean in the proportion of weight. Data source Krishnan (2005).

Amino acid		β-conglycinin (7S)			Glycinin (11S)				
		α'	α	β	Gy1	Gy2	Gy3	Gy4	Gy5
Cysteine	%	0.8	0.9	0.0	1.7	1.7	1.7	1.1	1.2
Methionine	%	0.3	0.2	0.0	1.3	1.5	1.1	0.4	0.6
Total	%	1.1	1.1	0.0	3.0	3.2	2.8	1.5	1.8
	g kg^{-1}		7.3				24.6		

1989; Fukushima, 1991). Subunits of β-conglycinin contain no more than 11 g kg^{-1} of sulfur amino acids (Table 8–1) (Harada et al., 1989; Krishnan, 2005).

Soybean-protein composition can vary depending on the cultivar and, to some extent, be influenced by the growth conditions in the field. Different sources of nitrogen affect the amount and composition of proteins, in particular, the concentration of the sulfur-poor β-subunit of β-conglycinin that could be enhanced with higher nitrogen supply (Paek et al., 1997). Likewise, supraoptimal nitrogen application increased the soybean protein content by 15 to 28% over that of the control (Nakasathien et al., 2000). Sulfur nutrition also affects the protein composition, namely the glycinin concentration of sulfur-deficient soybean seeds decreased by 40% of that of the control, whereas the concentration of the β-subunit of β-conglycinin increased threefold (Gayler and Sykes, 1985). A change in sulfur supply from deficient to sufficient levels increased the glycinin concentration of soybean seed from 30 to 60% of the amount of total storage proteins, whereas the concentration of the β-subunit of β-conglycinin decreased from 40 to 10% (Fig. 8–5) (Sexton et al., 1998a).

The level of seed protein increased with the increase in nitrogen supply, while the glycinin or sulfur level remained unchanged (Nakasathien et al., 2000). In contrast, seed-nitrogen level remained unchanged across a range of tissue sulfur levels so as to maintain the homeostasis of the total protein concentration in seed (Fig. 8–5 and 8–6) (Blagrove et al., 1976; Sexton et al., 1998a; Hitsuda et al., 2004). These phenomena probably depend on the relative abundance of sulfur and nitrogen metabolites available for seed development. The seed-protein composition is mostly influenced by individual effects of sulfur and nitrogen, although these components interact to a limited extent in the process of synthesis (Paek et al., 1997, 2000). Nakasathien et al. (2000) concluded that the seed-protein concentration was regulated by nitrogen availability for the developing seeds and that the availability of sulfur-containing amino acids in the seeds controlled the synthesis of storage protein subunits from extra nitrogen.

These assumptions are supported by the fact that the sulfur-to-nitrogen (S/N) ratio in soybean seeds was not related to the seed-protein concentration. The ratio reflects merely the nitrogen values, as seed nitrogen concentrations are much higher than sulfur concentrations, and the concentrations of sulfur-containing amino acids tend to remain constant (Radford et al., 1977; Nakasathien et al., 2000; Wilcox and Shibles, 2001). Krishnan et al. (2005) reported that a high nitrogen nutrition level lowered the sulfur-containing amino acid concentrations because of a decrease in the expression of a Browman-Brik protease inhibitor (BBI) gene. Since BBI is rich in cysteine, seed-protein quality, indicated by the

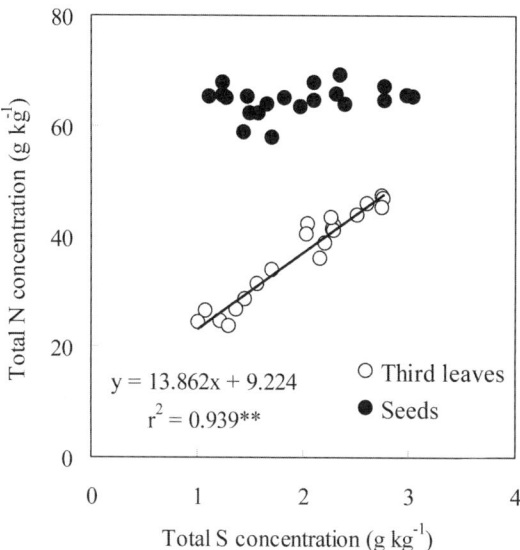

Fig. 8–6. Effect of sulfur application on sulfur and nitrogen concentrations in the third leaves (upper-most mature leaves) at flowering, and in seeds at harvest. Source: Hitsuda et al. (2004).

concentration of sulfur-containing amino acids, decreased when the protein concentration increased (Paek et al., 1997; Clarke and Wiseman, 2000). Nevertheless, the increase of the expression of the BBI gene to increase the levels of sulfur-containing amino acids is not desirable because it exerts an adverse effect on the performance of nonruminants fed with soybean meal (Wang et al., 2003). Moreover, the inverse relationship of the seed-protein concentration with yield and that with the oil concentration should be considered in soybean breeding aimed at high commercial quality (Table 8–2) (Nakasathien et al., 2000; Panthee et al., 2005).

Transgenic approaches to improve the soybean quality have not been fully successful, yet (Clarke and Wiseman, 2000; Krishnan, 2005). Improvement of sulfur assimilation of soybean plant is another option for the production of seeds with higher-protein quality. Sexton et al. (1998a, 1998b) concluded that sulfur was not remobilized to the seeds as efficiently as nitrogen, since the values of the harvest index for assimilated sulfur were consistently 15 to 20% lower than those for nitrogen. Sulfur concentrations in shoot and cotyledon of soybean tended to become stabilized, to some extent, despite the increase of sulfur application. However, the sulfur-poor β-subunit of β-conglycinin continued to synthesize

Table 8–2. Linear regression analysis and correlation coefficients between soybean yield and quality traits for 43 breeding lines averaged over three environments. Data source Wilcox and Shibles (2001).

Dependent vs. independent variables	Slope	Standard error of slope	Correlation coefficient
Protein vs. seed yield	−0.039	0.013	−0.45 **
Oil vs. seed yield	0.019	0.007	0.38 *
Sulfur vs. seed yield	<0.001	<0.001	0.23 ns†
Sulfur vs. protein	0.008	0.001	0.68 **
Protein vs. oil	−1.560	0.130	−0.88 **

† ns, nonsignificant.
* Significant at the 0.05 probability level.
** Significant at the 0.01 probability level.

seed storage proteins by 10 to 20% under sulfur-rich conditions. Therefore, it was assumed that the plant ability to reduce sulfate and to synthesize amino acids may limit the seed-protein quality. Naeve and Shibles (2005) suggested that the expansion of the sulfate-reductive capacity in seeds during the late-reproductive stage might contribute to improving soybean-seed quality.

Sulfur Deficiency in Soybean Cultivation

History

Sulfur deficiency had often been alleviated without prior recognition and has been sometimes confused with phosphorus deficiency because sulfur had been added as a component to other fertilizers, especially phosphoric materials. Furthermore, since soybean is a crop known for its low inputs, the importance of fertilization had been considered less compared with other crops. Soybean cultivation started to expand even in newly reclaimed infertile soils since the middle of the 1970s, and the application of gypsum was recommended for sodic soils (Keren et al., 1983; Sivapalan, 2003) or to highly weathered acidic soils (Ritchey et al., 1981; Sumner, 1993; Toma et al., 1999) because of its ameliorative effects. The need to apply sulfur was overlooked, though the beneficial effect of gypsum on crop growth may have been caused by sulfur supply to some extent (Alcordo and Rechcigl, 1993; Carvalho and Raij, 1997). Fertilizers with a higher purity that contain less sulfur are used more frequently than they were earlier to economize operation fees. Furthermore, it was recently noted that gypsum application is seldom required in no-till cropping systems, a method that has become popular in North and South America, because organic acids from crop residues in no-till fields ameliorate the acid subsoil by leaching of calcium and aluminum (Hue et al., 1986; Oliveira and Pavan, 1996; Miyazawa et al., 2001, 2002; Franchini et al., 2003). These factors have led to the reduction of the inadvertent sulfur supply to fields. Assessment of soil sulfur fertility is, therefore, required to avoid sulfur deficiency in crops.

Sulfur Deficiency Symptoms

Plant leaves from a higher position slowly turn yellow together with a marked decrease in their sulfur concentration once sulfur deficiency occurs (Fig. 8–3). The yellowing indicates a decrease in the leaf chlorophyll and protein levels because of the reduced synthesis of sulfur-containing amino acids, which results in a substantial decrease in photosynthesis (Nikiforova et al., 2005). In the most sulfur-deficient soybean leaves (0.8 g sulfur kg^{-1}), the CO_2-exchange rate and the dark respiration rate were only 30 and 50%, respectively, of those of control leaves (2.7 g sulfur kg^{-1}) (Sexton et al., 1997). As a result, the soybean plant stops growing as smaller leaves yellow and fewer branches are produced. Subsequently, brown spots appear on the edges of leaves and pods. Grains with a low yield are not fully ripened, and, in extreme cases, germinated seedlings are stunted and wilted (Hitsuda et al., 2004).

Yellowing of leaves is also indicative of other nutrient deficiencies, such as nitrogen, molybdenum, magnesium, manganese, or iron. The deficiency symptoms of these elements differ from those of sulfur but cannot easily be distinguished from one another once the symptoms advance. Moreover, plants occasionally experience nutritional stresses at the same time. Tissue analysis is, therefore, necessary to confirm the specific reason why the plants develop such symptoms in fields.

Sulfur Application for Soybean Production

Field Experiments

Sulfur requirements differ greatly depending on the kind of crop and in what country the crop is located. For example, crops can assimilate 0.1 to 25 kg sulfur to obtain 1 Mg of production, and 1 to 60 kg sulfur ha^{-1} to reach the average yield value. Crops need approximately 10 to 60 kg sulfur ha^{-1} to overcome sulfur deficiency (Sanchez, 1976; FAO Sulphur Network, 1992). In West Africa, for gramineous crops like millet [*Pennisetum glaucum* (L.) R. Br.], sorghum [*Sorghum bicolor* (L.) Moench], and maize (*Zea mays* L.), 5 to 10 kg sulfur ha^{-1} was sufficient to maintain maximum yields (Friesen, 1991). In Brazil, for sugarcane (*Saccharum officinarum* L.) and perennial grass production about 20 kg sulfur ha^{-1} is required (Malavolta et al., 1987). Although legumes generally require more sulfur to produce a unit weight of yield than nonleguminous crops, the amount of sulfur required for a unit area is intermediate among or lower than that of other crops (Table 8–3) (FAO Sulphur Network, 1992; Rocha and Malavolta, 1988; Lantmann and Castro, 2004). On the basis of numerous field trials, application of about 20 to 40 kg sulfur ha^{-1} is necessary to achieve high yields in pulse crops (FAO Sulphur Network, 1992). Field trials showed that the economically optimum soybean yield obtained with sulfur application ranged from 1.3 to 4.2 Mg ha^{-1} and that sulfur-deficient field conditions are widely varied in the soil type, texture, pH, and organic matter content. The deficiency occurs everywhere, in tropical, temperate, or cool regions, in dry or moist climates, and in fertile or highly leached soils. In spite of these vast differences, small sulfur applications (15–40 kg ha^{-1}) were sufficient for normal soybean growth (Table 8–4). A trial under hydroponic conditions indicated that the sulfur concentration required for seedling crops was so low that all the tested crops achieved optimum growth at a concentration of 2.0 mg sulfur L^{-1} in the culture solution (Hitsuda et al., 2005). A trace concentration of field sulfur likely reaches the required amount of sulfur for soybean growth because of the volume of soil available for the developing roots. Hence, sulfur deficiency is not often observed in fields, compared with pot trials.

Foliar Spray

A pot experiment using a Brazilian sandy Oxisol showed that foliar application of elemental sulfur (S^0) to soybean at the early-vegetative and initial-flowering stages increased the soybean yield and that the foliar application of 6 kg S^0 ha^{-1} was equivalent to a soil application of 20 kg S^0 ha^{-1} (Vitti et al., 2007). McGrath and Till (1992) sprayed ^{35}S-labeled S^0 on bean (*Phaseolus vulgaris* L.) leaves, and reported that S^0 was not absorbed through the leaf tissues. They assumed that S^0 fell onto the soil, underwent oxidation to sulfate-sulfur, and was absorbed by the roots. The sprayed S^0 is oxidized much faster than the coarse S^0 applied to the soil. Therefore, foliar spraying of S^0 at the stages where it is most required might result efficiently in yield increase in sulfur-deficient soils. Foliar spraying of sulfate-sulfur with nitrogen, phosphorus, and potassium at the early stage of soybean growth exerted a beneficial effect on yield at two out of 18 sites in the midwestern USA but seldom offset fertilization costs. The specific effect of sulfate-sulfur was

Table 8–3. Amount of sulfur absorbed by whole crop shoot to yield 1 Mg of harvested portion of the crop and to reach the country average yield per hectare.

Crop	Harvested part	Shoot S uptake for 1 Mg yield Source†			Crop	Average yield among countries‡	Total S uptake per hectare Source		
		I	II	III			I	II	III
		kg Mg⁻¹				Mg ha⁻¹	kg ha⁻¹		
Cotton	seed	–	25.0	–	Stylosanthes ¶	20.0 §	60	–	–
Rapeseed (*Brassica napus* L.)	seed	21.7	–	–	Tropical grasses	27.0 §	60	–	–
Coffee (*Coffea arabica* L.)	clean bean	13.5	1.5	–	Fodder beet	45.0 §	45	–	–
Greengram ¶ [*Vigna radiate* (L.) R. Wilczek]	grain	12.0	–	–	Turnip	45.0 §	45	–	–
Bean ¶	grain	10.4	25.0	–	Rapeseed	1.9	41	–	–
Tea [*Camilia sinensis* (L.) Kuntze]	made tea	10.0	–	–	Coastal bermuda	20.0 §	40	–	–
Chickpea ¶ (*Cicer arietinum* L.)	grain	8.7	–	–	Grass/clover ¶	12.0 §	35	–	–
Groundnut ¶ (*Arachis hypogaea* L.)	pod	7.5	3.0	–	Sugarcane	58.0	35	29	–
Pigeonpea ¶ [*Cajanus cajan* (L.) Huth]	grain	7.5	–	–	Sugarbeet	44.0	34	–	–
Millet	grain	7.3	–	–	Tomato	54.0	32	38	–
Soybean ¶	seed	6.7	8.0	8.2	Oilpalm	25.0 §	30	–	–
Wheat	grain	5.0	2.0	4.3	Cabbage	23.0	26	–	–
Tobacco (*Tabacum nicotiana* L.)	dry leaf	5.0	–	–	Maize (silage)	70.0 §	25	–	–
Sunflower	seed	5.0	–	8.4	Cotton	0.9	–	23	–
Sesame (*Sesamum indicum* L.)	seed	5.0	–	–	Red clover ¶	7.0 §	20	–	–
Castorbean (*Ricinus communis* L.)	seed	4.5	–	–	Alfalfa (lucerne) ¶	7.0 §	19	–	–
Maize	grain	4.2	7.0	2.6	Onion	29.0	17	29	–
Barley (*Hordeum vulgaris* L.)	grain	4.0	–	–	Maize (grain)	3.9	16	27	10
Sorghum	grain	3.8	1.5	–	Tea	1.6	16	–	–
Stylosanthes ¶ [*Stylosanthes guianensis* (Aublet) Sw.]	hay (7 cuts)	3.0	–	–	Coconut	4.8	15	–	–
Coconut (*Cocos nucifera* L.)	nut	3.0	–	–	Wheat	2.9	15	6	12
Lentil ¶ (*Lens culinaris* Medikus)	grain	3.0	–	–	Beans ¶	1.3	14	33	–
Cocoa (*Theobroma cacao* L.)	bean	–	3.0	–	Grape	8.6	13	2	–
Grass/clover § (*Trifolium* spp.) ¶	hay	2.9	–	–	Greengram ¶	1.0 §	12	–	–
Red clover§ (*Trifolium pratense* L.)	hay	2.9	–	–	Groundnut ¶	1.6	12	5	–
Alfalfa (lucerne) ¶	hay	2.8	–	–	Soybean ¶	1.7	11	14	14

continued.

Sulfur Management for Soybean Production

Table 8–3. Continued.

Crop	Harvested part	Shoot S uptake for 1 Mg yield Source†			Crop	Average yield among countries‡	Total S uptake per hectare Source		
		I	II	III			I	II	III
		kg Mg^{-1}				Mg ha^{-1}	kg ha^{-1}		
Tropical grasses	dry fodder	2.2	–	–	Sorghum	3.0	11	5	–
Coastal bermuda [*Cynodon dactylon* (L.) Pers.]	hay	2.0	3.0	–	Citrus	11.0	11	2	–
Rice (paddy)	grain	1.7	–	–	Barley	2.7	11	–	–
Grape (*Vitus* spp.)	fruit	1.5	0.2	–	Chickpea ¶	1.2	10	–	–
Oilpalm (*Elaeis guinensis* Jacq.)	fresh bunch	1.2	–	–	Tobacco	2.0§	10	–	–
Cabbage (*Brassica oleracea* L.)	head	1.1	–	–	Okra	20.0§	10	–	–
Fodder beet (*Beta vulgaris* L.)	plant	1.0	–	–	Papaya	50.0§	10	–	–
Turnip (*Brassica rapa* L.)	root	1.0	–	–	Millet	1.3	10	–	–
Citrus (*Citrus* spp.)	fruit	1.0	0.1	–	Potato	19.0	10	19	–
Sugarbeet (*Beta vulgaris* L.)	root	0.8	–	–	Pigeonpea ¶	1.2§	9	–	–
Sugarcane	cane	0.6	0.5	–	Castorbean	2.0§	9	–	–
Tomato (*Lycopersicon* spp.)	fruit	0.6	0.7	–	Coffee	0.7	9	1	–
Garlic (*Allium sativa* L.)	bulb	0.6	–	–	Pineapple	19.0	8	–	–
Onion (*Allium cepa* L.)	bulb	0.6	1.0	–	Banana	18.0	7	–	–
Okra [*Abelmoschus esculentus* (L.) Moench]	fruit	0.5	–	–	Sunflower	1.3	7	–	11
Potato (*Solanum tuberosum* L.)	tuber	0.5	1.0	–	Rice (paddy)	3.5	6	11	–
Cassava (*Manihot esculenta* Crantz)	tuber	0.5	0.4	–	Cassava	10.0	5	4	–
Pineapple [*Ananas comosus* (L.) Merr.]	fruit	0.4	–	–	Eggplant	30.0	5	–	–
Banana (*Musa* × *paradisiaca* L.)	fruit	0.4	–	–	Garlic	8.2	5	–	–
Maize	silage	0.4	–	–	Sesame	0.9	5	–	–
Papaya (*Carica papaya* L.)	fruit	0.2	–	–	Lentil ¶	1.0	3	–	–
Eggplant (*Solanum melongena* L.)	fruit	0.2	–	–	Cocoa	0.4	–	1	–

† Calculated from the following data: (I) FAO Sulphur Network (1992), (II) Rocha and Malavolta (1988), and (III) Lantmann and Castro (2004).
‡ The data without mark were calculated from the FAO Statistics Division data in 2005, and the data with § from FAO Sulphur Network (1992).
¶ Leguminous crops.

Table 8–4. Characteristics of topsoils to which sulfur application positively affected soybean growth.

Country	Soil type	Texture	O.M.	SO_4–S[‡]	pH [§]	Amount of S applied for optimum yield	Yield obtained	S source	Reference
			g kg^{-1}			kg S ha^{-1}	Mg ha^{-1}		
India	Vertisols	Silty clay–clay	5 [†]	–	7.9a	20	1.3	gypsum	Joshi and Billore (1998)
			7 [†]	–	8.3a	40	2.6		Ganeshamurthy (1998)
Bangladesh	Inceptisols	silt loam	17	–	6.7a	30	2.0	gypsum	Sarker et al. (2002)
Vietnam	Ultisols	–	5–7	–	5.0–5.5b	40	2.0–2.7	various	Bo and Thi (1997)
Canada	Spodosols	silty clay loam	21	–	6.9a	–	–	gypsum	Abbès et al. (1992)
USA	Ultisols	–	–	–	–	22	2.3	–	Kamprath and Jones (1986)
Brazil	Oxisols	silt	19	–	5.4c	30	–	–	Richart et al. (2006)
Argentina	Mollisols	loam	19	10.3	5.9a	15	3.0	gypsum or ammonium sulfate	Boem et al. (2007)
		sandy loam	27	11.4	5.6a	15	3.6		
		sandy loam	26	14.8	5.8a	15	4.2		
		silt loam	31	17.3	5.6a	15	3.5		

[†] Organic matter content was obtained by multiplying 1.724 with the total carbon content.
[‡] Extracted with 1 mol NH_4OAc (Lisle et al., 1994)
[¶] pH was determined with (a) H_2O, (b) 1 mol KCl, or (c) 0.01 mol $CaCl_2$.

not identified, compared with that of S^0 (Mallarino et al., 2001). It was reported that foliar fertilization of nitrogen, phosphorus, and potassium together with sulfate-sulfur during the seed-filling period had led to the increase of soybean yield, providing nutrition to seeds to compensate for root- and leaf-senescence (Garcia and Hanway, 1976). However, this phenomenon does not always occur, and the spray may even damage the foliage (Parker and Boswell, 1980). Changes in the nutrient-absorption rate by roots during the pod-filling stage were minimal (Vasilas et al., 1980). Root and leaf senescence was not causally related to the seed yield, since the yield did not increase despite the increase in the nutritional element concentrations in leaves during the pod-filling stage (Boote et al., 1978; Sesay and Shibles, 1980).

Latent Sulfur Deficiency

Determination of the critical value for a deficient element is necessary for both high-yield management and efficient use of fertilizers. Moreover, prevention of latent deficiency is required because crop yield markedly decreases once deficiency symptoms appear and because the product quality is not always high enough, although yield reduction is not identified. Extensive trials have revealed the effects of the absence of sulfur application on soybean yield. However, it is yet to be determined whether the plants were subjected to a latent sulfur deficiency before the application. If soybean yields, together with protein or oil concentration in seeds, increased by sulfur application in the presence of enough other essential elements (Malavolta et al., 1987; Dwivedi and Bapat, 1998; Saha et al., 2001), it could be considered that the plants did not grow normally because of severe soil sulfur deficiency. Meanwhile, if the protein or oil concentration of seeds increased by sulfur supply without concomitant increase of yield (Chowdhury et al., 1985; Caires et al., 2006), it is possible that the plants displayed a latent-sulfur deficiency, which prevented the full development of the genotype traits. On the basis of 43 soybean lines grown in three different environments, a positive relationship was observed between the seed-sulfur level and protein concentration, while the seed-sulfur level was not related to the seed yield (Table 8–2) (Wilcox and Shibles, 2001). This may imply that a number of fields were low in sulfur availability. Most producers, however, do not recognize the insufficiency before a decrease in yield or that sulfur nutrition is important in producing high-quality soybean seeds. A simple and reliable method of evaluating the sulfur-nutrition status in relation to soybean yield and seed quality needs to be developed.

Assessment of Sulfur Deficiency in Soybean

Difficulty in Diagnosing Deficiency for Sulfur Application

Sulfur application to field crops looks very simple, and numerous application trials have been conducted. However, many of them have remained only as a reference because of the difficulty in generalizing the data. Plants absorb soil sulfur as sulfate-sulfur, which easily leaches by precipitation and accumulates in lower soil layers (Friesen, 1991; Sweeney and Granade, 1993). In Brazilian Oxisols, the retention of sulfate-sulfur was highest in the 30- to 45-cm soil layer with the addition of 1200 mm of rain and irrigation water, but it was highest in the 45- to 60-cm layer when the added water increased to 2300 to 4600 mm, in spite of the equal amount of gypsum applied to each (Ritchey and Sousa, 1997). Seasonal changes in precipitation may lead to inconsistencies in the crop response to sulfur among the trials within the same area (Wihardjaka et al., 1999; Sawyer and Barker, 2002).

Soil texture also affects sulfate-sulfur leaching. The layers at a 30- to 45-cm depth accumulated the greatest amount of sulfate-sulfur in clay soil after 3 yr of gypsum application, while those at a 90- to 105-cm depth in a loamy soil accumulated the greatest amount (Sumner, 1990). The amount of sulfate-sulfur that accumulated in the surface layers was the lowest among the horizons studied, accounting for roughly 10 to 50% of the highest value (Caires et al., 2006; Ritchey and Sousa, 1997). A similar trend of sulfate-sulfur movement was observed in a field where sulfur-containing fertilizers had been used for many years (Hitsuda et al., 2004). Sulfur-stressed young plants may recover from the deficiency when roots reach the soil layers where sulfur accumulated (Coleman, 1966; Sanchez, 1976), and sulfur application can possibly be omitted in such fields. The variations in the sulfate-sulfur concentrations within the soil horizons are large, and the contribution of each horizon to plant growth cannot be easily determined. As a result, it is difficult to obtain a representative soil sample. Soil-sulfur assessment through routine soil analysis cannot be performed easily in cultivated fields. Sulfur deficiency has been reported to occur in crops when growth was enhanced by the application of high rates of nitrogen and phosphorus (Randall et al., 1981; Bansal et al., 1983). In Argentina, soybean growth likely responds more to sulfur application when the soil contains less than 3% organic matter; the field has been cultivated for more than 20 yr; and a yield of more than 3.5 Mg ha^{-1} can be achieved with sufficient nitrogen and phosphorus application (Ferraris, 2006). Crop responses to sulfur application are affected by so many factors that the results obtained under certain conditions are hardly applicable to other conditions. Thus, the difficulty in the assessment of soil-sulfur fertility in fields is considered to be a major problem in sulfur management.

Furthermore, sulfur analyses in soils and plant tissues are not sufficiently reliable. Crosland et al. (2001) compared concentrations of extractable sulfur in soils with total sulfur and nitrogen in plant tissues in the same samples that had been analyzed in 10 UK laboratories. The concentrations of plant-available sulfur in the soils extracted with KH_2PO_4 showed very high coefficients of variation (CVs) among the laboratories involved, in the range of 36 to 45%, presumably because of the difference in the determination methods applied. Calibration of the results for diagnostic purposes has been required to make the varied analyses comparable. On the other hand, plant analysis is generally more reliable for determining sulfur deficiency than soil testing. Results from interlaboratory plant analyses for total nitrogen determination were in reasonable agreement, with low CVs (2.7–5.4%), whereas the CVs for total sulfur were considerably higher (8.2–20.3%), presumably because of the methods of digestion or extraction applied. Consequently, the accuracy of the N/S ratio in plant tissues and the diagnosis of plant-sulfur deficiency determined on the basis of total-sulfur analysis may not be reliable. In contrast with the conventional method, Blake-Kalff et al. (2000, 2001) suggested that the determination of the malate/sulfate peak area ratio determined in a single analysis by ion chromatography was a reliable and practical indicator for sulfur deficiency; the ratio would be theoretically stable throughout plant development. Further studies should be performed to confirm how widely the method may be used to determine the sulfur status of plant tissues.

Soil Analysis

Soil samples (Vertisols, 0–15 cm layer) were collected from Central India, and 12 determination methods of available sulfur (mostly sulfate-sulfur) were examined in soybean plants grown in pots. Morgan reagent (1 mol NaOAc, pH 4.8), $Ca(H_2PO_4)_2$ (500 mg phosphorus L^{-1}), and KH_2PO_4 (500 mg phosphorus L^{-1}) solutions were found to be suitable to determine the available content of sulfur because of their higher correlation with indices used, namely relative yield, "A" value, and sulfur uptake of the control plants. Critical sulfate-sulfur concentration for 90% relative yield was 9 mg L^{-1} with Morgan reagent and 10 mg L^{-1} with both phosphate solutions (Bansal et al., 1983). The values were similar to those obtained from field trials with other crops, namely 12 mg L^{-1} for alfalfa (*Medicago sativa* L.) with Morgan reagent, 10 mg L^{-1} for alfalfa with $Ca(H_2PO_4)_2$ (500 mg phosphorus L^{-1}), and 8 mg L^{-1} for pasture with KH_2PO_4 (500 mg phosphorus L^{-1}) (Reisenauer, 1975). However, the results of the soil-sulfur concentration obtained from pot experiments are not always applicable in fields.

Soil-sulfur availability in soybean fields was evaluated on the basis of the soil SO_4–S concentration extracted with 0.01 M L^{-1} $Ca(H_2PO_4)_2$ (Embrapa-Soja, 2006). Soils were classified into two groups according to the texture (sandy and silt), sampling of two soil layers (0–20 and 20–40 cm depths), and levels of sulfate-sulfur fertility (low, medium, and high). The critical sulfate-sulfur concentration ranged from 2 to 9 mg kg^{-1} in the sandy soil (silt concentration ≤40%), and from 5 to 35 mg kg^{-1} in the silt soil (silt concentration >40%). Eighteen categories of soil sulfur fertility were identified by combining three sulfate-sulfur fertility levels in each soil layer with two types of soil texture, and sulfur application rates ranging from 0 to 80 kg sulfur ha^{-1} were proposed in the respective cases (Table 8–5).

Regardless of the categories, the supplementary application of 10 kg sulfur Mg^{-1} grain yield expected was recommended to ensure the fertility. Leaf analysis,

Table 8–5. Sulfur application rates based on soil sulfur fertility. Data source Embrapa-Soja (2006).

S fertility category		Soil depth (cm)				S application required	
		0–20	20–40	0–20	20–40		
		Sandy soil		Silty soil		Basal application	Fertility maintenance
		Silt concentration in soil					
		≤40%		>40%			
		mg S kg^{-1}†				kg ha^{-1}	
Low	low		<6		<20	80	10 kg ha^{-1} Mg^{-1} ha^{-1} seed yield expected
	medium	<2	6–9	<5	20–35	60	
	high		>9		>35	40	
Medium	low		<6		<20	60	
	medium	2–3	6–9	5–10	20–35	40	
	high		>9		>35	0	
High	low		<6		<20	40	
	medium	>3	6–9	>10	20–35	0	
	high		>9		>35	0	

† Soil S was extracted with 0.01 mol L^{-1} $Ca(H_2PO_4)_2$, and the concentration was determined by the turbidimetric method.

however, was suggested to complement the soil data if they were questionable. Obtaining sulfate-sulfur data through the recommended soil analysis is very laborious before they can be made available for the cultivation season, although the soybean sulfur requirements were met at relatively low levels. Therefore, such analysis does not seem practical for use by producers. Besides, sulfate-sulfur concentration in soil was sometimes the highest in the subsoil layers located at a depth of more than 40 cm (Sumner, 1990; Ritchey and Sousa, 1997), which may be the main source of sulfate-sulfur for plants. Occasionally, soybean plants did not significantly respond to sulfur application in a sandy soil despite the low sulfate-sulfur concentrations in the surface (0–20 cm; 1 mg sulfur kg^{-1}) and subsoil (20–41cm; 7 mg sulfur kg^{-1}) horizons (Kamprath and Jones, 1986). These inconsistencies in soybean responses to soil-sulfur concentration have been repeatedly observed in trials, indicating that soil-sulfur analysis cannot be easily applied under field conditions.

Leaf Analysis

Sulfur diagnostic indices have been examined in sulfur-responsive plants by excluding other limiting factors. Nutritional indices in a specific plant part for targeting a 90 to 95% maximum yield that were fitted to the Mitsucherlich function have been widely used to obtain the critical concentration (Munson and Nelson, 1990). Zhao et al. (1996) evaluated six diagnostic indices for sulfur on the basis of previous reports in wheat (*Triticum aestivum* L.) plants grown in pots as follows: (i) sulfur concentration of shoot, (ii) sulfate-sulfur concentration of shoot (Jones, 1963; Scaife and Burns, 1986), (iii) percentage of sulfate-sulfur-to-sulfur concentration in shoot (Spencer and Freney, 1980), (iv) glutathione concentration of uppermost, fully expanded leaves (Macnicol and Randall, 1987), (v) shoot N/S ratio (Dijkshoorn et al., 1960; Rasmussen et al., 1977), and (vi) chlorophyll-meter reading of uppermost, fully expanded leaves (Fox et al., 1994; Peltonen et al., 1995). The values of the indices were obtained at the stem extension stage (flag leaf just visible), and the relationship of the relative shoot-dry weight at maturity (grain and straw) with each index was examined. Either sulfur or sulfate-sulfur concentration was found to be a good indicator of sulfur deficiency because of the sharper demarcation from sufficiency to deficiency, compared with the other indices. Nevertheless, the determination of sulfur concentration is superior to that of sulfate-sulfur concentration, since tedious analysis procedures required for sulfate-sulfur determination, as well as for glutathione determination, are not practical as a routine service to producers (Blanchar, 1986; Zhao et al., 1996). The N/S ratio and the chlorophyll-meter reading indices were not accurate for diagnosing the sulfur-nutritional status in wheat or in soybean (Hitsuda et al., 2004). Besides, the chlorophyll-meter reading approach is not recommended because leaf yellowing can also be a nutritional disorder symptom due to a deficiency in other elements.

The response of young soybean plants to sulfur stress was examined relative to other crops; plants were grown under hydroponic conditions for 29 d after germination. The critical shoot-sulfur concentration for 75% of relative soybean shoot weight was 0.8 g sulfur kg^{-1}, a value similar to that of maize, and lower than that of sunflower (*Helianthus annuus* L.), field bean, wheat, rice (*Oryza sativa* L.), sorghum, and cotton (*Gossypium* spp.). The amount of sulfur absorbed in soybean shoots and roots was higher than that in wheat, rice, and sorghum but much lower than that in maize, sunflower, field bean, and cotton because of a slower initial growth rate. Crop tolerance to low external sulfur concentration was not related to the critical sulfur concentration in crop shoots or to the amount of sulfur absorbed for optimum growth, and the tolerance of soybean was intermediate among that of the tested crops (Hitsuda et al., 2005). In Indian Vertisols, the relative shoot-dry weight of soybean plants grown in pots was not related to the shoot-sulfur concentration in 36-d-old plants but was significantly related in 60-d-old plants. The critical sulfur concentration for more than 90% of relative shoot yield was 1.9 g sulfur kg^{-1} in the 60-d-old plants (Bansal et al., 1983). Although critical concentrations obtained from plants at the initial-vegetative stage may be useful for the preparation of starter-fertilizer application, the critical values at this stage will not be sufficient to obtain a satisfactory grain yield (Zhao et al., 1996).

In a study conducted in the USA to anticipate soybean seed yield, the nutritional status of leaf was calibrated by Mitscherlich function on the basis of the uppermost, fully expanded leaves (i.e., the last leaves to become mature) with petioles at the full-flowering stage (Peck, 1979; Tanaka et al., 1993). The proposed critical values based on this study have also been widely used in Brazil, though sampling of leaves without petiole at the initial flowering stage was adopted (Embrapa-Soja, 2006). The critical-sulfur concentration values of the last-mature soybean leaves proposed in five separate studies were in mutual agreement, despite the differences in varieties, cultivation conditions, the presence or absence of petioles, or considerable CVs in the analyses used. The optimum leaf-sulfur concentration for satisfactory yield ranged from 2.0 to 3.1 g kg^{-1}, and the concentration in the low ranges from 1.5 to 2.5 g sulfur kg^{-1} (Table 8–6). The third leaf from the top at the flowering stage is usually considered to be the last leaf to become mature in soybean plants. It was reported that the average sulfur concentration in the second leaf was about 9% higher than that in the third leaf, while the concentration in the fourth leaf was 4% lower (Hitsuda et al., 2004). Moreover, the sulfur concentration of the last mature soybean leaf decreased by 1 to 27% among six varieties during plant growth from the full-bloom stage to the pod stage (Fig. 8–2) (Fontanive et al., 1996). It is not always easy to identify the last-mature leaf in fields, moreover, to obtain a large number of leaf samplings at the same growth stage. The diagnosis by foliar analysis is, thus, associated with considerable variation because of the difficulty in sampling.

Seed Analysis

It is well known that the elemental concentration of harvested grain displays less variation than that of vegetative parts (Munson and Nelson, 1990). Although seeds are considered to be less affected by the nutrient status of the growth medium than leaves, they have hardly been utilized for diagnosing the nutritional status of crops (Ulrich, 1952). Nevertheless, Russell (1963) found that the nitrogen concentration in grain could be used for nitrogen-status assessment in wheat. Malhi et al. (2007) reported that the effects of the sulfur-nutritional status in the growth medium were more pronounced in seeds than in straw in *Brassica* oilseed crops, both in terms of yield and sulfur concentration. Grain-sulfur concentration in wheat and rice was highly correlated with the respective relative grain yield obtained under different growth conditions (Randall et al., 1981, 2003). Relative soybean-seed yield was better correlated with sulfur concentration in seeds ($R^2 = 0.744$) than that in the uppermost mature leaves at flowering (R^2

Table 8–6. Critical sulfur concentrations for nutritional diagnosis of soybean using shoots, leaves, or seeds.

			Sulfur deficiency category					
			Very deficient	Deficient	Low	Optimum	High	
			Appearance of visible symptoms	Yield decrease	Unstable both in yield and seed quality	Satisfactory both in yield and seed quality	Luxurious absorption	
					Relative yield			
			≤60%	60–80%	80–90%	90%≤		
Trial site	Plant part analyzed	Growth condition			g S kg^{-1}			Reference
	Young plant shoot							
Brazil	whole shoot (32 d old)	water culture	<0.8		–	–	–	Hitsuda et al. (2005)
	Shoot at flowering							
India	whole shoot (60 d old)	pot	–		<1.9	–	–	Bansal (1983)
	Leaf at flowering							
USA	the uppermost fully expanded trifoliate leaf prior to or during initial flowering	field	<1.5		1.5–2.5	2.5–3.5	–	Olsen's Agricultural Laboratory (1997)
USA, Brazil	the uppermost fully expanded trifoliate leaf without petiole at initial flowering	field	<1.5		1.5–2.0	2.0–4.0	>4.0	Peck (1979); Embrapa-Soja (2006)
Brazil	the uppermost fully expanded trifoliate leaf with petiole at full flowering	field	–		<2.0	2.0–3.1	>3.1	Kurihara (2004)
Brazil	the uppermost fully expanded trifoliate leaf without petiole at initial flowering	pot	<1.5	1.5–2.0	2.0–2.3	>2.3	–	Hitsuda et al. (2004)
	Seed							
Brazil	mature seed	pot	<1.5	1.5–2.0	2.0–2.3	>2.3	–	Hitsuda et al. (2004)

= 0.660) (Hitsuda et al., 2004). Fox et al. (1977) concluded that the use of seeds was preferable to that of vegetative materials for diagnosing the sulfur status of cowpea [*Vigna unguiculata* (L.) Walp.]. Nutritional stress is, thus, obviously reflected not only in leaves but also in seeds. Consequently, seeds can be used for assessing the sulfur-nutritional status of plants retrospectively. The results can then be used to determine the level of sulfur fertilizer application required for the following cropping. Furthermore, the sulfur status in soybean seeds should be evaluated in terms of the nutritional value of seed protein.

A trial in fields with moderate sulfur deficiency revealed that the vegetative growth of soybean did not differ among the sulfur treatments until the beginning of the seed stage but that the seed number had decreased at the time in the absence of sulfur supply. This implied that the amount of sulfur accumulated in the vegetative plant parts was not sufficient to support grain growth through remobilization leading to reduced grain yields (Boem et al., 2007). Likewise, the importance of the nutritional status before seed filling was confirmed in a trial under hydroponic conditions that could limit precisely the period of nutrient supply. It was found that the seed yield was very sensitive to sulfur deficiency occurring during the vegetative growth stage (second node- full pod) but not to the deficiency occurring at the reproductive stage (beginning seed to full maturity) (Sexton et al., 1998c). In the fields with sufficient fertility, the soybean growth rate between the beginning-bloom and full-seed stages was positively related to the seed number (Egli, 1998). The amount of sulfur absorbed in soybean shoots should become highest at the full-seed stage, as suggested by the plant dry weight (Fig. 8–1) (Hanway et al., 1984), and sulfur should be supplied to meet the demand at a particular time to obtain a satisfactory yield. Although there are variations depending on the sulfur-nutritional status of the plants until the full-pod stage, it appears that sulfur supply until the full-seed stage controls soybean yield. Since sulfur absorbed during the reproductive stage is mainly transformed into seed protein, strongly influencing the 11S/7S ratio (Sunarpi and Anderson, 1996, 1997a; Sexton et al., 1998c), the amount of sulfur supplied during this stage is important for the production of high-quality seeds. Accordingly, the diagnosis of soybean-sulfur deficiency based only on the leaf-sulfur concentration until the flowering stage, is not reliable for determining whether the supply of sulfur will be sufficient throughout the growth cycle to obtain high-quality seed yield.

To evaluate soybean-seed quality, as well as seed yield, the changes in the protein composition of seeds depending on the supply of external sulfur might be a more suitable indicator than the conventional indices used in the evaluation of field sulfur fertility (Gayler and Sykes, 1985). Hitsuda et al. (2004) cultivated soybean plants in pots with different levels of sulfur application throughout the growth cycle and compared the protein profile of the seeds from each sulfur treatment with that of the control seeds grown in a sulfur-sufficient field by sodium dodecyl sulfate-polyacrylamide gel electrophoresis (SDS-PAGE) analysis. In the plants with visible symptoms of sulfur deficiency, the seeds contained 1.5 g sulfur kg^{-1}, and the seed yield was 60% of that of the control. The electrophoresis analysis indicated that the critical seed-sulfur concentration for the deficiency in protein components was 2.0 g kg^{-1} when the yield was 80% of that of the control. The sulfur concentration was 2.3 g kg^{-1} or higher for more than 90% yield when the composition of the protein components was identical with that in the original seeds obtained under sufficient sulfur fertilization. The 90% yield as a critical

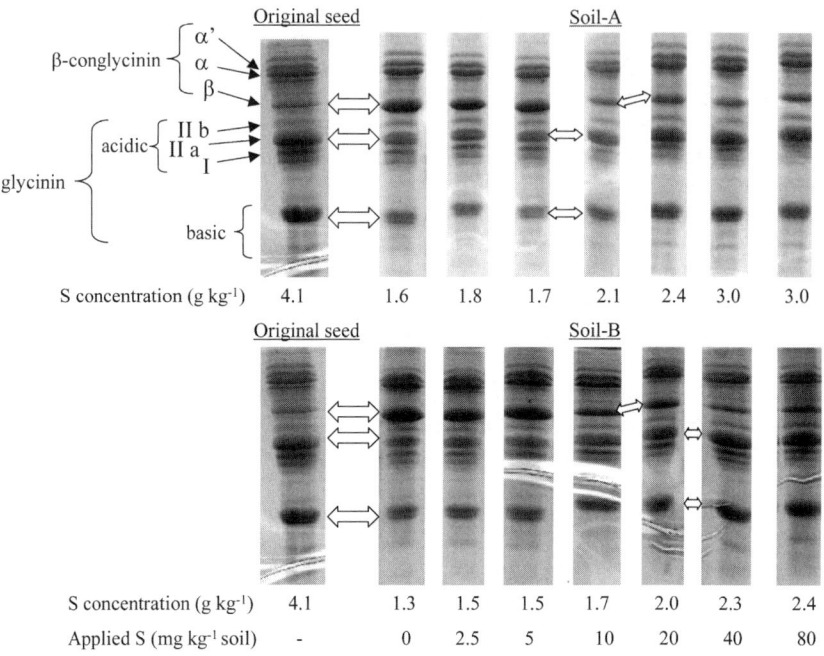

Fig. 8–7. Composition of the protein components of the seeds with different levels of sulfur application in the Soil-A and -B from Brazilian northeastern Oxisols. Arrows indicate corresponding subunits. In the seeds with 1.3–1.7 g sulfur kg^{-1}, the content of β-conglycinin (especially β-subunit) was higher and that of glycinin (acidic-IIa and basic subunits) was lower than the contents in the original seeds with 4.1 g sulfur kg^{-1}. The seeds with 2.1 g sulfur kg^{-1} in the Soil-A displayed a similar β-conglycinin profile to that of the original seeds, but still exhibited the sulfur-deficient glycinin profile. The seeds with 2.0 g sulfur kg^{-1} in the Soil-B showed a sulfur-deficient β-conglycinin profile, but had a sulfur-sufficient glycinin profile. Therefore, the amino acid composition in the seeds began to change from the deficient pattern to the sufficient one at around 2.0 g sulfur kg^{-1}. The seed sulfur concentration was 2.3 g kg^{-1} or higher in all the soils when the protein profile became identical with that of the original seeds. Source: Hitsuda et al. (2004).

value was appropriate from the viewpoint of the seed-protein composition. Plants displayed a latent deficiency when the seed sulfur concentration ranged between 2.0 and 2.3 g kg^{-1}, showing unstable yield and transition protein profiles from deficiency to sufficiency (Fig. 8–7). Consequently, the sulfur concentration in the seeds was classified as follows: deficient (sulfur < 1.5 g kg^{-1}), very low (1.5 g kg^{-1} ≤ sulfur < 2.0 g kg^{-1}), low (2.0 g kg^{-1} ≤ sulfur < 2.3 g kg^{-1}), and normal (2.3 g kg^{-1} ≤ sulfur) (Table 8–6). It was eventually concluded that sulfuric materials should be applied to keep the sulfur concentration above 2.3 g kg^{-1} in seeds. The difference in sulfur levels among the varieties differing in genetic background, which grew under different agronomic and climatic conditions, should be studied further.

While the value of the ratio between the sulfur concentrations of seed at full maturity and uppermost-mature leaves at flowering was 1:1 (Hitsuda et al., 2004), seed analysis was found to be preferable to leaf analysis for the following reasons: stable sulfur concentration, easy sampling, and sufficient time for planning fertilizer application for the subsequent cropping. Studies on the management of the sulfur-nutritional status in soybean to enhance the seed quality and yield should be further conducted.

Conclusions

We have presented various approaches to soybean crop management through sulfur application along with some of the genetic options to improve the seed-protein quality coupled with high levels of sulfur-containing amino acids without compromising yield levels. Reliable analysis of the sulfur status in soil is a prerequisite for determining whether sulfur should be applied to the soybean crop, but it remains difficult because of the heterogeneity of sulfur distribution among soil layers. Since seed-sulfur level reflects the field-sulfur status of soils, determination of the seed-sulfur concentration should be used as a diagnostic tool to determine the sulfur-fertility requirements of the field for soybean production to optimize the intrinsic capacity. In this way, sulfur sufficiency can be ensured to obtain maximum high-quality soybean yield.

References

Abbès, C., A. Karam, D. Isfan, and L.E. Parent. 1992. Sulfur fertilization of soybean. (In French, with English abstract.). Can. J. Plant Sci. 72:377–382.

Alcordo, I.S., and J.E. Rechcigl. 1993. Phosphogypsum in agriculture: A review. Adv. Agron. 49:55–118.

Anderson, J.W., and M.A. Fitzgerald. 2001. Physiological and metabolic origin of sulphur for the synthesis of seed storage proteins. J. Plant Physiol. 158:447–456.

Bansal, K.N., D.P. Motiramani, and A.R. Pal. 1983. Studies on sulphur in vertisols: I. Soil and plant tests for diagnosing sulphur deficiency in soybean (*Glycine max* (L.) Merr.). Plant Soil 70:133–140.

Beaton, J.D., and M. White. 1997. Occurrence and correction of sulfur deficiencies in the Asian and Pacific region: A review and update. Sulphur Agric. 20:31–46.

Blagrove, R.J., J.M. Gillespie, and P.J. Randall. 1976. Effect of sulphur supply on the seed globulin composition of *Lupinus angustifolius*. Aust. J. Plant Physiol. 3:173–184.

Blair, G.J. 1979. Sulfur in the tropics. Joint IFDC/Sulphur Institute Bulletin. IFDC, Muscle Shoals, AL.

Blake-Kalff, M.M.A., M.J. Hawkesford, F.J. Zhao, and S.P. McGrath. 2000. Diagnosing sulfur deficiency in field-grown oilseed rape (*Brassica napus* L.) and wheat (*Triticum aestivum* L.). Plant Soil 225:95–107.

Blake-Kalff, M.M.A., F.J. Zhao, M.J. Hawkesford, and S.P. McGrath. 2001. Using plant analysis to predict yield losses caused by sulphur deficiency. Ann. Appl. Biol. 138:123–127.

Blanchar, R.W. 1986. Measurement of sulfur in soils and plants. p. 455–490. *In* M.A. Tabatabai (ed.) Sulfur in agriculture. Agron. Monogr. 27. ASA, CSSA, and SSSA, Madison, WI.

Boem, F.H.G., P. Prystupa, and G. Ferraris. 2007. Seed number and yield determination in sulfur deficient soybean crops. J. Plant Nutr. 30:93–104.

Boote, K.J., R.N. Gallaher, W.K. Robertson, K. Hinson, and L.C. Hammond. 1978. Effect of foliar fertilization on photosynthesis, leaf nutrition, and yield of soybeans. Agron. J. 70:787–791.

Caires, E.F., S. Churka, F.J. Garbuio, R.A. Ferrari, and M.A. Morgano. 2006. Soybean yield and quality as a function of lime and gypsum applications. Sci. Agric. (Piracicaba, Brazil) 63: 370–379.

Carvalho, M.C.S., and B. van Raij. 1997. Calcium sulphate, phosphogypsum and calcium carbonate in the amelioration of acid subsoils for root growth. Plant Soil 192:37–48.

Ceccotti, S.P. 1996. Plant nutrient sulfur- a review of nutrient balance, environmental impact and fertilizers. Fert. Res. 43:117–125.

Chowdhury, I.R., K.B. Paul, F. Eivazi, and D. Bleich. 1985. Effects of foliar fertilization on yield, protein, oil and elemental composition of two soybean varieties. Commun. Soil Sci. Plant Anal. 16:681–692.

Clarke, E.J., and J. Wiseman. 2000. Developments in plant breeding for improved nutritional quality of soya beans I. Protein and amino acid content. J. Agric. Sci. (Cambridge) 134:111–124.

Coleman, R. 1966. The importance of sulfur as a plant nutrient in world crop production. Soil Sci. 101:230–239.

Craswell, E.T., and U. Karjalainen. 1990. Recent research on fertilizer problems in Asian agriculture. Fert. Res. 26:243–248.

Crosland, A.R., F.J. Zhao, and S.P. McGrath. 2001. Inter-laboratory comparison of sulphur and nitrogen analysis in plants and soils. Commun. Soil Sci. Plant Anal. 32:685–695.

De Oliveira, E.L., and M.A. Pavan. 1996. Control of soil acidity in no-tillage system for soybean production. Soil Tillage Res. 38:47–57.

Dijkshoorn, W., J.E.M. Lampe, and P.F.J. Van Burg. 1960. A method of diagnosing the sulphur nutrition status of herbage. Plant Soil 13:227–241.

Dwivedi, A.K., and P.N. Bapat. 1998. Sulphur-phosphorus interaction on the synthesis of nitrogenous fractions and oil in soybean. J. Indian Soc. Soil Sci. 46:254–257.

Egli, D.B. 1998. Seed biology and the yield of grain crops. CAB International, New York.

Embrapa-Soja (Soybean Research Center of the Brazilian Agricultural Research Corporation). 2006. Correção e manutenção da fertilidade do solo. (In Portuguese.) p. 41–63. In Sistemas de produção 11, Tecnologias de produção de soja- Região Central do Brasil, 2007. Embrapa-Soja, Londrina, PR, Brazil.

FAO Sulphur Network. 1992. The importance of sulphur for crop production. FAO, Rome.

FAOSTAT (Statistics Division. Food and Agriculture Organization of the United Nations). 2007. Production, Data archives [Online]. Available at http://faostat.fao.org/site/291/default.aspx; verified 9 April 2008. FAO, Rome.

Ferraris, G.N. 2006. Fertilización del cultivo de soja. (In Spanish.) In Nutrición de soja, Reunión de actualización técnica, 4 de octubre de 2006. Desarrollo Rural INTA-EEA, Pergamino, Argentina.

Fontanive, A.V., A.M. de la Horra, and M. Moretti. 1996. Foliar analysis of sulphur in different soybean cultivar stages and its relation to yield. Commun. Soil Sci. Plant Anal. 27:179–186.

Fox, R.H., W.P. Piekielek, and K.M. Macneal. 1994. Using a chlorophyll meter to predict nitrogen fertilizer needs of winter wheat. Commun. Soil Sci. Plant Anal. 25:171–181.

Fox, R.L., T. Kang, and D. Nangju. 1977. Sulfur requirements of cowpea and implications for production in the tropics. Agron. J. 69:201–205.

Franchini, J.C., C.B. Hoffmann-Campo, E. Torres, M. Miyazawa, and M.A. Pavan. 2003. Organic composition of green manure during growth and its effect on cation mobilization in an acid oxisol. Commun. Soil Sci. Plant Anal. 34:2045–2058.

Friesen, D.K. 1991. Fate and efficiency of sulfur fertilizer applied to food crops in West Africa. p. 59–68. In A.U. Mokwunye (ed.) Alleviating soil fertility constraints to increased crop production in West Africa. Kluwer Academic Publ., the Netherlands.

Fukushima, D. 1991. Recent progress of soybean protein foods: Chemistry, technology, and nutrition. Food Rev. Int. 7:323–351.

Ganeshamurthy, A.N. 1998. An evaluation of sulfur efficiency parameters in soybean and wheat cropping systems in relation to fertilizer sulfur on a Typic Haplustert. Aust. J. Agric. Res. 49:33–40.

Garcia, R.L., and J.J. Hanway. 1976. Foliar fertilization of soybeans during the seed-filling period. Agron. J. 68:653–657.

Gayler, K.R., and G.E. Sykes. 1985. Effect of nutritional stress on the storage proteins of soybeans. Plant Physiol. 78:582–585.

Hanway, J.J., E.J. Dunphy, G.L. Loberg, and R.M. Shibles. 1984. Dry weights and chemical composition of soybean plant parts throughout the growing season. J. Plant Nutr. 7:1453–1475.

Harada, J.J., S.J. Barker, and R.B. Goldberg. 1989. Soybean [beta]-conglycinin genes are clustered in several DNA regions and are regulated by transcriptional and posttranscriptional processes. Plant Cell 1:415–425.

Hitsuda, K., G.J. Sfredo, and D. Klepker. 2004. Diagnosis of sulfur deficiency in soybean using seeds. Soil Sci. Soc. Am. J. 68:1445–1451.

Hitsuda, K., M. Yamada, and D. Klepker. 2005. Sulfur requirement of eight crops at early stages of growth. Agron. J. 97:155–159.

Hue, N.V., G.R. Craddock, and F. Adams. 1986. Effect of organic acids on aluminum toxicity in subsoils. Soil Sci. Soc. Am. J. 50:28–34.

Jansson, H. 1995. Status of sulphur in soils and plants of thirty countries. World Soil Resources Reports 79. FAO, Rome.

Jones, M.B. 1963. Effect of sulfur applied and date of harvest on yield, sulfate sulfur concentration, and total sulfur uptake of five annual grassland species. Agron. J. 55:251–254.

Joshi, O.P., and S.D. Billore. 1998. Economic optima of sulphur fertilizer for soybean (*Glycine max*.). Indian J. Agric. Sci. 68:244–246.

Kamprath, E.J., and U.S. Jones. 1986. Plant response to sulfur in the southeastern United States. p. 323–343. *In* M.A. Tabatabai (ed.) Sulfur in agriculture. Agron. Monogr. 27. ASA, CSSA, and SSSA, Madison, WI.

Keren, R., I. Shainberg, H. Frenkel, and Y. Kalo. 1983. The effect of exchangeable sodium and gypsum on surface runoff from Loess soil. Soil Sci. Soc. Am. J. 47:1001–1004.

Knights, J.S., F.J. Zhao, B. Spiro, and S.P. McGrath. 2000. Long-term effects of land use and fertilizer treatments on sulfur cycling. J. Environ. Qual. 29:1867–1874.

Koshiyama, I. 1983. Storage proteins of soybean. p. 427–450. *In* W. Gottschalk and H.P. Müller (ed.) Seed proteins: Biochemistry, genetics, nutritive value. Martinus Nijhoff/Dr. W. Junk Publ., The Hague.

Krishnan, H.B. 2005. Engineering soybean for enhanced sulfur amino acid content. Crop Sci. 45:454–461.

Krishnan, H.B., J.O. Bennett, W. Kim, A.H. Krishnan, and T.P. Mawhinney. 2005. Nitrogen lowers the sulfur amino acid content of soybean (*Glycine max* [L.] Merr.) by regulating the accumulation of Bowman-Brik protease inhibitor. J. Agric. Food Chem. 53:6347–6354.

Kurihara, C.H. 2004. Demanda de nutrientes pela soja e diagnose de seu estado nutricional. (In Portuguese.) Ph.D. thesis, Federal Univ. Viçosa, Minas Gerais, Brazil.

Lantmann, A.F., and C. de Castro. 2004. Soil fertility and soybean nutrition management for maximum yield. (In Portuguese, with English abstract.) p. 1269–1274. *In* F. Moscardi et al. (ed.) VII World Soybean Research Conference, VI International Soybean Processing and Utilization Conference, III Congreso Brasileiro de Soja, Proceedings. 29 Feb.–5 Mar. 2004, Foz do Iguassu, PR, Brasil. Embrapa-Soja, Londrina, PR, Brazil.

Macnicol, P.K., and P.J. Randall. 1987. Changes in the levels of major sulfur metabolites and free amino-acids in pea cotyledons recovering from sulfur deficiency. Plant Physiol. 83:354–359.

Malavolta, E., G.C. Vitti, C.A. Rosolem, N.K. Fageria, and P.T.G. Guimarães. 1987. Sulphur responses of Brazilian crops. J. Plant Nutr. 10:2153–2158.

Malhi, S.S., Y. Gan, and J.P. Raney. 2007. Yield, seed quality, and sulfur uptake of Brassica oilseed crops in response to sulfur fertilization. Agron. J. 99:570–577.

Mallarino, A.P., M.U. Haq, D. Wittry, and M. Bermudez. 2001. Variation in soybean response to early season foliar fertilization among and within fields. Agron. J. 93:1220–1226.

McGrath, S.P., and R. Till. 1992. Sulphur uptake following foliar applications of elemental sulphur. J. Sci. Food Agric. 63:120.

Miyazawa, M., M.A. Pavan, and J.C. Franchini. 2002. Evaluation of plant residues on the mobility of surface applied lime. Braz. Arc. Biol. Technol. 45:251–256.

Miyazawa, M., M.A. Pavan, C.O. Ziglio, and J.C. Franchini. 2001. Reduction of exchangeable calcium and magnesium in soil with increasing pH. Braz. Arc. Biol. Technol. 44:149–153.

Munson, R.D., and W.L. Nelson. 1990. Principles and practices in plant analysis. p. 359–387. *In* R.L. Westerman (ed.) Soil testing and plant analysis. SSSA Book Ser. 3. 3rd ed. SSSA, Madison, WI.

Naeve, S.L., and R.M. Shibles. 2005. Distribution and mobilization of sulfur during soybean reproduction. Crop Sci. 45:2540–2551.

Nakasathien, S., D.W. Israel, R.F. Wilson, and P. Kwanyen. 2000. Regulation of seed protein concentration in soybean by supra-optimal nitrogen supply. Crop Sci. 40:1277–1284.

Nielsen, N.C., C.D. Dickinson, T.J. Cho, V.H. Thanh, B.J. Scallon, R.L. Fischer, T.L. Sims, G.N. Drews, and R.B. Goldberg. 1989. Characterization of the glycinin gene family in soybean. Plant Cell 1:313–328.

Nikiforova, V.J., J. Kopka, V. Tolstikov, O. Fiehn, L. Hopkins, M.J. Hawkesford, H. Hesse, and R. Hoefgen. 2005. Systems rebalancing of metabolism in response to sulfur deprivation, as revealed by metabolite analysis of Arabidopsis plants. Plant Physiol. 138:304–318.

Olsen's Agricultural Laboratory. 1997. Plant tissue interpretative guidelines. Olsen's Agricultural Laboratory, McCook, NE.

Paek, N.C., J. Imsande, R.C. Shoemaker, and R. Shibles. 1997. Nutritional control of soybean seed storage protein. Crop Sci. 37:498–503.

Paek, N.C., P.J. Sexton, S.L. Naeve, and R. Shibles. 2000. Differential accumulation of soybean seed storage protein subunits in response to sulfur and nitrogen nutrition sources. Plant Prod. Sci. 3:268–274.

Panthee, D.R., V.R. Pantalone, D.R. West, A.M. Saxton, and C.E. Sams. 2005. Quantitative trait loci for seed protein and oil concentration, and seed size in soybean. Crop Sci. 45:2015–2022.

Parker, M.B., and F.C. Boswell. 1980. Foliage injury, nutrient intake, and yield of soybeans as influenced by foliar fertilization. Agron. J. 72:110–113.

Peck, T.R. 1979. Plant analysis for production agriculture. p. 1–45. *In* Soil Plant Analysis Workshop, 7. Proceedings. Bridgetown, Barbados.

Peltonen, J., A. Virtanen, and E. Haggren. 1995. Using a chlorophyll meter to optimize nitrogen fertilizer application for intensively-managed small-grain cereals. J. Agron. Crop Sci. 174:309–318.

Radford, R.L., Jr., C. Chavengsaksongkram, and T. Hymowitz. 1977. Utilization of nitrogen to sulfur ratio for evaluating sulfur-containing amino acid concentrations in seed of *Glycine max* and *G. soja*. Crop Sci. 17:273–277.

Randall, P.J., J.R. Freney, and K. Spencer. 2003. Diagnosing sulfur deficiency in rice by grain analysis. Nutr. Cycling Agroecosyst. 65:211–219.

Randall, P.J., K. Spencer, and J.R. Freney. 1981. Sulfur and nitrogen fertilizer effects on wheat. I Concentrations of sulfur and nitrogen and the nitrogen to sulfur ratio in grain, in relation to the yield response. Aust. J. Agric. Res. 32:203–212.

Rasmussen, P.E., R.E. Raming, L.G. Ekin, and C.R. Rohde. 1977. Tissue analyses guidelines for diagnosing sulfur deficiency in white wheat. Plant Soil 46:153–163.

Reisenauer, H.M. 1975. Soil assays for the recognition of sulphur deficiency. p. 182–187. *In* K.D. McLachlan (ed.) Sulfur in Australasian agriculture. Sydney Univ. Press, Sydney.

Richart, A., M.C. Lana, L.R. Schulz, J.C. Bertoni, and A. de Lucca e Braccini. 2006. Phosphorous and sulfur availability for soybean in the presence of reactive natural phosphate, triple superphosphate and elemental sulfur. (In Portuguese, with English abstract.) R. Bras. Ci. Solo 30: 695–705.

Ritchey, K.D., and D.M.G. de Sousa. 1997. Use of gypsum in management of subsoil acidity in Oxisols. p. 165–178. *In* A.C. Moniz et al (ed.) Plant-soil interactions at low pH: Sustainable agriculture and forestry production, Proceedings of the fourth international symposium on plant-soil interactions at low pH, Belo Horizonte, Minas Gerais, Brazil. Bra. Soil Sci. Soc., Campinas, SP, Brazil.

Ritchey, K.D., D.M.G. Souza, E. Lobato, and O. Correa. 1981. Calcium leaching to increase rooting depth in a Brazilian savannah Oxisol. Agron. J. 72:40–44.

Rocha, M., and E. Malavolta. 1988. Perspectivas de demanda, comercialização e produção industrial de enxofre e micronutrientes para a agricultura. p. 277–309. (In Portuguese.) *In* C.M. Borkert and A.F. Lantmann (ed.) Enxofre e micronutrientes na agricultura brasileira. EMBRAPA-Soja, IAPAR, and Soci. Bras. Ci. Solo, Londrina, PR, Brazil.

Russell, J.S. 1963. Nitrogen content of wheat grain as an indication of potential yield response to nitrogen fertilizer. Aust. J. Exp. Agric. Anim. Husb. 3:319–325.

Saha, J.K., A.B. Singh, A.N. Ganeshamurthy, S. Kundu, and A.K. Biswas. 2001. Sulfur accumulation in Vertisols due to continuous gypsum application for six years and its effect on yield and biochemical constituents of soybean (Glycine max L. Merrill). J. Plant Nutr. Soil Sci. 164:317–320.

Sanchez, P.A. 1976. Properties and management of soils in the tropics. John Wiley & Sons, New York.

Sarker, S.K., M.A.H. Chowdhury, and H.M. Zakir. 2002. Sulphur and boron fertilization on yield quality and nutrition uptake by Bangladesh soybean-4. OnLine J. Biol. Sci. 2:729–733.

Sawyer, J.E., and D.W. Barker. 2002. Sulfur application to corn and soybean crops in Iowa. The 2002 Integrated Crop Management Conference. 4–5 Dec. 2002. Iowa State Univ., IA.

Scaife, A., and I.G. Burns. 1986. The sulphate-S/total S ratio in plants as an index of their sulphur status. Plant Soil 91:61–71.

Sesay, A., and R. Shibles. 1980. Mineral depletion and leaf senescence in soya bean as influenced by foliar nutrient application during seed filling. Ann. Bot. (London) 45:47–55.

Sexton, P.J., S.L. Naeve, N.C. Paek, and R. Shibles. 1998a. Sulfur availability, cotyledon nitrogen:sulfur ratio, and relative abundance of seed storage proteins of soybean. Crop Sci. 38:983–986.

Sexton, P.J., N.C. Paek, and R. Shibles. 1998b. Soybean sulfur and nitrogen balance under varying levels of available sulfur. Crop Sci. 38:975–982.

Sexton, P.J., N.C. Paek, and R.M. Shibles. 1998c. Effects of nitrogen source and timing of sulfur deficiency on seed yield and expression of 11S and 7S seed storage proteins of soybean. Field Crops Res. 59:1–8.

Sexton, P.J., W.D. Batchelor, and R. Shibles. 1997. Sulfur availability, rubisco content, and photosynthetic rate of soybean. Crop Sci. 37:1801–1806.

Sivapalan, S. 2003. Soybean responds to amelioration of sodic soils using polyacrylamides. Irrig. Aust. 18:45–47.

Smith, I.K., and A.L. Lang. 1988. Translocation of sulfate in soybean (*Glycine max* L. Merr). Plant Physiol. 86:798–802.

Spencer, K., and J.R. Freney. 1980. Assessing the sulfur status of field-grown wheat by plant analysis. Agron. J. 72.469–472.

Sumner, M.E. 1990. Gypsum as an ameliorant for the subsoil acidity syndrome. Publ. No. 01–024–090, Florida Inst. Phosphate Res., Bartow, FL.

Sumner, M.E. 1993. Gypsum and acid soils: The world scene. Adv. Agron. 51:1–32.

Sunarpi, and J.W. Anderson. 1996. Distribution and redistribution of sulfur supplied as [^{35}S]sulfate to roots during vegetative growth of soybean. Plant Physiol. 110: 1151–1157.

Sunarpi, and J.W. Anderson. 1997a. Allocation of S in generative growth of soybean. Plant Physiol. 114:687–693.

Sunarpi, and J.W. Anderson. 1997b. Inhibition of sulphur redistribution into new leaves of vegetative soybean by excision of the maturing leaf. Physiol. Plant 99: 538–545.

Sunarpi, and J.W. Anderson. 1998. Direct evidence for the involvement of the root in the redistribution of sulfur between leaves. J. Plant Nutr. 21: 1273–1286.

Sweeney, D.W., and G.V. Granade. 1993. Yield, nutrient, and soil sulfur response to ammonium sulfate fertilization of soybean cultivars. J. Plant Nutr. 16:1083–1098.

Tanaka, R.T., H.A.A. Mascarenhas, and C.M. Borkert. 1993. Nutrição mineral da soja. (In Portuguese.) p. 105–135. *In* N.E. Arantes and P.I. de Mello de Souza (ed.) Cultura da soja nos cerrados. Associação Brasileira para pesquisa da potassa e do fosfato, Piracicaba, SP, Brazil.

Tisdale, S.L., R.B. Reneau, Jr., and J.S. Platou. 1986. Atlas of sulfur deficiencies. p. 295–322. *In* M.A. Tabatabai (ed.) Sulfur in agriculture. Agron. Monogr. 27. ASA, CSSA, and SSSA, Madison, WI.

Toma, M., M.E. Sumner, G. Weeks, and M. Saigusa. 1999. Long-term effects of gypsum on crop yield and subsoil chemical properties. Soil Sci. Soc. Am. J. 39:891–895.

Ulrich, A. 1952. Physiological bases for assessing the nutritional requirements of plants. Annu. Rev. Plant Physiol. 3:207–228.

Van Bo, N., and N.T. Thi. 1997. Efficiency of S-containing fertilizer application for soybean grown in degraded soils in northern Vietnam. Sulphur Agric. 20:85–92.

Vasilas, B.L., J.O. Legg, and D.C. Wolf. 1980. Foliar fertilization of soybeans: Absorption and translocation of ^{15}N-labeled urea. Agron. J. 72:271–275.

Vitti, G.C., J.L. Favarin, L.A. Gallo, S.M. de Stefano Piedade, M.R.M. de Faria, and F. Cicarone. 2007. Foliar elementary sulfur assimilation by soybean. (In Portuguese, with English abstract.). Pesq. Agropec. Bras. 42:225–229.

Wang, T.L., C. Domoney, C.L. Hedley, R. Casey, and M.A. Grusak. 2003. Can we improve the nutritional quality of legumes seeds? Plant Physiol. 131:886–891.

Wihardjaka, A., Soeprapto, and C.P. Mamaril. 1999. Response of rainfed lowland rice and soybean to sulphur in light textured soils in Central Java (Indonesia). (In Indonesian, with English abstract.) Indone. J. Crop Sci. 14: 29–34.

Wilcox, J.R., and R.M. Shibles. 2001. Interrelationships among seed quality attributes in soybean. Crop Sci. 4:11–14.

Zhao, F.J., M.J. Hawkesford, A.G.S. Warrilow, S.P. McGrath, and D.T. Clarkson. 1996. Responses of two wheat varieties to sulphur addition and diagnosis of sulphur deficiency. Plant Soil 181:317–327.

9

Sulfur in a Fertilizer Program for Corn

George W. Rehm
University of Minnesota, St. Paul

John G. Clapp
Tessenderlo Kerley Inc., Greensboro, North Carolina

Abstract

The importance of sulfur in a fertilizer program for corn (*Zea mays* L.) has been the focus of diverse research projects conducted over several years at various locations. The requirement for this essential nutrient is not universal. Therefore, it would be desirable to have a useful and practical analytical procedure to predict the requirements. Analysis of either soil for sulfate-sulfur or plants for total sulfur concentration has not proven to be satisfactory in predicting the need for fertilizer sulfur. That challenge remains. Broadcast application of 28 kg sulfur ha^{-1} or use of 13.5 kg sulfur ha^{-1} in a band near the seed at planting has proven to be satisfactory in most production situations. Fertilizers containing sulfate-sulfur or thiosulfate are preferable. Measurements to define the effect of fertilizer sulfur on the quality of corn grown for forage have been inconsistent. Additional research on this topic is certainly justified. Because of (i) use of higher-analysis fertilizers, (ii) rapid adoption of conservation tillage, and (iii) reduced deposition of sulfate (SO_4) from atmospheric sources, a renewed evaluation of sulfur in fertilizer programs is justified.

For centuries, the importance of sulfur for crop production has been documented repeatedly. The most popular and early response citation was attributed to the application of gypsum by Ben Franklin. In the twentieth century, early documentation of the essential nature of this nutrient occurred in Minnesota in the 1920s followed by Nebraska in the 1950s. In other states, sulfur fertilization for crop production expanded in the 1960s continuing through the 1970s. The monograph edited by Tabatabai (1986) provides a comprehensive review of the earlier research with sulfur. Other chapters in this book will focus on the diverse role of sulfur in agriculture. This chapter will attempt to provide a summary of the more recent research that concentrates on the need for and management of sulfur in a fertilizer program for corn production.

Prediction of Sulfur Needs

Numerous research projects have focused on developing and/or evaluating measures of sulfur in soils and/or plants to predict the requirement for sulfur in a

---

Copyright © 2008. American Society of Agronomy, Crop Science Society of America, Soil Science Society of America, 677 S. Segoe Rd., Madison, WI 53711, USA. *Sulfur: A Missing Link between Soils, Crops, and Nutrition.* Agronomy Monograph 50.

fertilizer program for corn. Measurements of (i) extractable sulfate-sulfur in soil, (ii) total sulfur in plant material, and (iii) ratios of nitrogen and sulfur in plants are the indexes that have been used most frequently.

Soil Analysis

A measure of sulfate-sulfur extracted from soil (0–15 cm) has been used effectively in specific production environments to predict need of fertilizer sulfur for corn production (Fox et al., 1964; Kang and Osiname, 1976; Reneau, 1983; Rehm, 1984; Stecker et al., 1995). Extraction of soil with $Ca(H_2PO_4)_2$ (Hoeft et al., 1973) has been the most consistent analytical procedure used for predicting sulfur needs. This procedure extracts soluble as well as adsorbed sulfate-sulfur. In addition, this analytical procedure is easily adapted to routine analysis in public and commercial soil-testing laboratories and, thus, has become popular.

The reliability of this method for use in corn production over a wide geographic area with a wide variety of soil characteristics has been questioned because other major sources of plant-available sulfur are not accounted for by measures of only sulfate-sulfur in Illinois soil (Hoeft et al., 1985). In this Illinois study, this extractant reliably predicted a response to sulfur at four of five responding sites. However, it did not reliably predict those sites which did not respond to the use of fertilizer sulfur because 14 of the sites that did not respond to sulfur fertilizers had values of extractable sulfate-sulfur that were below levels normally considered to be deficient.

The inconsistency of analytical procedures that extract sulfate-sulfur from soil in predicting a need for fertilizer sulfur for corn can be attributed to the fact that 95 to 98% of the total sulfur in typical Midwest soils was found to be contained in soil-organic matter (Tabatabai and Bremner, 1972). There is some conversion of organic sulfur to sulfate-sulfur with mineralization. But the rate of mineralization is not constant for soils with diverse textures (O'Leary and Rehm, 1991). Therefore, analytical procedures measuring only sulfate-sulfur in solution or readily soluble sulfate-sulfur in soils have not been reliable in predicting fertilizer-sulfur need for corn. To be effective at predicting fertilizer requirements for sulfur, an analytical procedure should extract readily available sulfate-sulfur (soluble and absorbed) as well as the amount of organic sulfur that might become available. Development of a procedure that meets these requirements has been a challenge for researchers for many years. That challenge remains.

Plant Analysis

The concentration of total sulfur in plant tissue and the ratio of total nitrogen to total sulfur (N/S ratios) are measurements that have been used to evaluate the sulfur status of crops. In greenhouse studies, Kemper and Sorensen (1974) observed that an interaction of nitrogen and sulfur influenced corn growth on four different soils. For corn, successful use of N/S ratios in diagnosing sulfur deficiency has been reported by Stewart and Porter (1969), Terman et al. (1973), and Reneau (1983). These researchers reached the conclusion that an acceptable N/S ratio was 16:1. In contrast, Daigger and Fox (1971) Kang and Osiname (1976), and Rehm (1984) found that N/S ratios did not adequately predict the response of corn to fertilizer sulfur.

However, caution should be used in the interpretation of ratios (Sumner, 1978). When the N/S ratio is low, a response to nitrogen fertilization will be measured if nitrogen is limiting. A low N/S ratio may also be a consequence of uptake of large amounts of sulfur and, in this situation, a response to fertilizer sulfur may not be observed. When the N/S ratio is high, the reverse is true.

To be effective as a diagnostic tool, the N/S ratio should be calculated for corn plants early in the growing season. However, uptake of nitrogen and sulfur in the early stages of development is dynamically affected by hybrid and production environment. Therefore, it is doubtful if firm, adequate N/S ratios can be developed for young corn plants. The N/S ratio, if used, will probably be utilized as a crop-monitoring tool.

The ability of the N/S ratio as a diagnostic tool to describe the sulfur status of corn has been inconsistent and, therefore, has not gained wide acceptance. From a laboratory perspective, use of this ratio requires analysis for both nitrogen and sulfur, which increases cost.

A measurement of the sulfur concentration of corn ear leaf tissue has frequently been advocated as a diagnostic tool that can be used in corn production. The optimum range (1.0–3.0 g kg^{-1}) is wide (Jones and Eck, 1977) and can be influenced by a number of factors including hybrid and growing conditions. This optimum range is based on data collected from hybrids commonly grown in the 1960s and early 1970s. The optimum range may be different with modern hybrids and the new genetics. As development of new hybrids is a constant process, the optimum sulfur concentration may also be constantly changing. The use of the analysis for concentration of total sulfur in the ear leaf tissue has limited utility as a predictive tool. There has not been a research effort in recent years focused on defining a more specific critical level.

For plant analysis to be useful and effective as a predictive management tool, it is necessary to develop sufficiency standards for young plants. Because of the transient nature of sulfur in young plant tissue, definition of standards for rapidly growing plants early in the growing season has been a difficult task. Except for research described by Lockman (1969), there are few reports of the relationship between corn yield and sulfur concentration in young corn plants.

Management of Fertilizer Sulfur

There is general recognition that sulfur is not a universal requirement for optimum corn yields in all production environments. For situations where use of fertilizer sulfur is needed for optimum yield, there are several management decisions that are necessary. The major concern is usually selection of an economic rate. Then, there are questions about source, placement, time, and frequency of application.

The response of corn to fertilizer sulfur has been documented in several regions of the USA. However, responses to sulfur usage have been most prevalent on sandy soils of the Atlantic Coastal Plain (Bullock and Goodroad, 1989; Buttrey et al., 1986; Rabuffetti and Kamprath, 1977; Reneau, 1983) as well as some sandy soils in the North Central region (Hoeft et al., 1985; O'Leary and Rehm, 1990; Rehm, 1984). The majority of this research has focused on identifying the optimum rate of sulfur application while attempting to relate response to some measure of sulfate-sulfur in the surface soil (0 to 15 cm) or other soil properties.

Table 9–1. Yield of corn grown on a nonirrigated sandy soil as affected by rate of sulfur broadcast and incorporated before planting. Source is Fox et al. (1964).

Sulfur rate	Yield
kg ha^{-1}	Mg ha^{-1}†
0	3.32a
11	4.20b
22	4.08b
44	3.89b

† Treatment means followed by the same letter are not significantly different at the .05 confidence level.

Research reported by Reneau (1983) is typical of much of the response to sulfur observed on the Atlantic Coastal Plain. Optimum corn yields (90% of maximum) were achieved with the broadcast application of 18 to 28 kg sulfur ha^{-1} when sulfate-sulfur extracted by Ca(H$_2$PO$_4$)$_2$ was less than 2.3 mg kg^{-1}.

Early research with corn in Nebraska (Fox et al., 1964) showed that a broadcast rate of 11 kg sulfur ha^{-1} was adequate for nonirrigated corn grown on a soil with a loamy fine-sand texture (Table 9–1). The sulfate-sulfur extracted by Ca(H$_2$PO$_4$)$_2$ was less than 4 mg kg^{-1}. This response with sandy soils is typical of research with sulfur in the North Central USA. Subsequent Nebraska research (Rehm and Sorensen, 1977) showed that corn did not necessarily respond to sulfur fertilizers when grown on all sandy soils (Table 9–2). Two soil properties (organic-matter content, sulfate-sulfur extracted by Ca(H$_2$PO$_4$)$_2$) were considered to be important. Lower concentrations of extractable sulfate-sulfur were associated with the lower percentages of soil-organic matter.

Response of corn to fertilizer sulfur has been related to the rate of applied nitrogen. In studies reported by Rabuffetti and Kamprath (1977), there was no response to sulfur when the nitrogen rate was 56 and 112 kg^{-1}. At nitrogen rates of 168 and 224 kg ha^{-1}, however, a broadcast application of 44 kg sulfur ha^{-1} was adequate for optimum yield (Table 9–3).

However, the relationship of a response to fertilizer sulfur and nitrogen rate has not been consistent. A nitrogen-sulfur interaction was not measured at sites where corn responded to fertilizer sulfur in Minnesota (O'Leary and Rehm, 1990). A broadcast application of 22 kg sulfur ha^{-1} was optimum for nitrogen rates that

Table 9–2. Yield of corn grown on irrigated sandy soils as affected by application of broadcast sulfur. Source is Rehm and Sorensen (1977).

Year	County	S rate, kg ha^{-1}		Organic matter	Extractable SO$_4$–S
		0	22		
		Mg ha^{-1}†		g kg^{-1}	mg kg^{-1}‡
1973	Pierce	10.22a	11.48b	8.9	4.7
1973	Holt	10.35a	9.72a	14.0	5.6
1974	Pierce	8.97a	10.47b	9.7	3.8
1975	Pierce	7.03a	8.28b	8.5	2.7
1975	Holt	11.67a	11.79a	21.5	7.9

† Mean yields in each row followed by the same letter are not significantly different at the 0.05 confidence level.
‡ SO$_4$–S was extracted by the calcium phosphate procedure.

Table 9–3. Response of corn to rate of fertilizer sulfur in relation to rate of applied nitrogen. Source is Rabuffetti and Kamprath (1977).

Nitrogen rate	Sulfur rate, kg ha^{-1}			
	0	22	44	66
kg ha^{-1}	Mg ha^{-1}			
56	6.2	6.23	6.14	5.55
112	6.98	7.13	7.08	6.98
168	7.11	7.09	7.82	7.98
224	7.52	8.06	7.58	7.98

ranged from 0 to 252 kg ha^{-1}. Considering the results of research conducted on the Coastal Plain as well as in the Corn Belt, there is no requirement for the use of high rates of fertilizer sulfur. Broadcast applications of 22 to 44 kg sulfur ha^{-1} have been adequate for high yields. The absence of a consistent nitrogen-sulfur interaction leads to the conclusion that the rate of sulfur applied should not vary with the rate of nitrogen used. Although both nutrients are mobile in soils, the rate needed for each nutrient should be considered separately.

The research summarized in the previous paragraphs utilized broadcast applications of fertilizer sulfur. The placement of fertilizer sulfur in a band close to the seed at planting has also been evaluated (Rehm, 1984; 2005). A rate of 13.4 kg sulfur ha^{-1} placed in a band close to the seed at planting was adequate for optimum yield of irrigated corn grown on sandy soils at three sites in Nebraska. At sites where a response was measured, the organic-matter content was less than 20 g kg^{-1}, but there was no relationship of response to concentration of sulfate-sulfur extracted by the Ca(H$_2$PO$_4$)$_2$ procedure.

The results of the Minnesota trials were in close agreement with the conclusions reached in Nebraska (Rehm, 2005). There was a substantial response to the application of 21–0–0–24 applied in a band at planting for soils with a loamy-sand or sandy-loam texture (Table 9–4). Even though the amount of sulfate-sulfur extracted by the Ca(H$_2$PO$_4$)$_2$ procedure was similar, this type of response was not measured where soil texture was a silty-clay loam.

Table 9–4. Influence of 21–0–0–24 fertilizer applied in a band at planting on corn yield. Source is Rehm (2005).

Soil texture	Placement	Sulfur applied, kg ha^{-1}			
		0	6.7	13.4	20.2
		Mg ha^{-1}†			
Loamy fine sand	seed	10.52ab	11.15b	10.31a	10.76b
	starter	–	10.67a	10.55a	11.05b
Sandy loam	seed	8.90a	9.73b	10.25c	10.26c
	starter	–	9.52b	10.05c	10.73c
Silty clay loam	seed	11.64a	11.83a	12.54a	12.59a
	starter	–	11.56a	11.77a	11.61a

† Treatment means in each row followed by the same letter are not significantly different at the .05 confidence level. Yields for rates of applied sulfur should be compared to the control. Placements should be considered separately.

Table 9–5. Influence of source of fertilizer sulfur on yield of corn grown on irrigated sandy soils. Source is Rehm and Sorensen (1977).

Sulfur source	Yield (2-yr average)
	Mg ha^{-1}†
None	9.59a
Granular gypsum	11.16c
Potassium/magnesium sulfate	10.60b
Elemental sulfur plus bentonite clay	11.10c

† Treatment means followed by the same letter are not significantly different at the .05 confidence level.

Considering placement of fertilizer sulfur, the majority of the research conducted with soils that are responsive suggests that placement in a band near the seed is highly effective and rates can be reduced substantially when compared with a broadcast application. For example, a rate of 13 kg sulfur ha^{-1} in a band instead of a broadcast application of 28 kg sulfur ha^{-1} appears to be appropriate. The effectiveness of the banded application can be explained by the fact that sulfur is readily available in close proximity to a rapidly growing, but limited, root system of the young corn plant. In addition, release of sulfur from soil-organic matter early in the growing season is limited by lower-soil temperatures in most production environments and fertilizer sulfur close to the seed stimulates early growth.

Choice of source of fertilizer sulfur is an important decision for the corn producer when this nutrient is needed in a fertilizer program. Considering the dry sources, sulfur present in the form of sulfate-sulfur or elemental sulfur are the dominant choices in today's agriculture. When broadcast and incorporated before planting, use of sulfate-sulfur rather than elemental sulfur has produced yields slightly higher (Table 9–5). The fact that elemental sulfur must first be oxidized to sulfate-sulfur before it can be utilized is the most plausible explanation for the lower yields. The microbial process necessary to complete the oxidation reaction requires warm soil temperatures and time. The typical cooler temperatures at planting characteristic of most production environments could delay the early-season availability of sulfur supplied as elemental sulfur thereby restricting early-season growth and subsequent yield.

For corn producers who choose to use fluid fertilizers, the choices are either ammonium thiosulfate (12–0–0–26S) or potassium thiosulfate (0–0–21–17S). The thiosulfate ion is not absorbed by the corn plant. Studies show that thiosulfate-sulfur is rapidly converted to sulfate-sulfur. Laboratory studies have shown that oxidation of thiosulfate in potassium thiosulfate (0–0–21–17S) is nearly complete after 2 or 3 wk at temperatures of 5, 15, and 25°C (Goos and Johnson, 2001). The application of 12–0–0–26S in contact with the seed at planting can have a negative impact on corn emergence (Rehm, 2005). The use of 0–0–21–17S in this way has not been fully evaluated. When using either of these fluid materials in a band at planting, adjusting placement so that there is some soil between seed and fertilizer would be a good management practice. This is especially true for soils with a loamy-sand or sandy-loam texture.

As an anion, sulfate-sulfur, analogous to nitrate-nitrogen, is subject to leaching. This is especially true with sandy soils. Guidelines for nitrogen management on sandy soils call for split applications of fertilizer nitrogen. Logic suggests that split applications of fertilizer sulfur might also be appropriate. Minnesota trials,

Table 9–6. Corn yield as affected by time and frequency of sulfur application. Source is Rehm (1993).

| \multicolumn{4}{c}{Sulfur applied} | \multicolumn{2}{c}{Location} |
ST	PE	8 Leaf	TA‡	Benton Co.	Wadena Co.
		% of total			Mg ha^{-1}†
0	0	0	0	8.54a	7.57a
100	0	0	0	9.67b	8.92b
50	50	0	0	9.51b	9.10b
50	0	50	0	9.36b	9.20b
50	0	0	50	9.66b	9.26b
33.3	33.3	0	33.5	9.41b	9.21b
0	100	0	0	9.61b	8.99b

† Treatment means in each column followed by the same letter are not significantly different at the 0.05 confidence level.

‡ ST = starter band at planting; PE broadcast after planting but before emergence; 8 leaf = broadcast on soil surface at 8-leaf stage of development; TA = broadcast at time of tassel ling.

however, have shown that split applications did not improve yields (Table 9–6). With a loamy fine-sand texture at each site and low concentration of sulfate-sulfur extracted by $Ca(H_2PO_4)_2$, a response to fertilizer sulfur was expected and measured (Rehm, 1993). The results of this research led to the conclusion that, although there are several opportunities to apply sulfur to irrigated corn, application of fertilizer sulfur in a band at planting is adequate for optimum production.

Although the practice has not been a focus of any research efforts, it is possible to inject 12–0–0–26S or 0–0–21–17S into irrigation water. The combination of 28–0–0 and 12–0–0–26S injected early in the growing season may prove to be beneficial for irrigated corn grown on sandy soils.

Sulfur and Quality of Corn Silage

The previous discussion has focused on the impact of sulfur fertilization on corn grain production. Since sulfur is a component of three essential amino acids, it is logical to assume that application of this nutrient might also have some effect on the quality of corn when used for silage. However, there have been very few research projects with the objective of evaluating the impact of sulfur fertilization on corn grown as a forage crop.

Forage-feeding quality has been correlated to plant-sulfur concentration. Desirable concentrations of sulfur in forage has been established for ruminants (National Research Council, 1978, 1984). In feeding trials with lambs, Buttrey et al. (1986) reported that sulfur fertilization improved the digestibility of ensiled corn. Using in vitro analysis, O'Leary and Rehm (1990) reported that nitrogen fertilization increased the protein percentage of corn harvested for silage. Sulfur fertilization, however, had no significant effect on this parameter as well as the percentage of acid detergent fiber (ADF) and neutral detergent fiber (NDF). These parameters are frequently used as measures of forage quality. Results from this limited research are not consistent. In other research with forages, there is usually mutual agreement between in vivo and in vitro measures of digestibility. If corn

Renewed Evaluation of Sulfur Fertilization

In the past 3 to 5 yr, there has been a renewed interest in sulfur fertilization of corn. Much of this interest is stimulated by reduced deposition of sulfate (SO_4) from atmospheric sources (Table 9–7). Many of the measurements reported were taken from urban areas. Smaller depositions were recorded in rural areas. In the past, the atmosphere, soil organic matter, and phosphate fertilizers were primary sources of sulfur. There are miniscule concentrations of sulfur in modern fertilizers and the sulfur derived from atmospheric sources is reduced. Therefore, the renewed interest in sulfur fertilization is appropriate.

The continued adoption of conservation tillage is further justification for a renewed evaluation of sulfur fertilization of corn. With reduced soil disturbance, mineralization of soil-organic matter is reduced, thereby decreasing the amount of sulfur supplied by soil-organic matter. As a consequence, the requirement for sulfur in a fertilizer program may become more prevalent in conservation tillage systems. In Minnesota research, corn grown on a silt-loam soil with a low-organic matter content in a ridge-till planting system responded to the application of fertilizer sulfur (Rehm, 2005). The ridge-till system of planting involves planting followed by one cultivation. Therefore, soil disturbance and subsequent mineralization is minimized.

Recent evaluations of the use of sulfur fertilizers for corn production have been mixed. Clapp (2002) provided a summary for much of the recent research. Sulfur fertilization has improved corn yields in Kansas trials (Lamond et al., 2002). In Iowa, however, responses to sulfur fertilization have been inconsistent in recent years (Sawyer and Barker, 2002).

With the reduction in atmospheric deposition and increasing use of conservation tillage, there is a need for expanded research to evaluate the response of corn to fertilizer sulfur.

Table 9–7. Annual sulfur deposition at selected locations. Source is Clapp (2002), National Atmospheric Deposition Program.

Locations	Initial year	2000	Change
	kg SO_4 ha^{-1}		%
Davis, CA (1979) †	4.64	1.60	–66
Pullman, WA (1986)	2.12	1.46	–31
Muleshoe, TX (1986)	9.8	4.00	–59
Manhattan, KS (1983)	11.50	7.16	–38
Lamberton, MN (1979)	17.70	6.14	–65
Kalamazoo, MI (1980)	30.87	7.67	–75
Dixon Springs, IL (1979)	38.66	17.19	–56
Wooster, OH (1979)	34.80	20.07	–42
Quincy, FL (1984)	13.58	9.93	–27
Clinton, NC (1979)	21.48	16.51	–23
Penn State, PA (1984)	37.72	23.75	–37

† The year listed in parentheses is the year that initial measurements were taken.

References

Bullock, D.G., and L.L. Goodroad. 1989. Effect of sulfur rate, application method and source on yield and mineral content of corn. Commun. Soil Sci. Plant Anal. 20:1209–1217.

Buttrey, S.A., V.G. Allen, J.P. Fontenot, and R.B. Reneau, Jr. 1986. Effect of sulphur fertilization on chemical composition, ensiling characteristics, and utilization of corn silage by lambs. J. Anim. Sci. 62:1236–1245.

Clapp, J.G. 2002. Addition of sulfur to starter fertilizers. p. 39–43. *In* Proceedings of 2002 Fluid Forum. 17–19 Feb. 2002. Scottsdale, AZ.

Daigger, L.A., and R.L. Fox. 1971. Nitrogen and sulfur nutrition of sweet corn in relation to fertilization and water composition. Agron. J. 63:729–730.

Fox, R.L., R.A. Olson, and H.F. Rhoades. 1964. Evaluating the sulfur status of soils by plant and soil tests. Soil Sci. Soc. Am. Proc. 28:243–256.

Goos, R.J., and B.E. Johnson. 2001. Thiosulfate oxidation by three soils as influenced by temperature. Commun. Soil Sci. Plant Anal. 32:2841–2849.

Hoeft, R.G., J.E. Sawyer, R.M. Vanden Huevel, M.A. Schmitt, and G.S. Brinkman. 1985. Corn response to sulfur in Illinois soils. J. Fert. Issues 2:95–104.

Hoeft, R.G., L.M. Walsh, and D.R. Keeney. 1973. Evaluation of various extractants for available soil sulfur. Soil Sci. Soc. Am. Proc. 37:401–404.

Jones, J.B., and H.V. Eck. 1977. Plant analysis as an aid in fertilizing corn and grain sorghum. p. 349–364. *In* L.M. Walsh and J.D. Beaton (ed.) Soil testing and plant analysis. SSSA and ASA, Madison, WI.

Kang, B.T., and O.A. Osiname. 1976. Sulphur response of maize in western Nigeria. Agron. J. 68:333–336.

Kemper, D.W., and R.C. Sorensen. 1974. Comparative effects of nitrogen and sulfur fertilization and liming on three crops grown on four soils. Agron. J. 66:92–97.

Lamond, R.E., W.B. Gordon, B.J. Niehius, and C.J. Olsen. 2002. Flexibility in starter fertilizer application methods for conservation tillage crop production. p. 172–179. *In* Proceedings of 2002 Fluid Forum. 17–19 Feb. 2002. Scottsdale, AZ.

Lockman, R.B. 1969. Relationships between corn yields and nutrient concentration in seedling whole-plant samples. p. 97. *In* Agron. Abstr. ASA, Madison, WI.

National Research Council. 1978. Nutrient requirements of domestic animals. Nutrient requirements of dairy cattle. 5th ed. National Academy of Science, Washington, DC.

National Research Council. 1984. Nutrient requirements of domestic animals. Nutrient requirements of beef cattle. 6th ed. National Academy of Science, Washington, DC.

O'Leary, M.J., and G.W. Rehm. 1990. Nitrogen and sulfur effects on the yield and quality of corn grown for grain and silage. J. Prod. Agric. 3:135–140.

O'Leary, M.J., and G.W. Rehm. 1991. Evaluation of some soil and plant analysis procedures as predictors of the need for sulfur for corn production. Commun. Soil Sci. Plant Anal. 22:87–98.

Rabuffetti, A., and E.J. Kamprath. 1977. Yield, nitrogen and sulfur content of corn as affected by nitrogen and sulfur fertilization on Coastal Plain soils. Agron. J. 69:785–788.

Rehm, G.W. 1984. Source and rate of sulfur for corn production. J. Fert. Issues 1:99–103.

Rehm, G.W. 1993. Timing sulfur applications for corn (*Zea mays*, L.) production on irrigated sandy soil. Commun. Soil Sci. Plant Anal. 24:285–294.

Rehm, G.W. 2005. Sulfur management for corn grown with conservation tillage. Soil Sci. Soc. Am. J. 69:709–717.

Rehm, G.W., and R.C. Sorensen. 1977. Sulfur can boost corn yields on sandy soils. Nebraska Quart. Summer, 1977. p. 23–24.

Reneau, R.B., Jr. 1983. Corn response to sulfur application in Coastal Plain soils. Agron. J. 75:1036–1040.

Sawyer, J.A., and D.W. Barker. 2002. Sulfur application to corn and soybean in Iowa. *In* Proceedings of the Integrated Crop Management Conference. 4–5 Dec. 2002. Ames, IA.

Stecker, J.A., D.D. Buchholz, and P.W. Tracy. 1995. Fertilizer sulfur effects on corn yield plant sulfur concentration. J. Prod. Agric. 8:61–65.

Stewart, B.A., and L.K. Porter. 1969. Nitrogen-sulfur relationships in wheat (*Triticum aestivum* L.). Agron. J. 61:267–276.

Sumner, M.E. 1978. Interpretation of nutrient ratios in plant tissue. Commun. Soil Sci. Plant Anal. 9:335–345.

Tabatabai, M.A. (ed.). 1986. Sulfur in agriculture. Agron. Monogr. 27. ASA, CSSA, and SSSA, Madison, WI.

Tabatabai, M.A., and J.M. Bremner. 1972. Distribution of total and available sulfur in selected soils and soil profiles. Agron. J. 64:40–44.

Terman, G.L., S.E. Allen, and P.M. Giordano. 1973. Dry matter yield, nitrogen and sulfur concentration relationship and ratios in young corn plants. Agron. J. 65:633–636.

10

Sulfur Nutrition and Wheat Quality

Hamid A. Naeem
University of Manitoba, Winnipeg, Canada

Abstract

Global sulfur deficiency in agricultural lands is the result of gradual replacement of sulfur-containing fertilizers with high purity nitrogen fertilizers, introduction of high yielding crop cultivars, and since 1970s strict restrictions on sulfur dioxide emissions to reduce greenhouse gases. Wheat (*Triticum aestivum* L.) has a low sulfur requirement, and an addition of 15 to 20 kg ha^{-1} is usually sufficient for optimum yield and quality. Grain sulfur content of ≥1.2 mg g^{-1} and a nitrogen/sulfur ratio of ≤16:1 appears to be critical for optimum quality characteristics. The endosperm storage proteins or prolamins determine the end-use quality of wheat flour. They are classified on the basis of their solubility in different solvents, structure, and molecular size. There are three main groups based on the structure, namely sulfur-rich, sulfur-poor, and high molecular weight glutenin subunits (HMW-GS). Sulfur deficiency does not affect total grain protein but does affect accumulation of different protein groups during grain development. Amounts of sulfur-poor proteins such as ω-gliadins and HMW-GS increase at the expense of sulfur-rich protein groups such as α-, β-, and γ-gliadins and low molecular weight glutenin subunits (LMW-GS). HMW- and LMW-GS are polymeric proteins held together by disulfide bonds. Sulfur deficiency reduces the accumulation of LMW-GS, total amount of glutenins, and changes HMW-GS/LMW-GS ratio, thus a shift in molecular weight distribution of gluten proteins toward higher molecular weight. These quantitative changes in protein composition modify flour functional properties. Dough made from sulfur-deficient flour is stronger, has increased mixing requirements, reduced extensibility, and loaf volume. Sulfur deficient grains show higher pearling index. Nonprotein nitrogen significantly increases because of accumulation of excessive amounts of free asparagine and glutamine. Free asparagine serves as a precursor in the formation of acrylamide, a carcinogenic neurotoxin, during baking, roasting, and frying because of the Maillard reaction. Inadequate sulfur supply promotes accumulation of increased levels of asparagine thus posing a health risk. Recent studies have shown that accumulation of free asparagine in wheat grain depends on wheat cultivar, soil type, and growing season weather conditions. Sprouting significantly increases quantity of free asparagine in the grain. Ways to reduce free asparagine in the raw materials and formation of acrylamide in the processed food are summarized.

Plants absorb sulfur through roots as sulfate. It is then reduced to sulfite before incorporation in the amino acids, enzymes, coenzymes, and many other physiological functions. Supplies of soil sulfate are supplemented at varying rates during the growing season by accretions from the atmosphere, fertilizers, pesticides, and soil organic matter degradation. Plants also absorb sulfur directly from

---

Copyright © 2008. American Society of Agronomy, Crop Science Society of America, Soil Science Society of America, 677 S. Segoe Rd., Madison, WI 53711, USA. *Sulfur: A Missing Link between Soils, Crops, and Nutrition.* Agronomy Monograph 50.

the atmosphere. Thus, sulfur deficiency was not a limiting factor in the industrialized world as atmospheric deposition was enough to meet the crop requirements. It was during the late 1960s and early 1970s when widespread sulfur deficiency was observed in Europe and North America. This was due to the use of fertilizers that were low or contained no sulfur but were high in nitrogen such as urea and di-ammonium phosphate (Ceccotti et al., 1997), introduction of high yielding crop cultivars, and decrease in sulfur dioxide emissions because of environmental concerns. For example, the use of ammonium sulfate (percentage of total nitrogen use) to supplement nitrogen in Western Europe dropped from 7.2% in 1973 to 3% in 1991. Total nitrogen consumption worldwide doubled between 1974 and 1990, whereas total sulfur consumption remained static at about 10 million megagrams during the same period (Duke and Reisenauer, 1986). The yield of crops has almost doubled in the past two decades, resulting in excessive removal of nutrients including sulfur. The sulfur deficit is projected to be more than 13 million megagrams by the year 2020 (Messick and Brey, 2001, quoted by Pasricha and Abrol, 2003).

Wheat is widely grown and is the staple food for a vast majority of people across the globe. Analysis of a large number of wheat grain samples collected from the main wheat growing areas in Britain during 1981 and 1982 showed no signs of sulfur deficiency. Mean grain sulfur content of these samples was 1.72 mg g^{-1} (Byers et al., 1987b). However, the average grain sulfur content of samples collected in 1992 and 1993 was significantly lower 1.35 mg g^{-1} (Zhao et al., 1995). This decrease in grain sulfur increased nitrogen/sulfur (N/S) ratio from 12:1 in 1981–1982 to 16:1 in 1992–1993. Grain sulfur of 1.2 mg g^{-1} and an N/S ratio of 17:1 were suggested to be the threshold limits for sulfur deficiency in wheat (Randall et al., 1981). With these criteria, only a small number of samples were sulfur deficient in 1992–1993. Later field experiments indicated that N/S ratio of 17:1 was probably a good indicator of optimum sulfur supply for wheat yield. However, lower N/S ratio is required to affect the dough properties and bread making quality. Grain N/S ratio of 16:1 was suggested as the lower limit for optimum dough and bread making properties (Zhao et al., 1999c). Naeem and MacRitchie (2003) recently reviewed diagnosis of sulfur deficiency through chemical testing, its amelioration, and its effects on wheat production and dough quality.

A number of early studies have indicated that sulfur availability affected grain protein composition (Byers et al., 1987a; Moss et al., 1981; Wrigley et al., 1984). These studies also indicated that plants accumulated excessive amounts of free asparagine and glutamine when either sulfur supply was inadequate or excessive nitrogen was available. The significance of this observation remained unclear until the discovery that free asparagine served as a precursor in the formation of acrylamide (AA) during baking and the Maillard reaction (Tareke et al., 2002; Mottram et al., 2002). This finding stirred the food industry and the food chemists around the world. Thus, research interest in sulfur as a plant nutrient got a new thrust. Therefore, this chapter will focus on the effects of sulfur availability on wheat quality and its relevance to acrylamide formation.

Effect of Sulfur and Nitrogen Fertilizer on Wheat Yield

Crop yield potential is defined as "the yield of a cultivar when grown in environments to which it is adapted, nutrient and water non-limiting, and with pests,

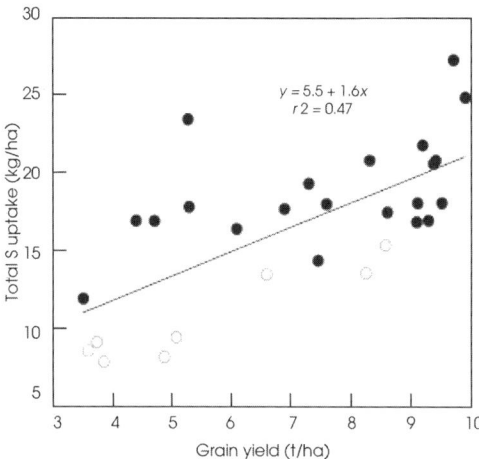

Fig. 10–1. Sulfate uptake of winter wheat at crop maturity in relation to grain yield in the UK. The designation of sulfur-deficient (○) or sulfur-sufficient (●) crop was defined by whether a significant yield increase was obtained in response to sulfur addition [Reprinted from Zhao et al. (1999a), with permission of Elsevier].

diseases, weeds, lodging and other stresses effectively controlled" (Evans and Fischer, 1999). Thus, sulfur requirement of wheat would be the minimum amount of sulfur required to achieve maximum yield (Randall et al., 1981). Sulfur requirement is even higher to achieve the optimum dough characteristics (Zhao et al., 1999a). Bread wheat cultivars are high in protein with relatively higher proportion of larger polymers, which are held together by disulfide bonds and therefore necessitate more sulfur supply. Soil type and nutrient status also affect sulfur requirement. As illustrated in Fig. 10–1, 15 to 20 kg ha^{-1} of sulfur are required for optimum wheat grain yield (Zhao et al., 1999c). However, sulfur requirement could be as high as 60 kg ha^{-1} under sulfur-deficient conditions for highly calcareous soils (Sakal et al., 1999, 2000).

A close link between sulfur and nitrogen metabolism has been known for many years. Deprivation of either sulfur or nitrogen leads to disorder in metabolism of the other (Prosser et al., 2001). Byers and Bolton (1979) conducted the classic study on the relationship of sulfur and nitrogen availability with four nitrogen and three sulfur levels. The wheat plants were grown in growth chambers where filtered air was used that contained 50% less sulfur compared with the atmospheric sulfur. Grain yield losses of 4.47- and 14-fold were reported in the absence of any sulfur supplementation in the first and second experiments when nitrogen supply was increased from 60 to 240 mg pot^{-1}. An increase in nitrogen supply from 60 to 180 mg pot^{-1} was associated with an increase in N/S ratio from 15.3 to 41. In greenhouse experiments, yield gains of more than 6-fold were also observed at a fixed nitrogen supply and a range of sulfur supplementation (Byers et al., 1987a). It is well known that sulfur becomes a limiting factor more quickly in greenhouses than in a field. Thus, yield increases and high N/S ratios would be expected in such experiments. A wide range of yield responses to sulfur fertilization was also observed under field conditions. Beaton and Soper (1986) reported yield gains from 16 to 345% at different locations in Canada. Withers et al. (1997) reported yield increases over 1 Mg (Fig. 10–2). These yield increases indicate extreme sulfur deficiency and do not reflect the average yield responses.

Wang et al. (2003) studied genomic response on the effect of low levels of nitrate in *Arabidopsis thaliana* (L.) Heynh. In addition to the expected nitrate-

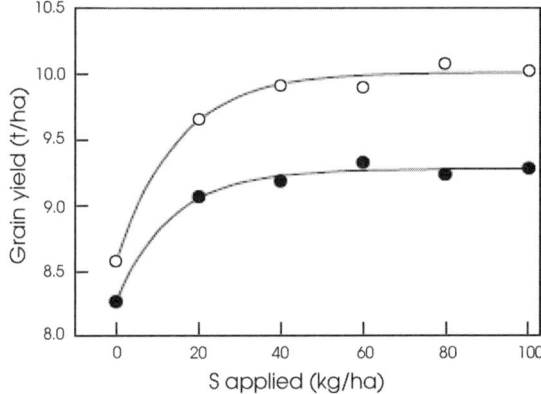

Fig. 10–2. Response of wheat yield to the addition of sulfur at Bridgets, UK. Two nitrogen levels were used: 180 kg N/ha (○) and 230 kg N/ha (●) [Reprinted from Withers et al. (1997) with permission from Elsevier].

responsive genes (e.g., those encoding for nitrate transporters, nitrate, and nitrite reductases, etc.) and a plethora of other genes, key sulfur genes were also influenced by nitrate level. These included two putative sulfate transporter genes and a 5-adenylylphosphosulfate reductase gene, which is part of the sulfate assimilation pathway.

Effect of Sulfur Availability on Wheat Protein Composition

Classification of Wheat Proteins

Wheat proteins are classified in a number of different ways depending on various physicochemical properties. The earliest classification was based on differential solubility characteristics (Osborne, 1907). The sequential extraction separates albumins–globulins (soluble in salt solution), prolamins (soluble in 70% aqueous ethanol), and glutenins (insoluble in either salt or 70% ethanol). Type of alcohol, composition of the extraction solvent, and extraction temperature affect extractability (Byers et al., 1983). The heterogeneous nature and overlapping composition of these fractions are the major drawbacks (Singh et al., 1990). A new classification is based on molecular size of the different groups of proteins (MacRitchie, 1992) that divides wheat proteins into three main groups on the basis of the size, albumin–globulins (20,000–30,000), gliadins (30,000–70,000), and glutenins (100,000 to many millions). Gliadins are single chain proteins further divided into four classes on the basis of mobility in acid–polyacrylamide gel electrophoresis. These four groups, in decreasing molecular size with some overlap, are omega (ω)-, gamma (γ)-, beta (β)- and alpha- (α) gliadins. The α- and β-gliadins are now believed to belong to essentially the same group. The glutenins are the polymeric proteins made up of subunits held together by inter- and intrachain disulphide (S–S) bonds, and divided into two main classes. The high molecular weight glutenin subunits (HMW-GS) have molecular weight in the range of 80,000 to 120,000, and low molecular weight glutenin subunits (LMW-GS) have molecular weight in the range of 40,000 to 55,000. This classification is particularly useful in explaining structure–function relationships.

Another classification is based on similarity in chemical structure and amino acid composition of different proteins, which might reflect their common genetic origin (Shewry et al., 1986). The main groups of prolamins are sulfur-rich, sulfur-poor, and high molecular weight (HMW) prolamins. The sulfur-poor prolamins are ω-gliadins and do not contain any cysteine. The sulfur-rich prolamins consist of LMW-GS, α-, β-, and γ-gliadins. The HMW prolamins correspond to HMW-GS. Each of the above classification has its own merits. The last classification based on structural similarities is more relevant in the current context to explain the effects of sulfur on protein composition.

Effect of Sulfur Availability on Gliadins

Sulfur availability influences synthesis and accumulation of endosperm storage proteins during grain development. Omega-gliadins significantly increased in proportion at the expense of other gliadins when sulfur supply was inadequate (Wrigley et al., 1980). Total flour protein extracts from Australian wheat cv. Olympic grown under different sulfur fertilization were analyzed by two-dimensional electrophoresis. Data revealed that sulfur supply modified protein composition quantitatively and not qualitatively. Neither new protein spots were found nor was any protein missing because of sulfur deficiency. Chemical properties such as isoelectric focusing point, charge, or size remained unaffected (Wrigley et al., 1984). Nonprotein nitrogen as a percentage of total nitrogen in grain is usually low (4%) under normal sulfur supply. This increased to 30% when grown under restricted sulfur supply (Byers et al., 1987a).

Effects of sulfur supplementation on wheat protein composition were recently quantified by means of reversed-phase high performance liquid chromatography (RP-HPLC) from German wheat cv. Star (Wieser et al., 2004). Wheat was grown in pots with two different soils and four different levels of sulfur supplementation. Sulfur availability had a relatively small effect on quantity of total gliadins (Table 10–1). However, significant quantitative changes in different gliadin types were evident (Fig. 10–3). The biggest increase was observed in quantities of ω5-type (Soil–I, 117%; Soil–II, 124%) and ω1, 2-type gliadins (Soil–I, 135%; Soil–II, 123%). A significant decrease in quantities of γ-gliadins was associated with sulfur deficiency, while α-gliadins were least affected because of variation in sulfur availability.

Table 10–1. Crude protein content and quantities/proportions of gluten proteins of whole meal flour.†

Flour	Crude protein‡	Gliadins					Glutenins				Ratio	
		Total	ω5	ω1,2	α	γ	Total	ωb	HMW	LMW	Gli/Glu	HMW/LMW
	%					AU/mg						
B1–S4	13.4	889	53	107	374	355	459	10	144	305	1.94	0.47
B1–S2	13.5	889	72	154	389	274	456	12	164	280	1.95	0.59
B1–S1	13.0	864	103	226	370	165	407	17	186	204	2.12	0.91
B1–S0	13.3	791	115	251	305	120	344	19	171	154	2.30	1.11
B2–S4	14.6	877	54	122	381	320	583	12	141	430	1.5	0.33
B2–S2	14.8	878	61	141	387	289	590	14	160	416	1.49	0.38
B2–S1	14.2	924	87	187	416	234	543	15	182	346	1.70	0.53
B2–S0	14.4	912	121	272	373	146	476	22	201	253	1.92	0.79

† Reprinted from Wieser et al. (2004) with permission.
‡ Crude protein, N×5.7, %.

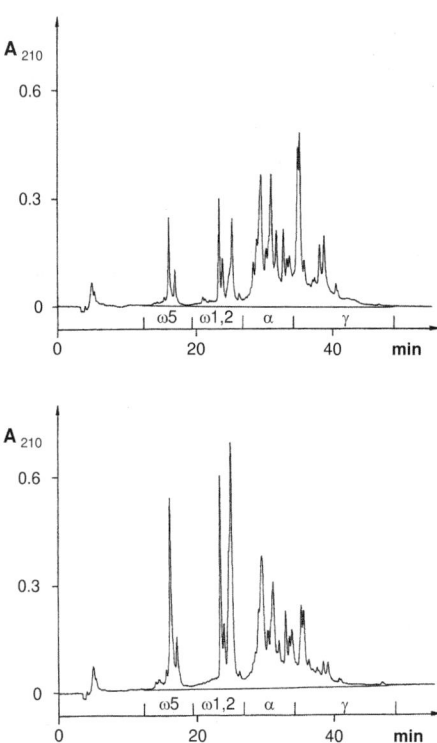

Fig. 10–3. RP-HPLC of gliadins from whole meal flours. Top panel: Sulfur-sufficient; Bottom panel: Sulfur-deficient [Reprinted from Wieser et al. (2004) with permission from Elsevier].

Effect of Sulfur Availability on LMW-GS and HMW-GS

Quantitatively, LMW-GS are two- to threefold more than HMW-GS in normal flour and thus is a major contributor to the gluten content. The electrophoretic analysis of flour samples from cv. Olympic indicated a close relationship between sulfur availability and quantities of LMW- and HMW-GS (Wrigley et al., 1984). Analysis of wheat proteins by size-exclusion HPLC (SE-HPLC) indicated significant decrease in quantity of total wheat glutenin grown under sulfur deficiency (MacRitchie and Gupta, 1993). This was due to a drastic decrease in the absolute quantities of LMW-GS (– 45%), while some quantitative increase in HMW-GS (+19%) was observed (Table 10–1). Changes in HMW- and LMW-GS with sulfur supply for cv. Star are illustrated in Fig. 10–4 (Wieser et al., 2004). Sulfur deficiency alters molecular weight distribution of gluten by changing the HMW-GS to LMW-GS ratio. This ratio plays a significant role in determining dough properties. A 1% change in percentage Peak 1 area of SE-HPLC due to HMW-GS would bring a change of 27 Brabender units (BU) in R_{max}, 0.5 cm in extensibility, and 0.6 min in mixograph dough development time (MDDT). A similar change due to LMW-GS would be associated with a change of 6 BU in R_{max}, 0.2 cm in extensibility and 0.1 min in MDDT (MacRitchie, 1992).

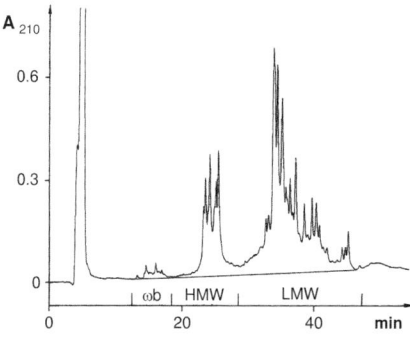

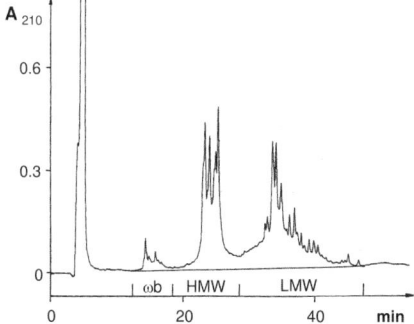

Fig. 10–4. RP-HPLC of glutenin subunits from whole meal flours. Top panel: Sulfur-sufficient; Bottom panel: Sulfur-deficient [Reprinted from Wieser et al. (2004) with permission from Elsevier].

Effect of Sulfur on Rheological Properties and Baking Performance

The distinctive property of wheat flour dough is its ability to be baked into leavened bread. This feature emanates from its proteins, which form a continuous network called gluten. The polymeric glutenins account for elasticity and monomeric gliadins for dough extensibility and viscosity. An inimitable balance of glutenins and gliadins is required for a vast majority of wheat products. For example, highly extensible dough is required for biscuits or cookies; more elastic and moderately extensible dough is suitable for bread making, whereas dough with intermediate properties performs better for the manufacture of noodles. It is well known that disulphide bonds (S–S) formed between the sulfhydryl (-SH) groups of cysteine residues play a vital role in determining the structural and functional characteristics of wheat proteins. Intrachain or no S–S bonds are formed within monomeric proteins. Interchain S–S bonds are formed within polymeric proteins that make glutenin proteins the largest natural polymers. These polymers contribute to the elasticity of the dough. Moss et al. (1981) conducted a comprehensive study on the influence of sulfur deficiency on dough rheology. Flours from three wheat cultivars, namely Shortim, a hard bread wheat, Olympic, a multipurpose soft wheat, and Egret, a biscuit wheat, were grown

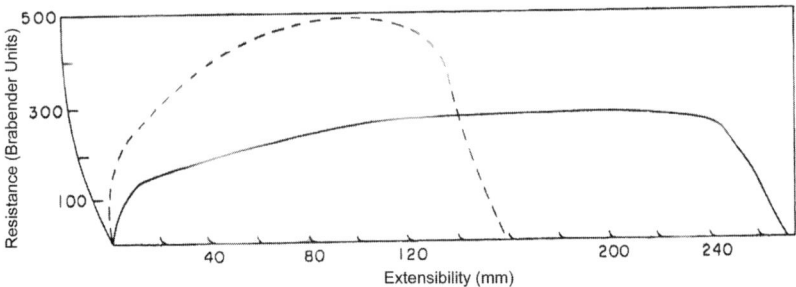

Fig. 10–5. Extensograph traces for the flour samples of different sulfur contents. The control flour (—) had a 0.146% sulfur, 1.82% nitrogen and N:S ratio 12.4; its extensibility (length) was 270 mm, resistance at 5 cm was 175 BU, and resistance ratio was 1.06. The low sulfur flour (- - -) had 0.089% sulfur, 1.72% nitrogen, and N/S ratio of 19.33; extensibility was 156 mm, resistance at 5 cm was 365 BU, and resistance ratio was 3.14 [Reprinted from Wrigley et al. (1984) with permission from Elsevier].

in the field at five sulfur and three nitrogen levels. Grain sulfur content varied from 0.08 to 0.18%. The response of all three wheat cultivars to sulfur fertilization showed a similar trend. Wheat grain grown under sulfur deficiency showed higher pearling index and decreased dough elasticity. The dough became harder and less extensible (Fig. 10–5) when flour sulfur levels dropped below a certain limit (Moss et al., 1981; Wrigley et al., 1984). This was reflected in correlations with extensograph measurements, which were negative with resistance and positive with extensibility. The mixograph peak dough development time decreased with increase in flour-sulfur contents. A positive correlation was reported between flour-sulfur content, cookie spread, and pup-loaf volume. A negative relationship was reported for loaf volume in a microbaking test. This can be a result of the testing methods. In the case of pup-loaves, mixing time is kept constant, while microbaking dough is mixed to peak-dough development time. This also points to the differences in testing methods for comparison of the end products.

Zhao et al. (1999a, 1999b) reported that sulfur supplementation increased dough extensibility and loaf volume but decreased dough resistance for the three cultivars, which varied considerably in elasticity and extensibility of dough. Grain sulfur and N/S ratio appeared to be better indicators of rheological and bread making properties than protein content. Wieser et al. (2007) also reported that flours with low sulfur content had poor technological properties such as short dough development time, hard and less extensible dough, high dynamic viscosity, and low bread volume.

Several workers later confirmed positive contribution of sulfur on bread loaf volume and crumb structure. Byers et al. (1987a) evaluated spring wheat cv. Timmo grown in the greenhouse using vermiculite as the culture medium with three sulfur nutrient supplementation. The N/S ratio of flours from the three sulfur treatments were 31, 17, and 16. Bread baked from low sulfur-flour failed to rise, and a rock-hard crust surrounded the unbaked core without crumb structure (Fig. 10–6). The N/S ratios of 31 (Byers et al., 1987a) or 32 (Wieser et al., 2007) are very unlikely to be observed in the field grown wheat (Zhao et al., 1999c).

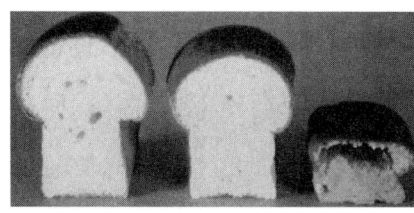

Treatment:	High S +S	High S	Low S
Flour S (mg/g):	1.9	1.7	0.8
Flour N/S ratio:	15.9	16.9	30.9
Loaf volume (mL):	1055	930	475

Fig. 10–6. Influence of sulfur supply on bread making quality of spring wheat. Data from Byers et al. (1987a) with permission of Assoc. Appl. Biol.

Effect of Sulfur Availability and Acrylamide Formation in Processed Foods

Changes in the proportions of different wheat protein groups and accumulation of free amino acids because of sulfur deficiency have been known for a long time. Similarly, increase in nonprotein nitrogen was up to six times higher and aspartic acid and asparagine levels were 2- to 3-fold higher in sulfur deficient wheat grains than in the control (Byers and Bolton, 1979; Byers et al., 1987a). Significance of the protein composition alterations on dough and bread making properties are well documented as discussed earlier. However, the significance of elevated levels of asparagine remained a mystery until the discovery of the adventitious formation of acrylamide (AA) in carbohydrate rich foods such as wheat and potatoes when subjected to baking, roasting, or frying at high temperatures (Tareke et al., 2002). Acrylamide is a neurotoxin and a potential carcinogenic chemical. Acrylamide concentrations of some processed foods are given in Table 10–2 (Friedman, 2003).

Thermal degradation of free asparagine in the presence of reducing sugars in the Maillard reaction (a nonenzymatic reaction responsible for aroma, flavor, and color of baked, fried, or roasted foods) was suggested as the key step in the formation of acrylamide (Becalski et al., 2003; Mottram et al., 2002; Stadler et al., 2002). Different pathways are proposed for the formation of AA in processed foods and

Table 10–2. Acrylamide concentration in some processed foods.†

Processed food	Acrylamide
	µg/kg
Corn chips, crisp	34–416
Bagels, breads, cakes, cookies, pretzels	70–430
Crispbread	800–1200
Cereals, breakfast	30–1346
Biscuits, crackers	30–3200
Potato chips, crisps	170–3700
Potato, French-fried	200–12,000

† Reprinted from Friedman (2003) with permission from Elsevier.

asparagine is the most widely accepted precursor of acrylamide. Processing temperature and type of reducing sugar are critical to the final concentration of acrylamide.

The asparagine model systems containing either fructose, glucose, glyceraldehyde, glycolaldehyde, 2,3-pentanedione, sorbitol, or sucrose were heated at 250 and 350°C to find the ability of different sugars to form AA (Yaylayan et al., 2003). Asparagine–glycolaldehyde system was more efficient in generating AA at 250°C than the asparagine–glucose model, while asparagine–sucrose produced more AA than asparagine–glucose at 350°C. In another study, equimolar solutions prepared in 0.05 M citrate buffer (pH 6.0) consisting of asparagine and either fructose, glucose, or sucrose were heated at four different temperatures from 140 to 200°C in a closed system to avoid any evaporation losses. Formation of AA proceeded quickly and was more temperature sensitive in asparagine–glucose combination than in the asparagine–fructose model (Claeys et al., 2005). Stadler et al. (2002) did not find significant differences in the yield of AA when D-fructose, D-galactose, lactose, and sucrose were heated to 180°C in model reactions with asparagine. Biedermann et al. (2003) used corn starch, dark wheat, and potato dried at 40°C as matrix materials to investigate the effect of different sugars (fructose, glucose, and sucrose) on the AA formation. Fructose was found to be twice as effective as glucose in the formation of AA when added to dry potato and heated at 150°C for 30 min. The formation of AA is a two-way process. Part of the AA formed is eliminated during the heating process. Fructose was also the most effective sugar in the AA elimination between 120 and 200°C. The apparent inconsistencies in the outcomes are due to the nature of reactants, physical state of the system whether dry or liquid, pH of buffer, the molar ratios, food matrix, and even the heating equipment, since all can influence AA formation (Claeys et al., 2005).

Elmore et al. (2005) examined the nexus between formation of AA and its precursors, namely free asparagine and reducing sugars in cakes made from potato flakes, whole meal wheat, and whole meal rye, cooked at 180°C. Major losses of asparagine, water, and total reducing sugars occurred between 5 and 20 min of baking accompanied by large increases in AA, which maximized in all three products between 25 and 30 min (Elmore et al., 2005). Acrylamide can also result from protein pyrolysis. Large amounts of AA were generated when wheat gluten balls free of low molecular weight precursors in the absence of reducing sugars were heated up to 240°C (Claus et al., 2006b). Discussion of various theories and reaction fundamentals involved in the formation of AA and the Maillard reaction are not intended here. Interested readers may consult recent reviews on the subject (Friedman, 2003; Yaylayan et al., 2003). The phenomenon of AA formation due to the Maillard reaction is new. There are only a few studies with respect to free amino acids as affected by genotypes, fertilization, and growing-season weather conditions. The limited information that is available is critically reviewed here.

Genotype

Claus et al. (2006a) studied the effect of genotype on the accumulation of free asparagine in nine bread wheat, two spelt wheat, and two rye cultivars over 2 yr. A wide range in asparagine concentration was observed in flour of bread wheat (4.9–25 mg 100 g^{-1}) and spelt-wheat (6.46–12.17 mg 100 g^{-1}). Asparagine quantities in the two rye cultivars were much higher than in either bread or spelt wheat (41.4–44.1 mg 100 g^{-1}). More work is required to validate these results.

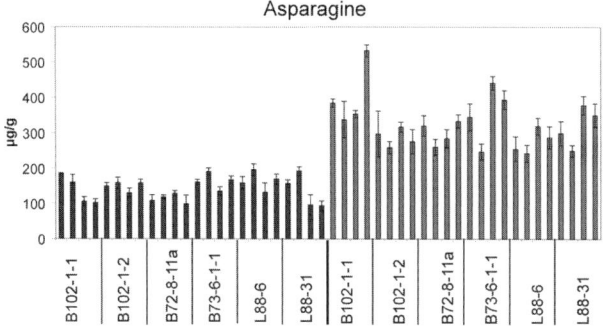

Fig. 10–7. Concentrations of asparagine in plants. In 2000, plants were grown in 48 plots, then asparagine concentrations were determined by gas chromatography-mass spectrometry (GC–MS). Light bars, Rothamsted Research (RRes); Dark bars, Long Ashton Research Station [Baker et al. (2006) with permission].

Growing-Season Weather

Baker et al. (2006) used metabolomic approach to compare three transgenic wheat lines expressing multiple copies of different HMW-GS and corresponding parental lines grown in replicated field trials at two locations (Long Ashton, near Bristol and Rothamsted, near London, UK) over 3 yr. Amino acid composition of samples grown in 2000 was determined by gas chromatography–mass spectrometry. No significant differences in amino acid composition between the parental and corresponding transgenic lines were observed. However, the samples grown at Rothamsted consistently revealed higher concentrations of asparagine (≈170 µg g^{-1} difference), aspartic acid (≈250 µg g^{-1} difference), glutamine (≈90 µg g^{-1} difference), and glutamic acid (≈70 µg g^{-1} difference) compared with the Long Ashton site (Fig. 10–7). In contrast, proline and γ-aminobutyric acid were higher for samples grown at Long Ashton. Mean sulfur contents on a dry-weight basis were 1.71 and 1.5 mg g^{-1}, for the samples grown at Rothamsted and Long Ashton, respectively. Similarly, percentage nitrogen in the grains was higher at Rothamsted (2.73%) compared with Long Ashton (Lea et al., 2007). Obviously, these results cannot be explained in terms of nitrogen and sulfur contents of grain as the N/S ratios at two sites were very similar (15.97 for Rothamsted and 16.36 for Long Ashton). Some other factors such as soil type and growing-season weather must have important implications that require further investigation (Lea et al., 2007).

High temperatures during the growing season up to 30°C are known to enhance bread making quality of wheat and temperatures above 30°C are detrimental to quality (Ciaffi et al., 1996; Johansson et al., 2004; Zhu and Khan, 2001; Claus et al., 2006a). Claus et al. (2006a) analyzed nine wheat and two rye cultivars grown during 2003 and 2004. A huge variation in sunshine duration (1580 h compared with 1296 h) and average daily temperature (15.2 compared with 13.2°C) from February to August was recorded for 2003 and 2004, respectively. Since samples were harvested from the same plots with same fertilization practices, higher amounts of asparagine in 2003 could be ascribed to variation in growing-season weather in 2 yr, specifically to higher temperatures in 2003. This is the only report so far in which detailed growing-season weather data was collected and related to the AA formation. The results indicate the importance of genotype, environ-

Effect of Sprouting

Wet weather shortly before harvest time can result in grain sprouting that translates into increased enzyme activity, degradation of starch and proteins, and adversely affects the end-use quality. The effect of α-amylase and proteases during sprouting is well documented in the literature. Depending on the wheat cultivar, sprouting could increase quantities of asparagine by up to 10-fold and AA formation from 54.5 to 273.4 µg kg^{-1} (Claus et al., 2006a). Increase of asparagine contents in germinating seeds of *Lupinus luteus* L. seedlings has been known since 1898 (Lea et al., 2007). Ten-fold increase in accumulation of asparagine was also observed in leaves of *Arabidopsis thaliana* when placed in the dark up to 6 d (Lin and Wu, 2004).

Effect of Sulfur and Nitrogen Supply on Asparagine Accumulation

Wheat grown under excess supply of nitrogen (Claus et al., 2006a) or restricted supply of sulfur produced grains with high N/S ratios and increased levels of free asparagine (Granvogl et al., 2007; Muttucumaru et al., 2006). Depending on wheat cultivar grown under restricted supply of sulfur, the concentration of free asparagine in the grains could increase up to 30-fold compared with that from samples grown under normal levels of sulfur supply (Muttucumaru et al., 2006). This huge increase in wheat asparagine was observed for greenhouse experiments in pots as well as under field conditions. Concentrations of some other amino acids also increased in sulfur-depleted flours. The increase in glutamine was specifically of the same order and magnitude as asparagine (Granvogl et al., 2007; Muttucumaru et al., 2006). Interestingly, Claus et al. (2006a) did not find significant increase in free asparagine because of sulfur deficiency that may be reflected in differences between genotypes and growing conditions.

Effect of Sulfur Supply on Amounts of Reducing Sugars

Carbohydrates, reducing sugars and their degradation products are vital reaction constituents along with asparagine to generate AA (Mottram et al., 2002; Stadler et al., 2002). Sulfur availability did not affect the quantity of total reducing sugars (Claus et al., 2006b). No significant difference was found in the amounts of fructose, glucose, maltose, and sucrose between grains obtained from plants grown either in the absence or under normal sulfur supply (Muttucumaru et al., 2006). However, Granvogl et al. (2007) reported a 2.7-fold increase in maltose concentration in sulfur-depleted flour samples.

Distribution of Acrylamide in Milling Streams

A range of milling fractions including sifted-flour, whole grain flour, fine and coarse shorts, bran, and germ from wheat and rye were analyzed for free asparagine content (Fredriksson et al., 2004). Free asparagine contents were higher in rye than in wheat in all the milling fractions analyzed. Sifted flour from wheat and rye had the lowest free asparagine (0.16 and 0.61 mg g^{-1}) followed by whole grain flour (0.50 and 1.07 mg g^{-1}). Bran contained considerably higher amounts of asparagine (1.48 and 2.90 mg g^{-1} for wheat and rye). The highest concentration of asparagine was found in wheat germ (4.94 mg g^{-1}). Low asparagine in white flour has recently been confirmed by a number of research groups (Granvogl et al., 2007; Mustafa et al., 2007). Claus et al. (2006a) prepared three flours representing low (0.63%), medium (0.74%), and high (1.35%) ash contents by mixing milling fractions from wheat and rye. Asparagine contents increased linearly with increase in ash contents for wheat (0.14–0.49 g kg^{-1}) and rye (0.60–0.83 mg g^{-1}) confirming previous findings that bran contained higher amounts of asparagine. It should be noted that flour ash content is directly related to extraction rate and is a measure of bran contamination.

Strategies to Reduce Acrylamide Formation

The amount of acrylamide produced is proportional to the amount of free asparagine present in raw materials. As discussed earlier, the quantity of asparagine depends on genotype, growing-season weather, and availability of sulfur to the crop. Thus, the problem is multifaceted and requires all the stakeholders to perform their roles effectively. This includes the plant breeders for development of genotypes with a special focus on genetics of free asparagine and reducing sugars, soil scientists to study the effect of soil types, agronomists to optimize growing conditions and sulfur supplementation, the farmer who actually will put all the available research knowledge together to raise the crop, and the food scientists to develop such processing methods that can minimize the amount of AA formed.

The Maillard reaction is responsible for development of wonderful flavor, breathtaking aroma, and eye-catching color in food products. Therefore, all those variables such as pH, temperature, time of processing etc. that affect the Maillard reaction will also influence AA formation. Some of the measures to control free asparagine contents of raw materials and quality of AA in processed food will include the following.

1. Wheat cultivars should be screened for asparagine contents grown at a large number of locations over multiple growing seasons under optimal nutrient availability.
2. Wheat cultivars should be bred with a special focus on the genes governing the synthesis and accumulation of free asparagine.
3. Educate farmers regarding quantitative and qualitative effects of sulfur deficiency on wheat yield and quality.
4. A premium may be paid for lower N/S ratio as a premium is paid for higher grain protein. Wheat with lower N/S ratio may be sold at higher price to potential customers.
5. Sulfur deficiency causes accumulation of augmented levels of asparagine in wheat. Therefore, necessary measures must be taken at government levels to help farmers in free soil tests and appropriate advice on fertilizer use.
6. Sprouting because of unfavorable weather conditions contributes to higher levels of asparagine in wheat. Therefore, weather damaged crop must not enter human food stream.
7. Hydrolysis of asparagine to aspartic acid and ammonia by enzymes such as asparaginase amidase (Friedman, 2003) needs further investigation.
8. Acetylating asparagine to *N*-acetylasparagine to prevent the formation of *N*-glycoside intermediates that form AA (Friedman, 2003).

9. Low pH also helps to minimize formation of AA and should be used if it suits the process (Jung et al., 2003).
10. Use of monovalent and divalent cations such as Na^+ and Ca^{2+} reduced formation of acrylamide in a model system. Divalent cations were more effective than monovalent cations (Gökmen and Senyuva, 2007).
11. Fermentation with yeast and long fermentation time effectively reduce free asparagine in the dough. Long fermentation (180+180 min) was more effective than short fermentation (15+15 min) (Fredriksson et al., 2004).
12. Application of glycine on the dough surface or its addition in the dough significantly reduced AA formation. However, this resulted in increased browning, which is undesirable (Fink et al., 2006).

Conclusions

Global sulfur deficiency has become wide spread over the last two decades. Atmospheric deposition of sulfur will continue to decrease especially in the industrialized countries because of strict sulfite emission control measures being implemented. A large number of countries now have agreed to be part of Kyoto Protocol, which is an amendment to the United Nations Framework Convention on Climate Change aimed at reducing greenhouse gases. However, the reverse is also true for the developing countries where recent industrialization without strict emission controls has been responsible for increased greenhouse gas emissions. Sulfur deficiency is associated with reduced wheat grain yield and loss of quality. Grain sulfur concentrations of 1.2 mg g^{-1} and N/S ratio of 17:1 are the lower limits for optimum grain yields, but higher sulfur contents and lower N/S ratio are required for better quality flour. Wheat quality losses due to sulfur deficiency include extended mixing time, increased strength, and lower extensibility in the dough that fails to rise when baked. These changes in dough properties originate from change in proportions of different protein groups and molecular weight distribution of glutenin proteins.

Sulfur deficiency impinges on plant physiology and results in the accumulation of increased levels of free amino acids, especially asparagine and glutamine. The high N/C ratio of asparagine makes it ideal for storage and transportation of nitrogen. Excessive quantities of free asparagine thus accumulate in a range of plant tissues, especially under stress conditions where plants are unable to support normal levels of protein synthesis. In addition to many unknown factors, sulfur-deficiency stress triggers accumulations of higher amounts of asparagine in cereal grains, which is responsible for the formation of AA during processing. The limited literature available strongly points to the genotypic differences, effect of soil type, and growing-season weather conditions such as total sunshine and temperature. Unraveling these differences and their interaction with the environment should be the focus in future research. Modern molecular techniques are available in wheat and other cereals to help us understand this phenomenon efficiently (Howarth et al., 2005; Shewry and Jones, 2005). The knowledge gathered on factors that determine asparagine accumulation will be helpful in genetic manipulation to produce cereals with low or no free asparagine. The raw materials with low or no free asparagine would be used to manufacture AA free products. Food chemists should study the kinetics of AA formation in individual processes used for the manufacture of various products before making any alteration.

References

Baker, J.M., N.D. Hawkins, J.L. Ward, A. Lovegrove, J.A. Napier, P.R. Shewry, and M.H. Beale. 2006. A metabolomic study of substantial equivalence of field-grown genetically modified wheat. Plant Biotech. J. 4:381–392.

Beaton, J.D., and R.J. Soper. 1986. Plant response to sulfur in western Canada. p. 375–403. In M.A. Tabatabai (ed.) Sulfur in agriculture. Agron. Monogr. 27. ASA, CSSA, and SSSA, Madison, WI.

Becalski, A., B.P.Y. Lau, D. Lewis, and S.W. Seamen. 2003. Acrylamide in foods: Occurrence, sources, and modeling. J. Agric. Food Chem. 51:802–808.

Biedermann, M., and K. Grob. 2003. Model studies on acrylamide formation in potato, wheat flour and corn starch; ways to reduce acrylamide contents in bakery ware. Mitt. Lebensm. Hyg. 94:406–422.

Byers, M., and J. Bolton. 1979. Effects of nitrogen and sulphur fertilizers on the yield, nitrogen and sulfur content, and amino acid composition of the grains of spring wheat. J. Sci. Food Agric. 34:447–462.

Byers, M., J. Franklin, and S.M. Smith. 1987a. The nitrogen and sulphur nutrition of wheat and its effect on the composition and baking quality of the grain. Aspects Appl. Biol. 15:337–344.

Byers, M., S.P. McGrath, and R. Webster. 1987b. A survey of the sulphur content of wheat grown in Britain. J. Sci. Food Agric. 38:151–160.

Byers, M., B.J. Miflin, and S.J. Smith. 1983. A quantitative comparison of the extraction of protein fractions from wheat grain by different solvents, and of polypeptide and amino acid composition of the alcohol-soluble proteins. J. Sci. Food Agric. 34:447–462.

Ceccotti, S.P., R.J. Morris, and D.L. Messick. 1997. A global overview of the sulphur situation: Industry's background, market trends, and commercial aspects of sulphur fertilizers. p. 175–202. In E. Schnug (ed.) Sulphur in agroecosystems. Kluwer Academic Publishers, the Netherlands.

Ciaffi, M., L. Tozzi, B. Borghi, M. Corbellini, and D. Lafiandra. 1996. Effect of heat shock during grain filling on the gluten protein composition of bread wheat. J. Cereal Sci. 24:91–100.

Claeys, W.L., K. De Vleeschouwer, and M.E. Hendrickx. 2005. Kinetics of acrylamide formation and elimination during heating of an asparagine-sugar model system. J. Agric. Food Chem. 53:9999–10005.

Claus, A., P. Schreiter, A. Weber, S. Graeff, W. Herrmann, W. Claupein, A. Schieber, and R. Carle. 2006a. Influence of agronomic factors and extraction rate on the acrylamide contents in yeast-leavened breads. J. Agric. Food Chem. 54:8968–8976.

Claus, A., G.M. Weisz, A. Schieber, and R. Carle. 2006b. Pyrolytic acrylamide formation from purified wheat and gluten-supplemented bread rolls. Mol. Nutr. Food Res. 50:87–93.

Duke, S.H., and H.M. Reisenauer. 1986. Roles and requirements of sulfur in plant nutrition. p. 123–168. In M.A. Tabatabai (ed.) Sulfur in agriculture. Agron. Monogr. 27. ASA, CSSA, SSSA, Madison, WI.

Elmore, J.S., G. Koutsidis, A.T. Dodson, D.S. Mottram, and B.L. Wedzicha. 2005. Measurement of acrylamide and its precursors in potato, wheat, and rye model systems. J. Agric. Food Chem. 53:1286–1293.

Evans, L.T., and R.A. Fischer. 1999. Yield potential: Its definition, measurement and significance. Crop Sci. 36:1544–1551.

Fink, M., R. Andersson, J. Rosén, and P. Åman. 2006. Effect of added asparagine and glycine on acrylamide content in yeast-leavened bread. Cereal Chem. 83:218–222.

Fredriksson, H., J. Tallving, J. Rosén, and P. Åman. 2004. Fermentation reduces free asparagine in dough and acrylamide content. Cereal Chem. 81:650–653.

Friedman, M. 2003. Chemistry, biochemistry, and safety of acrylamide. A review. J. Agric. Food Chem. 51:4504–4526.

Gökmen, V., and H.Z. Senyuva. 2007. Acrylamide formation is prevented by divalent cations during the Maillard reaction. Food Chem. 103:196–203.

Granvogl, M., H. Wieser, P. Koehler, S.V. Tucher, and P. Schieberle. 2007. Influence of sulfur fertilization on the amounts of free amino acids in wheat. Correlation with baking properties as well as with 3-aminopropionamide and acrylamide generation during baking. J. Agric. Food Chem. 55:4271–4277.

Howarth, J.R., J.N. Jacquet, A. Doherty, H.D. Jones, and M.E. Cannell. 2005. Molecular genetic analysis of silencing in two lines of *Triticum aestivum* transformed with the reporter gene construct pAHC25. Ann. Appl. Biol. 146:311–320.

Johansson, E., R. Kuktaite, A. Anderson, and M. Luisa. 2004. Protein polymer built-up during wheat grain development: Influences of temperature and nitrogen timing. J. Sci. Food Agric. 85:473–479.

Jung, M.Y., D.S. Choi, and J.W. Ju. 2003. A novel technique for limitation of acrylamide in fried and baked corn chips and in French fries. J. Food Sci. 68:1287–1290.

Lea, P.J., L. Sodek, M.A.J. Parry, P.R. Shewry, and N.G. Halford. 2007. Asparagine in plants. Ann. Appl. Biol. 150:1–26.

Lin, J.F., and S.H. Wu. 2004. Molecular events in senescing Arabidopsis leaves. Plant J. 39:612–628.

MacRitchie, F. 1992. Physicochemical properties of wheat proteins in relation to functionality. Adv. Food Nutr. Res. 36:1–87.

MacRitchie, F., and R.B. Gupta. 1993. Functionality-composition relationships of wheat flour as a result of variation in sulfur availability. Aust. J. Agric. Res. 44:1767–1774.

Messick, D.L., and C. de Brey. 2001. The global sulfur situation and outlook. *In* Proceeding of 51st Fertilizer Industry Round Table, St Peter Beach, FL.

Moss, H.J., C.W. Wrigley, F. MacRitchie, and P.J. Randall. 1981. Sulfur and nitrogen fertilizer effects on wheat. II. Influence on grain quality. Aust. J. Agric. Res. 32:213–226.

Mottram, D.S., B.L. Wedzicha, and A.T. Dodson. 2002. Acrylamide is formed in the Maillard reaction. Nature 419:448–449.

Mustafa, A., P. Åman, R. Anderson, and A. Kamal-Eldin. 2007. Analysis of free amino acids in cereal products. Food Chem. 105:317–324.

Muttucumaru, N., N.G. Halford, J.S. Elmore, A.T. Dodson, M. Parry, P.R. Shewry, and D.S. Mottram. 2006. Formation of high levels of acrylamide during the processing of flour derived from sulfate-deprived wheat. J. Agric. Food Chem. 54:8951–8955.

Naeem, H.A., and F., MacRitchie. 2003. Effect of sulfur nutrition on agronomic and quality attributes of wheat. p. 305–322. *In* Y.P. Abrol and A. Ahmad (ed.) Sulphur in plants. Kluwer Academic Publishers, the Netherlands.

Osborne, T.B. 1907. The proteins of wheat kernel. Carnegie Inst., Washington, DC.

Pasricha, N.S., and Y.P. Abrol. 2003. Food production and plant nutrient sulphur. p. 29–44. *In* Y.P. Abrol and A. Ahmad (ed.) Sulphur in plants. Kluwer Academic Publishers, the Netherlands.

Prosser, I.M., J.V. Purves, L.R. Saker, and D.T. Clarkson. 2001. Rapid disruption of nitrogen metabolism and nitrate transport in spinach plants deprived of sulphate. J. Exp. Bot. 52:113–121.

Randall, P.J., K. Spencer, and J.R. Freney. 1981. Sulphur nitrogen fertilizer effects on wheat. I. Concentration of sulfur and nitrogen, and the nitrogen to sulfur ratio in the grain. Aust. J. Agric. Res. 32:203–212.

Sakal, R., A.P. Singh, R.B. Sinha, N.S. Bhogal, and M.D. Ismail. 1999. Impact of sulphur fertilization in sustaining the productivity of rice-wheat cropping system. Fert. News 44:49–52.

Sakal, R., R.B. Sinha, A.P. Singh, N.S. Bhogal, and M.D. Ismail. 2000. Influence of sulphur on yield and mineral nutrition of crops in maize-wheat sequence. J. Indian Soc. Soil Sci. 48:325–329.

Shewry, P.R., and H.D. Jones. 2005. Transgenic wheat: Where do we stand after the first 12 years? Ann. Appl. Biol. 147:1–14.

Shewry, P.R., A.S. Tatham, J. Florde, M. Kreis, and B.J. Miflin. 1986. The classification and nomenclature of wheat gluten proteins: a reassessment. J. Cereal Sci. 4:97–106.

Singh, N.K., G.R. Donovan, I.L. Batey, and F. MacRitchie. 1990. Use of sonication and size-exclusion high performance liquid chromatography in the study of wheat proteins. I. Dissolution of total proteins in the absence of reducing agents. Cereal Chem. 67:150–161.

Stadler, R.H., I. Blank, N. Varga, F. Robert, J. Hau, P.A. Guy, M.C. Robert, and S. Riediker. 2002. Acrylamide from Maillard reaction products. Nature 419:449–450.

Tareke, E., P. Rydberg, P. Karlsson, S. Eriksson, and M. Toernqvist. 2002. Analysis of acrylamide, a carcinogen formed in heated foodstuffs. J. Agric. Food Chem. 50:4998–5006.

Wang, R., M. Okamoto, X. Xing, and N.M. Crawford. 2003. Microarray analysis of the nitrate response in *Arabidopsis* roots and shoots reveals over 1000 rapidly responding genes and new linkages to glucose, trehalose-6-phosphate, iron and sulfate metabolism. Plant Physiol. 132:556–567.

Wieser, H., R. Guster, and S.V. Tucher. 2004. Influence of sulphur fertilization on quantities and proportions of gluten protein types in wheat flour. J. Cereal Sci. 40:239–244.

Wieser, H., P. Koehler, and S.V. Tucher. 2007. Influence sulphur fertilization on the technological properties of wheat flour. p. 158–161. In G.L. Lookhart and P.K.W. Ng (ed.) Gluten proteins 2006. American Association of Cereal Chemists International, St. Paul, MN.

Withers, P.J.A., F.J. Zhao, S.P. McGrath, E.J. Evans, and A.H. Sinclair. 1997. Sulphur inputs for optimum yields of cereals. p. 191–198. In M.J. Gooding and P.R. Shewry (ed.) Aspects of applied biology 50, optimizing cereal inputs: Its scientific basis. The Assoc. Appl. Biologists, Wellesbourne, UK.

Wrigley, C.W., D.L. Du Cros, J.G. Fullington, and D.D. Kasarda. 1984. Changes in polypeptide composition and grain quality due to sulfur deficiency in wheat. J. Cereal Sci. 2:15–24.

Yaylayan, V.A., A. Wnorowski, and C.P. Locas. 2003. Why asparagine needs carbohydrates to generate acrylamide. J. Agric. Food Chem. 51:1753–1757.

Wrigley, C.W., D.L. Du Cros, M.J. Archer, P.G. Downie, and C.M. Roxburgh. 1980. The sulfur content of wheat endosperm proteins and its relevance to grain quality. Aust. J. Plant Physiol. 7:755–766.

Zhao, F.J., M.J. Hawkesford, and S.P. McGrath. 1999c. Sulphur assimilation and effects on yield and quality of wheat. J. Cereal Sci. 30:1–17.

Zhao, F.J., S.E. Salmon, P.J.A. Withers, E.J. Evans, J.M. Monaghan, P.R. Shewry, and S.P. McGrath. 1999a. Variation in the bread making quality and rheological properties of wheat in relation to sulphur nutrition under field conditions. J. Cereal Sci. 30:19–31.

Zhao, F.J., S.E. Salmon, P.J.A. Withers, J.M. Monaghan, E.J. Evans, P.R. Shewry, and S.P. McGrath. 1999c. Responses of bread making quality to sulphur in three wheat varieties. J. Sci. Food Agric. 79:1865–1874.

Zhao, F.J., S.P. McGrath, and A.R. Crosland. 1995. Changes in sulphur status of British wheat grain in the last decade, and its geographical distribution. J. Sci. Food Agric. 68:507–514.

Zhu, J., and K. Khan. 2001. Effects of genotype and environment on glutenin polymers and bread making quality. Cereal Chem. 78:125–130.

11

Sulfur and Marketable Yield of Potato

Alexander D. Pavlista
University of Nebraska, Panhandle Research and Extension Center, Scottsbluff

Abstract

Sulfur is considered a micronutrient of which relatively little is needed to overcome sulfur deficiency. However, in potato (*Solanum tuberosum* L.), large amounts of sulfur can be applied to increase marketable yield. In many potato-growing areas, marketable yield is reduced by tuber blemishes caused by *Streptomyces scabies* (Thaxt) Waks. & Henrici (common scab) and *Rhizoctonia solani* Kühn (black scurf). A reduction of common scab can be accomplished on slightly acidic and neutral soils by addition of large amounts of sulfur to lower the pH below the optimal range for *S. scabies* growth. However, a reduction of common scab and black scurf has been reported due to application of sulfur without affecting pH. A mechanism has been proposed for the latter observation by which sulfur and sulfate (SO_4^{2-}) are converted to volatile sulfur forms, especially hydrogen sulfide (H_2S), that act as biocides.

Sulfur deficiency in potato is characterized by the vine becoming light green or with leaves becoming yellow under severe deficiency (anonymous, 1982). It usually has been associated with soil that is sandy or has low organic matter (Stevenson et al., 2001). Relief from sulfur deficiency requires less than 50 kg sulfur ha^{-1} (anonymous, 1982; Caldwell et al., 1972; Kunkel, 1978, p. 81–87; Loman, 1977). Applying sulfur to deficient soil results in healthier leaves and increased sulfur in leaf tissue (Gupta and Sanderson, 1993) and improved nutritional value of tubers (Eppendorfer and Eggum, 1994). Most of the sulfur needed to overcome deficiency can be obtained from environmental sources (Haneklaus et al., 2003) such as the atmosphere (Schroeder, 1993), organic matter decomposition, rainfall, and irrigation water (Singh and Srivastava. 1993). Some nitrogen-phosphorus-potassium (NPK) fertilizers contain sulfur, such as superphosphate with 12% sulfur and about 20% P_2O_5 (anonymous, 1982). These levels of sulfur eliminate deficiency (Pavlista and Blumenthal, 2001). The objective of this review is to determine whether additional sulfur would improve meeting market yield requirements. The historical use of sulfur as a soil amendment was to reduce pH to control common scab, a tuber blemish for which tubers are culled and thereby reduce market yield (Loria et al., 1997; Pavlista, 1993). This review will present evidence for an effect by sulfur irrespective of soil pH on common scab (Mabbett, 2001; Pavlista, 1993) and another tuber blemish, black scurf, that also reduces market yield (Banville, 1989). This will include a discussion on heavy metal sulfates and conclude

---

Copyright © 2008. American Society of Agronomy, Crop Science Society of America, Soil Science Society of America, 677 S. Segoe Rd., Madison, WI 53711, USA. *Sulfur: A Missing Link between Soils, Crops, and Nutrition.* Agronomy Monograph 50.

with a proposed mechanism for sulfur and sulfate for suppression of common scab and black scurf to increase marketable yield.

Tuber Diseases

Common Scab

Common scab is primarily caused by the pathogen *S. scabies* (Afanasiev, 1937), which infects tubers through lenticels as tubers are in the early stage of growth (Lapwood and Hering, 1970). The mild symptoms are referred to as surface scab (Fig. 11-1a) (Stevenson et al., 2001). Pitted scab (Fig. 11-1b), once thought to be different, is caused by the same pathogen (Archuleta and Easton, 1981). Evidence suggests that the pathogen releases vivotoxins that result in the symptoms (Lawrence et al., 1990). Other species such as *Streptomyces acidiscabies* Lambert & Loria (Lambert and Loria, 1989) have been reported as pathogens in some regions. The pathogen in the soil causes common scab (Pavlista, 1996). Tubers with common scab are culled for all major markets, fresh (table stock), seed, potato chip, and French fry. In the chip and fry markets, surface scab can be removed in the peeling process but pitted scab is usually 1 cm deep, lowering yield of the final product (Pavlista and Ojala, 1997).

Black Scurf

The tuber blemish caused by infection with *R. solani* is called black scurf and is characterized by unwashable black spots on the skin (Fig. 11-1c) (Stevenson et al., 2001). This pathogen also causes the symptom stem canker, a discoloration to pinching of stem tissue (Fig. 11-1d) (Carling et al., 1989). Black scurf is associated with the pathogen in the soil, while stem canker occurs from both tuber-borne and soil-borne infection (Frank and Leach, 1980). Potatoes for the fresh market and those sold for seed are culled for black scurf; process markets are not affected as black scurf is easily removed with the peel.

Sulfur, pH, and the Common Scab

Early applications of sulfur to potato seed tubers and soil date back to the 1890s. Sulfur sublimate applied to seed tubers resulted in a yield increase and less harvested tubers with common scab (Garman, 1898, p. 9–17). Others at that time also reported that using sulfur as a soil amendment reduced common scab when potato was grown under good conditions (Halsted, 1895; Halsted, 1900, p. 326–345; Wheeler and Adams, 1897). The amount of sulfur needed was over 300 kg ha^{-1} and usually higher. An inverse correlation between increasing sulfur and decreasing scab existed (Sherbakoff, 1914). As a result of these studies, the suggestion came that sulfur reduced the soil pH and that this may reduce infection. A way to accurately measure soil pH was needed and 20 yr later, a pH meter to detect H$^+$ ions in the soil was invented (Gillespie, 1918; Gillespie and Hurst, 1918). Surveying 47 soils with pH ranging from 4.50 to 7.21, they concluded that when pH was less than 5.2, scabby tubers were rare and when it was above 5.6, scabby tubers were common at harvest. Soil type did not matter.

With these observations and the development of instrumentation, research began to identify the pH optimum of common scab. Sulfur at 336 to 1344 kg ha^{-1}

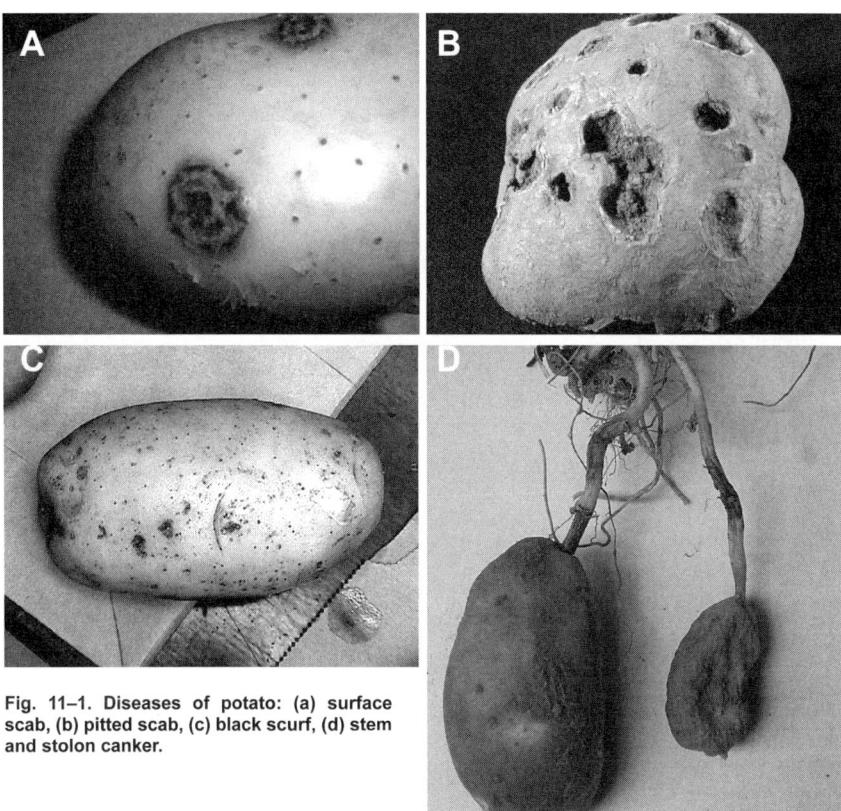

Fig. 11-1. Diseases of potato: (a) surface scab, (b) pitted scab, (c) black scurf, (d) stem and stolon canker.

was broadcast and incorporated into NJ soil, and pH, percentage of scabby tubers, and marketable yield were determined (Martin, 1920). Adding sulfur lowered pH and reduced common scab thereby increasing marketable yield. The percentage of harvested tubers with scab averaged 24% below pH 5.4 and averaged 57% above 5.4 (Fig. 11-2d). Martin's observations were unchanged when sulfofying inoculum was added (1921) or when soil moisture varied (1923). In culture, Sanford (1926) concluded that the pathogen *S. scabies* did not grow at pH below 5.3 and had an upper growth limit at pH 8.7. Optimal temperature for growth was 24 to 30°C, oxygen (air) was required, and growth was best at 80% relative humidity while not occurring above 90% RH. Isolates of the pathogen were collected across the USA with little difference found between them with regard to their sensitivity to pH (Schaal, 1940, 1943). Population of *S. scabies* in slightly acidic soils was reduced by further acidifying the soil with sulfur (Vlitos and Hooker, 1951). A survey of soils in western Nebraska indicated an optimal pH range for scabby tubers to be 5.9 to 8.0 (Goss, 1934) and another in New York concluded an optimal pH range of 5.5 to 7.4 (Blodgett and Howe, 1934). Tests were conducted with fertilizers to acidify soil (Cook and Nugent, 1939; Cook and Houghland, 1942). Studies were conducted around the world to acidify further slightly acid soil below the optimal pH range for common scab with elemental sulfur (e.g., Barnes and Chestnutt, 1966; El-Fayoumy and El-Gamal, 1998; Klikocka et al., 2005; McCreary, 1967).

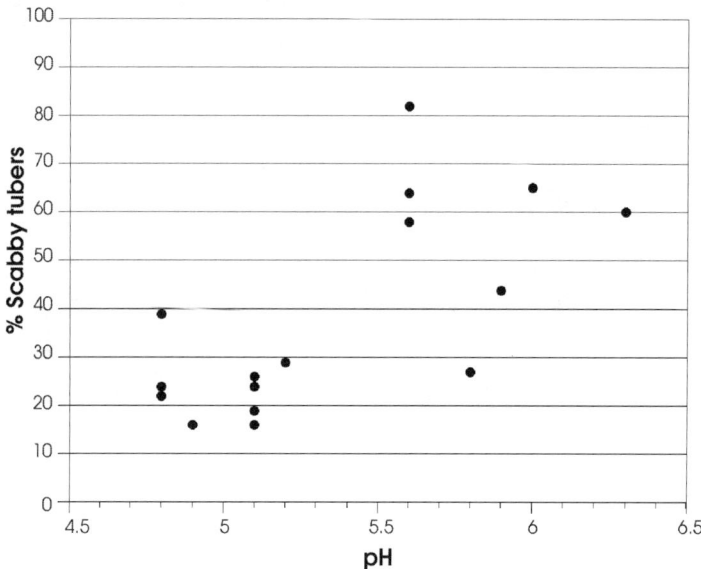

Fig. 11–2. Relation of soil pH as altered by sulfur on the incidence of scabby tubers, cultivars American Giant and Irish Cobbler (modified from Martin, 1920).

Wisconsin soil pH was lowered from 6.1 to as low as 4.2 by the addition of 3360 kg of elemental S per ha (Larson et al., 1938) and suggested that, for every added 1100 kg sulfur ha^{-1}, pH was lowered by 0.6–0.7. Studies were also conducted to raise the pH of alkaline soils with lime above the optimal pH range of common scab by which common scab was reduced and marketable yield was increased (Waterer, 2002).

Ammonium Sulfate on Common Scab Irrespective of Soil pH

In 1954, Emilsson and Gustafsson reviewed the literature up to that time on controlling common scab by altering soil pH. They noted that 190 kg sulfur ha^{-1} applied as ammonium sulfate (AS) to slightly alkaline soil could lower common scab and increase marketable yield. This was suggested earlier when soil pH remained in the optimal range for common scab, but the addition of sulfur still lowered the incidence of common scab and increased marketable yield (Duff and Welch, 1927; Martin, 1923). This also occurred in acid soil when incorporating 225 to 450 kg sulfur ha^{-1} reduced common scab incidence and increased marketable yield (Barnes, 1971; McCreary, 1967; Muncie et al., 1944).

Ammonium sulfate could replace elemental sulfur (Hooker and Kent, 1950). They reported that 130 kg sulfur ha^{-1} in AS decreased incidence in common scab, increased marketable yield, and increased total yield. Although increasing total yield, ammonium nitrate and ammonium carbonate did not affect common scab or change marketable yield. In 1943, Hooker and Kent (1950) were able to distinguish between a sulfur effect and a pH effect on common scab on the basis of

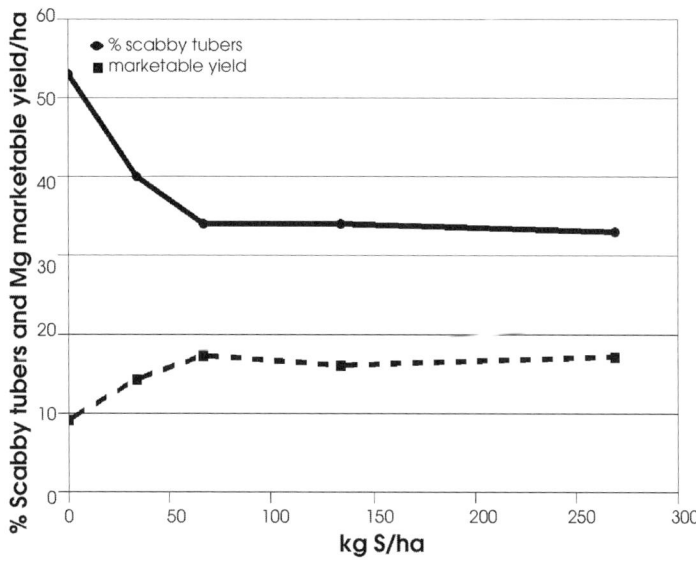

Fig. 11–3. Effect of furrow-incorporated ammonium sulfate on scabby potato and marketable yield, cultivar Irish Cobbler (modified from Table 5 in Hooker and Kent, 1950).

marketable yield. They also showed an AS rate response for common scab and marketable yield between 0 and 75 kg sulfur ha^{-1} when AS was incorporated into the furrow and the soil pH remained in the optimal range (Fig. 11–3). Many potato cultivars showed reduced common scab and increased marketable yield due to 134 kg sulfur ha^{-1} applied in AS.

On the basis of the pH range of common scab, when added sulfur acidifies slightly alkaline soil, the lowering of the pH to neutrality should increase common scab if the effect is only due to a change in soil pH. However, as early as 1951 (Vlitos and Hooker), it was reported that acidifying soil from about 7.9 to 7.3 with sulfur reduced the population of *S. scabies* and was verified in laboratory tests. Their conclusion was that regardless of pH, sulfur reduced *S. scabies* population. Under low irrigation especially, the addition of 125 kg sulfur ha^{-1} in AS reduced common scab and increased marketable yield only when AS was incorporated into the hill and not when it was applied broadcast preplanting (Hammerschmidt et al., 1986; Hammerschmidt and Vitosh, 1987).

More recently, sulfur at 56 and 112 kg ha^{-1} applied into the furrow in a wettable formulation, in AS or in ammonium thiosulfate (ATS), reduced common scab and wettable sulfur reduced black scurf while increasing yield (Pavlista, 2004). Applied early postemergence to the foliage, AS and ATS at these rates also reduced common scab and black scurf while increasing yield, while application to the foliage at tuber initiation reduced common scab. In agreement with Gray et al. (1961), preemergence applications had no effect at these rates with these compounds. Recent trials conducted in Europe on slightly acid soil reported that application of 50 kg sulfur ha^{-1} as elemental sulfur and K_2SO_4 increased yield and reduced the incidence of black scurf in 2001, but results were slight and erratic in 2002 (Klikocka et al., 2005). There was no consistent effect on common scab. The authors reported that as pH increased, the severity of black scurf was reduced, while the number of tubers with black scurf was increased. The severity and incidence of common scab was reduced by raising pH. Note that method and timing of sulfur application were not reported.

Recently an organic form of sulfur, ammonium lignosulfonate, has been reported to reduce common scab severity and increase marketable yield 3- to 7-fold (Soltani et al., 2002). The authors also reported that the incidence of early dying (*Verticillium dahlia* Kleb) was significantly decreased.

Metallic Sulfates, Common Scab, and Marketable Yield

An association of heavy metals with common scab was reviewed recently (Keinath and Loria, 1989). The review focused on alteration of the soil pH. Often the literature is viewed as indicating that heavy metals, especially calcium and manganese, directly affect common scab. Here, the effects of heavy metals will be analyzed with respect to the anion form in which they were applied with emphasis on sulfates.

Initial observations compared the effects of lime ($CaCO_3$) and gypsum ($CaSO_4$). Conclusions were that changes in common scab incidence was associated with pH changes and not because of calcium or sulfur (Blodgett and Cowan, 1935; Odland and Allbritten, 1950). Horsfall et al. (1954) suggested that by testing low levels of calcium and sulfur within the optimal pH range, common scab reduction by $CaSO_4$ may not be just a pH effect, and the increase in common scab by $CaCO_3$ was due to an increase of calcium in the tuber periderm (peel). The relation of higher soil calcium with increased common scab was reported also in Japan (Goto, 1985). However, calcium is not needed for growth of *S. scabies* in culture and Houghland and Cash (1956b) concluded that calcium in the peel may be a result of common scab. Later, Lambert and Manzer (1991) agreed that, in low pH soils, high-tissue calcium was a result of common scab, and the incidence of common scab was related to soil pH. Low levels of $CaSO_4$ and $Ca(OH)_2$ were added to acidic soil (pH 5.2) and the reduction in common scab by $CaSO_4$ was not attributed to pH while the promotion of common scab by $Ca(OH)_2$ was related to increased pH (Doyle and MacLean, 1960). No relation between common scab and Ca/K ratio was found (Doyle and MacLean, 1960; Houghland and Cash, 1956b). Comparing AS to $CaNO_3$, Huber and Watson (1970) showed that AS at 230 kg sulfur ha^{-1} increased marketable yield by reducing scab and that nitrogen had no effect. Studies with elemental sulfur, AS, gypsum ($CaSO_4$), and $Ca(NO_3)_2$ applied to alkaline soil (pH 7.5–7.8) showed that sulfur reduced common scab and increased marketable yield without affecting pH (Davis et al., 1974a, 1976a, 1976b). These studies also found a reduction of calcium in the potato peel and a decrease in the Ca/K and Ca/P ratios. With increasing irrigation, common scab was further reduced even in the presence of 230 kg sulfur ha^{-1} (Davis et al., 1974b).

Additional reports showed that $CaSO_4$, K_2SO_4, and AS applied at sulfur levels greater than 50 kg ha^{-1} in both acidic and alkaline soils increased marketable yield although tuber disease measurements were not recorded (Aulakh et al., 1977; Ramamurthy and Suseela-Devi, 1982; Simmons et al., 1988).

In greenhouse experiments, $MnSO_4$ was added around tubers of a scab-prone potato cultivar. A reduction in common scab with a correlated increase in scab-free tubers resulted (Mortvedt et al., 1961). However, in field trials, when $MnSO_4$

at 168 kg ha^{-1} (62 kg Mn ha^{-1}, 36 kg sulfur ha^{-1}) was applied into the furrow, the yield results were not significant but scab incidence was low. Additional field trials were conducted adding 504 kg MnSO$_4$ ha^{-1} (184 kg Mn ha^{-1}, 106 kg sulfur ha^{-1}) applied in-furrow and this rate significantly decreased common scab (Mortvedt et al., 1963). The authors concluded that the reduction in common scab was due to Mn, but an alternative conclusion would be that it is due to the S. From studies on AS, the rate of sulfur applied in-furrow needed to suppress common scab would be over 56 kg sulfur ha^{-1} (Hooker and Kent, 1950; Hammerschmidt et al., 1986; Pavlista, 2004). This could explain the results with MnSO$_4$. This is further supported by studies showing that 125 kg MnSO$_4$ ha^{-1} did not affect either common scab or black scurf (Barnes, 1971), and this threshold for activity seems to be generally accepted (Keinath and Loria, 1989). However, exceptions have been reported that lower rates could reduce common scab in acidic and neutral soils (McGregor and Wilson, 1964, 1966).

Potassium sulfate has been reported to reduce *R. solani* infection while KCl did not (Panique et al., 1997). Aluminum sulfate (250 kg sulfur ha^{-1}) was reported to be ineffective against common scab (Houghland and Cash, 1956a).

Hydrogen Sulfide

Hydrogen sulfide has been identified as fungitoxic in animals (Reiffenstein et al., 1992). A recent review identified sulfur and sulfur-containing compounds as biocides that reduce incidence and severity of many diseases of various plants (Table 1 in Haneklaus et al., 2007). These observations have led to the term sulfur-induced resistance (SIR) and the suggestion that volatile sulfur-compounds produced by plants exposed to sulfur or sulfate may be responsible (Schroeder, 1993). Most notable of these is H$_2$S (Haneklaus et al., 2007; Yang et al., 2006). The precise mechanism for the production of H$_2$S by plants against pathogens is not known, but several have been suggested (Haneklaus et al., 2007; Schmidt, 2005).

These reports and reviews represent a renewed interest in the role of sulfur in controlling pathogens of plants. However, Vlitos and Hooker (1951) had already suggested a role for H$_2$S in controlling common scab on potato. They cultured *S. scabies* on neutral media exposing colonies to H$_2$S and monitored colony growth and media pH (Fig. 11–4). The pH of the growth media was acidified into the optimal pH range for colony growth but instead colony survival was strongly inhibited and dry weight neared zero. Recently, Yang et al. (2006) reported a similar effect of H$_2$S on *R. solani*, the pathogen for black scurf and stem canker. One-hour exposure to as little as 2 μL L^{-1} H$_2$S significantly reduced colony diameter.

The conversion of sulfate to H$_2$S in the soil could explain the reduction of common scab and black scurf when AS or ATS is applied into the furrow (Hammerschmidt et al., 1986; Pavlista, 2004). Also, this would account for the ineffectiveness of preemergence application to the soil (Gray et al., 1961; Pavlista, 2004) as H$_2$S would escape into the atmosphere (Haneklaus et al., 2003). Furthermore, Pavlista (2004) reported that foliar applications of AS and ATS also reduced common scab and black scurf, but a mechanism was not formulated. The recent report of conversion of sulfate to H$_2$S or other volatile sulfur compounds in plants producing SIR (Haneklaus et al., 2007; Schmidt, 2005) suggests how foliar absorption of sulfate may have resulted in the observed reductions.

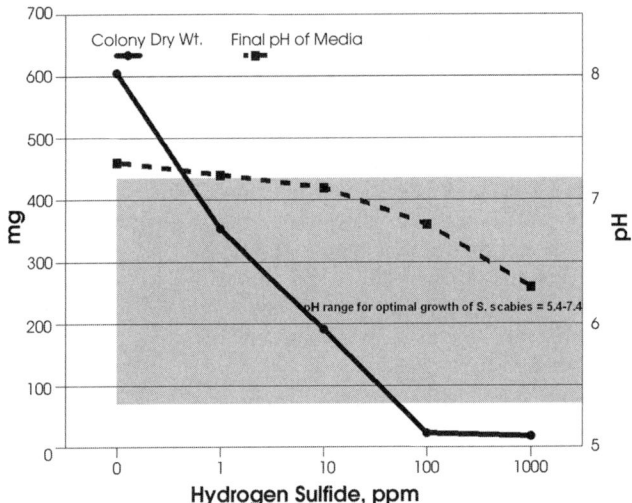

Fig. 11–4. Hydrogen sulfide-induced reduction of the growth of Streptomyces scabies in culture (modified from Vlitos and Hooker, 1951).

Conclusions

There seem to be two mechanisms for sulfur to reduce common scab and increase marketable yield of potato (Pavlista, 1993). With the application of high amounts of sulfur, soil pH is reduced. When applied to slightly acidic soil, above pH 5.4, it can reduce the pH below the optimal range for the common scab pathogen. The other mechanism involves the production of H_2S and possibly other volatile sulfur compounds to affect the pathogens for both common scab and black scurf. This mechanism would require less sulfur than needed for acidification, would involve plant metabolism accounting for SIR, and would explain the effect of foliar exposure to sulfur. Also suggested is that the increase in marketable yield that has been attributed to certain heavy metals may be due to the sulfate form applied.

References

Afanasiev, M.M. 1937. Comparative physiology of Actinomyces in relation to potato scab. Nebr. Agric. Exp. Sta. Res. Bull. 92.

Anonymous. 1982. Sulphur—The fourth major nutrient. The Sulphur Inst., Washington, DC.

Archuleta, J.G., and G.D. Easton. 1981. The cause of deep-pitted scab of potatoes. Am. Potato J. 58:385–392.

Aulakh, M.S., B. Singh, and B.R. Arora. 1977. Effect of sulphur fertilization on the yield and quality of potatoes (*Solanum tuberosum* L.). J. Indian Soc. Soil Sci. 25:182–185.

Banville, G.J. 1989. Yield losses and damage to potato plants caused by *Rhyzoctonia solani* Kuhn. Am. Potato J. 66:821–834.

Barnes, E.D. 1971. The effects of irrigation, manganese sulphate and sulphur applications on common scab of potato. Record Agric. Res. 20:35–44.

Barnes, E.D., and D.M.B. Chestnutt. 1966. Some aspects of the control of common scab in potatoes. Record Agric. Res. 15:35–42.

Blodgett, F.M., and E.K. Cowan. 1935. Relative effects of calcium and acidity of the soil on the occurrence of potato scab. Am. Potato J. 12:265–274.

Blodgett, F.M., and F.B. Howe. 1934. Factors influencing the occurrence of potato scab in New York. NY (Cornell) Agric. Exp. Sta. Bull. 581.

Caldwell, A.C., E.C. Seim, G.W. Rehm, and J. Grava. 1972. Sulfur studies in the field on a sulfur-deficient Minnesota soil. U. Minn. Tech. Bull. 284.

Carling, D.E., S.S. Leiner, and P.C. Westphale. 1989. Symptoms, signs, and yield reduction associated with Rhyzoctonia disease of potato induced by tuberborne inoculum of *Rhyzoctonia solani* AG-3. Am. Potato J. 66:693–701.

Cook, H.T., and G.V.C. Houghland. 1942. The severity of potato scab in relation to the use of neutralized and one-third neutralized fertilizers. Am. Potato J. 19:201–208.

Cook, H.T., and T.J. Nugent. 1939. The influence of acid-forming and non-acid-forming fertilizer on the development of potato scab. Am. Potato J. 16:1–5.

Davis, J.R., J.G. Garner, and R.H. Callihan. 1974a. Effects of gypsum, sulfur, Terraclor and Terraclor Super-X for potato scab control. Am. Potato J. 51:35–43.

Davis, J.R., R.E. McDole, and R.H. Callihan. 1976a. Fertilizer effects on common scab of potato and the relation to calcium and phosphate-phosphorus. Phytopathology 66:1236–1241.

Davis, J.R., G.M. McMaster, R.H. Callihan, J.G. Garner, and R.E. McDole. 1974b. The relationship of irrigation timing and soil treatments to control potato scab. Phytopathology 64:1404–1410.

Davis, J.R., G.M. McMaster, R.H. Callihan, F.H. Nissley, and J.J. Pavek. 1976b. Influence of soil moisture and fungicide treatments on common scab and mineral content of potatoes. Phytopathology 66:228–233.

Doyle, J.J., and A.A. MacLean. 1960. Relationship between Ca:K ratio, pH, and prevalence of potato scab. Can. J. Plant Sci. 40:616–619.

Duff, G.H., and C.G. Welch. 1927. Sulphur as a control agent for common scab of potato. Phytopathology 17:297–314.

El-Fayoumy, M.E., and A.M. El-Gamal. 1998. Effects of sulfur application rates on nutrients availability, uptake and potato quality and yield in calcareous soil. Egypt. J. Soil Sci. 38:271–286.

Emilsson, B., and N. Gustafsson. 1954. Studies on the control of common scab on the potato. Acta Agric. Scand. 4:33–62.

Eppendorfer, W.H., and B.O. Eggum. 1994. Sulphur deficiency of potatoes as reflected in chemical composition and in some measures of nutritive value. Norw. J. Agric. Sci. Supp. 15:127–134.

Frank, J.A., and S.S. Leach. 1980. Comparison of tuberborne and soilborne inoculum in the Rhyzoctonia disease of potato. Phytopathology 70:51–53.

Garman, H. 1898. Corrosive sublimate and flour of sulphur for potato scab. Experiments made in 1896. Kentucky Agric. Exp. Sta., Bull. 72.

Gillespie, L.J. 1918. The growth of the potato scab organism at various hydrogen ion concentrations as related to the comparative freedom of acid soils from the potato scab. Phytopathology 7:257–269.

Gillespie, L.J., and L.A. Hurst. 1918. Hydrogen ion concentration- soil type- common potato scab. Soil Sci. 6:219–236.

Goss, R.W. 1934. A survey of potato scab and Fusarium wilt in western Nebraska. Phytopathology 24:517–527.

Goto, K. 1985. Relationship between soil pH, available calcium and prevalence of potato scab. Soil Sci. Plant Nutr. 31:411–418.

Gray, E.G., J.D. Smith, and B.C. Knight. 1961. Some effects of soil treatments for control of common scab and black scurf of potato. Eur. Potato J. 4:277–278.

Gupta, U.C., and J.B. Sanderson. 1993. Effect of sulfur, calcium, and boron on tissue nutrient concentration and potato yield. J. Plant Nutr. 16:1013–1023.

Halsted, R.D. 1895. Field experiments with potato. NJ Agric. Exp. Sta. Bull. 112.

Halsted, B.D. 1900. Soil fumigation for potato and turnip diseases. NJ Agric. Exp. Sta. 20th Annu. Rep.

Hammerschmidt, R., R.W. Chase, and M.L. Vitosh. 1986. Control of scab and rhizoctonia diseases. Mich. Potato Industry News 25(3):4–5.

Hammerschmidt, R., and M.L. Vitosh. 1987. Control of potato scab. Mich. Potato Industry News 26(2):10–11.

Haneklaus, S., E. Bloem, and E. Schnug. 2003. The global sulphur cycle and its links to plant environment. p. 1–28. *In* Y.P. Abrol and A. Ahmad (ed.) Sulphur in plants. Kluwer Acad. Pub., Dordrecht, the Netherlands.

Haneklaus, S., E. Bloem, and E. Schnug. 2007. Sulfur and plant disease. p. 101–118. *In* L.E. Datnoff et al. (ed.) Mineral nutrition and plant disease. APS, St. Paul, MN.

Hooker, W.J., and G.C. Kent. 1950. Sulfur and certain soil amendments for potato scab control in the peat soils of northern Iowa. Am. Potato J. 27:343–365.

Horsfall, J.G., J.P. Hollis, and H.G.M. Jacobson. 1954. Calcium and potato scab. Phytopathology 44:19–24.

Houghland, G.V.C., and L.C. Cash. 1956a. Some physiological aspects of the potato scab problem. I. Acidity and aluminum. Am. Potato J. 33:86–91.

Houghland, G.V.C., and L.C. Cash. 1956b. Some physiological aspects of the potato scab problem. II. Calcium and calcium-potassium ratio. Am. Potato J. 33:235–241.

Huber, D.M., and R.D. Watson. 1970. Effect of organic amendment on soilborne plant pathogens. Phytopathology 60:22–26.

Keinath, A.P., and R. Loria. 1989. Management of common scab of potato with plant nutrients. p. 152–166. *In* A. W. Engelhard (ed.) Soilborne plant pathogens: Management of diseases with macro- and microelements. APS, St. Paul, MN.

Klikocka, H., S. Haneklaus, E. Bloem, and E. Schnug. 2005. Influence of sulfur fertilization on infection of potato tubers with *Rhyzoctonia solani* and *Streptomyces scabies*. J. Plant Nutr. 28:819–833.

Kunkel, R. 1978. Sulphur requirements of Russet Burbank potatoes in Washington's Columbia Basin. 17th Wash. Potato Conf. and Trade Fair, Proc.

Lambert, D.H., and R. Loria. 1989. *Streptomyces acidiscabies* sp. nov. Int. J. Syst. Bacteriol. 39:393–396.

Lambert, D.H., and F.E. Manzer. 1991. Relationship of calcium to potato scab. Phytopathology 81:632–636.

Lapwood, D.H., and T.F. Hering. 1970. Soil moisture and the infection of young potato tubers by *Streptomyces scabies* (common scab). Potato Res. 13:296–304.

Larson, R.H., A.R. Albert, and J.C. Walker. 1938. Soil reaction in relation to potato scab. Am. Potato J. 15:325–330.

Lawrence, C.H., M.C. Clark, and R.R. King. 1990. Induction of common scab symptoms in aseptically cultured potato tubers by the vivotoxin, thaxtomin. Phytopathology 80:606–608.

Loman, H. 1977. Sulphur fertilization experiments on Dutch sandy soils. Inst. Bodemvruchtbaarheid, Rapp. 8–77.

Loria, R., R.A. Bukhalid, B.A. Fry, and R.R. King. 1997. Plant pathogenicity in the genus *Streptomyces*. Plant Dis. 81:836–846.

Mabbett, T. 2001. Sulphur is a barrier to potato scab. Int. Pest Control 43:16–17.

Martin, W.H. 1920. The relation of sulphur to soil acidity and to the control of potato scab. Soil Sci. 9:393–408.

Martin, W.H. 1921. A comparison of inoculated and uninoculated sulfur for the control of potato scab. Soil Sci. 11:75–84.

Martin, W.H. 1923. Influence of soil moisture and acidity on the development of potato scab. Soil Sci. 16:69–73.

McCreary, C.W.R. 1967. The effect of sulphur application to the soil in the control of some tuber disorders. p. 303–308. *In* Brit. Insect Fung. Conf., Proc., 4th.

McGregor, A.J., and G.C.S. Wilson. 1964. The effect of applications of manganese sulphate to a neutral soil upon the yield of tubers and the incidence of common scab of potatoes. Plant Soil 20:59–64.

McGregor, A.J., and G.C.S. Wilson. 1966. The influence of manganese on the development of potato scab. Plant Soil 25:3–16.

Mortvedt, J.J., K.C. Berger, and H.M. Darling. 1963. Effect of manganese and copper on the growth of *Streptomyces scabies* and the incidence of potato scab. Am. Potato J. 40:96–102.

Mortvedt, J.J., M.H. Fleischfresser, K.C. Berger, and H.M. Darling. 1961. The relation of soluble manganese to the incidence of common scab in potatoes. Am. Potato J. 38:95–100.

Muncie, J.H., H.C. Moore, J. Tyson, and E.J. Wheeler. 1944. The effect of sulphur and acid fertilizer on incidence of potato scab. Am. Potato J. 21:293–304.

Odland, T.E., and H.G. Allbritten. 1950. Soil reaction and calcium supply as factors influencing the yield of potato and the occurrence of scab. Agron. J. 42:269–275.

Panique, E., K.A. Kelling, E.E. Schulte, D.E. Hero, W.R. Stevenson, and R.V. James. 1997. Potassium rate and source effects on potato yield, quality, and disease interaction. Am. Potato J. 74:379–398.

Pavlista, A.D. 1993. Control of common scab with sulfur and ammonium sulfate. Spudman 31(8):13,32,34.

Pavlista, A.D. 1996. How important is common scab in seed potato? Am. Potato J. 73:275–278.

Pavlista, A.D. 2004. Early-season applications of sulfur fertilizers increase potato yield and reduce tuber defects. Agron. J. 97:599–603.

Pavlista, A.D., and J.M. Blumenthal. 2001. Potatoes. p. 151–156. In R.B. Ferguson and K.M. DeGrout (ed.) Nutrient management for agronomic crops in Nebraska, Univ. Nebraska Coop. Ext. Circ. 155.

Pavlista, A.D., and J.C. Ojala. 1997. Potatoes: Chip and French fry processing. p. 237–284. In D.S. Smith et al. (ed.) Processing vegetables: Science and technology. Technomics Publ. Co. Inc., Lancaster, PA.

Ramamurthy, N., and L. Suseela-Devi. 1982. Effect of different sources of sulphur on the yield and quality of potato. J. Indian Soc. Soil Sci. 30:405–407.

Reiffenstein, R.J., W.C. Hulbert, and S.H. Roth. 1992. Toxicology of hydrogen sulfide. Annu. Rec. Pharmacol. Toxicol. 32:109–134.

Sanford, G.B. 1926. Some factors affecting the pathogenicity of Actinomyces scabies. Phytopathology 16:525–546.

Schaal, L.A. 1940. Variation in the tolerance of certain physiological races of *Actinomyces scabies* to hydrogen-ion concentrations. Phytopathology 30:699–700.

Schaal, L.A. 1943. Variation and physiologic specialization in the common scab fungus (*Actinomyces scabies*). J. Agric. Res. 69:169–186.

Schmidt, A. 2005. Metabolic background of H_2S release from plants. Landbauforschung Volkenrode (FAL Agric. Res.). Special Issue 283:121–130.

Schroeder, P. 1993. Plants as sources of atmospheric sulfur. p. 253–270. In L.J. DeKok et al. (ed.) Sulfur nutrition and assimilation in higher plants. SPB Academic Pub., The Hague.

Sherbakoff, C.C. 1914. Potato scab and sulfur disinfection. p. 709–743. NY (Cornell) Agric. Exp. Sta. Bull. 350.

Simmons, K.E., K.A. Kelling, R.P. Wolkowski, and A. Kelman. 1988. Effect of calcium source and application method on potato yield and cation composition. Agron. J. 80:13–21.

Singh, J.P., and O.P. Srivastava. 1993. Irrigation water as a source of sulphur and its critical concentration for potato (*Solanum tuberosum*) crop. Indian J. Agric. Sci. 63:237–239.

Soltani, N., K.L. Conn, P.A. Abbasi, and G. Lazarovits. 2002. Reduction of potato scab and verticillium wilt with ammonium lignosulfonate soil amendment in four Ontario potato fields. Can. J. Plant Pathol. 24:332–339.

Stevenson, W.R., R. Loria, G.D. Franc, and D.P. Weingartner. 2001. Compendium of potato diseases. 2nd ed. APS, St. Paul, MN.

Vlitos, A.J., and W.J. Hooker. 1951. The influence of sulfur on populations of *Streptomyces scabies* and other *Streptomycetes* in peat soil. Am. J. Bot. 38:678–683.

Wheeler, H.J. and Adams, G.E. 1897. On the use of flowers of sulfur and sulphate of ammonia as preventive of the potato scab in contaminated soils. RI Agric. Exp. Stn. 10th Annu. Rep.

Waterer, D. 2002. Impact of high soil pH on potato yields and grade losses to common scab. Can. J. Plant Sci. 82:583–586.

Yang, Z., S. Haneklehaus, L.J. DeKok, E. Schnug, and B.R. Singh. 2006. Effect of H_2S and dimethylsulfide (DMS) on growth and enzymatic activities of *Rhizoctonia solani* and its implications for sulfur-induced resistance (SIR) of agricultural crops. Phyton (Buenos Aires) 46:55–70.

12

Sulfur, Its Role in Onion Production and Related Alliums

George E. Boyhan
University of Georgia, Statesboro

Abstract

Unlike other crop species, the Alliums, which include onion (*Allium cepa* L.), garlic (*Allium sativa* L.), leek (*Allium ampeloprasum* L.), shallot (*Allium cepa* L. var. *ascalonicum*), and chives (*Allium schoenoprasum* L.), rely on sulfur for their unique flavor. A variety of sulfur compounds in these plants determines the flavor profile of these crops. This chapter reviews how pungency is tested, fertilizer demands of Allium crops, and how sulfur influences bulb quality and crop growth.

From a production standpoint, sulfur is a secondary nutrient grouped with calcium and magnesium in importance, behind nitrogen, phosphorus, and potassium. All other nutrients necessary to plant growth are referred to as micronutrients, indicative of the amounts required by plants.

Sulfur is a significant constituent in all living organisms primarily for the role it plays in amino acids, proteins, and, in particular, to those proteins that act as enzymes. Among the 20 amino acids that are the basic building blocks of proteins, cysteine and methionine contain sulfur in their side chains (Stryer, 1988). In addition to this, many proteins contain disulfide bonds, which are cross-links within or between different amino acid chains. This ability to form cross-links is an important factor in determining the primary structure of proteins (Stryer, 1988).

Sulfur plays an important role in onions and other Alliums because it directly influences their unique flavors. Sulfur is the primary element in many compounds that produce a specific flavor profile to a specific species of Allium. These compounds are essential in pungency or heat development in Alliums, particularly onions and garlic. Pungency has become an important flavor component particularly in mild, sweet onions where a lack of pungency is particularly desirable.

A tremendous amount of research has been done with onions and related Alliums on sulfur-containing compounds, their fate in the plant, factors affecting accumulation in both environmental and cultural practices, and the consequence for flavor, nutrition, disease resistance, and medical purposes to name a few. In addition, their genetic regulation has also been studied.

Copyright © 2008. American Society of Agronomy, Crop Science Society of America, Soil Science Society of America, 677 S. Segoe Rd., Madison, WI 53711, USA. *Sulfur: A Missing Link between Soils, Crops, and Nutrition.* Agronomy Monograph 50.

History and Adaptability

These same flavor compounds were probably instrumental in early adoption and worldwide distribution of Alliums through human cultivation (National Onion Association, 2007). Early civilizations relied heavily on a single commodity for the bulk of their dietary calories and some groups still do so today. Rice (*Oryza sativa* L.) in Asia, wheat (*Triticum aestivum* L.) in Europe, and corn (*Zea mays* L.) in the Americas are but some examples. This heavy reliance on a few commodities meant that these civilizations would have been interested in diversifying their meals, particularly as it relates to flavor. Onions and other Alliums are an ideal compliment because they offer flavor and perhaps more importantly storability, at least with the higher solid onion types and garlic.

An important characteristic of onions is their adaptability to different environments, which also helped their widespread distribution. Although onions are a cool season crop, they have adapted to growing from the tropics to the arctic by day-length response among different onion types (Brewster, 1990). All onions begin to bulb as days get longer. Short-day onions begin to form bulbs at 11 to 12 h and are adapted to tropical, subtropical regions, and lower parts of the temperate regions where they can be grown as an over-wintering crop, whereas intermediate (12–14 h) and long-day (14–17 h) onions are adapted to more northerly range of the temperate zone to the subarctic region and can be grown as an over-wintering crop in milder areas or late winter–early spring planted crop. This adaptability is also found in garlic with soft-neck and hard-neck varieties adapted to different regions south or north (Boyhan et al., 2000). This wide range of adaptability, storability, and flavor characteristics has resulted in wide cultivation for onions and to a lesser extent garlic and other Alliums.

Flavor Development

Onions and related Alliums' flavor is in large part attributed to various sulfur-containing compounds, which has a fairly well-understood process of development. For a comprehensive discussion of flavor development in Alliums, view Brewster (1994) and Lancaster and Boland (1990). It begins with a set of sulfur-containing precursors with the general name, S-alk(en)yl cysteine sulfoxides (Brewster, 1994). The general structure of which is

$$R - S(\uparrow O) - CH_2 - CH(NH_2)COOH$$

The L-cysteine sulfoxide (R) group can be one of several forms:
1. $CH_3-(+)-S-$methyl
2. $CH_3-CH_2-CH_2-(+)-S-$propyl
3. $CH_3-CH=CH-trans(+)-S-(1-$propenyl$)$
4. $CH_2=CH-CH-(+)-S-(2-$propenyl$)$

These amino acids are not part of any protein and are not volatile until cell disruption occurs. When cell tissue is macerated, alliinase enzyme that is held in cell vacuoles is released and comes in contact with these precursors in the cytoplasm and quickly converts these precursors into volatile flavor compounds. This is why with many onions, when you begin to chew the onion, you may experience a sweet taste followed in a few seconds by an ever-increasing tingling sensation. This is the lag as the enzyme is released from the vacuole and begins to convert the precursors into flavor compounds.

Alliinase actually converts the precursors into sulfenic acid with the concomitant release of ammonia and pyruvate. The sulfenic acid immediately rearranges into many different volatile, strong-smelling compounds. There are over 80 of these different volatile compounds that have been identified from Alliums (Brewster, 1994). The most important of these are the thiosulfinates, which have the general structure

$$R - S - \overset{\overset{O}{\uparrow}}{S} - R$$

In onions, when 1-propenyl sulfenic acid is among these compounds, it rearranges to thiopropanal S-oxide, which is the lachrymatory (tear producing) factor. In garlic, alliin or S-ally L-cysteine sulfoxide, which is the same as the fourth precursor listed above, is catalyzed by alliinase it forms allicin, which gives garlic its characteristic odor.

The relative amounts of the flavor precursors have a significant effect on the flavor profile of any specific Allium species (Table 12–1). Flavor characteristics of specific forms of these precursors have been described. The flavor characteristic of the oxidation of 1-propyl disulphide was that of a freshly cut onion, while the oxidation of methyl disulfide had an unpleasant cabbage-like odor (Freeman and Whenham, 1976). Other descriptions of flavor profiles from specific precursors have included green onion flavor, radish-like, brassica-like, sweet sulfur, and livery (Randle et al., 1994).

These volatile compounds probably developed as a defense mechanism against insects and diseases (Brewster, 1994). Extracts from onion and garlic have been shown to be fungicidal or fungistatic in vitro. In addition, extracts have been shown to have insecticidal and nematocidal properties.

Table 12–1. Relative amounts of flavor precursors found in various alliums. Adapted from Brewster (1994).

Species	Cysteine sulfoxides			
	S-methyl	S-propyl	S-(1-propenyl)	S-(2-propenyl)
Onion (A. cepa)	+	++	+++	0
Garlic (A. sativum)	++	+	0	+++
Leek (A. ampeloprasum)	++	++	+	0
Shallot (A. cepa var. ascalonicum)	++	++	+	0
Japanese bunching onion (A. fistulosum L.)	+	++	++	0
Elephant garlic (A. ampeloprasum)	++	+	0	+++
Chives (A. schoenoprasum)	+	+	++	0
Chinese chives (A. tuberosum Rottler ex Sprengel)	++	+	+	+++
Rakkyo (A. chinense G. Don)	++	+	++	0

Other Alliums have been studied for volatile accumulation (Table 12–1). In one such study of Welsh onions (*A. fistulosum*) over 60 volatile sulfur-containing compounds were identified, many of these thiosulfinates and thiosulfonates had not been previously identified (Kuo and Ho, 1992).

Pungency Testing

A relatively easy test was developed by Schwimmer and Weston (1961) to test for pyruvate, which is produced in a one-to-one ratio with these flavor compounds. This test (now called the pyruvate test) is used to determine pungency in Alliums. This test has been modified by several researchers to streamline the process, improve standardization, reduce reagents used, and improve pungency throughout (Anthon and Barrett, 2003; Boyhan et al., 1999; Randle and Bussard, 1993; Yoo and Pike, 1999; Yoo et al., 1995). This is not the only test developed to determine pungency in onions, others have been developed, but are not as widely used (Thomas et al., 1992). Abbey et al. (2001), for example, has used an electronic nose to discriminate pungency and flavor among different Alliums to overcome some of the limitations with pyruvate analysis.

This pyruvate test involves the maceration of onion or other Allium tissue. Juice from this tissue is allowed to incubate for approximately 10 min at room temperature to allow the alliinase to catalyze the flavor precursors. At this point a trichloroacetic acid solution is added to the juice to stop the reaction. Then 2,4-dinitrophenylhydrazine (2,4-DNPH) in a hydrochloric acid solution and water are added to the solution and incubated at 37°C for 10 min. The 2,4-DNPH reacts with carbonyls, which are present in pyruvate that has been generated in the alliinase catalyzed reaction in a one-to-one ratio with the volatile flavor compounds. Finally a sodium hydroxide solution is added which reacts to form a yellowish colored solution. The intensity of the color development is indicative of the pyruvate content and indirectly the amount of volatile flavor compounds. The light absorbance at 420 nm is measured and compared with a standard curve of sodium pyruvate. The results are expressed as micromoles per gram fresh weight ($\mu m\ gfw^{-1}$). Pyruvate, of course, is part of other biological processes, most notably in glycolysis and the Krebs cycle, which is part of respiration (Curtis, 1979). For this reason the background pyruvate content is often determined by heating an intact onion, which inactivates the alliinase enzyme before juice expression. With experience, researchers have determined that subtracting 0.4 to 1.0 $\mu m\ gfw^{-1}$ from the pyruvate analysis is a good approximation of the background pyruvate, so the determination of background pyruvate is not always done.

Values with the pyruvate test can range from under 5 $\mu m\ gfw^{-1}$ for short-day mild onions as found in the Vidalia growing region into double digits for higher solids long-day onions (Boyhan et al., 2005; Kopsell and Randle, 1997a). Garlic can have a range from the high 20s to over 90 $\mu m\ gfw^{-1}$ (Natale et al., 2005; Verma et al., 2004).

The pyruvate test has not been without controversy, however, in its application and repeatability. The test is used as a marketing tool among sweet onion producers to ensure their customers that their onions are indeed mild flavored. In addition, different regions of sweet onion production have used the test to claim the mildest onions. Havey et al. (2002) found considerable variation among laboratories utilizing this procedure. Randle has done a good job of attempting to standardize the test so that different laboratory results are comparable (Ga. Dep. of Agric., 2007; Randle and Bussard, 1993). He has even gone so far as to have the protocol, including pictures, written into the rules governing Vidalia onion testing in Georgia (Ga. Dep. of Agric., 2007). In 1 yr, we found a high degree of correlation (>0.80) in results from laboratories following Randle's procedures, but in another year the correlation was only about 0.50 (unpublished).

Wall and Corgan (1992) found a high degree of correlation between the pyruvate test and taste panel flavor perception. Correlation coefficients ranged from 0.79 to 0.95. It should be pointed out, however, that their work was with relatively small samples over a range of about 3.5 to over 7 µm gfw^{-1}, and it is unclear how well the test results would have been if the range were below 5 µm gfw^{-1}, which is typical with mild short-day onions. Crowther et al. (2005) also found a correlation between flavor and pyruvate, but again the pyruvate range was from 1.2 to 9.3 µm gfw^{-1}. In addition, they found differences in the prediction of flavor attractiveness with some varieties that did not completely correlate with the pyruvate levels. Doruchowski (1970) found a correlation between flavor and sulfur content among breeding material in variety Wolska. Comparisons of taste panel evaluations and pyruvate analysis in Georgia where onions are routinely produced with 5 µm gfw^{-1} or less (in some years much less) has shown only a weak correlation between pyruvate analysis and taste perception (unpublished data).

Sulfur Fertilization

Soils are the primary source of sulfur in plants, including onions and related Alliums. Because of the central role sulfur plays in flavor development in the Alliums, the soil nutrient status as it relates to sulfur can be very important. About 80% of the sulfur in soils is stored in organic compounds with well-drained mineral soils having the least amount of sulfur (Plaster, 1996, p. 402). Sulfur is readily leached from soils, particularly coarse textured soils because it does not adsorb onto soil particles in its iconic form sulfate (SO_4^{-2}). The primary source of sulfur in the soil is from the weathering of sulfate containing minerals such as calcium sulfate (gypsum) (Plaster, 1996, p. 402).

The amount of sulfur fertilizer that a soil requires is determined by several factors—soil texture, organic matter, leaching, sulfur content of irrigation water, previous fertilizer practices, and soil test results. Soil test results for sulfur are often unreliable, especially where sulfur levels are low (Brown, 2000). Since sulfur is readily leached in many soils, there may be a sizable amount of sulfur below 30.5 cm, which may precipitate with calcium forming gypsum (Brown, 2000). This does not usually help onions because they are relatively shallow rooted. Deep turning may bring these sulfur latent soils into the root zone of onions reducing the possibility of sulfur deficiency.

Sources of sulfur include elemental sulfur, ammonium sulfate, ammonium thiosulfate, magnesium sulfate (Epsom salts), potassium magnesium sulfate, potassium sulfate, calcium sulfate (gypsum or land plaster), ferric sulfate, and sulfuric acid. Many of these materials also supply other important plant nutrients, and in the case of elemental sulfur, it is used to reduce soil pH (Maynard and Hochmuth, 1997; Mitchell, 1987). With the use of elemental sulfur, it is recommended that it be finely ground with a screen size of less than 40 mesh (Oregon State University, 2004).

Soils used for onion production vary widely depending on the location. Washington State, for example, is one of the major onion producing regions in the USA. The soils onions are produced on in Washington are primarily in the Columbia Basin region of southeast Washington (Pelter and Sorensen, 2003). The soils in this region tend to be well-drained, sandy-loam soils. They produce both short- and intermediate-day, over-wintering onions along with spring planted long-day onions (Pelter and Sorensen, 2003).

Part of the Columbia Basin is in Oregon where they also produce these onions (Oregon State University, 2004). As in Washington, they produce both over-wintering, short-day mild onions, and spring-sown long-day Spanish onions. Sulfur recommendations in this area are from 45 to 67 kg ha^{-1} with the caveat that sweet onions should be fertilized on the low end of this scale to ensure mildness. A relatively new area of onion production occurs on the western side of the Cascade Mountains in the Willamette Valley (Hemphill, 2002). This production occurs on both mineral and muck soils. On the mineral soils, sulfur recommendations would follow those for eastern Oregon; however, on the muck soils, no additional sulfur is required.

In the Treasure Valley of southwestern Idaho and eastern Oregon, onion production does not generally suffer from sulfur deficiency. There is sufficient sulfur in most soils in this area (Brown, 2000). In addition, irrigation water in this area may be a source of sulfur. If, however, both soil and water sulfur levels are below 5 μg g^{-1}, fertilizer application of 45 kg ha^{-1} are recommended.

In California, onion production is primarily with short-day and intermediate-day onions, which are marketed as mild sweet onions, either as Grano or Granex types or Sweet Spanish (Voss and Mayberry, 1999). This means that soil sulfur levels are important because of their contribution to pyruvate levels. California is one of the largest producers of onions with production ranging from southern to central California.

California is also the major producing state for garlic. Garlic has a much higher level of pyruvate and therefore limiting sulfur to influence flavor is generally not done. The sulfur content of garlic, however, is important because it is a constituent in many compounds that are believed to have health benefits (Bloem et al., 2004).

New Mexico onion production recommends ammonium sulfate as the first nitrogen application preplant (Corgan et al., 2000). Other than this recommendation, sulfur is only recommended as sulfuric acid or gypsum based on tests of high levels of exchangeable sodium, which affect soil structure.

The production guide for Texas does not make specific recommendations for sulfur as a fertilizer, but it is recommended for disease control and as a herbicide as sulfuric acid (Hall et al., 2000). Finally, in New York most onion production is on organic soils with long-day onions, which have high solids, pungency, and readily store for long periods of time (Stivers, 1999). The organic soils of New York require very little fertilization and the soils are relatively high in sulfur. There are some onions grown on well-drained mineral soils that do require additional fertilizer.

In Georgia, application rates of sulfur can range from 11 to 67 kg ha^{-1} in soils and crops that require additional sulfur fertilizer (Kissel, 2003; Maynard and Hochmuth, 1997). In south Georgia, for onion production, it is recommended that 45 to 67 kg ha^{-1} of additional sulfur be applied to meet the crop's needs (Boyhan et al., 2001). This contrasts with recommendations for other vegetables in this region, which require only 11 kg ha^{-1} sulfur (Kissel, 2003).

In southeastern Georgia, sulfur application is usually accomplished with complete fertilizers that also contain sulfur. Products such 5–10–15 with 9% sulfur or 6–12–18 with 5% sulfur are commonly used. More recent work suggests that these formulations are inefficient and their use should be discontinued because they have such low levels of nitrogen and relatively high levels of phosphorus and potassium (Boyhan et al., 2007). A more realistic complete fertilizer to use that is common in the Vidalia region is 10–10–10 with 12% sulfur. This can be used to apply all the needed sulfur at transplanting. In any case, recommendations of sulfur fertilizer should be applied before the end of January to minimize pungency in spring harvested onions.

Leaf tissue sulfur levels for sweet onions is considered in the adequate range with 0.20 to 0.60% and is considered deficient below 0.20% (Maynard and Hochmuth, 1997). Sulfur deficiency in onions is not an uncommon occurrence and can occur during periods of heavy rain on light textured soils. Soil pH can also play a role in sulfur deficiency with the potential for low pH inducing deficiency. In addition, the nitrogen/sulfur ratio is considered important with recommendations of 5:1 to 15:1 or 10:1 to 17:1 (C. Owen Plank, unpublished; W. O. Chance, unpublished). Too high a ratio is supposed to induce a sulfur deficiency. This occurs when too much nitrogen fertilizer is used, particularly if the nitrogen is in the ammonia form (W.O. Chance, unpublished). This has been reported to cause sulfur deficiency, but in practical experience, it rarely does within standard nitrogen recommendations. In southeastern Georgia, growers are advised to back off on nitrogen fertilizer rather than increase sulfur fertilizer because sulfur is known to contribute to onion pungency.

Sulfur deficiency symptoms appear as chlorotic stripes in the leaf tissue. This should not be confused with chimeras that occasionally occur in onion fields where more dramatic green and yellow stripes occur. Other symptoms associated with sulfur deficiency are tip burn and stunted growth (Abbey et al., 2002).

Species and Variety Effects

The effect of sulfur fertility on Allium species and varieties has been shown in many studies. Abbey et al. (2002) found that onion was more responsive to sulfur than *A. fistulosum* and sulfur fertilization had differential effects on other plant constituents such as total soluble solids based on onion variety. Onion varieties could be distinguished from one another by volatile analysis. Varieties Jumbo and Lafort were found to be significantly different in headspace analysis by gas chromatography and gas chromatography–mass spectrometry of sulfur-containing volatiles (Kallio and Salorinne, 1990). In addition, differences in pyruvate analysis have been found between different varieties (Boyhan et al., 2005; Kopsell and Randle, 1997a). Varietal effect on the lachrymatory factor has also been studied in both short-day and long-day onions (Kopsell et al., 2002).

Investigations have been made on how different onion varieties respond to increasing solution sulfate levels (Randle et al., 1999). Savannah Sweet, a short-day mild onion, showed much more response (linear increase) to increasing sulfur fertility as measured by total sulfur with a lower intercept compared with Southport White Globe, a long-day hard onion, which had a lower slope and greater intercept (Randle et al., 1999). This indicates that long-day onions are more efficient at sulfur uptake at lower sulfur fertility.

Onion clones of 'TG 1015Y' were evaluated under low and high sulfur fertility (Hamilton et al., 1997). At low sulfur fertility, there was not much difference in pyruvate analysis, but at high sulfur fertility, there was considerable variation among the clones. In a related study with clones, it was found that genetic differences accounted for over 80% of differences in pyruvate levels (Yoo et al., 2006).

Pyruvate analysis has shown differences between onion type (day-length response) and onion varieties over time in storage (Kopsell and Randle, 1997a). Intermediate- and long-day onions showed a decrease in pyruvate, with long-term cold storage, while short-day onions had variable responses based on variety.

In Georgia, there has been an interest in how pungency and thus sulfur-containing compounds are affected by variety. Over 10 yr ago, a new type of short-day onion was introduced to the Vidalia onion growing region of southeast Georgia. The most dramatic distinguishing characteristic of these varieties was their earliness. They produced onions a full 2 wk earlier than any varieties grown in the region previously. These varieties, referred to as Japanese over-wintering onions, also exhibited greater foliar disease resistance and strong day-length response. At maturity, the tops would break over at the neck early and very uniformly (short time frame).

These early varieties quickly earned a reputation for being more pungent than the traditional onion varieties. This led to extensive testing comparing varieties by the pyruvate test as well as taste testing (Boyhan et al., 2006, 2005). This testing was not able to distinguish these early varieties from more traditional varieties grown in this region by either pyruvate analysis or taste test. It has been postulated that there is some other, as yet unidentified principle, responsible, but with taste tests not able to distinguish a difference, it may have been premature to label these short-day onions as unusually pungent.

When these Japanese over-wintering onions were first introduced, growers would harvest onions when 20 to 50% of the tops were down (broken over at the neck) (Boyhan et al., 2001). This was too early for these varieties, which should have 100% of their tops down before harvest. It has been shown that these onions will continue to increase in size presumably from water uptake after tops are down (Boyhan et al., 2004). Generally, higher water content and larger size are associated with mildness in short-day onions.

Sulfur, Bulb Quality, and Metabolic Pathways

Sulfur fertility has been shown to affect bulb firmness and dry weight in onion. Lancaster et al. (2001) showed a reduction in firmness, dry weight, and accumulation of sulfur in cell walls with lower sulfur fertility in hydroponic culture. Lower sulfur fertility under these conditions, however, did not affect total sulfur accumulation in bulbs but did limit growth.

Leaf tissue sulfur levels increased during early development but decreased during bulbing whether under a low or high sulfur fertility regimen. The decrease in leaf tissue under low sulfur fertility was more severe, but correlations between leaf tissue sulfur levels and final bulb sulfur levels were generally poor (Randle et al., 1993). It is well documented that soil sulfur levels will increase sulfur-containing precursors based on the pyruvate test (Randle, 1997). Randle et al. (1999) also found that increasing sulfate concentration in nutrient solutions resulted in increased sulfate, total sulfur, and onion volatiles as measured by pyruvate analysis in onion bulbs.

In more extensive research on onion sulfur and flavor precursors response to sulfur fertility, Randle et al. (1995) found that (+)-S-methyl-L-cysteine sulfoxide was the dominant flavor precursor under sulfur deficiency, while trans(+)-S-(1-propenyl)-L-cysteine sulfoxide was the dominant flavor precursor under high sulfur fertility.

Interactions of sulfur with other compounds have been studied especially as they affect uptake and sulfur metabolism. Bloem et al. (2004) investigated the effect of sulfur and nitrogen on the accumulation of alliin in onion and garlic and found increasing sulfur fertilizer increased alliin accumulation, but that nitrogen had little or no effect. In addition, they found that alliin was translocated to the bulbs during bulb enlargement and maturation, which could influence harvest if alliin content were an important criteria for flavor, nutritional, or medicinal reasons. Others have reported different results concerning nitrogen fertilizer rates and its affect on sulfur precursors. Randle (2000), for example, found that the accumulation of different sulfur precursors were variable depending on nitrogen fertilizer levels. He also found that pyruvate analysis increased with increasing nitrogen fertilizer in hydroponic solution, except for the highest concentration of 0.97 g L^{-1} nitrogen. In addition, methyl cysteine sulfoxide (cabbage-like and fresh onion flavor) increased with increasing nitrogen, while 1-propenyl cysteine sulfoxide (heat, mouth burn) increased only over the lowest nitrogen concentration and then decreased. Finally, propyl cysteine sulfoxide (fresh onion and sulfur flavor) increased with increasing nitrogen fertility. Evaluation of the interaction between nitrogen and sulfur fertility as it affects onion flavor showed, increasing sulfur enhanced trans-S-1-propenyl-L-cysteine sulfoxide precursor accumulation, while increasing nitrogen raised S-methyl-L-cysteine sulfoxide content (Coolong and Randle, 2003).

Studies of accumulation of sulfur in onion tissue and selenium uptake have been conducted (Kopsell and Randle, 1997b). Selenium, enhanced bulb sulfur levels, but pyruvate was reduced. It is conjectured that selenium interferes with metabolism of flavor precursors. In a later study, by these same researchers, it was found selenium lowered total bulb sulfur content (Kopsell and Randle, 1999). Barak and Goldman (1997) reported an antagonistic relationship between sulfate and selenate. Adding more selenate to solution did not continue to increase tissue selenium beyond a certain point, whereas increasing sulfate concentration continued to increase onion sulfur content and reduced selenium concentration in bulb tissue.

High sodium chloride levels have been shown to have deleterious effects on onion dry weight, but when applied in combination with calcium chloride, these effects could be ameliorated (Arvin and Kazemi-Pour, 2002). Onion sensitivity and response to various levels of sodium chloride has been studied. Others have found that increasing application of sodium chloride decreased bulb and leaf fresh weight as would be expected, but sulfur content as sulfate increased or remained the same, while overall sulfur content decreased (Chang and Randle, 2004; Chang and Randle, 2005). This suggested that less sulfur was entering the metabolic pathway of flavor precursors. Related to this, calcium chloride application on onions decreased sulfur accumulation, reduced pyruvate levels, and affected sulfur precursors (Randle, 2005). The influence of chloride in the application of calcium chloride in affecting sulfur accumulation and metabolism may be related to competition with uptake of sulfate rather than a specific need for chloride (T. Coolong, personal communication).

Sulfur and Other Effects

Soil temperature has a significant effect on tissue sulfur levels with onions grown at 12 or 21°C having similar concentrations, while onions grown at 34°C had significantly less sulfur (Coolong and Randle, 2006). Pyruvate analysis followed a similar trend.

As mentioned previously, volatile compounds generated by onion tissue maceration are associated with disease resistance. These compounds apparently can also be correlated with increased susceptibility to white rot (*Sclerotium cepivorum* Berk.). This susceptibility is associated with increased levels of 1- and 2-propenylcysteine sulfoxide (Hovius and Goldman, 2005). The fungicidal properties of sulfur-containing compounds is found in such Alliums as garlic, onion, leek, and shallot with the most effective compound being dimethyl disulfide (Arnault et al., 2005). In addition, disulfides from Alliums have been identified as a potential alternative to methyl bromide for soil fumigation (Arnault et al., 2004).

Sulfur-containing compounds in onions and related Alliums are known to have effects against phytophagus insects, but these same compounds act as attractants to certain insects. For example, onion fly [*Delia antiqua* (Meigen)] larvae have been shown to be attracted to several of the sulfur-containing compounds found in Allium tissue (Soni and Finch, 1979).

Traditional medicine has for centuries relied on herbal treatments prepared from Allium species (Ayaz and Alpsoy, 2007). This is particularly true of garlic, which has been used to treat cardiovascular disease, regulate blood pressure, lower blood sugar, and lower cholesterol to name but a few. A variety of compounds found in garlic have been associated with these therapeutic effects including many sulfur-containing compounds.

Genetic Regulation

There have been attempts to understand the underlying genetic control of sulfur metabolism in onions particularly as it relates to pungency. McCallum et al. (2007) in a cross of 'W202A' × 'Texas Grano 438' has found broad-sense heritability of 0.78 to 0.80 for pungency. PCR-based molecular markers were used to identify putative regions affecting onion pungency. They identified significant associations between pungency and soluble solids with marker intervals on chromosomes 3 and 5. In another study, a cloned cDNA coding for serine acetyltransferase and mapped to chromosome 7 has been shown to be induced with low sulfur supply (McManus et al., 2005). The alliinase enzyme controlling sulfur-containing volatile generation has been characterized for several different Allium species (Kaminishi et al., 2005). At least four isoenzymes of alliinase have been identified with different levels of activity based on species. Species included in the study were onion, garlic, wakegi (*A.* × *wakegi* Araki), rakkyo, Chinese chives, and Welsh onion. Rabinowitch and Currah (2002) have compiled a book of recent advances in understanding Alliums that includes more extensive discussions of gene regulation.

Conclusions

Alliums continue to be an important vegetable crop worldwide, in no small part because of the sulfur-containing compounds present. These compounds have been shown to be important in defense against insect and disease attack, product storability, flavor, health, and medicinal uses. Research continues on various fronts including fertility, physiology, biochemistry, and gene regulation.

References

Abbey, L., J. Aked, and D.C. Joyce. 2001. Discrimination amongst Alliums using an electronic nose. Ann. Appl. Biol. 139:337–342.

Abbey, L., D.C. Joyce, J. Aked, and B. Smith. 2002. Genotype, sulphur nutrition and soil type effects on growth and dry-matter production of spring onion. J. Hortic. Sci. Biotechnol. 77:340–345.

Anthon, G.E., and D.M. Barrett. 2003. Modified method for the determination of pyruvic acid and dinitrophenylhydrazine in the assessment of onion pungency. J. Sci. Food Agric. 83:1210–1213.

Arnault, I., I. André, S. Diwo-Allain, J. Auger, and F. Vey. 2005. Propriétés pesticides des alliacées: Biodésinfection des sols maraîchers au moyen d@oignon et poireau (Pesticidal properties of alliaceous bulb vegetables: Organic disinfection of market garden soils using onions and leeks). Phytoma (January) 40–43.

Arnault, I., N. Mondy, S. Diwo, and J. Auger. 2004. Soil behaviour of sulfur natural fumigants used as methyl bromide substitutes. Int. J. Environ. Anal. Chem. 84:75–82.

Arvin, M.J., and N. Kazemi-Pour. 2002. Effects of salinity and drought stresses on growth and chemical and biochemical compositions of 4 onion (*Allium cepa*) cultivars. J Sci. Tech. Agric. Nat. Res. 5:41–52.

Ayaz, E., and H.C. Alpsoy. 2007. Garlic (*Allium sativum*) and traditional medicine. [Article in Turkish] Sarimsak (*Allium sativum*) ve geleneksel tedavide kullanimi. Turkiye Parazitol Derg. 31:145–149.

Barak, P., and I.L. Goldman. 1997. Antagonistic relationship between selenate and sulfate uptake in onion (*Allium cepa*): Implications for the production of organosulfur and organoselenium compounds in plants. J. Agric. Food Chem. 45:1290–1294.

Bloem, E., S. Haneklaus, and E. Schnug. 2004. Influence of nitrogen and sulfur fertilization on the alliin content of onions and garlic. J. Plant Nutr. 27:1827–1839.

Boyhan, G.E., D.M. Granberry, and W.T. Kelley. 2001. Onion production guide. Univ. of Ga. Bull. 1198.

Boyhan, G.E., W.T. Kelley, and D.M. Granberry. 2000. Production and management of garlic, elephant garlic and leek. Circ. 852:1–4.

Boyhan, G.E., A.C. Purvis, W.C. Hurst, R.L. Torrance, and J.T. Paulk. 2004. Harvest date effect on yield and controlled atmosphere storagability of short-day onions. HortScience 39:1623–1629.

Boyhan, G.E., A.C. Purvis, W.M. Randle, R.L. Torrance, M.J.I. Cook, G. Hardison, R.H. Blackley, H. Paradice, C.R. Hill, and J.T. Paulk. 2005. Harvest and postharvest quality of short-day onions in variety trials in Georgia, 2000–03. Horttechnology 15:694–706.

Boyhan, G., B. Randle, A. Resurreccion, R. Shewfelt, R. Torrance, C. Hopkins, R. Hill, and T. Paulk. 2006. Mandated Vidalia onion variety trials; how well has it worked? HortScience 41:513 (Abstr.).

Boyhan, G.E., N.E. Schmidt, F.M. Woods, D.G. Himelrick, and W.M. Randle. 1999. Adaption of a spectrophotometric assay for pungency in onion to a microplate reader. J. Food Qual. 22:225–233.

Boyhan, G.E., R.L. Torrance, and C.R. Hill. 2007. Effects of nitrogen, phosphorus, and potassium rates and fertilizer sources on yield and leaf nutrient status of short-day onions. HortScience 42:653–660.

Brewster, J.L. 1990. Physiology of crop growth and bulbing. p. 53–88. In H.D. Rabinowitch and J.L. Brewster (ed.) Onions and allied crops volume I Botany, physiology, and genetics. CRC Press, Inc., Boca Raton, FL.

Brewster, J.L. 1994. Onions and other vegetable alliums. Vol. 36 3, p. 2. CAB Intl., Wallingford, UK.

Brown, B. 2000. Onions. University of Idaho CIS 1081:6.

Chang, P.T., and W.M. Randle. 2004. Sodium chloride in nutrient solutions can affect onion growth and flavor development. HortScience 39:1416–1420.

Chang, P.T., and W.M. Randle. 2005. Sodium chloride timing and length of exposure affect onion growth and flavor. J. Plant Nutr. 28:1755–1766.

Coolong, T.W., and W.M. Randle. 2003. Sulfur and nitrogen availability interact to affect the flavor biosynthetic pathway in onion. J. Am. Soc. Hortic. Sci. 128:776–783.

Coolong, T.W., and W.M. Randle. 2006. The influence of root zone temperature on growth and flavour precursors in *Allium cepa* L. J. Hortic. Sci. Biotechnol. 81:199–204.

Corgan, J., M. Wall, C. Cramer, T. Sammis, B. Lewis, and J. Schroeder. 2000. Bulb onion culture and management. New Mexico State Univ. Circ. 563:1–16.

Crowther, T., H.A. Collin, B. Smith, A.B. Tomsett, D. O'Connor, and M.G. Jones. 2005. Assessment of the flavour of fresh uncooked onions by taste-panels and analysis of flavour precursors, pyruvate and sugars. J. Sci. Food Agric. 85:112–120.

Curtis, H. 1979. Biology, p. 1043. Worth Publishers, Inc., New York.

Doruchowski, R.W. 1970. Variation of horticultural characters in some Polish onion varieties, and A, B and C inbreds used as parental forms in producing onion hybrids with heterosis. Biuletyn Warzywniczy. 11:287–306.

Freeman, G.G., and R.J. Whenham. 1976. Thiopropanal S-oxide alk(en)yl thiosulphinates and thiosulphonates: Simulation of flavour components of *Allium* species. Phytochemistry 15:187–190.

Ga. Dep. of Agric. 2007. georgia.gov- Vidalia Onion. Marketing Division, Atlanta. 13 Oct. 2007. http://agr.georgia.gov/00/channel_title/0,2094,38902732_81336415,00.html; verified 10 April 2008.

Hall, K.D., R.L. Holloway, and D. Smith. 2000. Texas crop profile onions. Texas A & M Univ. Publ. E-18.

Hamilton, B.K., L.M. Pike, and K.S. Yoo. 1997. Clonal variations of pungency, sugar content, and bulb weight of onions due to sulphur nutrition. Sci. Hortic. (Amsterdam) 71:131–136.

Havey, M.J., M. Cantwell, M.G. Jones, R.W. Jones, N.E. Schmidt, J. Uhlig, J.F. Watson, and K.S. Yoo. 2002. Significant variation exists among laboratories measuring onion bulb quality traits. HortScience 37:1086–1087.

Hemphill, D. 2002. Dry bulb onions—Western Oregon. Oregon State University. http://hort-devel-nwrec.hort.oregonstate.edu/onionb-w.html; verified 13 April 2008.

Hovius, M.H.Y., and I.L. Goldman. 2005. Flavor precursor [S-alk(en)yl-L-cysteine sulfoxide] concentration and composition in onion plant organs and predictability of field white rot reaction of onions. J. Am. Soc. Hortic. Sci. 130:196–202.

Kallio, H., and L. Salorinne. 1990. Comparison of onion varieties by headspace gas chromatography- mass spectrometry. J. Agric. Food Chem. 38:1560–1564.

Kaminishi, A., H. Doken, K. Nomura, and N. Kita. 2005. Divergence of alliinase among *Allium* species. p. 117–122.

Kissel, D. 2003. Soil test handbook for Georgia. Univ. of Ga., Athens.

Kopsell, D.E., and W.M. Randle. 1997a. Onion cultivars differ in pungency and bulb quality changes during storage. HortScience 32:1260–1263.

Kopsell, D.A., and W.M. Randle. 1997b. Short-day onion cultivars differ in bulb selenium and sulfur accumulation which can affect bulb pungency. Euphytica 96:385–390.

Kopsell, D.A., and W.M. Randle. 1999. Selenium affects the S-alk(en)yl cysteine sulfoxides among short-day onion cultivars. J. Am. Soc. Hortic. Sci. 124:307–311.

Kopsell, D.E., W.M. Randle, and N.E. Schmidt. 2002. Incubation time, cultivar, and storage duration affect onion lachrymatory factor quantification. HortScience 37:567–570.

Kuo, M.C., and C.T. Ho. 1992. Volatile constituents of the solvent extracts of Welsh onions (Allium-Fistulosum L Variety Maichuon) and scallions (a-Fistulosum L variety caespitosum). J. Agric. Food Chem. 40:1906–1910.

Lancaster, J.E., and M.J. Boland. 1990. Flavor biochemistry. p. 33–72. In H.D. Rabinowitch, and J.L. Brewster (ed.) Onions and allied crops Vol. 3. CRC Press, Boca Raton, FL.

Lancaster, J.E., J. Farrant, and M.L. Shaw. 2001. Sulfur nutrition affects cellular sulfur, dry weight distribution, and bulb quality in onion. J. Am. Soc. Hortic. Sci. 126:164–168.

Maynard, D.N., and G.J. Hochmuth. 1997. Knott's handbook for vegetable growers, p. 600. J. Wiley & Sons Inc., New York.

McCallum, J., M. Pither-Joyce, M. Shaw, F. Kenel, S. Davis, R. Butler, J. Scheffer, J. Jakse, and M.J. Havey. 2007. Genetic mapping of sulfur assimilation genes reveals a QTL for onion bulb pungency. Theor. Appl. Genet. 114:815–822.

McManus, M.T., S. Leung, A. Lambert, R.W. Scott, M. Pither-Joyce, B. Chen, and J. McCallum. 2005. Molecular and biochemical characterisation of a serine acetyltransferase of onion, *Allium cepa* (L.). Phytochemistry 66:1407–1416.

Mitchell, C.C. 1987. Nutrient Content of fertilizer materials. Auburn Univ. Circ. ANR-174:1–4.

Natale, P.J., A. Camargo, and C.R. Galmarini. 2005. Characterization of Argentine garlic cultivars by their pungency. Acta Hortic. 688:313–316.

National Onion Association. 2007. About onions: History. 822 7th St. Ste.510, Greeley, CO. http://www.onions-usa.org/about/history.asp; verified 10 April 2008.

Oregon State University. 2004. Dry bulb onions—Eastern Oregon. http://hort-devel-nwrec.hort.oregonstate.edu/onionb-e.html ; verified 13 April 2008.

Pelter, G.Q., and E.J. Sorensen. 2003. Crop profile for onions in Washington State, Washington State University. http://www.ipmcenters.org/cropprofiles/docs/WAonions.html; verified 10 April 2008.

Plaster, E.J. 1996. Soil science & management. Delmar Publishers, Albany, NY.

Rabinowitch, H.D., and L. Currah. 2002. Allium crop science: Recent advances. CABI Publ., New York.

Randle, W.M. 1997. Genetic and environmental effects influencing flavor in onion. Acta Hortic. 433:299–311.

Randle, W.M. 2000. Increasing nitrogen concentration in hydroponic solutions affects onion flavor and bulb quality. J. Am. Soc. Hortic. Sci. 125:254–259.

Randle, W.M. 2005. Advancements in understanding and manipulating *Allium* flavor: Calcium and chloride. 35–40. In Proceedings of the IVth International Symposium on Edible Alliaceae, Beijing, China. 21–26 April 2005.

Randle, W.M., E. Block, M.H. Littlejohn, D. Putman, and M.L. Bussard. 1994. Onion (*Allium cepa* L.) thiosulfinates respond to increasing sulfur fertility. J. Agric. Food Chem. 42:2085–2088.

Randle, W.M., and M.L. Bussard. 1993. Streamlining onion pungency analyses. HortScience 28:60.

Randle, W.M., M.L. Bussard, and D.F. Warnock. 1993. Ontogeny and sulfur fertility affect leaf sulfur in short-day onions. J. Am. Soc. Hortic. Sci. 118:762–765.

Randle, W.M., D.E. Kopsell, D.A. Kopsell, and R.L. Snyder. 1999. Total sulfur and sulfate accumulation in onion is affected by sulfur fertility. J. Plant Nutr. 22:45–51.

Randle, W.M., J.E. Lancaster, M.L. Shaw, K.H. Sutton, R.L. Hay, and M.L. Bussard. 1995. Quantifying onion flavor compounds responding to sulfur fertility- sulfur increases levels of alk(en)yl cysteine sulfoxides and biosynthetic intermediates. J. Am. Soc. Hortic. Sci. 120:1075–1081.

Schwimmer, S., and W. Weston. 1961. Enzymatic development of pyruvic acid in onion as a measure of pungency. J. Sci. Food Chem. 9:301–304.

Soni, S.K., and S. Finch. 1979. Laboratory evaluation of sulphur-bearing chemicals as attractants for larvae of the onion fly, Delia antiqua (Meigen) (Diptera: Anthomyiidae). Bull. Entomol. Res. 69:291–298.

Stivers, L. 1999. Crop profile for onions in New York. Cornell Cooperative Extension, Rochester, NY. October 16, 2007. http://pestdata.ncsu.edu/cropprofiles/docs/nyonions.html; verified 10 April 2008.

Stryer, L. 1988. Protein structure and function. p. 15–42. In Biochemistry. W.H. Freeman and Co., New York.

Thomas, D.J., K.L. Parkin, and P.W. Simon. 1992. Development of a simple pungency indicator test for onions. J. Sci. Food Agric. 60:499–504.

Verma, L.R., K.P.S. Chauhan, and D.K. Singh. 2004. Screening of garlic collections for dehydration. News Letter- National Horticultural Research and Development Foundation. 24:10–12.

Voss, R.E., and K.S. Mayberry. 1999. Fresh-market bulb onion production in California. University of California Vegetable Production Series Publication 7242:1–4.

Wall, M.M., and J.N. Corgan. 1992. Relationship between pyruvate analysis and flavor perception for onion pungency determination. HortScience 27:1029–1030.

Yoo, K., and L.M. Pike. 1999. Development of an automated system for pyruvic acid analysis in onion breeding. Sci. Hortic. (Amsterdam) 82:193–201.

Yoo, K.S., L. Pike, K. Crosby, R. Jones, and D. Leskovar. 2006. Differences in onion pungency due to cultivars, growth environment, and bulb sizes. Sci. Hortic. (Amsterdam) 110:144–149.

Yoo, K.S., L.M. Pike, and B.K. Hamilton. 1995. A simplified pyruvic acid analysis suitable for onion breeding programs. HortScience 30:1306.

13 S

Sulfur and the Production of Rice in Wetland and Dryland Ecosystems

Richard W. Bell
Murdoch University, Murdoch, WA, Australia

Abstract

Rice (*Oryza sativa* L.) is one of few crop species that can be productively grown in both wetland and dryland conditions. Yields of rice may have to rise by 40 to 60% over the next 20 to 30 yr because of the demand from increasing population. The requirement for sulfur will rise by a comparable level, and unless fertilizer inputs increase, more sulfur deficiency can be expected, threatening the attainment of yield targets. Management of the sulfur nutrition of rice varies depending on whether the soil is submerged throughout the crop cycle as in paddy cultivation, aerated in the root zone as in upland production systems, or intermittently flooded as in rain-fed lowland, aerobic, and water-deficit irrigation production systems. Under upland conditions, which generally use different cultivars than those grown in wetland conditions, rice attains greater rooting depth than in submerged soils. Sulfur nutrition of rice under upland conditions has much in common with that of other field crops. In submerged soil, access to sulfur is limited by the shallow root system, with >90% of roots confined to the top 20 cm of the soil. Low redox potential causes reduction of sulfate to sulfides, some of which are toxic (H_2S), and others low in solubility (FeS, ZnS). Moreover, the slower mineralization of organically bound sulfur decreases availability of sulfur to rice in submerged soils. Hence, sulfur deficiency has increased in prevalence in wetland rice. Negative sulfur budgets have been estimated in several countries where rice is grown without sulfur fertilizers, especially when residue is removed or burned. Further research is needed to develop optimal sulfur nutrient management for the emerging water-saving production systems where soil redox potential fluctuates over time, affecting the mineralization of organically bound sulfur and changing the stability of reduced or oxidized forms of sulfur.

Rice is the staple grain for half the world's population (Greenland, 1997). Globally, rice yields averaged 4 Mg ha^{-1} in 2004, doubling the yield in 1966 (Cassman, 2006). On the basis of growing populations in the regions where rice is the main staple, it was estimated that total rice production should escalate a further 38% by 2030 (Surridge, 2004). Most of the production increase will have to come from yield increases. However, yield increases will be offset by loss of paddy land due to urbanization and transport infrastructure, decreases in water allocation to rice production, and the impact of global climate change. Cassman (2006) estimates that cereal yields by 2025 will have to increase 59% compared with 1995 levels

---

Copyright © 2008. American Society of Agronomy, Crop Science Society of America, Soil Science Society of America, 677 S. Segoe Rd., Madison, WI 53711, USA. *Sulfur: A Missing Link between Soils, Crops, and Nutrition.* Agronomy Monograph 50.

to supply the rising demand for these commodities, after allowing for losses of land to biofuels and urbanization. Similar increases would apply to rice yield also. Increases in yield will be a challenge for breeders to develop cultivars with higher yield potential, agronomists to develop packages of technologies that allow realization of potential, and farmers to apply efficient practices to achieve economic returns for their crops in the face of rising fuel prices, increased cost of fertilizers, and the uncertain effects of climate change.

Higher yielding rice crops will require greater nutrient supply including increased sulfur. Dobermann et al. (1998) estimate that the total annual sulfur demand for irrigated rice in 2025 will increase 65 to 70% above 1990 requirements. Without substantial increases in sulfur fertilizer application, sulfur deficiency will become more prevalent over time. In addition to increasing demand from higher yielding rice crops, there are already several other factors that are contributing to more widespread sulfur deficiency: increased use of low sulfur analysis fertilizers, depletion of soil sulfur supplies, soil erosion, and leaching of sulfate-sulfur.

Unlike most crop species, rice is able to grow productively in both wetland and dry land conditions. Generally, different cultivars are grown under wetland and dry land conditions. The so-called upland rice cultivars for dry land conditions are generally deeper rooted and require greater tolerance of water stress than cultivars grown in wetland soils. The reasons for growing rice under submerged soil conditions relate mostly to ease of weed control and minimizing risk of water stress during the growing season. In addition, the traditional growing of rice on floodplains in submerged soils exploited the nutrient enrichment of soils from the annual floodwaters. Submerged soils also improve the supply of phosphorus to crops and minimize extreme pH conditions in the root zone (Greenland, 1997).

Submerging soils to prepare rice fields for transplanting requires 300 to 700 mm of water (Greenland, 1997). Moreover, maintaining paddy fields in a submerged condition consumes substantial water during crop growth. Depending on transpiration rate, percolation rate, and the amount of through flow of water in fields, 1020 to 2860 mm is required for a 100-d transplanted crop (Greenland, 1997). Submerged rice uses 4000 kL of water Mg^{-1} of grain produced compared with 1000 kL Mg^{-1} for wheat, *Triticum aestivum* L. (Greenland, 1997). In favorable environments, the water requirement is provided by rainfall or floodwater. In the many parts of the world where irrigation is required for rice production, the increased competition for water resources has reached the point where rice production can no longer be assured of current water supplies, much less increased allocations for expansion of irrigation. This has led to efforts toward rice production systems that require less water (Bouman, 2001), such as aerobic rice and water-deficit irrigation for rice. The emergence of each of these rice production systems will have an impact on the optimal methods for supplying sulfur needs of rice crops.

Sulfur deficiency in rice was reported by Fox and Blair (1986) to occur in Bangladesh, Brazil, Burma, India, Indonesia, Nigeria, Papua New Guinea, Philippines, Sri Lanka, southern China, and Thailand. Subsequently sulfur deficiency has also been reported in lowland rain-fed rice grown in Laos (Linquist et al., 1998), Cambodia (Seng et al., 2001), and Pakistan (Dobermann and Fairhurst, 2000). According to Dobermann and Fairhurst (2000), sulfur deficiency often occurs in upland rice but is also found in lowland rice areas. As discussed below, the risk factors associated with sulfur deficiency often differ between upland and lowland soils. Prevalence of sulfur deficiency has increased in rice because of increased cropping intensity and increased use of low sulfur fertilizers such as urea, rather than ammonium sulfate or superphosphate (Fox and Blair, 1986). Research on management of sulfur nutrition of rice has examined optimum rates, types, methods, and timing of sulfur fertilizer application. Effects of soil water regime on mineralization of organically bound sulfur and on the availability of native-soil sulfur and fertilizer sulfur in submerged and upland soils has also been investigated. The derived sulfur fertilization strategies are discussed below.

Sulfur in Aerobic Soils

Upland rice is grown in predominantly aerobic soils, and for this rice ecosystem, much of the reviews by Schoenau and Malhi and by Dick et al. (2008, this publication) on sulfur in soils is entirely relevant and hence only brief details will be provided here. Factors that increase sulfur deficiency risk in upland soils include sandy textures, leaching rainfall patterns, high iron and aluminum oxides (including volcanic soils with allophanic mineralogy), especially when combined with low soil pH, and low organic matter in the soil (Dobermann and Fairhurst, 2000).

Sulfur in aerobic soils occurs in the soil solution, sorbed on variable charge surfaces as sulfate, and in organic matter as sulfate esters, and bound in recalcitrant organic compounds. Sulfate is the primary soluble sulfur species in aerobic soils and the form that is adsorbed on soil colloids (Bohn et al., 1986). Generally, 80 to 98% of soil sulfur is organically bound, of which sulfate esters are a major fraction (Freney, 1986). Sulfur in organic matter is divided into ester-bound sulfur, carbon-bonded sulfur, and nonreducible organic sulfur (Freney et al., 1975). The abundance of each of these sulfur pools in humus appears to vary with land use. In two Chinese padi soils, there was a low abundance of ester-bound sulfur compared with that in arable soils, and this was attributed to formation of reduced forms of organically bound sulfur under the reduced chemical environment of the paddy soils (Zhao et al., 2006). By contrast, in a range of aerated arable soils, 46 to 53% of the organically bound sulfur in the soil humus was associated with esters (Zhao et al., 2006).

The soil solution and sorbed forms of sulfur appear to be most readily absorbed by plants; however, the organically bound sulfur provides a pool of sulfur that continually replenishes available sulfur supply for plant uptake (Blair et al., 1991b). Both carbon-bonded and ester-bonded sulfur in organic matter form plant-available pools, but their prevalence appears to vary with soil type and land use (Zhao et al., 2006). By contrast, sulfur in carbonate forms was not correlated with plant uptake (Shan et al., 1997).

Organic matter is the main pool of soil sulfur; hence, factors controlling the turnover of organic matter generally have profound effects on sulfur forms and availability. About 1 to 5% of soil-organic sulfur is mineralized per year (Schoenau and Malhi, 2008, this publication). However, immobilization and mineralization of sulfur occur simultaneously in soil (Freney, 1986), so that the pattern of sulfate release in a particular soil may be site specific. Agricultural soils generally have carbon/nitrogen/sulfur (C/N/S) ratios of 130:10:1.3 (Freney, 1986). Sulfate mineralization occurs when C/S ratio in organic material is <200, but at C/S ratio >400 sulfate immobilization is likely (Reddy et al., 2002). The general pattern of sulfate release following addition of organic matter to soils involves initial immobilization, a steady increase in sulfate over time comprising a rapid initial increase followed by a slow increase, and a slow decline in rates of release (Freney, 1986). Aulakh et al. (2002) reported that rates of sulfate release were most rapid at 0 to 14 d after first wetting the soil and then declined during subsequent incubation periods (14–28 and 28–42 d).

Sulfate is sorbed nonspecifically on anion exchange sites on the surfaces of variable charge colloids (Barrow, 1985). Sorption is weaker than for phosphate or molybdate but stronger than for nitrate or chloride. Sulfate sorption increases as anion exchange capacity increases. Hence, sulfate sorption increases in acid soils and is minimal in neutral to alkaline soils (Bohn et al., 1986). Iron and aluminum oxides and oxyhydroxides are the main minerals on which surface sorption of sulfate occurs; however, edges of aluminosilicate minerals are also important for sulfate sorption in some soils. Fox (1974) reports that the sulfate adsorption varies with mineralogy of the clay-sized fraction in tropical soils in the following sequence (highest to lowest): amorphous hydrated oxides > crystalline oxides > kaolin clays > 2:1 clays. Hence, highly weathered soils rich in iron and aluminum oxides and oxyhydroxides tend to have greater sulfate sorption especially when this mineralogy coincides with low soil pH. Sulfate sorption increases with increasing depth in the soil profile are correlated with lower pH and increased oxyhydroxides levels. Sulfate sorption increases in subsoil of soils with oxic B horizons (Bohn et al., 1986).

Sulfur in Wetland Soils

Wetland soils experience a range of water regimes from those that are continuously submerged to those that are intermittently waterlogged. In terms of rice production, the main categories of wetland soils are deepwater rice where during the growing season soils may be covered by several meters depth of water, irrigated rice where the paddy field is submerged to a depth of up to 50 cm for most if not all the growing season, and rain-fed lowland soils where the depth and duration of submergence is extremely varied during the growing season (Wade et al., 1999). However, almost all of these wetland soils experience some drying and aeration of soils, either between harvesting of one crop and the planting of the next in triple- or double-cropped irrigated systems or during the fallow period as in deep-water rice environments or during crop growth as in the rain-fed lowlands. The timing and duration of the drying or reoxidation phase of wetland soils may be important in determining the forms and availability of sulfur.

Submergence has profound effects on soil oxygen supply in the root zone as well as consequences for biological activity in soils and the chemistry of soil nutrients including sulfur. Possibly the most significant effect of submergence is that it limits root depth (Kirk, 2004). Rice has relatively large proportions of the cortex occupied by aerenchyma regardless of the root oxygen supply, but under anoxia, porosity levels increase to 30 to 40% of root volume (Colmer, 2003). The gas channels that develop in the cortex as a result of aerenchyma formation allow oxygen transport down the root axis from above ground to the root tips. The extent of radial oxygen loss from axial roots and fine lateral roots determines how much oxygen reaches the root tip and controls the maximum root depth that can be sustained by the plant in submerged soil (Armstrong, 1979). Oxygen loss into the rhizosphere creates a zone of oxidized soil within a predominantly reduced soil. The rhizosphere therefore supports different chemical reactions to those in the bulk soil. Sulfate-sulfur, while potentially unstable in the bulk soil, is chemically stable in the oxidized rhizosphere. Hu et al. (2003) found that ratios of inorganic sulfate-sulfur in the rhizosphere relative to the nonrhizosphere varied between 1.3 and 3.1, indicating an enrichment of sulfate-sulfur in the rhizosphere. Wind and Conrad (1997) reported that sulfate concentrations were 50 to 60% higher within 3 mm of the rice root surface than >4 mm from the root surface in a flooded soil. Coupled with the greater abundance of sulfate in the rhizosphere, Sheid et al. (2004) found higher abundance of sulfate reducing bacteria.

Rice in submerged soils has an unusually shallow root system. Generally, 70% or more of the roots are in the 0- to 10-cm layer, 90% in the 0- to 20-cm layer, and very few roots penetrate below 40 cm (Sharma et al., 1994). In contrast, upland rice can have rooting depths between 70 and 80 cm (Morita and Abe, 1996). The shallow rooting behavior of lowland rice is clearly controlled partly by genetics and partly by the soils physical and chemical properties. While a plow pan in paddy fields generally restricts root penetration below 15 to 20 cm, deeper penetration would not be possible in submerged soils because of inadequate oxygen supply to root tips. Hence the sulfur stored deep in the soil profile that might be accessed by dryland crops including upland rice is inaccessible to submerged rice. Hence, the sulfur supply to rice in submerged conditions depends on forms and availability in the surface 15 to 20 cm of the profile.

Sulfur in submerged soils occurs in the soil solution, sorbed on variable charge surfaces as sulfate, as sulfate esters and bound in more recalcitrant forms of organic matter, and as sulfides of iron and other metals such as manganese and zinc. Sulfate is the primary soluble sulfur species in aerobic soils but is unstable at low redox potential that favors hydrogen sulfide formation (Ponnamperuma, 1981). Levels of sulfate have been reported to decline from 600 mg L^{-1} to close to zero within 4 wk of submergence (Ponnamperuma, 1981). In another report, total sulfide in solution formed after 6 wk under submergence accounted for 85 to 90% of the ^{35}S-labeled sulfate supplied (Dobermann and Fairhurst, 2000). Provided active iron levels are sufficient, H_2S reacts to form FeS as the main sulfide in the soil. However, sulfides of manganese and zinc also form at low redox potential, and this may cause deficiency of zinc in particular (Dobermann and Fairhurst, 2000). Generally, the solubility of CuS is too high for it to form in reduced paddy soils (Ponnamperuma, 1981).

Sulfide formation proceeds at −150 to −200 mV and pH 6.5 to 7 in soils (Takai, 1978 cited by Suzuki, 1995). However, other reports suggest different threshold redox potentials for sulfide formation. Dobermann and Fairhurst (2000) suggest that at redox potential <50 mV at pH 7, sulfate is reduced to H_2S and there is some indication that reduction of sulfate can occur up to +50 mV (Lefroy et al., 1993). Sahrawat (2004) reports that sulfide formation proceeds at −120 to−180 mV. Yu et al. (2007) found no relationship between S^{2-} concentration in solution and Eh over the range −250 to + 600 mV. To some extent the differences between studies can be ascribed to soil pH differences since the redox potential at which sulfide forms varies with soil pH (Bohn et al., 1986). Reduction at pH 4 can proceed at 0 mV whereas at the more typical pH of flooded soils, pH 6, the threshold redox value is −150 to −200 mV. The low redox potentials that favor sulfide formation occur after prolonged flooding and are accentuated by low active iron status in soils, poorly drained organic soils, and acid sulfate soils. However, the soil redox environment is heterogeneous (Murase and Kimura, 1997), and even when the bulk soil is oxic, microsites in the soil may have low enough redox potential for sulfide formation to proceed (Kirk, 2004) and variation in soil soluble Mn, Fe, and Zn concentrations will determine the forms and stability of sulfides formed (Yu et al., 2007).

Oxidation of organic matter in general, and sulfur mineralization in particular, is suppressed under submerged conditions. Aulakh et al. (2002) recorded no mineralization of organically bound sulfur in soil-organic matter in three submerged soils over a 42-d incubation period. Indeed, incubation under submerged conditions caused an increase in organically bound sulfur and decrease in sulfate-sulfur, suggesting immobilization of sulfate-sulfur. Similar results were obtained during incubation for 42 d in three flooded Chinese rice soils (Zhou et al., 2002). Under flooded conditions, there is slower net mineralization of organically bound sulfur added as straw to the soil (Lefroy et al., 1994). However, at the level of 0.086% sulfur in rice straw, there was net mineralization of sulfur, suggesting that the critical sulfur level for mineralization was lower than this level. Mineralization of sulfur in straw was accelerated by the simultaneous addition of sulfate-sulfur as gypsum suggesting a priming effect of sulfate-sulfur for decomposition. Since the level of readily available carbon in the soil may also influence soil redox potential, this too could affect levels of sulfate-sulfur remaining in soil. However, in studies with addition of straw or ashed rice straw to flooded soil, the latter was the more effective sulfur source for rice uptake (Lefroy et al., 1994). This led Lefroy et al. (1994) to conclude that under the conditions of their experiment, there was limited sulfate reduction to sulfide, suggesting that in this case redox potential was not low enough to reduce sulfate.

The decline in availability of nitrogen from mineralization of organic matter in long-term flooded soils is attributed to the increased prevalence of recalcitrant forms of organic matter (Olk et al., 2000). Since organically bound sulfur is the major sulfur source in most soils, this has potential implications for sulfur supply to rice as well. However, as yet no investigation has been undertaken on whether this is a concern for sulfur supply to rice crops in long-term submerged soils (D. Olk, 2008, personal communication).

Varied sulfate sorption with soil pH that has been studied in dry land soils (Barrow, 1985) also occurs in submerged soils (Lefroy et al., 1993). Since submergence of soils causes pH to shift toward neutrality, it is likely that sulfate sorption in most flooded soils will be low. Indeed with flooding of acid soils, it has been reported that sulfate levels initially rise with increased pH and then decline over time as sulfate is reduced (Ponnamperuma, 1981). However, the oxidized rhizosphere adds a further complexity because of its lower pH as do the iron plaques, which can form around rice roots (Armstrong, 1964). At high levels of sulfur supply, the formation of iron plaque is increased, and the accumulation of increased sulfur in the rhizosphere suggests that sulfate is adsorbed on the iron oxyhydroxides that make up iron plaque (Hu et al., 2007).

Sulfur plays a significant role in the acid sulfate soils, a particular form of submerged soils used for rice (Moorman and van Breemen, 1978). Acid sulfate or potential acid sulfate soils have developed in shallow coastal or estuarine sedimentary environments in the last 10,000 yr. Because they occur in submerged environments, they have often been developed for rice cultivation. Large areas of acid sulfate soils are used for rice cultivation in the Mekong Delta of Vietnam and the Chao Prya Delta of Thailand. The acid sulfate soils contain high levels of reactive sulfides that are chemically stable when submerged. When the development of the acid sulfate soils involved drainage to lower the water table, the oxidation of sulfides in the soil produces highly acid soils that result in unproductive rice crops or require careful treatment to restore them as rice growing soils. Upon drainage of the soil, oxidation of the sulfides occurs rapidly and generates extreme acidity. Soil pH as low as three has been reported in some acid sulfate soils. Even when reflooded for rice production, the pH rises slowly and may never attain the near neutral pH typical of submerged rice soils (Kirk, 2004).

Sulfur Nutrition in Rice

Uptake

Sulfate ion is the ionic species absorbed by roots from the soil solution (Barber, 1984). Minimum sulfate concentrations in irrigation water for unrestricted sulfur uptake are in the range 2.7 to 6 mg L^{-1} (Blair et al., 1991b). By comparison, Hitsuda et al. (2005) suggests that the minimum solution sulfur concentration for maximum growth of rice is 2 mg L^{-1}. Blair and Lefroy (1998) indicate that sulfur concentrations had to exceed 8.1 mg L^{-1} in irrigation water before yields of rice exceeded those in the control plots.

Under low external sulfur, plants increase the rate of sulfur uptake by producing more sulfate transporters in the plasma membrane of roots (Hawkesford et al., 1995). Low levels of glutathione in the phloem appear to be the signal for increased synthesis of sulfur transporters when plant sulfur supply is low (Leustek and Saito, 1999). It is not known whether this happens in rice, especially in anaerobic soils, although glutathione is the main reduced form of sulfur in the phloem of rice.

Uptake of gaseous SO_2 and H_2S also occurs through leaves. Most of the uptake is through stomates, and, hence, the rate of uptake fluctuates diurnally with stomatal opening and closing. At low atmospheric concentrations, plants respond positively to SO_2 that can substitute for low soil sulfate levels (Suzuki, 1995). However, at adequate sulfate levels in the root medium, the contribution of SO_2 to plant sulfur content is small. At high concentrations (>50–120 mg SO_2 per m^3), long-term depression in photosynthesis occurs (Marschner, 1995, p. 889; Saalbach, 1984). Gaseous H_2S in the canopy is toxic to plants, but the sensitivity of rice is not known. Losses of volatile H_2S can also occur from plants through stomates.

While the bulk of a submerged soil is chemically reduced and has low oxygen concentration, leakage of oxygen from the aerenchyma produces an oxidized rhizosphere (Kirk, 2004). This has profound effects on the chemical forms of many nutrients (nitrogen, phosphorus, manganese, iron, and sulfur) (Stubner et al., 1998). In the oxidized rhizosphere, sulfides are readily oxidized to sulfate by an abundant and diverse population of sulfur-oxidizing bacteria. Ratios of sulfate in the rhizosphere to the nonrhizosphere of rice indicate an enrichment of sulfate in the rhizosphere soil (Hu et al., 2003). By contrast, the ester-bonded and carbon-bonded sulfur fractions decreased by 75 and 30%, respectively, in the rhizosphere of rice. Samosir et al. (1993) reported that rice cv. C4–63 with high O_2 diffusion rates from roots had greater sulfur uptake on a submerged sulfur deficient Aquic Haplustalf than cv. cultivar Pulu Bolong with low O_2 diffusion rates from roots.

Radial oxygen loss (ROL) into the rhizosphere causes oxidation of sulfides to sulfate and helps to alleviate sulfur deficiency when soil-sulfur levels are low.

Hydrogen sulfide toxicity is a potential constraint to rice root growth and nutrient uptake under low redox potential. The disorder associated with H_2S toxicity is known in Japan as "Akiochi" and in the USA as "straighthead." Hydrogen sulfide reduces root respiration and causes coarse, blackened, and sparse roots. In addition, interveinal chlorosis of emerging leaves and increased occurrence of leaf diseases are common. Release of O_2 into the rhizosphere of rice roots will oxidize sulfur, and, hence, roots with high ROL will tolerate higher levels of H_2S than roots with low ROL (Dobermann and Fairhurst, 2000). Decreased concentrations of phosphorus, potassium, magnesium, calcium, manganese, and silicon are commonly associated with H_2S toxicity in rice presumably because of damage to root systems that impairs nutrient uptake (Dobermann and Fairhurst, 2000).

Distribution in Plants

Sulfate absorbed by roots is loaded into the xylem and transported as sulfate ion. While most of the plant sulfur is in assimilated forms, some sulfur is stored in vacuoles as sulfate (Marschner, 1995, p. 889). Both reduced (cysteine, methionine, glutathione) and sulfate forms of sulfur are transported in the phloem. Rennenberg et al. (1979) found that glutathione is the main form of reduced sulfur transported in the phloem, and this was confirmed to be the case in rice also (Kuzuhara et al., 2000).

There is some evidence that loading of sulfur in rice grain is a limiting factor to increasing grain sulfur content (Sheehy, 1996; Hagan et al., 2003). Low sulfur levels were reported in the phloem of rice plants relative to that required in glutelin, one of the main storage proteins in rice grain (Sheehy, 1996). In transgenic rice expressing the sunflower seed albumin (SSA) gene, which produces a protein rich in methionine and cysteine, increases in the sulfur-rich SSA were obtained in grain, but there was no net increase in grain sulfur (Hagan et al., 2003). It was concluded that the sulfur-rich SSA was synthesized with endogenous sulfur in the grain and hence was produced at the expense of other sulfur-containing proteins in the grain. Islam et al. (2005) found that levels of three classes of glutelin proteins and a globulin were depressed in the transgenic plants expressing increased SSA. Hence, the extra sulfur sink in the grain did not increase sulfur loading of the grain of the transgenic rice. This suggests a limitation in sulfur transport or unloading of sulfur into grain of rice, unlike transgenic lines of several other species that had increases in both grain SSA and total grain sulfur.

Remobilization of organically bound sulfur is necessary for the phloem transport of sulfate-sulfur. Islam et al. (2005) suggests that the remobilization of sulfur in rice grain is under transcriptional control. Sulfur was classified by Marschner (1995, p. 889) as having high phloem mobility. The appearance of sulfur-deficiency symptoms in young leaves is ascribed by Marschner (1995, p. 889) to inadequate remobilization of sulfur. However, Loneragan et al. (1976) suggests that sulfur is variably mobile in the phloem because nitrogen status of the plant affects the remobilization of sulfur from mature and old leaves. Under nitrogen deficiency, the breakdown of protein in old leaves favors the release and remobilization of sulfur to growing parts of the plant (Loneragan et al., 1976; Robson and Pitman, 1983). Under adequate nitrogen, sulfur is less readily remobilized. Hence, sulfur-deficiency symptoms can appear in older leaves when plants are also nitrogen deficient, but the symptoms appear in young leaves when nitrogen supply is adequate. However, there appears to be no field report of such differences in symptom expression for sulfur-deficient rice plants depending on nitrogen supply.

Forms, Functions, and Internal Requirements for Sulfur

Sulfur occurs in a diverse range of forms necessary for plant function. Sulfur is a constituent of the amino acids methionine and cysteine, and its requirement in protein and coenzymes containing these amino acids is the main biochemical role for sulfur in plants. In addition to its structural role in the amino acids of protein and enzymes, sulfur affects the tertiary structure and hence function of polypeptides that may be important in protecting cells from heat or drought stress (Marschner, 1995, p. 889). Thioredoxin compounds are involved in cellular redox reactions, particularly in the chloroplasts. The vitamins thiamine (Vitamin B1) and biotin (Vitamin H) contain sulfur. Other sulfur compounds include glutathiones that are important in stress tolerance of plants by detoxifying superoxide species. Glutathione in turn is a precursor to the phytochelatin group of compounds that complex with heavy metals and are important in determining plant tolerance to heavy metals. Finally, sulfolipids occur in cell membranes and are particularly abundant in thylakoid membranes of the chloroplast (Marschner, 1995, p. 889).

Cysteine is the primary reduced sulfur compound formed in plants (Hofgen, 2008, this volume). The rate of sulfate reduction and incorporation into cysteine is greater in shoots than roots. All other organically bound forms of plant sulfur are derived from cysteine by various pathways (Marschner, 1995, p. 889).

The internal sulfur requirement for plants in general is 1 to 5 mg sulfur g^{-1} in dry weight (Marschner, 1995, p. 889). Rice like other cereals requires only 1 to 2 mg sulfur g^{-1} in shoots. Rice grain contains 1 to 3 mg sulfur g^{-1} (Dobermann et al., 1998). The nitrogen/sulfur (N/S) ratio of protein in cereals is generally about 30.

Because of its effects on levels of the amino acids methionine and cysteine, sulfur affects the nutritional quality of rice grain for humans and of animal feed derived from rice grain or rice bran. Rice grain grown under low-soil sulfur supply has reduced levels of methionine and cysteine.

At moderate rates of sulfur application that corrected sulfur deficiency on a low sulfur soil, Suwanarit et al. (1997) found increased aroma, softness, whiteness, stickiness, and glassiness of boiled milled grains from the aromatic rice Khaw Dauk Mali 105. However, at sulfur levels that exceeded the optimum rate, these quality parameters declined. Srivastava and Singh (2007) found that milling and hulling percentages, per thousand grain weight, length/breadth ratio of grain, cooked grain breadth, and aroma scores were significantly and positively correlated with the N/S ratio of rice grain, while grain breadth had a significant inverse relationship with N/S ratio in grain. N/S ratios in the range 3.8 to 3.9 were associated with the best combination of grain quality parameters.

Sulfur Management for Rice Crops

Sulfur Budgets and Cycling

A detailed sulfur cycle for rice grown in submerged soils was proposed by Lefroy et al. (1992). It outlines the main pools of sulfur in the soil–plant–animal system and the transfers among pools (Fig. 13–1).

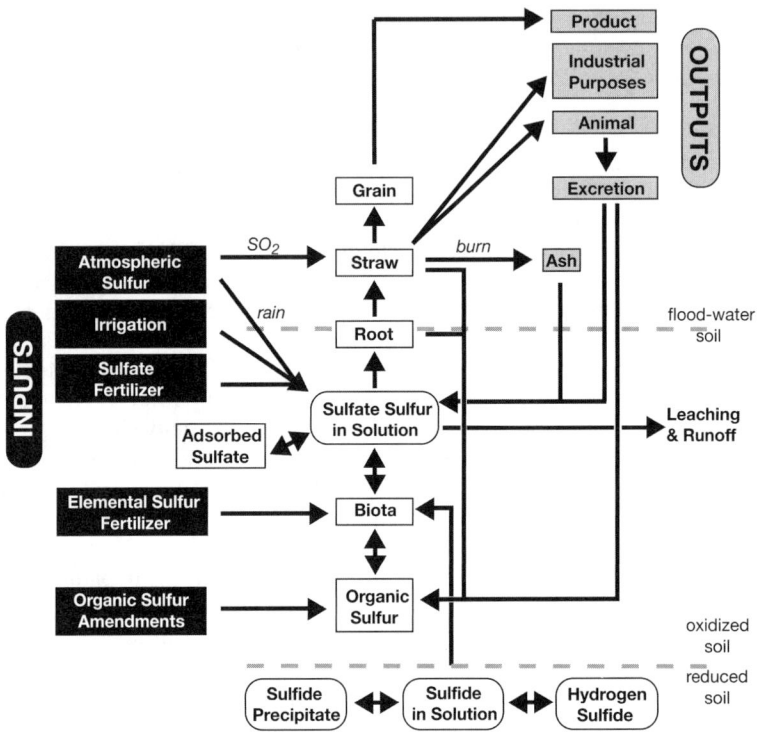

Fig. 13–1. Schema for sulfur cycling in paddy field growing rice under submerged conditions. (Adapted from Blair and Lefroy, 1998). Note the spatial arrangement of the flood-water, oxidized soil and reduced soil layers is not to scale. Oxidized soil occurs as a thin layer at the interface with flood water and in the rhizosphere where radial oxygen loss occurs from rice roots. For simplicity inputs are shown entering the oxidized soil, but in reality they may also enter the reduced soil layers.

Rainfall may be a significant source of sulfur input for rice crops (Blair and Lefroy, 1998; Lefroy et al., 1991, 1992). In Bangladesh, 3.5 kg sulfur ha^{-1} yr^{-1} has been recorded in rainfall (Adebin Mian et al., 1991), but Blair and Lefroy (1998) indicate that the range of values is large (0.4–47 kg ha^{-1} yr^{-1}). In data collected from China, Malaysia, Philippines, and Sri Lanka, the range of values for sulfur accretion in rainfall was 5 to 29 kg ha^{-1} yr^{-1} (Lefroy et al., 1992). In areas close to sites of industrial development, air pollution may result in rates of sulfur accretion in rainfall that fully satisfy crop requirements (Blair and Lefroy, 1998). Increasing industrialization in Asia may be contributing to the alleviation of sulfur deficiency in rice crops grown in adjacent areas. Sulfur accretion in rainfall is higher close to the coast but may still not be sufficient to supply crop sulfur requirements when soil sulfur is very low (Blair and Lefroy, 1998).

Irrigation water is also a significant input of sulfur for rice crops (Greenland, 1997). In general, more sulfur was supplied in irrigation water than in rainfall for rice crops (Dobermann et al., 1998). From studies in Bangladesh, Thailand, and the Philippines, sulfur input in irrigation water ranged from 35 to 86 kg ha^{-1} yr^{-1} with 713 to 4000 mm of water applied (Greenland, 1997). According to Blair et al. (1991b), sulfur response in rice was noted when sulfur in irrigation water was <2.8 mg L^{-1}, but when sulfate was >6.4 mg sulfur L^{-1}, enough sulfur was supplied to produce a 4.5 Mg ha^{-1} rice yield (Tisdale et al., 1986). Lowland rice can recover 35 to 50% of the sulfate in irrigation water (IRRI, 1977; Blair and Lefroy, 1998).

Sediment deposited on flooded fields may contribute significant sulfur to paddy rice, but no estimates were provided in Greenland's (1997) compilation of data on nutrients added in sediment. Sediment deposition rates vary from 1 to 10 Mg ha^{-1} yr^{-1}. On the basis of the mean sulfur content (106 mg sulfur kg^{-1}) found in a range of tropical soils as an estimate of sediment sulfur concentration, sediment could supply 0.1 to 1 kg sulfur ha^{-1} yr^{-1}. However, these rates may be an underestimate since sediment is likely to be sulfur-enriched relative to bulk soil since erosion removes mostly surface layers that are higher in organic matter.

When no fertilizer sulfur is applied to rice, the sulfur balance may be negative or positive depending on grain yield and residue fate (Table 13–1). In northeast Thailand, where the rice grain yields are low and the crop is entirely rain fed, the sulfur balance was negative if straw was removed or burned but positive if the residue was returned to the field. However, when sulfur fertilizer was added to the low yield system in northeast Thailand, sulfur balance was positive. Rainfall was a major sulfur input in the unfertilized fields, indicating that failure to account for this source may produce erroneous conclusions about sulfur balance in a low-productivity, rain-fed rice system. In the low-yield, rain-fed rice system, stubble contained 40 to 50% of the uptake of sulfur by the crop (Dobermann and Fairhurst, 2000). The fate of the stubble has a major bearing on sulfur budgets. Removal of the stubble from the field for use as fuel or cattle feed changes the sulfur balance from positive to negative. Similarly, burning the stubble results in a 40 to 60% loss of sulfur as volatile compounds and produces a negative sulfur balance. However, if burning is widespread in a region, it is likely that a significant proportion of the volatile sulfur is returned to the rice fields in rain so there may be limited net loss.

In irrigated systems, sulfur balance is likely to reflect extra inputs from irrigation water, offset by increased removal in grain. In the Indian example below (Table 13–2), the control crop and the crop fertilized with nitrogen, phosphorus, and potassium (but not sulfur) had apparent sulfur deficits of 5 to 14 kg sulfur ha^{-1}, but this may be misleading because the rainfall and irrigation water inputs have not been accounted for. Even without accounting for inputs from rainfall or irrigation, the crops fertilized with sulfur had large surpluses of sulfur, suggesting that rates of addition could be decreased substantially without loss of yield or compromising sustainability of the system. However, since the budget does not outline the amounts in stubble, the fate of the stubble, or report on the losses from leaching, some caution needs to be exercised in drawing conclusions about the wisdom of altering sulfur fertilizer rates on the basis of simple sulfur budget calculations.

Significant levels of dissolved sulfur are found in the standing water after fertilizer application. Lateral flow of standing water across fields may produce significant sulfur losses (Satrusajang et al., 1991). However, such losses of sulfur are likely to be episodic, occurring when recent sulfur fertilizer application coincides with flooding. Information on such episodic sulfur losses is rarely captured in sulfur cycling studies.

Presumably leaching losses under upland rice are equivalent to those under other field crops on the same soils. However, few studies have attempted to quantify sulfur leaching in rice paddy fields. Leaching losses of sulfur were estimated to be equivalent to 20 kg sulfur ha^{-1} in rice paddy fields in Korea (Shin and Han, 1988). In this system, the sulfur balance was negative because the 5 kg sulfur ha^{-1} yr^{-1} in rainfall and 15 kg sulfur ha^{-1} yr^{-1} in irrigation water were offset by the equivalent loss by leaching. In a study in Bangladesh, the estimated sulfur leached from a paddy field was 1.6 kg sulfur ha^{-1} yr^{-1} (Adebian Mian et al., 1991).

Little is known about the turnover of sulfur in crop residues and organic materials added to rice soils. However, as noted above, Lefroy et al. (1994) found reduced mineralization of sulfur from straw added in submerged soils compared with those maintained at 60% of pore water saturation. Indeed they concluded that straw, regardless of sulfur concentrations, was ineffective as a sulfur source for rice in submerged soils. By contrast, the same straw, subject to ashing, readily supplied sulfur to the rice plant in flooded soil, even though 40 to 60% of the sulfur in the original straw had been lost by volatilization. Nevertheless, Lefroy et al. (1994) still concluded that retention of the straw in the field was preferable to burning for the long-term sulfur supply and the cycling of sulfur in the agroecosystem. Li et al. (2001a) found that more S was taken up from mineralization of soil organically bound S by rice from flooded soils than by maize (*Zea mays* L.) from soils maintained at field capacity. The greater release of organically bound S for uptake by rice was attributed in part to the oxidized rhizosphere of rice roots that allows aerobic mineralization to proceed in the flooded soils. However, Li et al. (2001a) also note that the flooded soil was drained and aerated between each of four growth cycles and that soil aeration may have accelerated the release of mineral S.

Fertilizer Sulfur Responses

The need to supply sulfur fertilizer should be determined by whether fertilizer responses have been observed or whether diagnosis and prognosis by plant and soil analysis indicate a likelihood of deficiency. Further evidence of the need for fertilizer sulfur addition can be provided by negative sulfur budgets. However, it should be noted that nearby industrial and urban centers, inputs from rainfall, and irrigation water during the growing season may fully satisfy crop requirements even if soil levels indicate marginal sulfur supply.

Where moderate sulfur deficiency occurs, or low yield levels are expected, only 10 kg sulfur ha^{-1} yr^{-1} is required to correct deficiency (Dobermann and Fairhurst, 2000). However, such low levels provide limited residual benefit for future crops. Rates of 15 to 20 kg sulfur ha^{-1} yr^{-1} have been reported to provide residual sulfur to future crops. On severely deficient sites, 20 to 40 kg sulfur ha^{-1} yr^{-1} is required to overcome deficiency (Dobermann et al., 1998).

Since rice roots absorb sulfate-sulfur, sulfate fertilizers (e.g., gypsum), are immediately available while elemental sulfur fertilizer must first be oxidized before the sulfur is available. Hence the rate of oxidation of elemental sulfur is the main factor limiting its effectiveness as a sulfur source. Particle size of elemental sulfur determines the rate of oxidation (McCaskill and Blair, 1989). Lower rates of oxidation were also reported in acid than neutral and alkaline soils (Zhou et al., 2002). However, if elemental sulfur is broadcast on the surface of submerged soils, the rate of oxidation is generally not limiting. Indeed broadcasting elemental sulfur 2 wk before transplanting, has on occasion been reported to be less effective than application at transplanting if the rapid oxidation of elemental sulfur is followed by rapid leaching from the root zone (Blair and Lefroy, 1998).

Since oxidation of elemental sulfur only occurs in aerated zones of the soil, it should be placed close to the soil surface for rapid oxidation (Samosir et al., 1993; Chaitep et al., 1994). Samosir et al. (1993) reported that at maturity the recovery of sulfur by rice from sulfur fertilizers was as follows: 51% from surface applied elemental sulfur, 42% from surface applied sulfate-sulfur, 30% from deep sulfate-sulfur, and only 5% from deep applied elemental sulfur. The slower oxidation of deeply placed elemental sulfur in some cases has been reported to increase the residual value of the remaining sulfur for future crops in the rotation (Chien et al., 1988). The oxidation rate of elemental sulfur has been shown to vary among commercial formulations available in the southern USA (Slaton et al., 2001). The variation in oxidation rates of elemental sulfur formulations is important not only in determining the fertilizer rates and methods of application for optimal sulfur nutrition but on alkaline rice soils may also be significant for determining the acidification of soils and its impact on availability of P and Zn.

Elemental sulfur may be oxidized in the rhizosphere as well as at the oxidized surface of submerged soils (Zhou et al., 2002). However, deeply placed elemental sulfur generally has reduced effectiveness as a sulfur fertilizer for submerged rice. This suggests that oxidation of elemental sulfur particles in the oxic rhizosphere is a less efficient means of making sulfur available for uptake than oxidation at the soil surface. This can be ascribed to inadequate contact between elemental sulfur particles and the rhizosphere at depth and decreased root density with depth. However, Samosir et al. (1993) demonstrated that a cultivar of rice with higher O_2 diffusion rates from roots, which would make for a more strongly oxidized rhizosphere, had higher rates of sulfur uptake. Moreover, Zhou et al. (2002) showed that rates of oxidation of elemental sulfur over a 42-d period were also increased in submerged soils by 40 to 50% by the presence of live rice roots in the soil. This indicates the significance of rhizosphere oxidation rates for sulfur uptake by rice and of genotypic variation in this property. As yet no research has explored the prospects of using genotypic variation in sulfur acquisition to manage sulfur nutrition of rice in submerged soils.

Sulfur fertilizer is most effective when applied at sowing or transplanting (Blair and Lefroy, 1998). However, rapid leaching conditions at sowing may reduce the effectiveness of S fertilizers. For example, application of elemental sulfur, 2 wk before transplanting had reduced effectiveness, suggesting that in this case oxidation of elemental sulfur occurred rapidly but was followed sulfur leaching (Blair and Lefroy, 1998). Delaying sulfate-sulfur application until 2 wk after transplanting produced rapid recovery of crops from sulfur-deficiency symptoms (Dobermann and Fairhurst, 2000). However, delaying sulfur fertilizer application until maximum tillering, decreased yield by one third (Blair and Lefroy, 1998).

Blair and Lefroy (1998) suggest that since sulfur is best placed close to the soil surface at planting, like phosphate fertilizer, efforts should be made to combine the supply of sulfur with the main phosphate fertilizer applied at sowing. Recent work by Yasmin et al. (2007) reported on the effectiveness of sulfur-coating of urea, di-ammonium phosphate (DAP), mono-ammonium phosphate (MAP), and triple superphosphate (TSP), relative to gypsum when applied as a surface or deep-placed application to nonflooded and to flooded soil. All sulfur sources were equally effective in stimulating growth and sulfur uptake, although curiously gypsum in this trial was not different from the control in the flooded soils. Elemental sulfur and sulfur-coated urea resulted in the highest recovery of fertilizer sulfur in the plant (47 and 46%, respectively) followed by sulfur-coated DAP, sulfur-coated TSP, and gypsum. Recovery of fertilizer sulfur by rice was lower under nonflooded soil conditions compared with surface or deep application to submerged soils. Release of sulfur from sulfur-coated urea was highest when applied on the surface under flooded conditions, compared with other coated fertilizers. Hence, these results and those of Dana et al. (1994) suggest that supplying sulfur as a coating on basal phosphorus fertilizers may be effective. However, further evaluation is needed on a range of soils as well as validation from field trials before drawing firm conclusions about the value of sulfur-coated macronutrient fertilizers to deliver sulfur to rice crops in submerged soils.

Diagnosis of Sulfur Deficiency

The accurate diagnosis of sulfur deficiency in standing rice crops can be achieved by recognition of plant symptoms, by plant analysis, or by crop response to sulfur fertilizer addition.

Sulfur-deficient plants are pale green, especially in young leaves (Dobermann and Fairhurst, 2000). Other symptoms of sulfur deficiency include reduced plant height, reduced tiller number and fewer spikelets per panicle, delayed plant maturity by 1 to 2 wk, and decreased resistance to cold injury (Suzuki, 1995; Dobermann and Fairhurst, 2000). Sulfur deficiency may appear in the nursery as well as in the main field.

Nitrogen deficiency in rice, which normally occurs in old leaves, can be distinguished in rice from sulfur deficiency that occurs in young leaves or the whole shoot (Suzuki, 1995; Dobermann and Fairhurst, 2000). Moreover, sulfur deficiency does not produce senescence of old leaves like nitrogen deficiency and in the young leaves produces a paler green appearance. Nevertheless, for reasons discussed above, relating to variable sulfur mobility, it is not uncommon for sulfur deficiency to be misdiagnosed as nitrogen deficiency, especially if plants are deficient in both elements. Suzuki (1995) concluded that it may be difficult to distinguish sulfur deficiency in rice fields from nitrogen deficiency. Moreover, differences in symptom expression may occur among cultivars. Symptoms are most useful when the deficiency is quite severe but not in cases of moderate deficiency that produce unrecognizable symptoms.

Plant analysis can help to confirm the cause of symptoms and also to identify sulfur-deficient crops that do not express symptoms. The key to the use of plant analysis is to establish, for a defined plant part, the relationship between sulfur concentration in that part at a specific growth stage and the plant growth at the time of leaf sampling. Provided a well-defined relationship exists, a critical concentration or critical range of sulfur concentrations can be defined and used for diagnosis of deficiency. In the absence of well-defined critical concentrations, comparing leaf sulfur concentrations in plants with symptoms and those without symptoms in the same field can be useful in establishing whether sulfur is the likely cause.

The most recently matured leaf blade, called the Y-leaf, is widely used for diagnosis of nutrient disorders in rice (Westfall et al., 1990). Considerable calibration has established the Y-leaf to be the more reliable indicator for sulfur deficiency. Critical levels of sulfur vary with the plant part sampled and age of plant at sampling (Tables 13–2 and 13–3). The values reported by Dobermann and Fairhurst (2000) probably represent the current best estimates of critical values

Table 13–1. Sulfur budgets for rice crops in Thailand and India reflecting low and high input cropping systems, respectively. Source is Blair et al. (1991b).

	Thailand†		India‡		
	Control	Fertilized	Control	NPK–S§	NPK+S
	kg S ha^{-1}				
Inputs	3.6	11.6	0	0	82
Fertilizer	0	8	0	0	82
Rain	3.6	3.6	nd¶	nd	nd
Uptake	4.2	5.6	5.0	14.0	18.0
Product	1.5	1.2	nd	nd	nd
Stubble	1.5	2.4	nd	nd	nd
Residue	1.2	2.0	nd	nd	nd
Balance			−5.0	−14.0	+64
Residue returned	+0.9	+8.4	nd	nd	nd
Residue removed	−0.6	+6.0	nd	nd	nd

† Budget for a single rice crop.
‡ Budget for double cropped rice over 14 yr on a Tropaquent.
§ N = nitrogen, P = phosphorus, K = potassium, S = sulfur.
¶ nd = not determined.

Table 13–2. Normal ranges and critical levels of sulfur in plant tissue. Source is Dobermann and Fairhurst (2000).

Growth stage	Plant part	Optimum	Critical level for deficiency
		mg g^{-1}	
Tillering	Y leaf		<1.6
Tillering	Shoot	1.5–3.0	<1.1
Flowering	Flag leaf	1.0–1.5	<1.0
Flowering	Shoot		<0.7
Maturity	Straw		<0.6

Table 13–3. Critical levels of sulfur in plant parts for prediction of sulfur deficiency in rice.

Plant part	Relative yield	Critical concentration		Source
		S	N/S ratio	
		mg g^{-1}		
Y leaf–flowering	95	1.5	–	Osiname and Kang (1975)
Grain	95	1.2	20–22	Osiname and Kang (1975)
Straw	95	1.0	16–19	Osiname and Kang (1975)
Whole shoot–active tillering	90	2.3	–	Fox and Blair (1986)
Whole shoot–max tillering	90	1.4	–	Fox and Blair (1986)
Grain	90	1.1	–	Fox and Blair (1986)
Whole shoot–29 d	90	1.1–1.3		Hitsuda et al. (2005)
Whole shoot–30 d, tillering	90	1.7–2.2		Brar et al. (1982)
Leaf blades at tillering	100	1.6	29	Yoshida and Chaudhry (1979)
Leaf blades at flowering	100	1.0	24	Yoshida and Chaudhry (1979)
Shoot at tillering	90	1.0	–	Suzuki (1995)

for confirming sulfur deficiency by plant analysis. However, critical values may also vary depending on whether plant analysis is being used to detect an existing deficiency (diagnosis) or predict a possible future deficiency (prognosis). It is not always clear in the studies reported in Tables 13–2 and 13–3 whether the values were calibrated for diagnosis or prognosis.

Prediction of Sulfur Deficiency

While the diagnosis of sulfur deficiency in standing crops is important, the timing of the appearance of symptoms determines how useful the diagnosis will turn out to be. In the time that elapses between first recognition of a symptom of sulfur deficiency, confirmation of the diagnosis, application of a treatment, and a crop response, irreversible loss of yield may have occurred. Hence, it is more important for rice producers to be able to predict potential deficiency so that corrective action can be taken before loss of yield or yield potential has occurred. The accurate prognosis of future sulfur deficiency in rice crops can be achieved by plant and soil analysis. As described above, the key to the use of plant or soil analysis is to establish for a defined plant part or soil depth the relationship between sulfur concentration and rice grain yield. Provided a well-defined relationship exists, a critical concentration or critical range of sulfur concentrations can be defined and used for prognosis of deficiency. Soil analysis has the advantage over plant analysis in that soil samples can be taken before sowing and a decision made at sowing whether to add sulfur fertilizer or not. By contrast, predicting sulfur deficiency by plant analysis is based on sampling 3 to 8 wk following sowing. The elapse of time between sample collection and the receipt of plant analysis results may prevent timely intervention. The accuracy of the prediction depends on whether the sampled rice crop experiences expected growth conditions and assumes no major change in the sulfur supply after sampling. If other nutrient deficiencies (such as nitrogen deficiency), or growth constraints like drought intervene in growth, or roots encounter high subsoil sulfur, then the predicted sulfur deficiency may not be expressed.

Critical concentrations for prognosis of sulfur deficiency in rice decline with time (Tables 13–2 and 13–3). Hence, accurate prediction of sulfur deficiency depends on using critical concentrations appropriate for the time of leaf sampling as well as the plant part specified.

Soil Analysis

Relatively few studies have attempted to establish critical sulfur concentrations in soil for rice (Table 13–4). Johnson and Fixen (1990) suggest that <10 mg sulfur kg^{-1} by $Ca(H_2PO_4)$ extraction is critical on the basis of several studies with rice. Dobermann and Fairhurst (2000) summarize the reported critical values that

Table 13–4. Critical levels of sulfur in soil for prediction of sulfur deficiency in rice.

Extractant	Critical level	Source
	mg S kg^{-1}	
0.05 M HCl	5	Islam and Ponnamperuma (1982)
0.25 M KCl heated to 40°C	6	Dobermann and Fairhurst (2000)
0.01 M $Ca(H_2PO_4)$	9	Islam and Ponnamperuma (1982)
0.01 M $Ca(H_2PO_4)$	8–9	Suzuki (1995)

vary depending on the sulfur extractant used (Table 13–4). Where a significant proportion of sulfur taken up is derived from the ester sulfur pool, the KCl-40 extractant that extracts from the ester-sulfur pool (Blair et al., 1991a) may be the most reliable test for assessing levels of plant available sulfur in soils (Blair and Lefroy, 1998). However, there has been inadequate evaluation as yet of its reliability in predicting sulfur responses in rice, in either wetland or dry land conditions. Moreover, there are conflicting reports in the literature on the importance of the ester-bound sulfur for plant uptake. Zhao et al. (2006) suggest that in padi soils, ester sulfur is less prevalent than reduced forms of organically bound sulfur. Short-term gross mineralization of sulfur in the study by Zhao et al. (2006) was strongly correlated with reduced organic sulfur forms rather than with ester-bound sulfur.

Recent studies have examined the utility of a plant root simulator probe for determining extractable sulfur by an anion exchange resin strip encased in a plastic applicator (Li et al., 2001b). Available sulfur extracted from 18 Chinese soils by the resin strips after 24-h and 2-wk burial times was significantly correlated with sulfur extracted by 0.01 mol $Ca(H_2PO_4)_2$ L^{-1} ($r^2 = 0.73^{***}$ and $r^2 = 0.60^{***}$). There was also a significant correlation between sulfur availability by the resin strips at the two burial times and rice plant sulfur uptake ($r^2 = 0.77^{***}$, $r^2 = 0.55^{***}$) and for the 24-h burial time with relative dry matter yield of rice (0.69***). The anion exchange resin strip may be a suitable alternative method in the prediction of sulfur supply in rice soils, but further evaluation is needed, especially under field conditions.

Increased sulfur levels in the subsoil may decrease the reliability of prediction of sulfur deficiency in upland rice from soil analysis of the topsoil. Subsoil sulfur pools may be important for upland rice since it has roots deep enough to access this sulfur. However, as discussed above, subsoil sulfur pools in submerged soils are unlikely to be of significant use to the shallow-rooted rice crops and hence topsoil sulfur levels are likely to be an adequate predictor of sulfur deficiency for wetland rice.

Future Considerations

While most rice is grown in submerged soils where chemical and biological processes that affect sulfur forms and behavior in soil differ from those in aerobic soils (see above), there is considerable interest in expanding aerobic rice and water-deficit irrigation systems to reduce the water requirement for producing rice. In the aerobic rice system, rice is flood irrigated every 4 to 7 d, resulting in cycles of soil saturation followed by drainage. In the water-deficit irrigation system, water levels are maintained between the soil surface and 20-cm depth. In the study by Belder et al. (2005), water levels cycled between 0 and 14 cm, with soil water potential falling below −10 kPa at the end of each drainage period in the cycle. Hence, the root zone was saturated except for a surface layer that was cycling between saturation and aerated conditions. Finally, in the rain-fed lowland ecosystem for rice, water levels fluctuate throughout the growth cycle so that soils may experience one or more cycles of loss of soil water saturation (Zeigler and Puckridge, 1995; Wade et al., 1999; Bell et al., 2001). The timing and duration of these episodes may be predictable in some areas but not in others. The factors affecting sulfur supply for the wetland rice production systems that do not involve continuous soil submergence are understood less than for either the submerged or upland rice systems. However, given the different processes affecting sulfur forms and mineralization in oxic versus anoxic soils, and in the rhizosphere of roots grown in submerged versus aerated soils, changes in soil water and oxygen regime are likely to have significant bearing on sulfur nutrition of rice. Elemental sulfur is likely to be an effective sulfur source when surface applied in these water-saving rice production systems. Mineralization of organically bound sulfur, whether from the soil-organic matter or from crop residues, is likely to be accelerated compared with submerged soils (Aulakh et al., 2002). Insoluble sulfides of iron and zinc are less likely to form and toxicity from H_2S is less likely since redox potentials may remain too high for the formation of these compounds. However, the effect of frequent changes in root-zone oxygen supply on root-sulfur uptake efficiency remain unclear. There is evidence that changes in root-oxygen supply can induce quite rapid changes in phosphorus uptake in rice (Insalud et al., 2006), but no equivalent information exists for sulfur uptake.

References

Abedin Mian, M.J., H.P. Blume, Z.H. Bhuiya, and M. Eaqub. 1991. Water and nutrient dynamics of a paddy soil of Bangladesh. Z. Pflanzenernahr. Bodenk. 154:93–99.

Armstrong, W. 1964. Oxygen diffusion from the roots of some British bog plants. Nature 204:801–802.

Armstrong, W. 1979. Aeration in higher plants. p. 225–332. *In* H. W. Woolhouse (ed.) Advances in botanical research Vol. 7. Academic Press, London.

Aulakh, M.L., R.C. Jaggi, and R. Sharma. 2002. Mineralisation-immobilisation of soil organic sulfur and oxidation of elemental sulfur in subtropical soils under flooded and nonflooded conditions. Biol. Fertil. Soils 35:197–203.

Barber, S.A. 1984. Soil nutrient bioavailability. A mechanistic approach. J. Wiley and Sons, New York.

Barrow, N.J. 1985. Reactions of anions and cations with variable-charge soils. Adv. Agron. 38:183–230.

Belder, P., B.A.M. Bouman, J.H.J. Spiertz, S. Peng, A.R. Castaneda, and R.M. Visperas. 2005. Crop performance, nitrogen and water use in flooded and aerobic rice. Plant Soil 273:167–182.

Bell, R.W., C. Ros, and V. Seng. 2001. Improving the efficiency and sustainability of fertiliser use in drought- and submergence-prone rainfed lowlands in Southeast Asia. p. 155- 169. *In* S. Fukai. and J. Basnayake (ed.) Increased Lowland Rice Production in the Mekong Region. Proceedings of an International Workshop, Vientiane, Laos. 30 Oct,–1 Nov. 2000. ACIAR Proceedings 101.

Blair, G.J., N. Chinoim, R.D.B. Lefroy, G.C. Anderson, and G.J. Crocker. 1991a. A sulfur soil test for pastures and crops. Aust. J. Soil Res. 29:619–626.

Blair G.J., and R. Lefroy. 1998. Sulfur and carbon research in rice production systems Field Crops Res. 56: 177–181.

Blair, G.J., R.D.B. Lefroy, W. Chaitep, D. Santoso, S. Samosir, M. Dana, and N. Chinoim. 1991b. Matching sulfur fertilisers to plant production systems. p. 156–157. *In* S. Portch (ed.) International symposium on the role of sulfur, magnesium and micronutrients in balanced plant nutrition. Phosphate and Potash Institute, Hong Kong.

Bohn, H.L., N.J. Barrow, S.S.S. Rajan, and R.L. Parfitt. 1986. Reactions of inorganic sulfur in soils. p. 233–249. *In* M.A. Tabatabai (ed.) Sulfur in agriculture. Agron. Monogr. 27. ASA, CSSA, SSSA, Madison, WI.

Bouman, B.A.M. 2001. Water-efficient management strategies in rice production. Int. Rice Res. Notes 26:17–22.

Brar, M.S., C.L. Arora, and P.N. Takkar. 1982. Critical values and adequate nutrient ranges in rice. J. Indian Soc. Soil Sci. 30:562–566.

Cassman, K.G. 2006. Ecological intensification of agriculture and implications for improved water and nutrient management. *In* International Symposium on Fertigation, Sept 20–24, Beijing, China.

Chaitep, W., R.D.B. Lefroy, and G.J. Blair. 1994. Effect of placement and source of sulfur in flooded and non-flooded rice cropping systems. Aust. J. Agric. Res. 45:1547–1556.

Chien, S.H., D.K. Friesen, and B.W. Hamilton. 1988. Effect of application method on availability of elemental sulfur in cropping sequences. Soil Sci. Soc. Am. J. 52:165–169.

Colmer, T.D. 2003. Aerenchyma and an inducible barrier to radial oxygen loss facilitate root aeration in upland, paddy and deep-water rice (*Oryza sativa* L.). Ann. Bot. (London) 91:301–309.

Dana, M., R.D.B. Lefroy, and G.J. Blair. 1994. A glasshouse evaluation of sulfur fertilizer sources for crops and pastures. I. Flooded and non-flooded rice. Aust. J. Agric. Res. 45:1497–1515.

Dick, W.A., D. Kost, and L. Chen. 2008. Availability of sulfur to crops from soil and other sources. p. 59–82. *In* J. Jez (ed.) Sulfur: A missing link between soils, crops, and nutrition. Agron. Monogr. 50. ASA, CSSA, SSSA, Madison, WI.

Dobermann, A., K.G. Cassman, C.P. Mamaril, and J.E. Sheehy. 1998. Management of phosphorus, potassium, and sulfur in intensive, irrigated lowland rice. Field Crops Res. 56:113–138.

Dobermann, A., and T. Fairhurst. 2000. Rice: Nutritional disorders and nutrient management. Potash and Phosphate Institute and Potash and Phosphate Institute of Canada, Singapore: International Rice Research Institute, Makati City (Philippines), 192 p.

Fox, R.L. 1974. Examples of anion and cation adsorption by soils of Tropical America. Trop. Agric. 51:2000–2009.

Fox, R.L., and G.J. Blair. 1986. Plant responses to sulfur in tropical soils. p. 405–434. *In* M.A. Tabatabai (ed.) Sulfur in agriculture. Agron. Monogr. 27. ASA, CSSA, SSSA, Madison, WI.

Freney, J.R. 1986. Forms and reactions of organic sulfur compounds in soils. p. 207–232. *In* M.A. Tabatabai (ed.) Sulfur in agriculture. Agron. Monogr. 27. ASA, CSSA, and SSSA, Madison, WI.

Freney, J.R., G.E. Melville, and C.H. Williams. 1975. Soil organic matter fractions as sources of plant available sulphur. Soil Biol. Biochem. 7:217–221.

Greenland, D.J. 1997. The sustainability of rice farming. CAB International Wallingford, UK.

Hagan, N.D., N. Upadhyaya, L.M. Tabe, and T.V.J. Higgins. 2003. The redistribution of protein sulfur in transgenic rice expressing a gene for a foreign, sulfur-rich protein. Plant J. 34:1–11.

Hawkesford, M.J., A. Schneider, A.R. Belcher, and D.T. Clarkson. 1995. Regulation of enzymes in the sulfur-assimilatory pathway. Z. Pflanzenernahr. Bodenkunde 158:55–57.

Hitsuda, K., M. Yamada, and D. Klepker. 2005. Sulfur requirement of eight crops at early stages of growth. Agron. J. 97:155–159.

Hoefgen, R., and H. Hesse. 2008. Sulfur and cysteine metabolism. p. 83–104. *In* J. Jez (ed.) Sulfur: A missing link between soils, crops, and nutrition. Agron. Monogr. 50. ASA, CSSA, SSSA, Madison, WI.

Hu, Z., S. Haneklaus, S. Wang, C. Xu, Z. Cao, and E. Schnug. 2003. Comparison of mineralization and distribution of soil sulfur fractions in the rhizosphere of oilseed rape and rice. Commun. Soil Sci. Plant Anal. 34:2243–2257.

Hu, Z.Y., Y.G. Zhu, M. Li, L.G. Zhang, Z.H. Cao, and F.A. Smith. 2007. Sulfur (S) induced enhancement of iron plaque formation in the rhizosphere reduces arsenic accumulation in rice (*Oryza sativa* L.) seedlings. Environ. Pollut. 147:387–393.

Insalud, N., R.W. Bell, T.D. Colmer, and B. Rerkasem. 2006. Morphological and physiological responses of rice (*Oryza sativa* L.) to limited phosphorus supply in aerated and stagnant solution culture. Ann. Bot. (London) 98:995–1004.

IRRI. 1977. Preliminary report: First international trials of nitrogen fertilizer efficiency in rice. 1975–76. International Rice Research Institute, Los Banos The Philippines.

Islam, M.M., and F.N. Ponnamperuma. 1982. Soil and plant tests for available sulfur in wetland rice soils. Plant Soil 68:98–113.

Islam, N., N.M. Upadhyaya, P.M. Campbell, R. Akhurst, N. Hagan, and T.J.V. Higgins. 2005. Decreased accumulation of glutelin types in rice grains constitutively expressing a sunflower seed albumin gene. Phytochemistry 66:2534–2539.

Johnson, G.H., and P. Fixen. 1990. Testing soils for sulfur, boron, molybdenum and chlorine. p. 265–273. In R.L. Westermann (ed.) Soil testing and plant analysis 3rd ed. SSSA Book Series No. 3. SSSA, Madison, WI.

Kirk, G.J.D. 2004. The biogeochemistry of submerged soils. J. Wiley & Sons, Chichester. p. 291.

Kuzuhara, Y., A. Isobe, M. Awazuhara, T. Fujiwara, and H. Hayashi. 2000. Glutathione levels in phloem sap of rice plant under sulfur-deficient conditions. Soil Sci. Plant Nutr. 46:266–270.

Lefroy, R.D.B., W. Chaitep, and G.J. Blair. 1994. Release of sulfur from rice residues under flooded and non-flooded soil conditions. Aust. J. Agric. Res. 45:657–667.

Lefroy, R.D.B., C.P. Mamaril, G.J. Blair, and P.B. Gonzales. 1992. Sulphur cycling in rice wetlands. p. 279–299. In R.W. Howarth et al (ed.) Sulphur cycling on the continents. John John Wiley & Sons, New York.

Lefroy, R.D.B., S.R. Samosir, and G.J. Blair. 1993. The dynamics of sulfur, phosphorus and iron in flooded soils as affected by changes in Eh and pH. Aust. J. Soil Res. 31:493–508.

Lefroy, R.D.B., D. Santoso, and M. Ismunadji. 1991. Incidental sulfur inputs in rainfall and irrigation water. p. 101–104. In G. Blair and R. Lefroy (ed.) Sulfur fertilizer policy for lowland and upland rice cropping systems in Indonesia. Australian Centre for International Agricultural Research, Canberra.

Leustek, T., and K. Saito. 1999. Sulfate transport and assimilation in plants. Plant Physiol. 120:637–643.

Li, S., B. Lin, and W. Zhou. 2001a. Soil organic sulfur mineralisation in the presence of growing plants under aerobic or waterlogged condictions. Soil Biol. Biochem. 33:721–727.

Li, S., B. Lin, and W. Zhou. 2001b. Soil sulfur supply assessment using anion exchange resin strip-plant root simulator probe. Commun. Soil Sci. Plant Anal. 32:711–722.

Linquist, B., P. Sengxua, A. Whitbread, J. Schiller, and P. Lathvilayvong. 1998. Evaluating nutrient deficiencies and management strategies for lowland rice in Lao PDR. p. 59–73. In J.K. Ladha et al. (ed.) Rainfed lowland rice: Advances in nutrient management research. Proceedings of the International Workshop on Nutrient Management Research in Rainfed Lowlands, 12–15 October 1998, Ubon Ratchatani, Thailand. International Rice Research Institute, Los Baños, Philippines.

Loneragan, J.F., K. Snowball, and A.D. Robson. 1976. Remobilisation of nutrients and its significance in plant nutrition. p. 463–469. In I.F. Wardlaw and J.B. Passioura (ed.) Transport and transfer processes in plants. Academic Press, London.

Marschner, H. 1995. Mineral nutrition of higher plants, 2nd ed. Academic Press, New York.

McCaskill, M.R., and G.J. Blair. 1989. A model for the release of sulfur from elemental S and superphosphate. Fert. Res. 19:77–84.

Moorman, F.R., and N. van Breemen. 1978. Rice: Soil, water, land. International Rice Research Institute, Los Banos, The Philippines. p. 185.

Morita, S., and J. Abe. 1996. Development of root systems in wheat and rice. p. 199–209. In O. Ito et al. (ed.) Dynamics of roots and nitrogen in cropping systems of the semi-arid tropics. Japan International Research Centre for Agricultural Sciences, Tsukuba, Japan.

Murase, J., and M. Kimura. 1997. Anaerobic reoxidation of Mn^{2+}, Fe^{2+}, S^0 and S^{2-} in submerged paddy soils. Biol. Fertil. Soils 25:302–306.

Olk, D.C., G. Brunetti, and N. Senesi. 2000. Decrease in humification of organic matter with intensified lowland rice cropping: A wet chemical and spectroscopic investigation. Soil Sci. Soc. Am. J. 64:1337–1347.

Osiname, O.A., and B.T. Kang. 1975. Responses of rice to sulfur application under upland conditions. Commun. Soil Sci. Plant Anal. 6:585–598.

Ponnamperuma, F.N. 1981. Some aspects of the physical chemistry of paddy soils. p. 59–94. *In* Institute of Soil Science, Academic Sinica (ed.) Proceedings of the symposium on paddy soil Vol. 1. Springer-Verlag, Berlin.

Reddy, K.S., M. Singh, A. Swarup, A.S. Rao, and K.N. Singh. 2002. Sulfur mineralization in two soils amended with organic manures, crop residues, and green manures. J. Plant Nutr. Soil Sci. 165:167–171.

Rennenberg, H., K. Schmidtz, and L. Bergmann. 1979. Long distance transports of sulfur in *Nicotiana tabacum*. Planta 147:57–62.

Robson, A.D., and M.G. Pitman. 1983. Interactions between nutrients in higher plants. p. 147–180. *In* A. Lauchli and R.L. Bieleski (ed.) Encyclopaedia of plant physiology. New Series Vol. 15A. Springer Verlag, Berlin.

Saalbach, E. 1984. The significance of atmospheric sulfur compounds for the supply of agricultural crops. Ann. Bot. (London) 58:147–156.

Sahrawat, K.L. 2004. Ammonium production in submerged soils and sediments: The role of reducible iron. Commun. Soil Science Plant Nutr. 35:399–411.

Samosir, S.S.R., G.J. Blair, and R.D.B. Lefroy. 1993. Effects of placement of elemental sulfur and sulfate on the growth of two rice varieties under flooded conditions. Aust. J. Agric. Res. 44:1775–1788.

Satrusajang, A., P. Snitwongse, R.J. Buresh, and D.K. Friesen. 1991. Nitrogen-15 and sulfur-35 balances for fertilizers applied to transplanted rainfed lowland rice. Fert. Res. 28:55–65.

Scheid, D., S. Stubner, and R. Conrad. 2004. Identification of rice root associated nitrate, sulfate and ferric iron reducing bacteria during root decomposition. FEMS Microbiol. Ecol. 50:101–110.

Schoenau, J.J., and S.S. Malhi. 2008. Sulfur forms and cycling processes in soil and their relationship to sulfur fertility. p. 1–10. *In* J. Jez (ed.) Sulfur: A missing link between soils, crops, and nutrition. Agron. Monogr. 50. ASA, CSSA, SSSA, Madison, WI.

Seng, V., C. Ros, R.W. Bell, P.F. White, and S. Hin. 2001. Nutrient requirements for lowland rice in Cambodia. p. 169- 178. *In* S. Fukai and J. Basnayake (ed.) Increased lowland rice production in the Mekong Region. Proceedings of an international workshop, Vientiane, Laos. 30 Oct.–1 Nov. 2000. ACIAR Proceedings 101.

Shan, X.-Q., B. Chen, T.H. Zhang, F.L. Li, B Wen, and J. Qian. 1997. Relationship between sulfur speciation in soils and plant availability. Sci. Total Environ. 199:237–246.

Sharma, P.K. G., Pantuwan, K.T. Ingram, and S.K. De Datta. 1994. Rainfed lowland rice roots: Soil and hydrological effects. p. 55–66. *In* G.J.D. Kirk (ed.) Rice roots: Nutrient and water use. International Rice Research Institute, Manila, Philippines.

Sheehy, J.E. 1996. Models, mechanisms and yield. p. 306–310. *In* R. Ishii and T. Horie (ed.) Crop research in Asia: Achievements and perspective. University of Tokyo, Japan.

Shin, J.S., and K.H. Han. 1988. Sulfur balance in Korean agriculture. p. 29–34. *In* Proceedings of the International Symposium on Sulfur in Korean Agriculture. Korean Society for Soil Science and the Sulfur Institute, Washington, DC.

Slaton, N.A., R.J. Norman, and J.T. Gilmour. 2001. Oxidation rates of commercial elemental sulfur products applied to an alkaline silt loam from Arkansas. Soil Sci. Soc. Am. J. 65:239–243.

Srivastava, P.C., and U.S. Singh. 2007. Effect of graded levels of nitrogen and sulfur and their interaction on yields and quality of aromatic rice. J. Plant Nutr. 30:811–828.

Stubner, S., T. Wind, and R. Conrad. 1998. Sulfur oxidation in rice field soil: Activity, enumeration, isolation and characterization of thiosulfate-oxidizing bacteria. Syst. Appl. Microbiol. 21:569–578.

Surridge, C. 2004. Feast or famine? Nature 428:360–361.

Suwanarit, A., S. Kreetapirom, S. Buranakarn, P. Suriyapromchai, W. Varanyanond, and P. Tungtrakul. 1997. Effects of sulfur fertilizer on grain qualities of Khaw Dauk Mali-105 rice. Kasetsart J. Nat. Sci. 31:305–313.

Suzuki, A. 1995. Metabolism and physiology of sulfur. p. 395–401. *In* T. Matsuo et al. (ed.) Science of the rice plant. Vol. 2. Physiology. Food and Agriculture Policy Research Center, Tokyo.

Tisdale, S.L., R.B. Reneau, and J.S. Platou. 1986. Atlas of sulfur deficiencies. p. 295–322. *In* M.A. Tabatabai (ed.) Sulfur in agriculture. Agron. Monogr. 27. ASA, CSSA, and SSSA, Madison, WI.

Wade, L.J., S. Fukai, B.K. Samson, A. Ali, and M.A. Mazid. 1999. Rainfed lowland rice: Physical environment and cultivar requirements. Field Crops Res. 64:3–12.

Westfall, D.G., D.A. Whitney, and D.M. Brandon. 1990. Plant analysis as an aid in fertilizing small grains. p. 495–519. *In* R.L. Westermann (ed.) Soil testing and plant analysis 3rd ed. SSSA Book Ser. 3. SSSA, Madison, WI.

Wind, T., and R. Conrad. 1997. Localization of sulfate reduction in planted and unplanted rice field soil. Biogeochemistry 37:253–278.

Yasmin, N., G.J. Blair, and R. Till. 2007. Effect of elemental sulfur, gypsum, and elemental sulfur coated fertilizers, on the availability of sulfur to rice. J. Plant Nutr. 30:79–91.

Yoshida, S., and M.R. Chaudhry. 1979. Sulfur nutrition of rice. Soil Sci. Plant Nutr. 25:121–134.

Yu, K., F. Bohme, J. Rinklebe, H.-U. Neue, and R.D. DeLaune. 2007. Major biogeochemical processes in soils- a mocrocosm incubation from reducing to oxidising condition. Soil Sci. Soc. Am. J. 71:1406–1417.

Zeigler, R.S., and D.W. Puckridge. 1995. Improving sustainable productivity in rice-based rainfed lowland systems of South and South-East Asia. GeoJournal 35:307–324.

Zhao, F.J., J. Lehmann, D. Solomon, M.A. Fox, and S.P. McGrath. 2006. Sulphur speciation and turnover in soils: Evidence from sulphur K-edge XANES spectroscopy and isotope dilution studies. Soil Biol. Biochem. 38:1000–1007.

Zhou, W., M. Wan, P. He, S. Li, and B. Lin. 2002. Oxidation of elemental sulfur in paddy soils as influenced by flooded condition and plant growth in pot experiment. Biol. Fertil. Soils 36:384–389.

14

Evaluation of the Relative Significance of Sulfur and Other Essential Mineral Elements in Oilseed Rape, Cereals, and Sugar Beet Production

Ewald Schnug and Silvia Haneklaus
Julius-Kühn Institute, Braunschweig, Germany

Abstract

This chapter provides a current overview of mathematical approaches for assessing the nutritional status of a crop, to supply verified algorithms for interpretation and comparative weighting of plant analytical data, and to present a procedure for calculating the sulfur fertilizer demand of oilseed rape (*Brassica napus* L.). In each case, special attention is given to sulfur nutrition.

Arnon (1950) determined three criteria that must be fulfilled before a mineral element can be recognized as essential for the growth of higher plants: (i) the element must be pivotal for completing the life cycle; (ii) the element must have specific functions in plant metabolism and a limited supply must yield characteristic macroscopic deficiency symptoms; and (iii) the element must be directly involved in plant nutrition. Besides carbon, hydrogen, and oxygen, six major (nitrogen, phosphorus, sulfur, potassium, calcium, and magnesium) and seven minor elements (iron, manganese, zinc, copper, chloride, boron, and molybdenum) are essential for the growth of higher plants.

The plant nutrient supply affects yield, quality, plant health, and environment. The supply of plants with mineral elements has a direct impact on the content of primary and secondary compounds and their mineral composition (Barker and Pilbeam, 2006). Nutrient induced resistance mechanisms and allelopathies may be used to develop effective countermeasures against weeds, pests, and diseases (Datnoff et al., 2007). Last but not least, nutrient losses to the environment cannot be avoided but can be minimized, for instance by a balanced supply with nitrogen and sulfur as sulfur deficiency reduces significantly the nitrogen utilization efficiency (Schnug, 1991). Here, the efficacy of new technologies such as precision agriculture for a site-specific, variable rate input of fertilizers is intrinsic; however, it is not yet profitable to implement the technology (Haneklaus and Schnug, 2006).

Important threshold markers for the nutrient supply are (i) the symptomatological value, which reflects the nutrient concentration below which deficiency

---

Copyright © 2008. American Society of Agronomy, Crop Science Society of America, Soil Science Society of America, 677 S. Segoe Rd., Madison, WI 53711, USA. *Sulfur: A Missing Link between Soils, Crops, and Nutrition.* Agronomy Monograph 50.

symptoms become visible; (ii) the critical nutrient value, which stands for the nutrient concentration above which the plant is sufficiently supplied for achieving the maximum potential yield or yield reduced by 5, 10, and 20%; and (iii) the toxicological value, which indicates the nutrient concentration above which toxicity symptoms can be observed. A comprehensive overview of crop-specific deficiency and sufficiency ranges for plant nutrients are given in Reuter and Robinson (1997) and Barker and Pilbeam (2006).

Critical plant nutrient values are dependent on crop type, yield level, sampling date, sampled plant organ, and chemical form of the nutrient (Finck, 1979; Schnug and Haneklaus, 1998). Soil samples can be taken throughout the year (Finck, 1979), but for comparison of soil and plant analysis, the same sampling date and location is recommended (Haneklaus et al., 1995).

Liebig's "Law of the Minimum" (1855) is the basis for plant nutrition. It states that the exploitation of the genetically fixed yield potential of crops is limited by the variable which is insufficiently supplied to the greatest extent. This theory neglects interactions between growth factors but proved to be sufficiently accurate to estimate the single response of the nutrient (Lark, 2001). A quite important disadvantage of critical nutrient values or even ranges of critical values (tissue concentration for 95% of maximum yield) (Mills and Benton-Jones, 1991) or *no effect values* (NEV) (tissue concentration for maximum yield or the concentration above which yields fail to response) (Finck, 1970) is that they do not respect the nonlinear relationship between growth factors and yield postulated by Mitscherlich (Wallace, 1990). This problem can be overcome efficiently by the determination of upper boundary line functions and the subsequent calculation of optimum nutrient values and ranges within which highest yields can be achieved (Haneklaus et al., 2006a).

It is the objective of this chapter to provide an up-to-date overview of mathematical approaches for assessing the nutritional status of a crop, to supply verified algorithms for interpretation and comparative weighting of plant analytical data, and to present a procedure for calculating the sulfur fertilizer demand of oilseed rape. In each case, special attention is paid each time to sulfur.

Algorithms for the Deduction of Critical Ranges in the Nutrient Supply

Optimum or critical nutrient concentrations for plant and soil data are generally deduced from fertilizer response trials in greenhouse and field experiments by regression analysis (Bergmann, 1993, Finck, 1979). Methods based on regression analysis such as the "broken stick method" (Hudson, 1966; Spencer and Freney, 1980) and "vector analysis" (Timmer and Armstrong, 1987) investigate mathematical but not necessarily causal interactions between nutrient content and yield. The reason is the dictate of minimizing the sum of squared distances, which aims exclusively at determining a function that fits best across the data set; another reason is the inadequacy of employing linear interpretation models for the nonlinear relationship between growth factors and yield (Wallace, 1990). Another major criticism of critical values for the interpretation of tissue analysis is the small experimental basis, which often consists of not more than a single experiment (Vielemeyer et al., 1983). Yet a further shortcoming of these procedures is that a similar nutrient supply can be realized by comparatively wide nutrient ranges and nutrient ratios (Smith, 1962). This results finally in significantly different optima for the nutritional status in different experiments (Bergmann, 1993). Hence, the accuracy of critical values and optimum ranges (Baier, 1995; Bergmann, 1992; Mills and Benton-Jones, 1991), and NEV (Finck, 1970) is questionable.

An exemplary compilation and attribution of individual data and their evaluation revealed ultimately for sulfur response trials that it is not possible to assign a general validity of critical nutrient ranges, let alone threshold values and this not only for individual crop plants, but also plant families (Haneklaus et al., 2006a). A clear differentiation of the sulfur supply can be determined only after assigning plant groups based on morphogenetic and physiological features and restricting the evaluation of the sulfur supply irrespective of the sampled plant part during vegetative growth to three major categories—deficient, adequate, and high (Haneklaus et al., 2006a).

The ideal basis for establishing critical values for the interpretation of tissue analysis are large sets of yield data and nutrient concentrations in defined plant organs, which cover a wide range of growth factor combinations. These data may include samples from field surveys, field trials, or pot trials if the reference yield for 100% was obtained always under optimum growth conditions. Such data sets will show a characteristic pattern when plotted in a Cartesian coordinate system if the nutrient concentrations cover the entire range from deficiency to surplus: in the lower right, the points relating nutrient concentration to crop yield will be frayed, but in the upper left, points will arrange to a characteristic bow-shaped bulk. The line describing the highest yields observed over the range of nutrient values measured is known as the boundary line since it lies on the upper edge of the body of the data (Webb, 1972). Thus, boundary lines describe the "pure effect of a nutrient" on crop yield under ceteris paribus conditions (Evanylo and Sumner, 1987; Moeller-Nielsen and Frijs-Nielsen, 1976; Walworth et al., 1986). A major criticism of using boundary lines is that they are drawn manually and thus prone to subjective influence. With BOLIDES (BOundary LIne DEvelopment System), it became feasible to calculate upper boundary line functions by defined mathematical routines (Schnug et al., 1995, 1996). Noteworthy is that critical nutrient ranges of the sulfur supply that have been deduced this way for various cultures fit harmoniously in a specifically compiled data set from innumerable experiments (Haneklaus et al., 2006a).

Boundary Line Development Systems (BOLIDES)

The BOLIDES is a mathematical approach which was elaborated to determine upper boundary line functions and to evaluate optimum nutrient values and ranges, respectively. BOLIDES is based on a five-step algorithm (Fig. 14–1; Schnug et al., 1996). For the identification of outliers, cell sizes are defined for nutrient and yield values together with an optional number of data points per cell (Fig. 14–1a). The cell size can be chosen variably with proposed values for X (nutrient contents) corresponding to the standard deviations and for Y (yield) with the coefficient of variation. If another variable than the plant-available nutrient content in the soil, for instance a stable soil feature such as organic matter or clay content, has a significant effect on the response to the nutrient, its presence is indicated by two or more distinct concentrations of points, each with its own boundary line response to the nutrient (Fig. 14–1b). The data can be classified on

(a) Identification of outliers

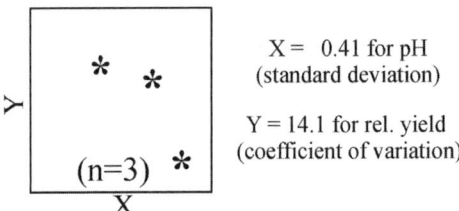

(b) Discrimination against a third variable

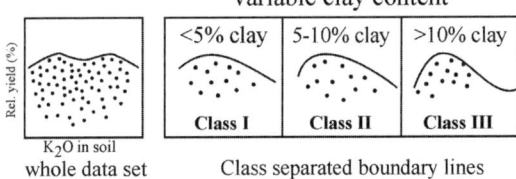

(c) Calculation of step functions

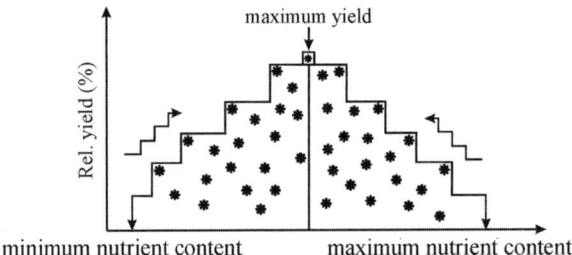

(d) Determination of the upper boundary line and calculation of optimum nutrient values and ranges

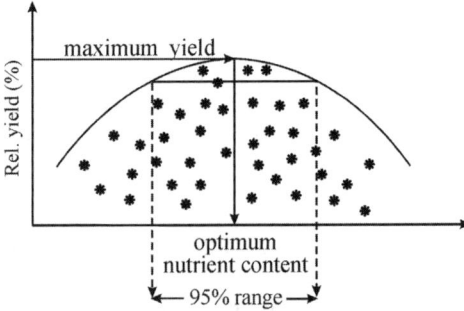

Fig. 14–1. Structure of BOLIDES for the determination of upper boundary line functions and optimum nutrient values and ranges in plants and soils (adapted from Haneklaus and Schnug, 1998).

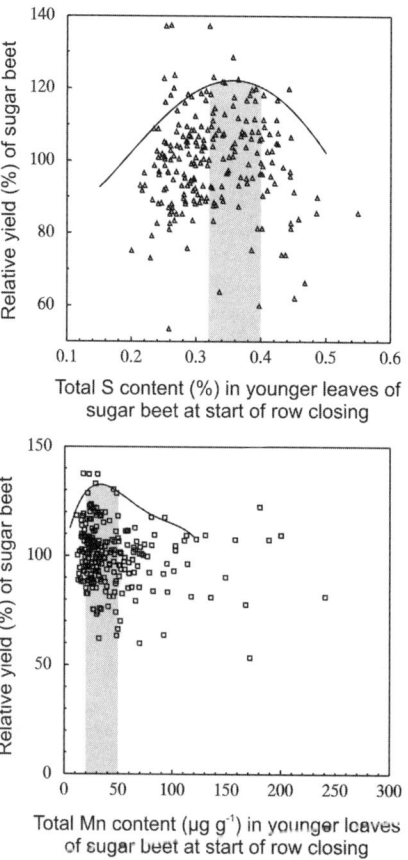

Fig. 14–2. Upper boundary lines calculated by BOLIDES for the derivation of optimum nutrient ranges of the total sulfur (S, left) and manganese (Mn, right) content in younger, fully differentiated leaves at start of row closing.

the basis of this third variable and the boundary line can be determined separately for each class.

Next, a boundary step function is calculated with each class starting from the minimum nutrient content up to the point of maximum yield as well as from the maximum nutrient content up to the maximum yield (Fig. 14–1c). Then the boundary line, usually a fourth order polynomial function, is fitted according to the least square method (Fig. 14–1d). The first derivative of the fitted polynomial gives predicted yield response to fertilization in relation to the nutrient content (Fig. 14–1). The last step is the classification of the nutrient supply (Haneklaus and Schnug, 1998) to determine optimum nutrient levels and/or optimum nutrient ranges. The optimum nutrient value corresponds with the zero of the first derivative of the upper boundary line and the sign of the second derivative at this point. For the determination of the optimum ranges, i.e., the range of nutrient concentration which gives 95% of the maximum yield, standard, numerical root-finding procedures are used for real polynomials of degree four with constant coefficients (Fig. 14–1).

The advantage of upper boundary line functions is that they describe the yield response to variation in the test parameter where all other factors are, within the constraints of the fields concerned, as close as possible to nonlimiting. The lack of mathematical procedures, however, restricted the acceptance and use of this approach, but this has been overcome by BOLIDES, a procedure that proved to provide precise and reproducible results (Schnug et al., 1995, 1996; Haneklaus et al., 2006a). In Fig. 14–2, two upper boundary line functions are shown, one for a macro and one for a micronutrient. Both the upper boundary lines for sulfur and Mn show a characteristic course in so far as for macronutrients generally a wider optimum range can be found than for micronutrients where the equivalent span is much more confined. This indicates that the risk of yield losses by an insufficient and excessive supply is distinctly higher for micro than for macronutrients.

Upper boundary line functions, their range of validity, and optimum nutrient ranges have been summarized for selected essential macro and micronutrients and crops (oilseed rape, cereals and sugar beet, *Beta vulgaris* L.) in Tables 14–1 to 14–3.

In regard to sugar beet, the question arises whether optimum nutrient ranges vary in relation to the reference value of productivity, namely root and sugar yield. The upper boundary line analysis revealed that relevant differences in the optimum nutrient ranges exist only for Mg and Na (Table 14–4). For these elements, the lower concentration for a sufficient supply is about 50% higher than that required for maximizing crop productivity (Table 14–4). This implies that on marginal sites a sufficient supply should be ensured by appropriate fertilization practices.

A comprehensive overview of crop-specific deficiency and sufficiency ranges of the sulfur supply has been elaborated before by Haneklaus et al. (2006b), and the results are summarized in Table 14–5. Severe to moderate sulfur deficiency is indicated generally by total sulfur concentrations of less than 1.2, 1.7, and 3.5 mg g^{-1} sulfur in cereals, sugar beet, and oilseed rape, respectively (Table 14–5). Correspondingly, an adequate sulfur supply is reflected by total sulfur concentrations between 3.0 and 6.5 mg g^{-1} sulfur. With view to the nitrogen/sulfur (N/S) ratio values of 16 to 20 reflect a sufficient sulfur supply (Haneklaus et al., 2006a). In literature, sulfur concentrations which impair crop performance are rare for sulfur. An excessive sulfur supply, which diminishes crop yield, can be expected if sugar beet, cereals, and oilseed rape contain more than 4.5, 7.5, and 14 mg g^{-1} sulfur (d.w.), respectively, in fully differentiated leaves or whole above-ground biomass (cereals) at row closing and stem extension (Table 14–5).

Comparing these threshold values with median values calculated from literature data (Haneklaus et al., 2006a), it is striking that total sulfur concentrations, which can be found when macroscopic symptoms are visible, are in good agreement. The same applies for threshold concentrations, indicating a sufficient sulfur supply of cereals and sugar beet, while for oilseed rape, significantly higher values were determined. The reason is most likely that the potential yield of oilseed rape crops was distinctly lower in the majority of other studies (Haneklaus et al., 2006a). Momentarily, the physiological background of the specific toxicity of sulfate is unknown, but some speculations about regulatory mechanisms were suggested by Haneklaus et al. (2006b). A disproportionate sulfur supply may derange crosstalk between sulfur and Ca metabolism in the plant in such way that it hampers sulfur homeostasis and thus induces toxic effects. This theory is

Table 14–1. Algorithms and validated ranges of upper boundary lines for unifactorial relationships between nutrient concentrations and relative seed yield of oilseed rape. [Y = relative yield (%); X = nutrient concentration in mg/g (d.w.) for macro and mg/kg (d.w.) for micronutrients.]

Nutrient	Oilseed rape					
	Upper boundary line function (linear)	Validated range	Upper boundary line function (nonlinear)	Validated range		NEV†
N	$6.07*X - 121.43$	20–34	$2.5*X + 0.01$	4	-40	40
P	$43*X - 64.5$	1.5–3.5	$2.0*X + 16$	3.5	-4.2	4.2
S	$-166.97*1/X + 125.7$	1.32–6.5	$-166.97*1/X + 125.7$	1.32	-6.5	6.5
K	$18.89*X - 283.33$	15–19.5	$1685.2*1/X - 1.55*X + 202.07$	19.5	-35	35
Ca	$65*X - 650$	10–11	$-1012.3*1/X - 1.04*X + 169.56$	11	-22.5	22.5
Mg	$211.11*X - 168.89$	0.9–1.25	$20*X + 70$	1.25	-1.5	1.5
Fe	$1.49*X - 29.8$	20–77	$-14262.5*1/X - 1.22*X + 364.07$	77	-100	100
Mn	$18.5*X - 185$	10–14	$-1080*1/X - 0.97*X + 164.86$	14	-30	30
Zn	$7.17*X - 107.5$	15–27	$-4762.1*1/X - 3.1*X + 346.5$	27	-33	33
Cu	$52.78*X - 105.56$	2.0–3.8	$7.14*X + 67.86$	3.8	-4.5	4.5
Cl	$2.8*X - 70$	25–50	$0.6*X + 40$	50	-100	100
B	$13.75*X - 123.75$	9.0–13	$-1804.9*1/X - 1.81*X + 217$	13	-25	25
Mo	$800*X - 40$	0.05–0.1	$-10.842*1/X - 74.45*X + 156.24$	0.1	-0.3	0.3

† NEV, no effect value.

Table 14–2. Algorithms and validated ranges of upper boundary lines for unifactorial relationships between nutrient concentrations and relative grain yield of cereals. [Y = relative yield (%); X = nutrient concentration in mg/g (d.w.) for macro and mg/kg (d.w.) for micronutrients.]

Nutrient	Cereals				
	Upper boundary line function (linear)	Validated range	Upper boundary line function (nonlinear)	Validated range	NEV†
N	12*X − 150	12.5–19.5	−1476.2*1/X − 1.21*X + 184.1	19.5 −35	35
P	168*X − 366	2.0–2.5	−336.54*1/X − 24.16*X + 280.39	2.5 −4	4
S	120.9*X − 145.09	1.2–1.75	−187.09*1/X − 12.21*X + 195.18	1.75 −4	4
K	21*X − 336	16–20	−1743*1/X − 1.52*X + 202.62	20 −35	35
Ca	200*X − 200	1.4–1.8	−145.18*1/X − 11.91*X + 183.02	1.8 −4	4
Mg	460*X − 230	0.5–0.7	20*X + 78	0.7 −1.1	1.1
Fe	4.2*X − 105	25–45	1.07*X + 36	45 −60	60
Mn	17.6*X − 158.4	9–14	−634.3*1/X − 0.85*X + 146.26	14 −28	28
Zn	56*X − 672	12–13.5	−664.38*1/X − 0.56*X + 140.45	13.5 −25	25
Cu	168*X − 235.2	1.4–1.9	−81.89*1/X − 4*X + 135.9	1.9 −4	4
Cl	2.8*X − 70	25–50	0.6*X + 40	50 −100	100
B	50*X − 50	1.0–1.4	−7.54*(X*X*X) + 165.22*X − 191.9	1.4 −3	3
Mo	1600*X − 80	0.05–0.075	−2847.3*(X*X) + 1258.6*X − 38.02	0.075 −0.2	0.2

† NEV, no effect value.

Sulfur in Oilseed Rape, Cereals, and Sugar Beet Production

Table 14–3. Algorithms and validated ranges of upper boundary lines for unifactorial relationships between nutrient concentrations and relative root yield of sugar beet. Adapted from Haneklaus and Schnug 1998. [Y = relative yield (%); X = nutrient concentration in % for macro (d.w.) and mg/kg (d.w.) for micronutrients.]

Nutrient	Sugar Beet		
	Upper boundary line function (nonlinear)	Validated range	NEV†
N	$Y = -3.594e - 4*X^4 + 68.5866*X^3 - 495.62*X^2 + 1606.484*X - 1842.222$	3.2 – 4.6	4.6
P	$Y = 190.650*X^4 - 391.398*X^3 + 56.694*X^2 + 154.292*X + 69.332$	0.15 – 0.45	0.45
S	$Y = 4807.537*X^4 - 7332.276*X^3 + 3175.999*X^2 - 340.317*X + 94.413$	0.15 – 0.35	0.35
K	$Y = 0.294*X^4 - 4.835*X^3 + 23.364*X^2 - 25.360*X + 85.575$	2 – 4.2	4.2
Ca	$Y = -48.633*X^4 + 171.211*X^3 - 249.140*X^2 + 145.851*X + 97.507$	0.1 – 0.42	0.42
Mg	$Y = -327*X^4 + 4773.236*X^3 - 2584.860*X^2 + 566.015*X + 90.34099.72$	0.08 – 0.18	0.18
Fe	$Y = -1.389e - 13*X^6 + 2.700e - 10*X^5 - 2.068e - 7*X^4 + 7.923e - 5*X^3 - 0.0159*X^2 + 1.548*X + 78.630$	50 – 100	100
Mn	$Y = -1.670e - 6*X^4 + 5.170e - 4*X^3 - 0.0565*X^2 + 2.272*X + 102.497$	5 – 30	30
Zn	$Y = 5.599e - 6*X^4 - 5.807e - 4*X^3 - 0.0298*X^2 + 4.315*X + 26.334$	20 – 50	50
Cu	$Y = 2.144e - 4*X^4 - 0.010*X^3 + 0.0032*X^2 + 4.067*X + 86.802$	3 – 15	15
B	$Y = 3.868e - 5*X^4 - 0.00485*X^3 + 0.169*X^2 - 1.105*X + 109.09$	15 – 30	30
Na	$Y = -1745.977*X^6 + 7010.894*X^5 - 11023.249*X^4 + 8593.433*X^3 - 3491.474*X^2 + 691.674*X + 73.923$	0.05 – 0.2	0.2

† NEV, no effect value.

Table 14–4. Comparison of optimum nutrient ranges for root and sugar yield of sugar beet. [Nutrient concentration in percentage (d.w.) for macro and mg/kg (d.w.) for micronutrients.]

Yield	Nutrient												
	N	S	N/S	P	K	Ca	Mg	Fe	Mn	Zn	Cu	B	Na
Root	4.2–5.4	0.32–0.40	13.5–18	0.45–0.65	3.6–6.0	0.35–0.8	0.12–0.32	80–200	20–50	40–60	10–20	24–40	0.2–0.77†
Sugar	4.8–5.6	0.32–0.40	14–18	0.4–0.65	4.0–6.0	0.35–0.7	0.22–0.40	100–230	20–50	45–60	13–19	24–40	0.4–0.65

† If K varies between 3.6 and 6.0%.

Table 14–5. Threshold values for total sulfur concentrations (mg g^{-1} sulfur, d.w.) in younger leaves of oilseed rape and sugar beet and whole above-ground biomass of cereals at start of stem extension and canopy closing. Adapted from Haneklaus et al. (2006b).

Crop	Deficiency	Sufficiency		Excess
	Symptomatological threshold	Lower critical value (–5% yield)	Maximum yield†	Upper critical value (–10% yield)
Cereals	<1.2	3.2	4.0	>7.5
Rape	<2.8‡ and <3.5§	5.5	6.5	>14.0
Sugar beet	<1.7	3.0	3.5	>4.5

† Seed (oilseed rape), grain (cereals), root and sugar (sugar beet) yield.
‡ Single low varieties.
§ Double low varieties.

compatible with experiments performed by Cerda et al. (1984) which showed that an excessive sulfur supply to tomatoes (*Lycopersicon esculentum* Mill.) induced calcium deficiency as evidenced by blossom end rot.

Professional Interpretation Program for Plant Analysis (PIPPA)

Professional Interpretation Program for Plant Analysis (PIPPA) is a program which makes use of upper boundary lines and enables the interpretation of plant analysis data of oilseed rape, cereals, and sugar beet crops. It is based on databases containing more than 6000 observations for the three major cash crops and summarizing more than 30 yr of research and experimentation in plant analysis. Much experimental work was conducted in close cooperation with official advisory boards, plant breeders, agrochemical manufacturers, and practical farmers in Germany, Denmark, and the UK. The collection of data began in 1973 for cereals, 1980 for oilseed rape, and 1990 for sugar beet (Schnug, 1990; Schnug and Haneklaus, 1992).

Recommended Sampling Procedures

The BBCH code was used for defining growth stages for the sampling of vegetative plant tissue (Lancashire et al., 1991). The entire plant tissue of cereals was taken after emergence of the first node (BBCH 30–31) by cutting the plants about 1 cm above ground to avoid contamination of the sample by adherent soil particles. If required, withered leaf material was removed from the sample. The optimum time for sampling oilseed rape starts at stem extension (BBCH 30–31) and continues until flowering. Younger, fully differentiated leaves of the upper third were taken for analysis. In case of sugar beet, younger, fully differentiated leaves were taken at start of row closing (BBCH 31–32) until closing of rows.

It is important to note that appropriate tools for cutting the material is pivotal to avoid contamination by contact, particularly with view to analyzing micronutrients. Before taking samples on production fields, it is necessary to ensure that no liquid fertilizers have been applied foliarly. Haneklaus and Schnug (2006) provide a detailed description for how to take representative samples in the field.

PIPPA is the first and only interpretation program, which acknowledges Liebig's "Law of the Minimum" (Liebig's Law, see Brock, 1997) and Mitscherlich's

Law of diminishing returns (Mitscherlich, 1909). Liebig's Law states that the growth of a plant is controlled not by the sum of all available resources but by the scarcest resource (popularly imaged as a barrel with stanchions of different length and the shortest stanchion representing the most stringent growth limiting nutrient). This example reveals that increasing the amount of nutrients plentifully does not necessarily increase plant growth. Only by increasing the one, most scarce element in relation to demand, will improve growth of a plant. Liebig's Law states that growth only occurs at the rate permitted by the most limiting factor. Many other interpretation programs (e.g., the Polish "INFOPLAN" see Cupial, 2005, and Wróbe, 2004) violate both laws because they assume a linear relationship between nutrient status and crop productivity, when the percentage of a given nutrient concentration in the plant in relation to the critical value is used for calculating the degree of deficiency.

PIPPA employs upper boundary lines for the essential mineral plant nutrients obtained by BOLIDES to evaluate ceteris paribus to which extent an individual nutrient limits the yield of cereals, oilseed rape, and sugar beet crops. PIPPA employs the upper boundary lines for Mo and Cl of oilseed rape, also for sugar beet, as long as analytical data for the latter crop are missing. Those upper boundary lines represent the application of Mitscherlich's Law on diminishing returns to the effect of increasing the supply of a particular plant nutrient on biomass production. Mitscherlich's Law states that in a production system (for example a field crop) with variable inputs (for example the fertilizer supply) beyond some defined point, each additional unit of variable input yields less and less additional output (crop yield).

PIPPA operates the whole range of essential plant nutrients (N, P, S, K, Ca, Mg, Cl, B, Mo, Fe, Mn, Zn, Cu, and Na for sugar beet). PIPPA calculates first the expected relative yield level for each nutrient concentration under ceteris paribus conditions by means of the boundary line functions (Tables 14–1 to 14–3). Then PIPPA ranks the nutrients according to their yield limiting power. Accordingly, fertilizer recommendation schemes should start supplying the elements in the same order, since according to Liebig's Law, it is essential to eliminate the strongest yield limiting element first to improve crop growth. PIPPA includes a routine which enables a reliable estimation of the sulfur fertilizer demand of oilseed rape on the basis of plant analysis data.

From Plant Analysis to Fertilizer Recommendation

A missing link between soils, crops, and nutritional status are decision making tools on how to calculate fertilizer rates on the basis of the results of plant analysis. PIPPA closes this gap at least for the essential plant nutrient sulfur. PIPPA contains a routine which enables the deduction of a reliable recommendation for soil-based sulfur fertilization of a (winter) oilseed rape crop from plant analysis data. Besides fertilizer recommendations, the validity of which have been proven under field conditions, PIPPA provides an estimate of crop quality, economical return, and nutrient utilization. Input, internally calculated and output variables of the "S" subroutine of PIPPA are listed in Table 14–6.

The program calculates the sulfur demand of oilseed rape that is required to raise the actual sulfur concentration in the leaf tissue to that of the critical nutrient value and optimum range, respectively. In case of oilseed rape, sulfur

Table 14–6. Input, internally calculated, and output variables of the sulfur routine of PIPPA3.

Input variables

"Symbol for national currency→ :"
"conversion factor ECU to national currency→:"
"fertilizer effect on sulfur content (mg/g/kg S)→:"
"potential maximum seed yield (dt/ha)†————→:"
"rapeseed price without premium ("SYM$"/dt)————→:"
"rapeseed price with premium ("SYM$"/dt)————→:"
"S-fertilizer price 'free root' ("SYM$"/kg S)————→:"
"quality threshold (μmol/g)————→:"
"S-content in younger leaves (mg/g)————→:"
"percentage volunteers (%)————→:"
"GSL-content of the sawn rapeseed (μmol/g)————→:"
"GSL-content of the volunteers (μmol/g)————→:"

Internally calculated variables‡

$D = (1 - A)*B + C*A$
$F = E - D$
$G = (F/(7*A) + 1.3))$
$I = G/H$
$L = (-166.97/K + 125.7)*O$
$M = ((-166.97/(K + G) + 125.7))*O$
$N = M - L$
$R = (Q*N) - (S*I)$
$T = 6.5 - K$
$W = T/H$
$U = O*100 - ((-166.97/K + 125.7)*O)$
$X = C + ((W*H)*8.3)$
$Y = B + ((W*H)*1.3)$
$Z = (A*X) + ((1 - A)*Y)$
$R1 = (U*P1) - (W*S)$
ENTO = $((D + 43.87)/14.99)*L/10$
ENTOT = $((0.66*K + 6.73)*L/10) + ((0.57*K + 0.79)*L/10)$
ENTMAX = $(((Z + 43.87)/14.99)*U/10) + $ ENTO
NUTZMAX = (ENTMAX − ENTO)/$W*100$
ENMAT = $(((0.66*6.5 + 6.73)*U/10) + ((0.57*6.5 + 0.79)*U/10)) + $ ENTOT
NUMAT = (ENMAT − ENTOT)/$W*100$
ENTQUA = $(((E + 43.87)/14.99)*N/10) + $ ENTO
NTTZQUA = (ENTQUA − ENTO)/$I*100$
ENQUAT = $((((0.66*(K + G)) + 6.73)*N/10) + ((((0.57*(K + G)) + 0.79)*N/10)) + $ ENTOT
NUQUAT = (ENQUAT − ENTOT)/$I*100$

Output variables

"GSL-content in harvest without fertilization (μmol/g) → :";D
"amount S-fertilizer needed (kg/ha S)————→:";W
"monetary yield increase ("SYM$"/ha)————→: ";R1
"utilization of fertilizer sulfur (netto) (%)→: ";NUTZMAX
"utilization of fertilizer sulfur (gross) (%)→ : ";NUMAT
"GSL-content in harvest (μmol/g)————→: ";Z
"utilization of fertilizer sulfur (net) (%)→: ";NUTZQUA
"utilization of fertilizer sulfur (gross) (%)→: ";NUQUAT

† dt is decatonnes (0.1 Mg).
‡ For references see Schnug (1988) and Schnug and Haneklaus (1988).

fertilization before flowering (sufficient water for dissolving and transporting the nutrient in the soil matrix provided) is still efficient to achieve full crop productivity. In contrast, this is not feasible for cereals (Haneklaus et al., 1995).

PIPPA calculates the recommended sulfur fertilizer dose in a two step process: First, the required increase of total sulfur in younger, fully differentiated leaves is calculated and thereafter, the fertilizer sulfur rate that is required to achieve this increase. For this, PIPPA calls for the following data entries.

1. The response to the fertilizer in terms of the sulfur concentration (mg g^{-1}) in the leaf tissue, expressed by the increase of the total sulfur concentration by 1 kg ha^{-1} of sulfur applied as sulfate to the soil. This value is site-specific and depends on humidity conditions, the main risk being leaching at the time of application. This value can easily be retrieved from sulfur fertilizer response experiments. The default value of 0.2, for instance, is valid under northern European conditions where sulfur is applied together with the first nitrogen dressing in spring.

2. The potential maximum seed yield is a site-specific value. The default value of 50 dt ha^{-1} (5 Mg ha^{-1}) for instance is valid under conditions of northern European.

3. The rapeseed price without premium, which is the price if no extra payment for achieving quality standards in low seed glucosinolate content is received. This is usually the world market price for rapeseed and correspondingly the "rapeseed price with premium."

4. The sulfur-fertilizer price free root, which are the costs for 1 kg sulfur plus the expenditures for application. The default value is 0.25 € kg^{-1} sulfur.

In addition, PIPPA estimates the glucosinolate (GSL) content, which can be expected in the present rapeseed crop. As the seed glucosinolate content depends on the quality of the variety sown and the number of volunteers (plants emerging from seeds left on the field from formerly cultivated rapeseed crops), PIPPA requires the following input: (i) the percentage of volunteers (%), (ii) the GSL content of the sown rapeseed, which varies usually between 8 and 12 μmol g^{-1} for a double low variety, and (iii) the GSL content of the volunteers, which is for instance 100 to 120 μmol g^{-1} for a typical old single low variety in a S-rich environment, and 60 to 80 μmol g^{-1} in a S-deficient area.

This information provided, PIPPA will calculate the economically most successful fertilizing strategy for sulfur—either going for maximum yield by violating the quality threshold or balancing sulfur fertilization to meet the quality threshold by not gaining maximum seed yield to obtain maximum economic return.

Thus, in addition to an evaluation and ranking of the supply of individual plant nutrients, PIPPA provides practical information such as the amount of sulfur required for fertilization, the expected monetary yield increase (net value, costs for sulfur fertilization subtracted), gross fertilizer efficiency (all parts of the crop removed from the field), net fertilizer efficiency (only seeds removed from the field), and the expected GSL content of the harvested rapeseed crop.

References

Arnon, D.I. 1950. Criteria of essentiality of inorganic micronutrients for plants. Loytsia 3:31–38.

Baier, I. 1995. Computer program for foliar fertilization. p. 23–28. *In* Proc. IAOPN Symposium. Cairo.

Barker, A.V., and D.J. Pilbeam. 2006. Handbook of plant nutrition. CRC Press, Boca Raton, FL.

Bergmann, W. 1992. Nutritional disorders of plants-visual and analytical diagnosis. Gustav Fischer Verlag, Jena, Germany.

Bergmann, W. 1993. Ernaehrungsstoerungen bei Kulturpflanzen. 3. Aufl., Gustav Fischer Verlag, Jena. Germany.

Brock, W.H. 1997. Justus von Liebig: The chemical gatekeeper, Cambridge Univ. Press, London.

Cerda, A., V. Martinez, M. Caro, and F.G. Fernandez. 1984. Effect of sulfur deficiency and excess on yield and sulfur accumulation in tomato plants. J. Plant Nutr. 7:1529–1543.

Cupial, M. 2005. Program Wspomagajacy Nawosenie mineralne, Nawozy 2". *In* Ŝynieria Rolnicza 14/2005. http://ir.ptir.org/artykuly/pl/74/IR(74)_1235_pl.pdf; verified 28 March 2008.

Datnoff, L., W. Elmer, and D. Huber. 2007. Mineral nutrition and plant diseases. APS Press, St. Paul, MN.

Evanylo, G.K., and M.E. Sumner. 1987. Utilization of the boundary line approach in the development of soil nutrient norms for soybean production. Commun. Soil Sci. Plant Anal. 18:1355–1377.

Finck, A. 1970. Die Pflanzenanalyse als Hilfsmittel zur Ermittlung des Duengerbedarfes. Sonderdruck aus Chemie und Landw. Produktion 183–188.

Finck, A. 1979. Duenger und Duengung. NewYork.

Haneklaus, S., E. Bloem, E. Schnug, L. De Kok, and I. Stulen. 2006a. Sulphur. p. 183–238. *In* A.V. Barker and D.J. Pilbeam (ed.) Handbook of plant nutrition. CRC Press, Boca Raton, FL.

Haneklaus, S., E. Bloem, and E. Schnug. 2006b. Sulphur interactions in crop ecosystems. p. 17–58. *In* M.J. Hawkesford and L.L. DeKok (ed.) Sulfur in plants—An ecological perspective. Springer, Dordrecht, the Netherlands.

Haneklaus, S., J. Fleckenstein, and E. Schnug. 1995. Comparative studies of plant and soil analysis for the evaluation of the sulphur status of oilseed rape and wheat. J. Plant Nutr. Soil Sci. 158:109–112.

Haneklaus, S., and E. Schnug. 1998. Evaluation of critical values for soil and plant analysis of sugar beets by means of boundary lines applied to field survey data. Aspects Appl. Biol. 52:87–94.

Haneklaus, S., and E. Schnug. 2006. Site-specific nutrient management: Objectives, current status, and future research needs. p. 91–151. *In* A. Srinivasan (ed.) Handbook of precision agriculture. Food Products Press, New York.

Hudson, D.J. 1966. Fitting segmented curves whose join points have to be estimated. J. Am. Statist. Assoc. 61:1097–1129.

Lancashire, P. D., H. Bleiholder, P. Langelüddecke, R. Stauss, T. van den Boom, E. Weber, and A. Witzen-Berger. 1991. An uniformdecimal code for growth stages of crops and weeds. Ann. Appl. Biol. 119:561–601.

Lark, R.M. 2001. Some tools for parsimonius modelling of within-field variation of soil and crop systems. Soil Tillage Res. 58:99–111.

Mills, H.A., and J. Benton Jones Jr. 1991. Plant analysis handbook II. MicroMacro Publishing, Athens, GA.

Mitscherlich, E.A. 1909. Das Gesetz des Minimums und das Gesetz des abnehmenden Bodenertrages. Landwirtschaftliche Jahrbücher Bd. 38:537–552.

Moeller-Nielsen, P., and H. Frijs-Nielsen. 1976. Evaluation and control of the nutritional status of cereals. II. Pure effects of a nutrient. Plant Soil 45:339–351.

Reuter, D.J., and J.B. Robinson. 1997. Plant analysis—An interpretation manual. CSIRO Publishing, Collingwood, Australia.

Schnug, E. 1988. Quantitative und qualitative Aspekte der Diagnose und Therapie der Schwefelversorgung von Raps (Brassica napus L.) unter besonderer Berücksichtigung glucosinolatarmer Sorten. Habilitationsschrift (Dsc thesis) Agrarwiss. Fakultät der Christian-Albrechts-Universität zu Kiel Dezember.

Schnug, E. 1990. PIPPA- a professional interpretation program for plant analysis data of oilseed rape and cereals. Proc. Inaugural Congress of the European Society of Agronomy Paris 1: 35.

Schnug, E. 1991. Sulphur nutritional status of European crops and consequences for agriculture. Sulphur Agric. 15:7–12.

Schnug, E., and S. Haneklaus. 1988. The sulphur concentration as a standard for the total glucosinolate content of rapeseed and meal and its determiation byX-ray fluorescence spectroscopy (X-RF method). J. Sci. Food Agric. 45:243–254.

Schnug, E., and S. Haneklaus. 1992. PIPPA: Un programme d'interprétation des analyses de plantes pour le colza et les céréales. Perspec. Agricol. Supp. 171:30–33.

Schnug, E., and S. Haneklaus. 1998. Diagnosis of sulphur nutrition. p. 1–38. In E. Schung (ed.) Sulphur in agro-ecosystems. Kluwer Academic Publ., Dordrecht, the Netherlands.

Schnug, E., J. Heym, and F. Achwan. 1996. Establishing critical values for soil and plant analysis by means of the boundary line development system (BOLIDES). Commun. Soil Sci. Plant Anal. 27:2739–2748.

Schnug, E., J. Heym, and D.P.L. Murphy. 1995. Boundary line determination technique (BOLIDES). p. 899–908. In Site-specific management for agricultural systems. ASA, CSSA, and SSSA, Madison, WI.

Smith, P.F. 1962. Mineral analysis of plant tissues. Annu. Rev. Plant Physiol. 13:81–108.

Spencer, K., and J.R. Freney. 1980. Assessing the sulfur status of field- grown wheat by plant analysis. Agron. J. 72:469–472.

Timmer, V.R., and G. Armstrong. 1987. Diagnosing nutritional status of containerized tree seedlings: Comparative plant analyses. Soil Sci. Soc. Am. J. 51:1082–1086.

Vielemeyer, H.-P., P. Neubert, I. Hundt, G. Vanselow, and P. Weissert. 1983. Ein neues Verfahren zur Ableitung von Pflanzenanalyse-Grenzwerten fuer die Einschaetzung des Ernaehrungszustandes landwirtschaftlicher Kultur-pflanzen. Arch. Acker- u. Pflanzenbau u. Bodenkde 27:445–453.

Wallace, A. 1990. Crop improvement through multidisciplinary approaches to different types of stresses-law of the maximum. J. Plant Nutr. 13:313–325.

Walworth, J.L., W.S. Letzsch, and M.E. Sumner. 1986. Use of boundary lines in establishing diagnostic norms. Soil Sci. Soc. Am. J. 50, 123–128.

Webb, R.A. 1972. Use of the boundary line in the analysis of biological data. J. Hortic. Sci. 47:309–319.

Wróbel. Sulfur 2004. Potrzeby nawożenia roślin uprawnych mikroelementamiXII Krajowe Seminarium "Stosowanie agrochemikaliów" (19–20.10.04)- materiały szkoleniowe, Puławy, Wydawnictwo IUNG, s.3–22, 2004 (pol)- rys.5, tab.9, bibliogr.15 Dokonano oceny; http://www.iung.pulawy.pl/_Oferty.html (verified 28 Mar. 2008); http://www.iung.pulawy.pl/WWW/naw/Naw3.html (verified 28 Mar. 2008).

15

Improving the Sulfur-Containing Amino Acids of Soybean to Enhance its Nutritional Value in Animal Feed

Hari B. Krishnan
USDA-ARS and University of Missouri, Columbia

Abstract

Soybean [*Glycine max* (L.) Merr.] is an economical source of quality protein in the formulation of animal feeds. Domestic swine, poultry, beef, and dairy industries consume a total of 28 million megagrams of soybean meal annually. Though soybean is an excellent source of protein, increasing the quantity of sulfur-containing amino acids could further enhance their nutritive value. This chapter deals with recent developments in enhancing the levels of sulfur-containing amino acids and other essential amino acids in soybean. Introduction of methionine-rich heterologous proteins, elevating the expression of endogenous methionine-rich proteins, and expression of synthetic proteins containing a high percentage of essential amino acids are some of the approaches employed for improving the nutritional quality of the soybean seed. Though the prospects for increasing the sulfur-containing amino acid content of soybean appear promising, a thorough understanding of the intricacies of the sulfur assimilatory pathway and its regulation in soybean will aid in improving the quality of soybean seed protein.

Importance of Soybean in Animal Feed

One of the unique qualities of soybean is its high content of protein and oil. Commercial soybean cultivars grown in the USA contain approximately 38% protein and 18% oil. Nearly 85% of the world's soybeans are processed into soybean meal and oil, and when compared with other protein-rich crops, soybean is by far the cheapest source of quality protein-rich meal.

Soy meal is the single most important source of animal feed in the USA and Europe (Fig. 15–1). More than 100 million bushels of U.S. soybean are utilized for feed in domestic swine and poultry production, which represents about 37% of total U.S. soybean production (http://www.soygrowers.com/; verified 12 April 2008). As global population increases steadily, the demand for food continues to rise. Demand for supplementary sources such as seafood is growing as well, resulting in an increase in soy meal utilization in aquaculture in addition to agriculture. It is estimated aquaculture will account for 8 to 10 million megagrams of soy meal use by the next decade. Thus, soybean is a vital part of the food supply

---

Copyright © 2008. American Society of Agronomy, Crop Science Society of America, Soil Science Society of America, 677 S. Segoe Rd., Madison, WI 53711, USA. *Sulfur: A Missing Link between Soils, Crops, and Nutrition.* Agronomy Monograph 50.

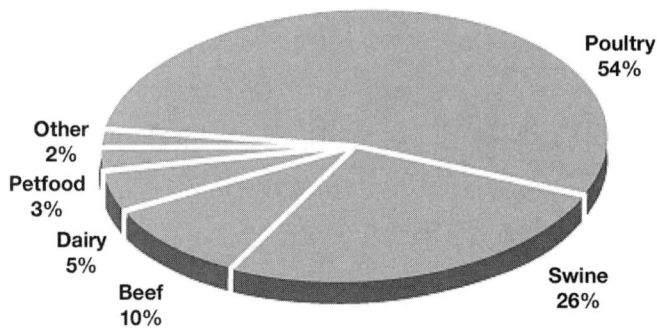

Fig. 15–1. U.S. soybean use by livestock. Figure based on the American Soybean Association Statistics (2006).

chain. It is also obvious that the continued use of soybean in animal feed will be crucial to the success of the soybean industry. However, the continued dominance of soybean use in the animal feed industry will depend on soybean trait improvement that meets livestock nutritional requirements and promotes feeding efficiencies while lowering the environmental impacts of livestock production.

Improving the Essential Amino Acid Content of Soybean Will Increase the Value of Soy Meal in the Animal Feed Industry

Soy meal is the preferred protein of choice in poultry and swine industry throughout the world. The relatively high protein content, along with a superior amino acid profile compared with other plant sources, makes soy meal an ideal protein supplement in the livestock industry. However, rapidly growing swine and poultry have a higher dietary requirement for sulfur-containing amino acids than is provided by grain-soybean meal rations. Synthetic lysine and methionine, the primary limiting amino acids in swine and poultry rations, are regularly supplemented to build balanced animal rations. Although the cost of these two synthetic amino acids is minimal, feed manufacturers and producers spend approximately 100 million dollars annually on the methionine analog alone (Imsande, 2001). After lysine and methionine requirements are met, other amino acids such as arginine, isoleucine, tryptophan, and valine can become limiting. However, the higher cost of some synthetic analogs minimizes their use in swine and poultry rations (Kerley and Allee, 2003). Hence, soybean varieties expressing adequate amounts of the essential amino acids necessary to meet the nutritional demands of growing livestock and poultry would significantly increase the utility and value of soybean seed meal as a feed ingredient.

Soybean Seed Composition and Improvment

Commercially grown soybean does not accumulate sufficient methionine and cysteine to meet the needs of growing swine and poultry. Researchers are attempting to increase the amounts of essential amino acids in seed proteins and many reviews on improving the nutritionally quality of legume seeds have been published (Muntz et al., 1998; Tabe and Higgins, 1998; Wang et al., 2003; Hagan et al., 2003; Williams, 2003; Sun and Liu, 2004; Krishnan, 2005; Beauregard and Hefford, 2006). In soybean, two groups of salt-soluble proteins, glycinin (11S) and β-conglycinin (7S), account for 70% of the soybean seed-storage proteins (Nielsen, 1996; Krishnan, 2000). These proteins together account for approximately 50% of the total seed protein; hence, they are largely responsible for the nutritive value and quality of soybean seed protein (Beilinson et al., 2002).

The 11S glycinin is synthesized as precursor protein; then it is posttranslationally processed to yield acidic and basic subunits that are held together by a single disulfide bond (Beachy et al., 1981; Barton et al., 1981; Staswick et al., 1984). They are isolated from seeds as hexamers with an estimated molecular mass of 320 to 375 kDa (Wolf and Briggs, 1958; Bradley et al., 1975). Because of their nutritional value, glycinins have been characterized extensively. Five genes encoding the major glycinins ($Gy1$, $Gy2$, $Gy3$, $Gy4$, and $Gy5$) have been cloned and grouped into two families on the basis of the identity of their amino acid sequences (Staswick et al., 1984; Nielsen et al., 1989; Xue et al., 1992). Group 1 glycinin includes $Gy1$, $Gy2$, and $Gy3$ genes, while Group 2 includes $Gy4$ and $Gy5$ genes. In addition, a glycinin pseudogene, $gy6$, and a functional gene, $Gy7$, have been identified and partially characterized (Beilinson et al., 2002). The $Gy7$ gene, which does not fit either the Group 1 or the Group 2 glycinin families, appears to be expressed at considerably low levels when compared with Group 1 and Group 2 members (Beilinson et al., 2002). Interestingly, $Gy7$ gene encodes a protein that has a relatively higher content of sulfur amino acids than the products from other glycinin genes.

The other abundant storage protein of soybean, 7S β-conglycinin, is isolated from seed as a trimer having a molecular mass of 150 to 175 kDa (Thanh and Shibasaki, 1976, 1978). It is composed of three subunits designated α′, α, and β that have molecular masses of 76, 72, and 53 kDa, respectively (Coates et al., 1985; Tierney et al., 1987). The α′ and α subunits of β-conglycinin contain an extension region that is absent in the β-subunit. The three subunits exhibit high similarity in their amino acid sequence of their coding regions. β-Conglycinin is a glycoprotein and is encoded by a multigene family containing 15 to 20 genes (Harada et al., 1989).

A comparison of sulfur amino acid contents reveals that glycinin contains relatively more cysteine and methionine than the β-conglycinin. The β-subunit of β-conglycinin is considered as nutritionally poor since it contains no methionine or cysteine residues. The accumulation of this subunit can be enhanced either by excess nitrogen or by sulfur deficiency (Gayler and Sykes, 1985; Paek et al., 1997, 2000; Imsande and Schmidt, 1998; Imsande, 2003). It is speculated that an increase in the accumulation of β-conglycinin will lower the cysteine and methionine content of soybean seed protein and, thus, its nutritive quality.

In addition to the 11S glycinin, other proteins somewhat rich in sulfur amino acids are found in soybean seed but in lower abundance (Choi et al., 1995). Albumin or water-soluble seed proteins, including the Bowman-Birk protease inhibitors, can account for 4 to 6% of the total seed protein and are rich in cysteine (Nielsen, 1996). Methionine-rich proteins have also been identified in the albumin fraction (Kho and de Lumen, 1988; George and de Lumen, 1991). Other seed storage proteins include lectins, lipoxygenases, and ureases. These proteins have been attributed to play a role in various physiological functions such as germination, plant defense, and growth and development.

It is important to note here that in addition to improving the quality and quantity of soybean protein, much work is currently being done to increase the quality of soybean oil and carbohydrates and the remaining seed components (phospholipids, isoflavones, saponins, phytic acid, minerals, and vitamins). In the USA, the oil component of soybean is predominantly used for human consumption. Oil with a higher content of oleic acid, a monounsaturated fatty acid, has increased its heat stability and further minimizes the need for hydrogenation, a process generating undesirable trans-fatty acids. Another major constituent of soybean seed, carbohydrates (including starch, sucrose, and a variety of other sugars) are earmarked for improvement. Two of these sugars, raffinose and stachyose, have a low digestibility in monogastric animals, resulting in gastric distress and a reduction in energy derivable from metabolic processes. Research examining modification of these characteristics is under way (Cromwell et al., 2000; Spencer et al., 2000) and would make soybean more desirable to the animal feed industry (Kerley and Allee, 2003). In addition to improving seed quality and quantity to meet the demand for increased animal production efficiency, stringent future waste management regulations may require modifications of soybean protein, oil, and/or carbohydrates in an effort to reduce nitrogen and phosphorus effluent from these feed components.

Improvement of Sulfur Amino Acid Content of Soybean through Plant Breeding

During the past few decades, the primary focus of soybean breeders has been yield improvement, with only a minor emphasis on improving the quality of protein composition. Recently, this situation has changed as breeders have concentrated on generating high yielding soybean cultivars also with desirable seed composition. Substantial progress has been made in increasing the total protein content even though increases in protein are often associated with decreased yield (Brim and Burton, 1979; Burton et al., 1982; Wilcox and Shibles, 2001). However, the work of Wilcox and Cavins (1995) has shown that protein content could be increased without sacrificing yield. Regardless of increases in seed protein, the amount of sulfur amino acids has remained constant (Burton et al., 1982; Wilcox and Shibles, 2001). Some of the high-protein soybean lines have been shown to improve nutritional value (Edwards et al., 2000), while others have resulted in significant variations in both the essential and nonessential amino acids (Zarkadas et al., 1993, 1994; Serretti et al., 1994). A comparison of soybean meals produced from conventional and high-protein soybean revealed digestible lysine, methionine, cysteine, threonine, and valine along with true metabolizable energy were higher in high-protein lines. Similarly, it was reported that broilers fed with Prolina soybean meal (a high-protein soybean line) resulted in the best overall performance compared with birds fed commercial variety soybean meal under heat stress conditions (Lenfestey, 2005). Thus, the high-protein soybeans may have considerable advantages over conventional soybean meal as a feed ingredient for poultry (Edwards et al., 2000).

Selective soybean breeding for the purpose of increasing methionine content has been hampered by the lack of variability of sulfur amino acid content among soybean cultivars (Kuiken and Lyman, 1949; Krober, 1956). Imsande (2001) mutated soybean seeds with ethyl methanesulfonate (EMS) and was able to select several methionine over-producing genetic lines of soybean. One mutant exhibited a 20% higher methionine and cysteine content than did the parental lines (Imsande, 2001). Recently, a soybean germplasm line was released that combined high yield and increased protein concentration (Panthee and Pantalone, 2006). Interestingly, this line (TN04–5321) also had increased total sulfur containing amino acids. Thus, increasing the sulfur amino acid content of soybean by standard breeding appears to be feasible.

The 7S globulins contain fewer sulfur amino acids than their 11S counterparts; thus, the ratio of these proteins influences nutritional quality of soybean seed proteins (Paek et al., 1997, 2000). Variation exists in the mean protein ratio of 11S and 7S globulins among different soybean cultivars. The ratio, which varies from 1.6 to 2.5, has an important role in determining the functional properties of food products made from soybean (Cai and Chang, 1999). Excess nitrogen increases the accumulation of sulfur-poor β-subunit of β-conglycinin while lowering the accretion of glycinin (Paek et al., 1997, 2000). The preferential increase in the accumulation of β-subunit of β-conglycinin alters the 11S:7S ratio resulting in poorer protein quality. It has been suggested that down regulating or eliminating the expression of the 7S proteins could enhance the nutritional quality of soybean seed proteins. In *Glycine soja* Sieb. and Zucc., the wild ancestor of *G. max*, a significant variation in the 11S:7S ratio has been reported (Kwanyuen et al., 1997).

Breeders for improvement of protein quality in cultivated soybean have exploited this valuable genetic resource offered by wild soybean. Several mutant lines from U.S. and Japanese germplasm collections have been identified that either lack individual or entire 7S or 11S subunits (Kitamura and Kaizuma, 1981; Ladin et al., 1984; Tsukada et al., 1986; Takahashi et al., 1994, 1996, 2003). A naturally occurring, single dominant mutant *Scg-1* (suppressor of β-conglycinin) has been found in a Japanese wild soybean, which is devoid of β-conglycinin (Hajika et al., 1996; Teraishi et al., 2001). Plant breeders have incorporated seed protein mutations into domestic cultivars with no deleterious consequences observed on agronomic parameters. Interestingly, dry seed nitrogen content of a soybean line that lacked both glycinin and β-conglycinin was similar to the wild-type cultivar (Takahashi et al., 2003). A similar approach was taken by Canadian researchers who have developed 20 null soybean genotypes that lacked different components of 7S and 11S storage proteins (Poysa et al., 2006; Zarkadas et al., 2007). These mutants, which failed to accumulate the major seed storage proteins, accumulated significantly higher amounts of free amino acids. It is interesting to note that the total amino acid content of seeds showed little variation among the mutants and the wild-type soybean. Since the 7S globulins contain three to four times fewer sulfur amino acids than 11S glycinin, one would expect that sulfur amino acid content of soybean mutants lacking β-conglycinin would be significantly greater. However, the methionine and cysteine content of soybean mutant lines lacking β-conglycinin were not different from that of the wild type. This observation raises doubts on the validity of claims that suggests the sulfur amino acid content of soybean can be improved by lowering the accumulation of β-conglycinin.

Expression of Heterologous Genes Encoding Sulfur-Rich Proteins in Soybean

Genetic engineering has provided additional avenues for improving the sulfur amino acid content of soybean. One common approach employed by several investigators is to express seed proteins that are rich in the sulfur amino acids in soybean. Brazil nut (*Bertholletia excelsa* H.B.K.), sunflower (*Helianthus annuus* L.), and corn (*Zea mays* L.) each contain proteins that are rich in the sulfur amino acids. The water-soluble 2S albumins from Brazil nut and sunflower (SFA8) are each comprised of approximately 25% cysteine and methionine (Altenbach et al., 1987; Kortt et al., 1991). The δ-zeins, the hydrophobic corn proteins, contain a similar percentage of the sulfur amino acids (Kirihara et al., 1988; Chui and Falco, 1995; Kim and Krishnan, 2003). An early attempt to introduce a heterologous high methionine protein into the soybean involved the use of the 2S albumin from the Brazil nut (Townsend and Thomas, 1994). Although the 2S albumin was expressed successfully and the methionine content improved, the introduced protein was found to be an allergen; thus, development of this approach was not rigorously pursued (Nordlee et al., 1996). In another study, a 15-kDa zein was expressed in soybean. Analyses of these transgenic soybeans showed the genetically engineered plants to have a 12 to 20% increase in methionine and 15 to 35% increase in cysteine compared with nontransformed lines (Dinkins et al., 2001).

Subsequently a novel 11-kDa δ-zein in which the cysteine and methionine account for 25% of the total amino acids was expressed in soybean (Kim and Krishnan, 2004). Analysis of the alcohol-soluble protein fraction of this transgenic soybean showed a 1.5- to 1.7-fold increase in the methionine content in comparison to that of nontransgenic soybean. However, the overall methionine content of the seed flour was not increased when compared with the control seed flour. Interestingly, the accumulation of the 11-kDa δ-zein was confined to cells that were situated between seed vascular tissue and storage parenchyma cells (Kim and Krishnan, 2004). The reason for restricted accumulation of the 11-kDa δ-zein to an area near the vascular tissue of the seed has not been determined, but is possibly related to the availability and transport of methionine. Similarly, a 27-kDa γ-zein was expressed in soybean (Li et al., 2005). Even though the transgenic soybean accumulated the 27-kDa γ-zein (0.59–0.81 mg per gram of soy flour), the overall increase in cysteine and methionine was only 2 to 3% compared with nontransformed soybean (Li et al., 2005). It is evident from the above-mentioned studies that, although sulfur-rich proteins from the other plant species can be successfully expressed in soybean, only modest increases in the methionine and cysteine contents have resulted.

Modification of Endogenous Proteins

Another approach to increase the protein quality involves the introduction of sulfur amino acids in the abundant seed storage proteins of soybean (Nielsen et al., 1989). A comparison of the amino acid sequences of Group 1 and Group 2 glycinin reveal a region that is characterized by an extensive natural variation in its sequence. This region has been designated as the hypervariable region (HVR) and has been targeted for insertion of sulfur amino acids (Nielsen et al., 1990). Nielsen and his coworkers (1995) inserted multiple methionine residues in the HVR of the *Gy4* gene. This modified glycinin gene was subsequently expressed in tobacco (*Nicotiana tabacum* L.). Even though the mRNA transcript of the gene was detected, the putative methionine-enriched protein did not accumulate, presumably because of degradation in the storage vacuoles (Nielsen et al., 1995). Ostensibly, the inserted methionine residues interfered with the correct folding of the protein, thus initiating proteolysis. Subsequently, methionine residues were introduced into different regions of the *Gy1* gene to determine whether other sites in the protein were amenable to alteration (Utsumi et al., 1994). Insertions proximal to the HVR C-terminal region and to the C terminus of the β-chain did not interfere with holoprotein formation. Glycinins enriched in methionine were found to accumulate in transgenic tobacco seeds and potato (*Solanum tuberosum* L.) tubers, indicating that the correct folding and assembly of modified protein occurred in these transgenic plants (Gidamis et al., 1995; Takaiwa et al., 1995).

Introducing additional methionine residues to specific regions of the glycinin thus appears feasible. Recently, a modified glycinin gene containing four contiguous methionine residues inserted in the coding region of the protein was expressed in transgenic soybean plants (El-Shemy et al., 2007). Two-dimensional SDS-PAGE analysis of seed proteins revealed that the transgenic soybean accumulated higher amounts of glycinin when compared with untransformed soybean. Unfortunately, the amino acid composition of transgenic soybean was not investigated in this study, and it is not known if the accumulation of the modified glycinin had elevated the sulfur amino acid content. Since only one member of the glycinin gene family has been altered, the overall increase in the methionine content will most likely be insufficient to meet the requirements of animal feeds.

An examination of patents issued during the last decade reveals renowned efforts of international seed companies to increase the levels of essential amino acids for improvement of the nutritional value of soybean. The primary limiting amino acids in corn–soybean based rations are lysine and methionine. Threonine, isoleucine, tryptophan, valine, and arginine have been identified as *second-tier* limiting amino acids in animal feeds. The latter amino acids become limiting when sufficient lysine and methionine are present for optimal growth. Researchers are attempting to improve the content of some of the second-tier amino acids in soybean through genetic engineering. Researchers from the private sector have attempted to elevate the isoleucine content of soybean by introducing 10 to 20 isoleucine residues into the coding region of the β-conglycinin gene. In vitro assays revealed that introduction of multiple isoleucine residues did not interfere with the correct folding of the β-conglycinin. Transgenic soybean with high-level expression of isoleucine-enriched β-conglycinin has been successfully generated (Rapp et al., 2003).

Other strategies that are being attempted to increase protein quality are enhanced expression of native soybean proteins that contain substantial quantities of sulfur amino acids. A 2S albumin (Gm2S-1) of soybean, which is rich in methionine (Revilleza et al., 1996), has been identified as a potential candidate for overexpression to improve the nutritional quality of soybean (de Lumen et al., 1999).

Soybean Gm2S-1 is synthesized as a precursor protein and posttranstionally processed to give rise to small and large subunits. The 43-amino acid small subunit contains the cell adhesion motif Arg-Gly-Asp (RGD) followed by eight Asp residues and a 77-amino acid large subunit that is rich in methionine (Galvez et al., 1997). The small subunit, which is named lunasin, may play a role as a novel cancer chemo-preventive agent (Jeong et al., 2003). Thus, overexpression of this unique protein has not only the potential to enhance the sulfur amino acid content but also increase the levels of the bioactive peptide in soybean seeds. Even though overexpressing the 2S albumin may be a viable strategy to improve the nutritional quality, one should be cautious with this approach since some of the 2S albumins are potential allergens. Interestingly, the 2S albumins from soybean are stable to heat and chemical treatments, which are hallmarks of plant allergens (Lin et al., 2004). However, a recent study has shown that soybean 2S albumins are not major allergens, as demonstrated by lack of IgE-specific to soybean 2S albumin in the sera of 23 individuals with known soybean allergy (Lin et al., 2006).

In addition to the five major glycinin genes (*Gy1* to *Gy5*), another functional glycinin gene (*Gy7*) has been identified (Beilinson et al., 2002). The product of Gy7 contains more sulfur amino acid than the other glycinins. Unlike the other abundant glycinins, this sulfur-rich glycinin has not been extensively studied. It has been reported that *Gy7* mRNA accounts for only a small proportion of the total glycinin message and the product of *Gy7* may account for less than 5% of the seed protein (Beilinson et al., 2002). Overexpression *Gy7* gene in soybean can be a possible option to increase the sulfur amino acid content. Another class of native proteins, the Bowman-Birke protease inhibitors, is rich in cysteine. Even though high accumulation of this protein may improve the sulfur amino acid content, it may also have a deleterious effect on animal nutrition since Bowman-Birk protease inhibitors are known to interfere with the action of digestive enzymes in monogastric animals.

Synthetic Protein as a Source of Sulfur Amino Acids

Another approach being employed to increase the essential amino acid of seed crops is to engineer synthetic proteins with desired amino acid composition. Initially, synthetic genes consisting of repeating segments coding for essential amino acids were synthesized. Proteins encoded by such synthetic genes were often unstable and were subject to proteolytic degradation (Beauregard et al., 1995). A better understanding of the rules that govern protein folding and topology has enabled scientists to rationally design synthetic proteins.

By exploiting emerging protein folding principles a synthetic protein Milk Bundle-1 (MB1) enriched in essential amino acids was constructed. Four essential amino acids methionine, threonine, lysine, and leucine accounted for 57% of the synthetic protein. This protein was successfully expressed in *E. coli* and attempts are being made to express in crop plants (Beauregard and Hefford, 2006). A synthetic gene encoding a storage protein (ASP1) was designed whose structure was based on zeins (Fig. 15–2), the storage proteins of maize (Kim et al., 1992). Eighty percent of this designer protein is made up of essential amino acids consisting of four, 20 amino acid helical-repeating monomers (Fig. 15–2). Several Glu-Lys salt bridges stabilize the amphipathic helical region of the ASP1. The four monomers of the ASP1 tetramer are connected by three β-turn (Gly-Pro-Gly-Arg) sequences that provide structural stability to the mature protein (Kim et al., 1992). ASP1 has been successfully expressed in tobacco (Kim et al., 1992), sweet potato, [*Ipomoea batatas* (L.) Lam.] (Prakash and Egnin, 1997) and rice, *Oryza sativa* L. (Potrykus, 2003) and the resulting transgenic plants revealed an increase in the essential amino acid content. The ASP1 gene was also introduced in cassava (*Manihot esculenta* Crantz), a staple food of more than 500 million people in the tropics, to increase the nutritional quality of storage roots (Zhang et al., 2003). Even though ASP1 accumulated in different levels in transgenic cassava, the overall protein

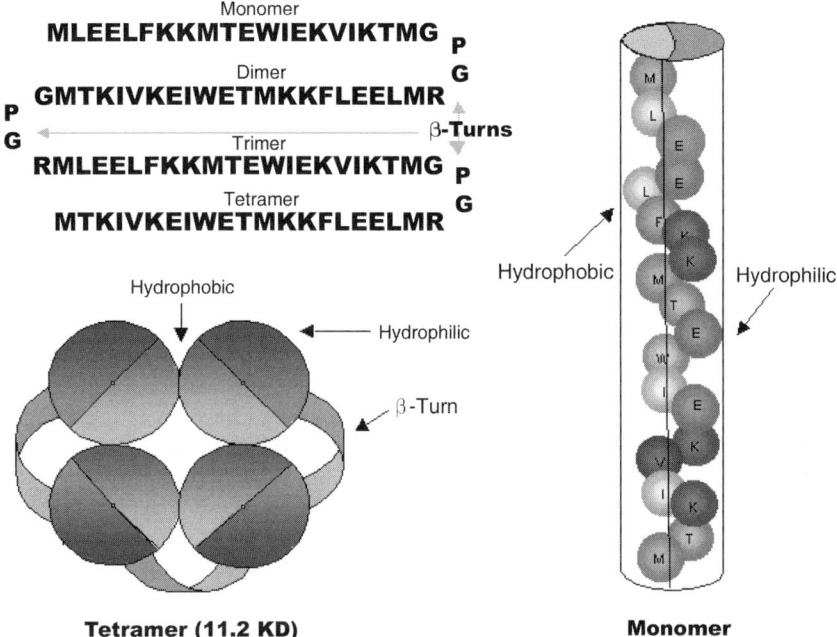

Fig. 15–2. Amino acid sequence and conformation of an artificial storage protein (ASP1). Note 80% of this synthetic protein is made up of essential amino acids consisting of four 20 amino acid helical-repeating monomers (figure courtesy of Peng Zhang).

content and amino acid composition of leaves of transgenic lines and wild-type control plants were not significantly different (Zhang et al., 2003). However, recent studies have demonstrated when the expression of ASP1 was controlled by p54 (a storage root specific promoter), a 2-fold increase in the protein content was observed in field-grown transgenic cassava (Peng Zhang, 2007, ETH Zurich, personal communication). Since ASP1 contains a significant amount of methionine, this synthetic protein could possibly be expressed in soybean and other legumes to improve their sulfur amino acid content.

Prospects

Although sulfur-rich proteins from other plant species have been successfully expressed in soybean, only modest increases in the methionine and cysteine contents have resulted. It is estimated that the existing levels of sulfur amino acids in soybean needs to be doubled to meet the nutritional requirements of monogastric animals. Analyses indicate that accumulation of methionine-rich heterologous proteins comes at the expense of sulfur-rich native proteins. Moreover, it appears that there is a threshold limitation in the availability of cysteine and methionine in developing seeds. A promising approach that has potential to double the sulfur amino acid content of soybean involves metabolic engineering of sulfur assimilatory pathway enzymes in combination with production of sulfur-rich sink proteins.

ATP sulfurylase, APS reductase, O-acetylserine-(thiol)lyase (OAS-TL), and serine acetyl transferase (SAT), all key enzymes in sulfur assimilation pathway, have been cloned and their expression in developing soybean seeds have been examined (Chronis and Krishnan, 2003, 2004; Phartiyal et al., 2006, 2007). SAT and OAS-TL are subject to multiple levels of regulation (Leustek et al., 2000; Noji and Saito, 2002; Francois et al., 2006), including product feedback inhibition. Engineering SAT and OAS-TL, which exhibit reduced sensitivity to this type of inhibition, would facilitate increased production of the sulfur amino acids in developing soybean seeds. There is a precedent for the success of this approach (Falco et al., 1995; Rapp et al., 2003). When feedback-insensitive aspartokinase and dihydrodipicolinic acid synthase, enzymes in lysine pathway, were expressed in soybean, a 5-fold increase in the accumulation of the amino acid was noted (Falco et al., 1995). Likewise, a modified form of anthranilate synthase, an enzyme in the tryptophan pathway, expressed in transgenic soybean permitted a 20- to 30-fold increase in the accumulation of the amino acid (Rapp et al., 2003).

Undoubtedly, a greater understanding of biochemical pathways involved in amino acid biosynthesis and protein accumulation will facilitate progress in the development of soybean varieties containing amino acids that provide optimal nutrition for swine and poultry. However, generations of a commercially viable high sulfur soybean must take into consideration the potential allergenicity of introduced proteins, the stability and subcellular localization of introduced proteins, and preservation of good agronomic characteristics. The improved amino acid profile should lead to improvements in feed efficiency and diminished nitrogen effluent from excess protein in the feeds needed to provide the minimum amino acid content.

Acknowledgments
The author wishes to thank Nathan Oehrle for critical review of the manuscript.

References

Altenbach, S.B., K.W. Pearson, F.W. Leung, and S.S.M. Sun. 1987. Cloning and sequence analysis of a cDNA encoding a Brazil nut protein exceptionally rich in methionine. Plant Mol. Biol. 8:239–250.

Barton, K.A., J.F. Thompson, J.T. Madison, R. Rosenthal, N.P. Jarvis, and R.N. Beachy. 1981. The biosynthesis and processing of high molecular weight precursors of soybean glycinin subunits. J. Biol. Chem. 257:6089–6095.

Beachy, R.N., N.P. Jarvis, and K.A. Barton. 1981. *In vivo* and *in vitro* biosynthesis of subunits of the soybean 7S storage protein. J. Mol. Appl. Gen. 1:19–27.

Beauregard, M., C. Dupont, R.M. Teather, and M.A. Hefford. 1995. Design, expression and initial characterization of MB-1, a *de novo* protein enriched in essential amino acids. Biotechnology (New York) 13:974–981.

Beauregard, M., and M.A. Hefford. 2006. Enhancement of essential amino acid contents in crops by genetic engineering and protein design. Plant Biotechnol. J. 4:561–574.

Beilinson, V., Z. Chen, R.C. Shoemaker, R.L. Fisher, R.B. Goldberg, and N.C. Nielsen. 2002. Genomic organization of glycinin genes in soybean. Theor. Appl. Genet. 104:1132–1140.

Bradley, R.A., D. Atkinson, H. Hauser, D. Oldani, J.P. Green, and J.M. Stubbs. 1975. The structure, physical and chemical properties of the soybean protein glycinin. Biochim. Biophys. Acta 412:214–228.

Brim, C.A., and J.W. Burton. 1979. Recurrent selection in soybeans. II. Selection for increased percent protein in seeds. Crop Sci. 19:494–498.

Burton, J.W., A.E. Purcell, and W.M. Walter, Jr. 1982. Methionine concentration in soybean protein from populations selected for increased seed protein. Crop Sci. 22:430–432.

Cai, T., and K.-C. Chang. 1999. Processing effect on soybean storage proteins and their relationship with tofu quality. J. Agric. Food Chem. 47:720–727.

Choi, Y., J.H. Ahn, Y.D. Choi, and J.S. Lee. 1995. Tissue-specific and developmental regulation of a gene encoding a low molecular weight protein in soybean seeds. Mol. Gen. Genet. 246:266–268.

Chronis, D., and H.B. Krishnan. 2003. Sulfur assimilation in soybean: Molecular cloning and characterization of o-acetylserine (thiol) lyase (cysteine synthase). Crop Sci. 43:1819–1827.

Chronis, D., and H.B. Krishnan. 2004. Sulfur assimilation in soybean (*Glycine max* [L.] Merr.): Molecular cloning and characterization of a cytosolic isoform of serine acetyltransferase. Planta 218:417–426.

Chui, C.F., and S.C. Falco. 1995. A new methionine-rich seed storage protein from maize. Plant Physiol. 107:291.

Coates, J.B., J.S. Medeiros, V.H. Thanh, and N.C. Nielsen. 1985. Characterization of the subunits of β-conglycinin. Arch. Biochem. Biophys. 243:184–194.

Cromwell, G., S. Traylor, M. Lindermann, H. Stilborn, and T. Sauber. 2000. Bioavailability of phosphorus in low oligosaccharide, low-phytate soybean meal for chicks. Poultry Sci. Supp. 79:127.

Dinkins, R.D., M.S.S. Reddy, C.A. Meurer, B. Yan, H. Trick, F. Thibaud-Nissen, J.J. Finer, W.A. Parrott, and G.B. Collins. 2001. Increased sulfur amino acids in soybean plants overexpressing the maize 15 kDa zein protein. In Vitro Cell. Dev. Biol. Plant 37:742–747.

de Lumen, B.O., A.F. Galvez, M.J. Revilleza, and D.C. Krenz. 1999. Molecular strategies to improve the nutritional quality of legume proteins. Adv. Exp. Med. Biol. 464:117–126.

Edwards, H.M., M.W. Douglas, C.M. Parsons, and D.H. Baker. 2000. Protein and energy evaluation of soybean meals processed from genetically modified high-protein soybeans. Poult. Sci. 79:525–527.

El-Shemy, H.A., M.M. Khalafalla, K. Fujita, and M. Ishimoto. 2007. Improvement of protein quality in transgenic soybean plants. Biol. Plant. 51:277–284.

Falco, S.C., T. Guida, J. Mauvais, C. Sanders, R.T. Ward, and P. Webber. 1995. Transgenic canola and soybean seeds with increased lysine. Biotechnology 13:577–582.

Francois, J.A., S. Kumaran, and J.M. Jez. 2006. Structural basis for interaction of O-acetylserine sulfhydrylase and serine acetyltransferase in the Arabidopsis cysteine synthase complex. Plant Cell 18:3647–3655.

Galvez, A.F., M.J.R. Revilleza, and B.O. de Lumen. 1997. A novel-methionine-rich protein from soybean cotyledon: Cloning and characterization of cDNA. Plant Physiol. 114:1567.

Gayler, K.R., and G.E. Sykes. 1985. Effects of nutritional stress on the storage proteins of soybeans. Plant Physiol. 78:582–585.

George, A.A., and B.O. de Lumen. 1991. A novel methionine-rich protein in soybean: Identification, amino acid composition and N-terminus sequence. J. Agric. Food Chem. 39:224–227.

Gidamis, A.B., P. Wright, Z.U. Haque, T. Katsube, M. Kito, and S. Utsumi. 1995. Modification tolerability of soybean proglycinin. Biosci. Biotechnol. Biochem. 59:1593–1595.

Hagan, N.D., L.M. Tabe, L. Molvig, and T.J.V. Higgins. 2003. Modifying the amino acid composition of grains using gene technology. p. 305–308. *In* I.K. Vasil (ed.) Plant biotechnology 2002 and beyond. Kluwer Academic Publishers, Dordrecht, the Netherlands.

Hajika, M., M. Takahashi, S. Sakai, and M. Igita. 1996. A new genotype of 7S globulin (β-conglycinin) detected in wild soybean (*Glycine soja* Sieb. Et Zucc.). Breed. Sci. 46:385–386.

Harada, J.J., S.J. Baker, and R.B. Goldberg. 1989. Soybean β-conglycinin genes are clustered in several DNA regions and are regulated by transcriptional and post-transcriptional processes. Plant Cell 1:415–425.

Imsande, J. 2001. Selection of soybean mutants with increased concentrations of seed methionine and cysteine. Crop Sci. 41:510–515.

Imsande, J. 2003. Sulfur nutrition and legume seed quality. p. 295–304. In Y.P. Abrol and A. Ahmad (ed.) Sulfur in plants. Kluwer Academic Publishers, Dordrecht, the Netherlands.

Imsande, J., and J.M. Schmidt. 1998. Effect of N source during soybean seed filling on nitrogen and sulfur assimilation and remobilization. Plant Soil 202:41–47.

Jeong, H.J., H.J. Park, Y. Lam, and B.O. de Lumen. 2003. Characterization of lunasin isolated from soybean. J. Agric. Food Chem. 51:7901–7906.

Kerley, M.S., and G.L. Allee. 2003. Modifications in soybean seed composition to enhance animal feed use and value: Moving from a dietary ingredient to a functional dietary component. AgBioForum 6:14–17.

Kho, C.J., and B.O. de Lumen. 1988. Identification and isolation of methionine and cysteine rich protein fraction in soybean seed. Plant Foods Hum. Nutr. 38:287–296.

Kim, J.H., S. Cetiner, and J.M. Jaynes. 1992. Enhancing the nutritional quality of crop plants: Design, construction and expression of an artificial plant storage protein gene. p. 1–36. In D. Bhatnagar and T.E. Cleveland (ed.) Molecular approaches to improving food quality and safety. AVI, New York.

Kim, W.S., and H.B. Krishnan. 2003. Allelic variation and differential expression of methionine-rich δ-zeins in maize inbred lines B73 and W23a1. Planta 217:66–74.

Kim, W.-S., and H.B. Krishnan. 2004. Expression of an 11 kD methionine-rich δ-zein in transgenic soybean results in the formation of two types of novel protein bodies in transitional cells situated between the vascular tissue and storage parenchyma cells. Plant Biotechnol. J. 2:199–210.

Kirihara, J.A., J.B. Petri, and J. Messing. 1988. Isolation and sequence of a gene encoding a methionine-rich 10-kDa zein protein from maize. Gene 71:359–370.

Kitamura, K., and N. Kaizuma. 1981. Mutant strains with low level of subunits of 7S globulin in soybean (Glycine max Merr.) seed. Jpn. J. Breed. 31:353–359.

Kortt, A.A., J.B. Caldwell, G.G. Lilley, and T.J.V. Higgins. 1991. Amino acid and cDNA sequence of a methionine-rich 2S protein from sunflower seed (Helianthus annuus L.). Eur. J. Biochem. 195:329–334.

Krishnan, H.B. 2000. Biochemistry and molecular biology of soybean seed storage proteins. J. New Seeds 2:1–25.

Krishnan, H.B. 2005. Engineeering soybean for enhanced sulfur amino acid content. Crop Sci. 45:454–461.

Krober, O.A. 1956. Methionine content of soybeans as influenced by location and season. J. Agric. Food Chem. 4:254–257.

Kuiken, A.A., and C.A. Lyman. 1949. Essential amino acid composition of soybean meals prepared from twenty strains of soybeans. J. Biol. Chem. 177:29–36.

Kwanyuen, P., V.R. Pantalone, J.W. Burton, and R.F. Wilson. 1997. A new approach to genetic alteration of soybean protein composition and quality. J. Am. Oil Chem. Soc. 74:983–987.

Ladin, B.F., J.J. Doyle, and R.N. Beachy. 1984. Molecular characterization of a deletion mutation affecting the α'-subunit of β-conglycinin of soybean. J. Mol. Appl. Genet. 2:372–380.

Lenfestey, B.A. 2005. An evaluation of the feeding value of Prolina soybean meal in male broiler chicken diets by altering dietary protein, amino acids, and metabolizable energy. MS thesis. North Carolina State University.

Leustek, T., M.N. Martin, J.-A. Bick, and J.P. Davies. 2000. Pathways and regulation of sulfur metabolism revealed through molecular and genetic studies. Annu. Rev. Plant Physiol. Plant Mol. Biol. 51:141–165.

Li, Z., S. Meyer, J.S. Essig, Y. Liu, M.A. Schapaugh, S. Muthukrishnan, B.E. Hainline, and H.N. Trick. 2005. High-level expression of maize γ-zein in transgenic soybean (Glycine max). Mol. Breed. 16:11–20.

Lin, J., R. Fido, P. Shewry, D.B. Archer, and M.J.C. Alcocer. 2004. The expression and processing of two recombinant 2S albumins from soybean (*Glycine max*) in the yeast *Pichia pastoris*. Biochim. Biophys. Acta 1698:203–212.

Lin, J., P.R. Shewry, D.B. Archer, K. Beyer, B. Niggemann, H. Haas, P. Wilson, and M.J. Alcocer. 2006. The potential allergenicity of two 2S albumins from soybean (*Glycine max*): A protein microarray approach. Int. Arch. Allergy Immunol. 141:91–102.

Muntz, K., V. Christov, G. Saalbach, I. Saalbach, D. Waddell, T. Pickardt, O. Schieder, and T. Wustenhagen. 1998. Genetic engineering for high methionine grain legumes. Nahrung 42:125–127.

Nielsen, N.C. 1996. Soybean seed composition. p. 127–163. *In* D.P.S. Verma and R.C. Shoemaker (ed.) Soybean: Genetics, molecular biology and biotechnology. CAB International, Wallingford, UK.

Nielsen, N.C., C.D. Dickinson, T.J. Cho, V.H. Thanh, B.J. Scallon, R.L. Fischer, T.L. Sims, G.N. Drews, and R.B. Goldberg. 1989. Characterization of the glycinin gene family. Plant Cell 1:313–328.

Nielsen, N.C., R. Jung, Y.-W. Nam, T.W. Beaman, L.O. Oliveira, and R. Bassüner. 1995. Synthesis and assembly of 11S globulins. J. Plant Physiol. 145:641–647.

Nielsen, N.C., M.P. Scott, and W.J.P. Lago. 1990. Assembly properties of modified subunit family. p. 635–640. *In* R. Hermann and B. Larkins (ed.) NATO Advanced Study Institute on Plant Molecular Biology. Plenum Press, New York.

Noji, M., and K. Saito. 2002. Molecular and biochemical analysis of serine acetyltransferase and cysteine synthase towards sulfur metabolism engineering. Amino Acids 3:231–243.

Nordlee, J.A., S.L. Taylor, J.A. Townsend, L.A. Thomas, and R.K. Bush. 1996. Identification of Brazil-nut allergen in transgenic soybeans. N. Engl. J. Med. 334:688–692.

Paek, N.C., J. Imsande, R.C. Shoemaker, and R. Shibles. 1997. Nutritional control of soybean seed storage protein. Crop Sci. 37:498–503.

Paek, N.C., P.J. Sexton, S.L. Naeve, and R. Shibles. 2000. Differential accumulation of soybean seed storage protein subunits in response to sulfur and nitrogen nutritional sources. Plant Prod. Sci. 3:268–274.

Panthee, D.R., and V.R. Pantalone. 2006. Registration of soybean germplasm lines TN03–350 and TN04–5321 with improved protein concentration and quality. Crop Sci. 46:2328–2329.

Phartiyal, P., W.-S. Kim, R.E. Cahoon, J.M. Jez, and H.B. Krishnan. 2006. Soybean ATP sulfurylase, a homodimeric enzyme involved in sulfur assimilation, is abundantly expressed in roots and induced by cold treatment. Arch. Biochem. Biophys. 450:20–29.

Phartiyal, P., W-S. Kim, R.E. Cahoon, J.M. Jez, and H.B. Krishnan. 2007. The role of 5′-adenylylsulfate reductase in the sulfur assimilation pathway of soybean: Molecular cloning, kinetic characterization, and gene expression. Phytochemistry 69:356–364.

Potrykus, I. 2003. Nutritional improvement of rice to reduce malnutrition in developing countries. p. 401–406. *In* I.K. Vasil (ed.) Plant biotechnology 2002 and beyond. Kluwer Academic Publishers, Dordrecht, the Netherlands.

Poysa, V., L. Woodrow, and K. Yu. 2006. Effect of soy protein subunit composition on tofu quality. Food Res. Int. 39:309–317.

Prakash, C.S., and M. Egnin. 1997. Engineered sweetpotato (*Ipomoea batatas*) plants with synthetic storage protein gene show high protein and essential amino acid levels. *In* J.F.D. Dean (ed.) Abstracts of 5th International Congress of Plant Molecular Biology. Kluwer Academic Publishers, Singapore.

Rapp, B.W., L. Weaver, Q. Wang, L. Crow, J. Ream, B. Hill, W. Brown, T. Oulmassov, and K. Gruys. 2003. Enhancing the nutritional value of soybean. Plant Biology 2003, Annual meeting of the American Society of Plant Biologists, abstract 38001.

Revilleza, M.J., A.F. Galvez, D.C. Krenz, and B.O. de Lumen. 1996. An 8 kDa methionine-rich protein from soybean (*Glycine max*) cotyledon: Identification, purification, and N-terminal sequence. J. Agric. Food Chem. 44:2930–2935.

Serretti, C., W.T. Schapaugh, and R.C. Leffel. 1994. Amino acid profile of high protein soybean. Crop Sci. 34:207–209.

Spencer, J., G. Allee, J. Frank, and T. Sauber. 2000. Nutrient retention and growth performance of pigs fed diets formulated with low-phytate corn and/or low-phytate/low oligosaccharides soybean meal. J. Anim. Sci. Supp. 78:73.

Staswick, P.E., M.A. Hermodson, and N.C. Nielsen. 1984. Identification of the cystines which link the acidic and basic components of the glycinin subunits. J. Biol. Chem. 259:13431–13435.

Sun, S.S.M., and Q. Liu. 2004. Transgenic approaches to improve the nutritional quality of plant proteins. In Vitro Cell. Dev. Biol. Plant 40:155–162.

Tabe, L.M., and T.J.V. Higgins. 1998. Engineering plant protein composition for improved nutrition. Trends Plant Sci. 3:282–286.

Takahashi, K., Y. Mizuno, S. Yumoto, K. Kitamura, and S. Nakamura. 1996. Inheritance of the α-subunit of β-conglycinin in soybean (*Glycine max* (L.) Merrill) line induced by gamma ray irradiation. Breed. Sci. 46:251–255.

Takahashi, M., Y. Uematsu, K. Kashiwaba, K. Yagasaki, M. Hajika, R. Matsunaga, K. Komatsu, and M. Ishimoto. 2003. Accumulation of high levels of free amino acids in soybean seeds through integration of mutations conferring seed protein deficiency. Planta 217:577–586.

Takaiwa, F., T. Katsube, S. Kitagawa, T. Hisago, M. Kito, and S. Utsumi. 1995. High level accumulation of soybean glycinin in vacuole-derived protein bodies in endosperm tissue of transgenic tobacco seed. Plant Sci. 111:39–49.

Teraishi, M., M. Takahashi, M. Hajika, R. Matsunaga, Y. Uematsu, and M. Ishimoto. 2001. Suppression of soybean β-conglycinin genes by a dominant gene, *Scg*-1. Theor. Appl. Genet. 103:1266–1272.

Thanh, V.H., and K. Shibasaki. 1976. Major proteins of soybean seeds. a straightforward fractionation and their characterization. J. Agric. Food Chem. 24:1117–1121.

Thanh, V.H., and K. Shibasaki. 1978. Major proteins of soybean seeds. Subunit structure of β-conglycinin. J. Agric. Food Chem. 26:692–695.

Tierney, M.L., E.A. Bray, R.D. Allen, Y. Ma, R.F. Drong, J. Slightom, and R.N. Beachy. 1987. Isolation and characterization of a genomic clone encoding the β-subunit of β-conglycinin. Planta 172:356–363.

Townsend, J.A., and L.A. Thomas. 1994. Factors which influence the *Agrobacterium*-mediated transformation of soybean. J. Cell. Biochem. Supp. 18A:78.

Tsukada, Y., K. Kitamura, K. Harada, and N. Kaizuma. 1986. Genetic analysis of subunits of two major storage proteins (β-conglycinin and glycinin) in soybean seeds. Jpn. J. Breed. 36:390–400.

Utsumi, S., S. Kitagawa, T. Katsube, T. Higasa, M. Kito, F. Takaiwa, and T. Ishige. 1994. Expression and accumulation of normal and modified soybean glycinin in potato tubers. Plant Sci. 102:181–188.

Wang, T.L., C. Domoney, C.L. Hedley, R. Casey, and M.A. Grusak. 2003. Can we improve the nutritional quality of legume seeds? Plant Physiol. 131:886–891.

Wilcox, J.R., and J.F. Cavins. 1995. Backcrossing high seed protein to a soybean cultivar. Crop Sci. 35:1036–1041.

Wilcox, J.R., and R.M. Shibles. 2001. Interrelationships among seed quality attributes in soybean. Crop Sci. 41:11–14.

Wolf, W.J., and D.R. Briggs. 1958. Studies on the cold-insoluble fraction of the water extractable soybean proteins. II. Factors influencing conformation changes in the 11S component. Arch. Biochem. Biophys. 76:377–393.

Williams, P.E. 2003. Engineering plants for animal feed for improved nutritional value. Proc. Nutr. Soc. 62:301–309.

Xue, Z.-T., M.-L. Xu, W. Shen, N.-L. Zhuang, W.-M. Hu, and S.C. Shen. 1992. Characterization of a *Gy4* glycinin gene from *Glycine max* cv. Forrest. Plant Mol. Biol. 18:897–908.

Zarkadas, C.G., C. Gagnon, V. Poysa, S. Khanizadeh, E.R. Cober, V. Chang, and S. Gleddie. 2007. Protein quality and identification of the storage protein subunits of tofu and null soybean genotypes, using amino acid analysis, one- and two-dimensional gel electrophoresis, and tandem mass spectrometry. Food Res. Int. 40:111–128.

Zarkadas, C.G., Z. Yu, H.D. Voldeng, H.J. Hope, A. Minero-Amador, and J.A. Rochemont. 1994. Comparison of the protein-bound and free amino acid contents of two Northern adapted soybean cultivars. J. Agric. Food Chem. 42:21–33.

Zarkadas, C.G., Z. Yu, H.D. Voldeng, and A. Minero-Amador. 1993. Assessment of the protein quality of a new high-protein soybean cultivar by amino acid analysis. J. Agric. Food Chem. 41:616–623.

Zhang, P., J.M. Jaynes, I. Potrykus, W. Gruissem, and J. Puonti-Kaerlas. 2003. Transfer and expression of an artifical storage protein (ASP1) gene in cassava (*Manihot esculenta* Crantz). Transgenic Res. 12:243–250.

16

Methionine Metabolism in Plants

Rachel Amir and **Yael Hacham**
MIGAL, Galilee Technology Center, Kiryat-Shmona, Israel.

Abstract

Methionine is a nutritionally essential, sulfur-containing amino acid whose low level in plants diminishes their value as a source of dietary protein for humans and animals. Methionine is also a fundamental metabolite in plant cells since through its first metabolite, S-adenosylmethionine (SAM), it controls the level of several key metabolites such as ethylene, polyamines, and biotin. SAM is also the primary methyl group donor that regulates different processes in plants. Despite its nutritional and regulatory significance, the factors regulating its synthesis and catabolism in plants are not fully known. Moreover, although methionine-associated metabolites play a major role in plant metabolism and growth, only little is known about how the methionine metabolism interacts with other metabolic networks regulating different processes in plants. In recent years, genetic molecular biology techniques have been used to increase and decrease the expression levels of several genes encoded to enzymes in the methionine metabolism. In this review, we summarize recent progress made in the molecular characterization of these genes. We also focus on specific examples where a deeper understanding of the regulation of metabolic networks in plants is needed for a tailor-made improvement of the methionine metabolism, with minimal interference in plant growth and productivity. We describe several different manipulations of methionine metabolism pathways and their effects on plant methionine content. These studies have resulted in the identification of steps important for the regulation of flux through the pathways and for the production of transgenic plants having increased free and protein-bound methionine. Similarly, the expression of methionine-rich storage proteins has resulted in significant improvements in methionine level of some target plant organs. These molecular approaches have provided new insights into the control of methionine level in plants and, in many cases, have resulted in significant improvements in the nutritional value of plants.

The Importance of Methionine in Plant Metabolism

The sulfur-containing amino acid methionine is an essential amino acid whose level limits the nutritional value of crop plants (Galili et al., 2005). Yet, aside from its nutritional importance, methionine is also a fundamental metabolite in plant cells. Apart from its role as a protein constituent and its central role in the initiation of mRNA translation, methionine indirectly regulates a variety of cellular processes as the precursor of SAM, which is the primary biological methyl group donor. The substrate of SAM-dependent methyltransferases participates in

---

Copyright © 2008. American Society of Agronomy, Crop Science Society of America, Soil Science Society of America, 677 S. Segoe Rd., Madison, WI 53711, USA. *Sulfur: A Missing Link between Soils, Crops, and Nutrition.* Agronomy Monograph 50.

both primary and secondary metabolism. Examples include lipids, DNA, RNA, proteins, pectin, alkaloids, phytosterols, osmoprotectants, reactions required for chlorophyll synthesis, for lignins and suberins synthesis, in flavonoids, hydroxycinnamic acids, stilbenes, and other aromatic as well as volatile fragrance and aroma compounds (Kagan and Clarke, 1994; Roje, 2006). Hence, as a donor for methyl groups, methionine through SAM regulates essential cellular processes such as cell division, synthesis of cell wall, synthesis of chlorophyll, and membrane synthesis (Roje, 2006). In higher plants, SAM is also the precursor for the hormone ethylene, which regulates developmental stages (Yang and Hoffman, 1984; Miyazaki and Yang, 1987; Yang et al., 1990; Matilla, 2000).

Besides, SAM is the source of the propylamino group in the synthesis of the polyamines, spermidine and spermine, which play crucial roles in many aspects of plant growth, including cell proliferation and differentiation, apoptosis, homeostasis, and gene expression (Cassol and Mattoo, 2003; Kaur-Sawhney et al., 2003). SAM is also the precursor for the metal ion chelating compounds, nicotinamide and phytosiderophores, and is also a precursor of the enzyme cofactor, biotin (Ravanel et al., 1998; Droux et al., 2000; Droux, 2004; Hesse et al., 2004; Roje, 2006) (Fig. 16–1). In addition, SAM posttranslationally regulates the levels of threonine and methionine in plant cells (Curien et al., 1998; Ravanel et al., 1998; Chiba et al., 2003). Methionine itself serves as a donor for many secondary metabolites such as glucosinolates, which are involved in pathogen and insect defense and are produced predominantly in plants belonging to the Brassicaceae family (Gigolashvili et al., 2007; Hirai et al., 2007). Finally, methionine leads to the synthesis of S-methylmethionine (SMM), which is considered to be the mobile and storage form of methionine (Mudd and Datko, 1990), the regulator of SAM level in cells (Ranocha et al., 2001; Kocsis et al., 2003), and a precursor to other secondary metabolites (Hanson et al., 1994).

Because of the biochemical and nutritional importance of methionine, considerable efforts were invested in studying the factors regulating its level and metabolism in plants, and highlights of these efforts are described in the following sections.

Methionine Biosynthesis Pathway and its Regulation

Methionine Biosynthesis

Methionine is synthesized in higher plants from a combination of three different moieties and, consequently, by three different pathways: (i) the carbon–amino skeleton is derived from aspartate (as are threonine, lysine, and isoleucine); (ii) the sulfur moiety is derived from cysteine; and (iii) the methyl group is derived from N-methyltetrahydrofolate (Fig. 16–1). Three enzymes operate methionine synthesis in plants: (i) the first unique enzyme of the methionine biosynthesis pathway, cystathionine γ-synthase (CGS), combines the carbon–amino skeleton with the sulfur moiety derived from cysteine (Giovanelli et al., 1980; Ravanel et al., 1995; Kim and Leustek, 1996); (ii) the second enzyme, cystathionine β-lyase, catalyzes the β-cleavage of cystathionine to homocysteine; and (iii) the last enzyme, methionine synthase, methylates homocysteine to methionine by means of N-methyltetrahydrofolate as a methyl group donor (Abdel-Kader, 2000; Droux, 2004; Azevedo et al., 2006). While the first two enzymes are localized in plastids, the last enzyme, methionine synthase, functions both in plastid and cytosol (Ravanel et al., 2004).

Methionine Metabolism in Plants

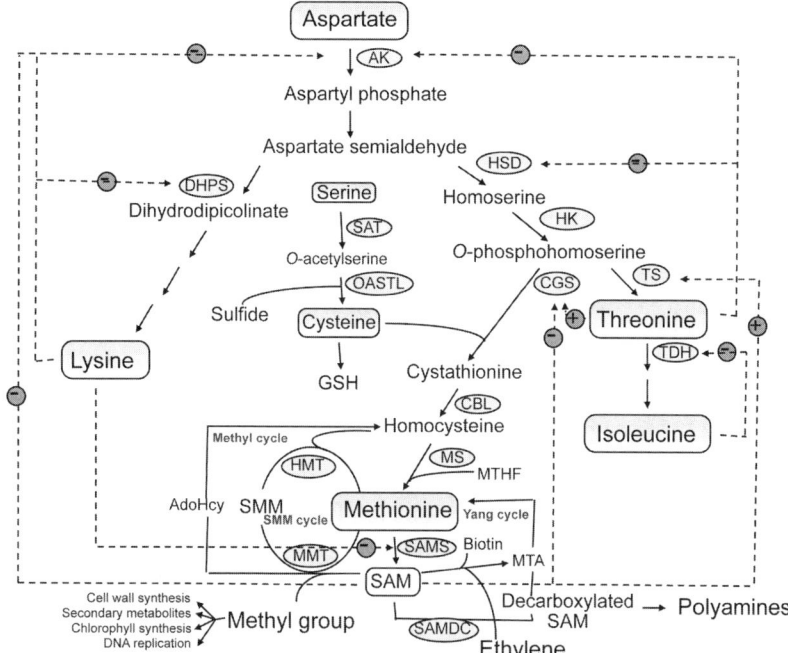

Fig. 16–1. Metabolic networks regulating sulfur, cysteine, and methionine metabolism. Only some of the enzymes and metabolites are specified. Dashed arrows with a "minus" sign represent either feedback inhibition loops or repression of gene expression. The dashed and dotted arrow with the "plus" sign represents the stimulation of gene expression or enzyme activity. Abbreviations: AK, Asp kinase; DHPS, dihydrodipicolinate synthase; HSD, homoserine dehydrogenase; HK, homoserine kinase; TS, threonine synthase; TDH, threonine dehydratase; OPH, O-phosphohomoserine; CGS, cystathionine γ-synthase (marked in red); CBL, cystathionine β-lyase; MS, methionine synthase; SAM, S-adenosyl Met; SAMS, SAM synthase; AdoHcys, adenosylhomocysteine; SMM, S-methylMet; MMT, methionine S-methyltransferase; HMT, homocysteine S-methyltransferase; MTHF, methyltetrahydrofolate; MTA, methylthioribose; SAT, serine acetyltransferase; OAS, O-acetyl serine; OASTL, O-acetylserine (thiol)lyase; GSH, glutathione; SAMDC, SAM decarboxylase; MTRK, methylthioribose kinase.

The Regulatory Role of Cystathionine γ-Synthase

In contrast to other regulatory enzymes in the aspartate biosynthesis pathway, whose activities are regulated by the feedback-inhibition mechanism mediated by their products, CGS activity is not feedback-inhibited by methionine or methionine metabolites (Ravanel et al., 1998a, 1998b). However, it was recently found that the methionine downstream product, SAM, negatively regulates the transcript level of *Arabidopsis* CGS (AtCGS) via a posttranscriptional mechanism (Chiba et al., 1999, 2003; Onouchi et al., 2004, 2005). This regulation occurs in the N-terminal region of AtCGS comprised of ~100 amino acids (without its plastid transit peptide), which does not exist in bacterial enzymes. This region is not essential for the catalytic activity of AtCGS but influences the levels of methionine and its metabolites in plants (Hacham et al., 2002). *Arabidopsis* mutants that accumulated methionine at a 40-fold higher rate than wild-type plants showed a set of mutations in a conserved subdomain within the N-terminal region (termed MTO1) (Chiba et al., 1999; Inba et al., 1994). A detailed analysis of this MTO1 domain reveals that SAM induces a temporal arrest in the translation elongation process during AtCGS mRNA translation of this domain. As a result of the translation arrest, mRNA degradation occurs upstream of the stalled ribosome, resulting in the production of the 5′-truncated RNA species (Onouchi et al., 2005). These studies suggest that methionine synthesis in the *Arabidopsis* plant cells is controlled by the amount of AtCGS transcript.

Notwithstanding, despite the regulatory posttranscriptional function of the MTO1 region, no inverse correlation between high methionine levels and low AtCGS mRNA levels were evident in transgenic *Arabidopsis* plants constitutively expressing the endogenous AtCGS (Kim et al., 2002). Moreover, overexpression of AtCGS in tobacco, *Nicotiana tabacum* L. (Hacham et al., 2002, 2006), potato, *Solanum tuberosum* L. (Di et al., 2003), and alfalfa, *Medicago sativum* L., plants (Avraham et al., 2005) showed a positive correlation between AtCGS level and methionine content instead of the expected negative correlation. Whether these observations are due to variations in MTO1 machinery between different *Arabidopsis* tissues and different plant species or the result of the overexpression itself must still be elucidated.

Similar to *Arabidopsis* CGS, it was recently shown by using feeding experiments that tomato, *Lycoperiscon esculentum* Mill., CGS having the MTO1 domain is also sensitive to high methionine application (Katz et al., 2006). However, the expression level of potato CGS1, which shares 97% identity with tomato CGS, is not affected by high methionine application, although the potato GCS1 gene differs from the tomato CGS in only six amino acids (one of them within the MTO1 region) (Hesse and Hofgen, 2003; Kreft et al., 2003). Moreover, overexpression of potato CGS in potato plants, even though it significantly increased enzyme activity, does not lead to a higher methionine level, implying that CGS in these plants does not control methionine level (Kreft et al., 2003). This observation suggests that potato is missing some elements required for posttranscriptional CGS regulation to occur and that regulation through the MTO1 region is insufficient. Therefore, either other cis elements outside the MTO1 region or additional trans-acting elements are apparently involved in the regulation of the CGS transcript level.

Indeed, it was recently found that an additional domain exists within the *N*-terminal region of the AtCGS very close to the MTO1 domain. This domain is comprised of 90 or 87 nucleotides, and omission of this domain maintains the reading frame of the protein of native AtCGS. Transgenic tobacco plants that overexpress from AtCGS lacking this domain show that this form enables plants to accumulate much higher levels of methionine compared with plants expressing full-length AtCGS (Hacham et al., 2006). Furthermore, feeding experiments revealed that this deleted form of AtCGS is not subject to feedback regulation by Met, as reported for the full-length transcript (Chiba et al., 2003). A native deleted form of AtCGS was found in all organs of *Arabidopsis* plants together with the full-length AtCGS. Since only one gene for CGS exists in *Arabidopsis* and the two forms of CGS are identical (except the one lacking the 90 nucleotide domain), it was suggested that the deleted form of AtCGS created from the full-length AtCGS, in still an unidentified process (Hacham et al., 2006).

As expected of the major role played by methionine in plant metabolism, the level of AtCGS is tightly regulated. In addition to the effect of SAM, which reduced the AtCGS transcript level (Chiba et al., 2003; Onouchi et al., 2005), threonine in *Arabidopsis* (Avraham and Amir, 2005), and the SAM-derived hormone ethylene in tomato (Katz et al., 2006), enhanced the transcript level of CGS. Other metabolites may also regulate the level of CGS or its activity, since it was recently shown that in cells starved for folates for a prolonged period, the N-terminal region of CGS was removed by proteolytic cleavage leaving the enzyme active (Loizeau et al., 2007).

Unlike CGS, which plays a major regulatory role in methionine synthesis at least in *Arabidopsis*, no direct evidence shows that the content of the second and third enzymes of the methionine biosynthesis pathway, cystathionine β-lyase and methionine synthase, play a regulatory role in determining methionine content in plants. Consistent with this assumption, overexpression of these two enzymes did not lead to an increase in flux toward methionine synthesis in potatoes (Maimann et al., 2001; Nikiforova et al., 2002; Hesse and Hofgen, 2003).

The Role of Threonine Synthase

Several lines of evidence have shown that methionine biosynthesis is also regulated by the competition between CGS, the first enzyme unique to the methionine pathway, and threonine synthase (TS), the last enzyme in the threonine pathway, for their common substrate O-phosphohomoserine (OPH) (Fig. 16-1) (Amir et al., 2002; Hesse et al., 2004). In vitro activity measurements indicate that TS in plants has 250- to 500-fold higher affinity for OPH compared with CGS, causing reduced OPH availability for methionine synthesis (Ravanel et al., 1998; Curien et al., 1998). However, a modeling analysis recently suggested that TS and AtCGS have a similar kinetic efficiency for OPH, but OPH is used more by TS because of the higher concentration of this enzyme compared with CGS (Curien et al., 2003). It was found that the TS level in *Arabidopsis* was 7-fold higher than the CGS level, causing the flux toward threonine synthesis to be 4-fold higher than the flux toward methionine (Curien et al., 2003).

Analyses of transgenic and mutant plants have also shown that the protein levels of TS and CGS are important for determining the distribution of OPH between the two pathways (reviewed by Amir et al., 2002; Hesse et al., 2004). For example, when the CGS level in transgenic *Arabidopsis* plants was reduced by antisense approaches, threonine concentration increased 3- to 8-fold relative to wild-type plants, while free methionine concentrations were slightly reduced (Gakiere et al., 2000; Kim et al., 2002). On the other hand, reduction in TS activity caused a 16-fold reduction in free threonine and a 22-fold increase in free methionine concentration in rosette leaves of the mutant compared with wild type (Bartlem et al., 2000). Similarly, a reduction in TS levels determined by an antisense approach in potato and *Arabidopsis* plants caused a significant increase in methionine level (Avraham and Amir, 2005; Zeh et al., 2001). These results imply that a reduction in TS causes either an increased upstream flux of the aspartate family pathway toward methionine or a reduced rate of methionine catabolism to SAM (Galili et al., 2005).

If only the ratio between CGS and TS affects the partition of OPH between threonine and methionine biosynthesis, it is expected that when TS is reduced, the methionine level will increase proportionally, and vice versa. However, antisense reduction in TS activity in potato and *Arabidopsis* plants caused a much stronger molar increase in methionine levels than the molar decrease in threonine levels (Avraham and Amir, 2005; Zeh et al., 2001). In addition, transgenic tobacco, alfalfa, and *Arabidopsis* plants that overexpressed AtCGS had a significantly higher methionine level that was not accompanied by a reduction in threonine levels (Avraham et al., 2005; Hacham et al., 2002, 2006). Moreover, *Arabidopsis* plants expressing the antisense form of AtCGS showed a significant increase in OPH levels (21-fold), while threonine increased only 4-fold (Gakiere et al., 2000). Similar results were obtained in potato plants expressing the antisense form of CGS, where the OPH level increased significantly while the threonine level remained unchanged (Gakiere et al., 2000; Kreft et al., 2003). These results show that even when OPH is available for threonine synthesis, TS does not utilize it efficiently when the CGS level is low.

SAM, the first catabolic product of Met, is apparently one of the factors that control the competitive fluxes of threonine and methionine biosynthesis. SAM negatively regulates the expression level of CGS (Chiba et al., 2003) and also positively regulates TS activity by an allosteric mechanism that increases the affinity of TS to OPH, as found by in vitro studies (Curien et al., 1996, 1998). Therefore, one option to explain why the level of threonine does not increase in plants expressing the antisense form of AtCGS is that in these plants, the SAM level is low and thus cannot lead to higher TS activity and consequently to a higher level of threonine. This effect of SAM on the transcript level of AtCGS and on TS activity implies that methionine regulates its own synthesis through SAM. When a high level of Met, and consequently SAM, is produced, the CGS level is downregulated and TS activity is increased, causing reduced OPH availability for methionine synthesis (Curien et al., 1998). However, other evidence suggests that this scenario does not exist under in vivo conditions. Analyses of some mutants have demonstrated that TS activity and threonine levels are not affected by SAM content. For example, it was found that in the *mto1* mutant where the CGS gene was mutated, free methionine increased 40-fold and SAM increased 3-fold, whereas the free threonine level was not altered compared with wild-type plants (Bartlem et al., 2000; Inba et al., 1994). In the *mto3* mutant having a mutation in the SAM synthase 3 gene, methionine increased more than 200-fold and SAM decreased by 35%, but the free threonine was again not affected (Shen et al., 2002). Moreover, in the *mmt* mutant, an opposite relationship occurred between the level of SAM and threonine content, since this mutant has a higher level of SAM but a lower level of threonine (Kocsis et al., 2003). These results are intriguing, as they imply that SAM may not lead to activation of TS in vivo.

The importance of the OPH level in methionine and threonine synthesis was recently shown in wild-type and transgenic *Arabidopsis* plants overexpressing homoserine kinase that were fed with homoserine one metabolite upstream to OPH (Fig. 16–1). The levels of both amino acids significantly increased in these plants (Lee et al., 2005). Moreover, a marked and significant increase in methionine content (a 180-fold increase above the level found in wild-type plants) was obtained when *Arabidopsis* plants overexpressing the AtCGS were fed with homoserine (Lee et al., 2005). These results suggest that under physiological conditions, the AtCGS and TS are substrate limited. This probably occurs because of the close regulation of the first unique enzyme of the aspartate family, aspartate kinase (AK), and homoserine dehydrogenase activities that limit OPH production. All in all, the results described above indicate that methionine content is tightly regulated by the flux of the carbon–amino skeleton toward its synthesis and that a strategy to increase methionine content might involve the coexpression of CGS along with the feedback-insensitive mutant form of AK.

Regulation of Carbon–Amino Flow into Methionine: The Role of Aspartate Kinase

Several studies indicate that the carbon–amino skeleton flux toward the threonine branch of the aspartate amino acid family is controlled by AK (reviewed by Azevedo et al., 2006). The activity of AK isozymes regulated in plants by the feedback-inhibition mechanism mediated by threonine and lysine. Mutants possessing feedback-insensitive AK isozymes have shown that the threonine level was over-accumulated, while the methionine content, whose biosynthesis pathway diverges from this branch, was not significantly changed (Frankard et al., 1991). In agreement with these studies, overexpressing *E. coli* feedback-insensitive AK enzyme constitutively in transgenic tobacco (Shaul and Galili, 1992), *Arabidopsis* (Ben Tzvi-Tzchori et al., 1996), and alfalfa plants (Galili et al., 2000) results in a significant overproduction of free threonine, while methionine content does not differ significantly.

A slight but significant elevation in methionine content was found when this bacterial enzyme was expressed in a seed-specific manner (Karchi et al., 1993). To study further the regulatory role of the carbon–amino flux in methionine synthesis, we crossed between tobacco plants overexpressing the bacterial AK (Shaul and Galili, 1992) and those overexpressing the full-length AtCGS (Hacham et al., 2002). Plants coexpressing these two genes have significantly higher methionine and threonine levels compared with levels found in wild-type plants, but the methionine level does not increase beyond that found in plants expressing the full-length AtCGS alone (Hacham et al., 2008). This finding contradicted that suggested by Lee et al. (2005) whereby the overexpression of CGS coupled with the feedback-insensitive form of AK can contribute to methionine production. However, our result could be explained through the feedback-inhibition regulation mediated by SAM on the transcript level of AtCGS when methionine increases beyond a certain threshold (Chiba et al., 2003).

To test this assumption, plants expressing the bacterial AK were crossed with plants expressing mutated forms of AtCGS in which the N-terminal region of AtCGS, or the 90-nucleotide domain, was deleted. These two forms of AtCGS are methionine–SAM insensitive. Indeed, significantly higher methionine contents accumulated in the newly produced plants compared with plants expressing only these forms of AtCGS (about 4.5-fold higher), while the level compared with wild-type plants increased to about 110- to 190-fold higher. The levels of threonine were doubled in these plants compared with wild-type plants (Hacham et al., 2008). These results suggest that the carbon–amino skeleton limits methionine synthesis, but this can hardly be seen under normal growth conditions since the regulatory role of methionine–SAM on the expression level of AtCGS has a relatively stronger effect. The results obtained in this study also suggest new ways of producing transgenic crop plants containing increased levels of methionine and threonine (an important essential amino acid that limits the nutritional value of cereals) and hence having improved nutritional quality.

Regulation of Sulfur Flow from Cysteine toward Methionine Synthesis

While it is well known that the carbon–amino skeleton level plays a significant role in methionine synthesis, it is still unclear if the sulfur level, and more specifically the cysteine content, limits methionine synthesis in plants. Plants having higher levels of cysteine because of manipulations of the cysteine biosynthesis pathway, such as overexpression of genes encoding to its two last biosynthesis enzymes, serine acetyl transferase or O-acetylserine (thiol) lyase, lead to higher cysteine and glutathione levels (Saito et al., 1994; Blaszczyk et al., 1999; Harms et al., 2000; Noji et al., 2001; Nikiforova et al., 2002; Wirtz and Hell, 2003; Matityahu et al., 2005; Stiller et al., 2007); however, the level of methionine in these transgenic plants was not significantly altered (Matityahu et al., 2005). This suggests that under natural conditions, cysteine was channeled toward glutathionine synthesis, whereas the incorporation of cysteine into methionine synthesis is tightly controlled, most probably by the level and activity of CGS. The differences between methionine and glutathione levels can also be explained by the rapid catabolism of methionine to SAM and its metabolites (Giovanelli et al., 1980, 1985), while the rate of glutathione catabolism is much slower (Ohkama-Ohtsu et al., 2007).

The observation that plants possess higher levels of Met-emitting compounds containing sulfur, such as methanethiol, dimethylsulfide, and carbon disulfide (Hacham et al., 2002; Boerjan et al., 1994), was also suggested because the sulfur compounds and cysteine content do not limit methionine synthesis in plants. To further test this assumption, we recently crossed plants overexpressing yeast O-acetyl(thiol)lyase (yOASTL) (Matityahu et al., 2005) and those overexpressing the AtCGS (Hacham et al., 2006). Plants overexpressing the yeast gene both in the cytosol and the chloroplasts exhibit high levels of cysteine and glutathione in their leaves (Matityahu et al., 2005). Plants overexpressing the chloroplast form of yOASTL and the AtCGS (which is naturally active in the chloroplasts) showed slightly but significantly higher amounts of methionine and the methionine metabolite, SMM, compared with plants overexpressing the AtCGS alone. However, the levels of methionine and SMM are not altered in plants overexpressing the AtCGS and the yeast protein targeted to the cytosol compared with the levels found in their corresponding parents (R. Amir, unpublished results). Taken together, the results suggested that in the chloroplasts, where the first enzyme for methionine is localized, higher levels of cysteine could contribute to methionine synthesis when a high expression level of CGS exists.

Additional evidence suggests that the relationship between cysteine and sulfur availability to methionine content is quite complicated. Unexpectedly, it was found that methionine levels are not grossly affected during a broad time range of sulfate deprivation (Nikiforova et al., 2005, 2006). The plants were able to keep the methionine levels constant, which are always low in plants. The level of SAM, however, was depleted under those conditions, while the levels of threonine and isoleucine (that compete with methionine for the carbon–amino skeleton) increased. These results suggest that factors other than sulfur content are involved in methionine homoeostasis in plants during sulfur starvation. One factor that might assist in maintaining the methionine homoeostasis in plants could be the methionine recycle pathways that recycle the Met-moieties to regenerate methionine (described below). Indeed, it was found that during sulfur starvation, two enzymes involved in the methyl cycle (one of these recycle pathways, Fig. 16–1), namely, SAM synthase and adenosylhomocysteine hydrolase, are induced (Nikiforova et al., 2003).

Regulation of Methyl Flow into Methionine

While the first two enzymes of methionine synthesis, CGS and cystathionine β-lyase, are localized in the chloroplasts, the third and last enzyme of methionine synthesis, methionine synthase, which adds the methyl moiety to homocysteine and thus produces Met, is localized in the cytosol (Eichel et al., 1995; Eckermann et al., 2000). It is suggested that the cytosolic form of methionine synthase plays a major role in regenerating the methyl group from the methyl cycle (Fig. 16-1) after the transmethylation reactions. This suggests that homocysteine, which was produced by cystathionine β-lyase, transfers from the chloroplast to the cytosol where methionine synthase is active. However, recent studies have shown that three forms of methionine synthase are found in *Arabidopsis*, one of which is present in the chloroplasts, suggesting that methionine can also be produced in the chloroplasts (Ravanel et al., 2004). This finding supports previous observations that isolated chloroplasts incubated with ether radioactive aspartate or sulfur can synthesize radiolabeled methionine (Mills, 1980).

Biochemical studies imply that methionine synthase does not play any regulatory or limiting role in methionine biosynthesis (Zeh et al., 2002). This assumption is supported by the massive turnover of methionine in methyl transfer reactions through SAM and the methyl cycle, without net consumption of methionine.

The Regulatory Roles of Lysine and Threonine in Determining the Methionine Levels in Plants

Two major branches exist in the aspartate family of amino acids. One leads to the lysine biosynthesis pathway, while the other leads to threonine, methionine, and isoleucine synthesis (Fig. 16-1). The regulation of the carbon–amino skeleton flux within these two branches is quite complicated, and metabolites from one branch affect the flux toward the other branch (Azevedo et al., 2006). It is well known that the application of both threonine and lysine leads to methionine starvation since they both inhibit the activity of AK (Thompson et al., 1982; Lee et al., 2005). However, we have recently found that threonine and lysine, when applied separately, increase the level of methionine. Since the methionine biosynthesis pathway diverges from the threonine branch of the aspartate family, we first elucidated the relationship between threonine and methionine synthesis. Using transgenic plants and feeding experiments, we have recently demonstrated that high levels of threonine enhanced the expression levels of AtCGS and thus led to an increase in methionine content (Avraham and Amir, 2005; Hacham et al., 2008). However, the nature of this regulation is not yet clear, since studies using the in vitro transcription–translation system suggest that threonine does not affect the level of AtCGS directly, rather through its effect on other metabolites whose nature is not yet revealed (Hacham et al., 2008).

The lysine biosynthetic pathway competes with the Thr-Met branch for the carbon–amino skeleton. Therefore, to elucidate the relationship between these biosynthetic branches and to study the factors that regulate methionine synthesis, we next crossed transgenic tobacco plants overexpressing AtCGS that exhibit higher levels of Met and those overexpressing the feedback-insensitive bacterial enzyme dihydrodipicolinate synthase (bDHPS) that contain a significantly higher level of lysine. As CGS is located downstream in the Thr–Met branch (Fig. 16-1), it was expected that the level of methionine would be reduced in the background of the overexpression of bDHPS. Unexpectedly however, methionine levels were significantly elevated in plants coexpressing both transgenes compared with those expressing AtCGS alone, while the level of lysine remained the same as those overexpressing bDHPS alone (Hacham et al., 2007). The increased levels of methionine and SMM correlated with the elevation in mRNA and protein levels of AtCGS and with the reduced mRNA level in genes encoding for SAM synthase, which converts methionine to SAM. Taking these results into account, we propose the following scheme for the crosstalk between lysine and methionine biosynthesis pathways (Fig. 16–1, dotted line). A high level of lysine brings about a reduction in the amount of enzyme SAM synthase because of a reduction in the amount of transcripts encoding this enzyme. This leads to a reduction in the amount of SAM, which negatively regulates the amount of AtCGS transcript (Chiba et al., 2003). As a result, the expression level of AtCGS is increased and consequently the level of methionine (Fig. 16–1) (Hacham et al., 2007). The methionine level can also be enhanced by a reduction in flux toward SAM and its metabolites (Giovanelli et al., 1985). Taken together, this mechanism ensures a fine balance between the amounts of lysine and methionine through an indirect biochemical crosstalk mechanism involving the regulation of the expression of SAM synthase.

These findings further explain the changes in lysine and methionine contents previously obtained in seeds of three sets of *Arabidopsis* transgenic plants that exhibit significantly higher levels of lysine, resulting from seed-specific expression of bDHPS and RNAi of lysine-ketoglutarate reductase–saccharopine dehydrogenase, the catabolic enzyme of lysine (Zhu and Galili, 2003, 2004). In these seeds, in addition to a significant 80-fold increase in lysine content, the methionine level increased significantly, up to 51-fold compared with wild-type seeds (Zhu and Galili, 2003, 2004). A positive correlation between high lysine and methionine levels was also found in transgenic barley plants that constitutively express the bDHPS. These plants exhibited a 14-fold increase in free lysine and an 8-fold increase in free methionine (Brinch-Pedersen et al., 1996). These results show that in seeds, as well as in vegetative tissues, a higher level of lysine enhances the production of methionine or reduces its catabolism.

We assume that because of the major role of methionine in plant metabolism, the lysine and threonine that compete with methionine for the carbon–amino skeleton regulate the methionine level in such a way that if mutation occurs, which leads to higher levels of these amino acids, the methionine level will not significantly change, since, when their content increases, the level of methionine increases as well. Notably, it was recently found that methionine also regulates the level of the fourth amino acid in this family, isoleucine. This amino acid, which is produced from α-ketobutyrate, is a product of the methionine catabolism that is produced when high levels of methionine are found in plants (Rebeille et al., 2006). These results imply that relations within the aspartate family of amino acids are much more complicated, and members of the family control the levels of other members in this family. From a biotechnological point of view, the results described here show that significantly higher methionine content can appear with significantly higher levels of lysine and separately with significantly higher levels threonine. The latter two amino acids are essential amino acids that limit the nutritional value of cereal grains, where a low level of methionine is also found (Galili et al., 2005). Thus, these studies present new ways of manipulating the levels of these amino acids in plants to increase their nutritional value.

The Regulatory Role of Methionine Catabolic Pathways

The Role of SAM Synthase in Controlling Methionine Content

The results described above showed that lysine regulates the level of SAM synthase and thus controls the methionine level, focusing the regulatory role of SAM synthase in controlling the methionine levels in plants. Indeed, several mutants and transgenic plants demonstrated that a reduction in the expression level of SAM synthase leads to significantly higher levels of methionine in plants. For example, the *Arabidopsis* mutant (*mto3*) that exhibits reduced SAM synthase 3 activity had an over 200-fold methionine content compared with wild-type plants (Goto et al., 2002; Shen et al., 2002). This high level of methionine was associated with a low level of SAM, and as a result, the lignin content, which is one of the major metabolic sinks for SAM, decreased by 22% compared with wild-type plants (Shen et al., 2002). A more severe phenotype appeared in SAM synthase silencing *Arabidopsis* plants. In these plants, the level of methionine increased 250-fold compared with wild-type plants, but severe abnormalities were observed (Kim et al., 2002). A severe, abnormal phenotype also appeared in tobacco plants where the SAM synthase level was significantly reduced (Boerjan et al., 1994). This suggested that a reduction in the expression level of SAM synthase could significantly increase the level of Met, but because of the importance of SAM in plant metabolism, such a reduction led to severe, abnormal phenotypes. Overexpression of SAM synthase in *Arabidopsis* does not lead to altered levels of methionine and SMM; although, as expected, the activity of CGS is reduced in this mutant, most probably because of the high level of SAM in these plants (Kim et al., 2002).

The Regulatory Roles of the Methionine Recycling Pathways

Three recycling pathways can regenerate the moieties that comprise methionine following the formation of SAM metabolites (Fig. 16–1). These recycling pathways are (i) the methyl cycle, in which the methyl group of methionine–SAM is donated to methyl transfer reactions, leaving *S*-adenosylhomocysteine, which recycles to methionine through homocysteine and the activity of the cytosolic form of methionine synthase that combines the methyl group to homocysteine to regenerate methionine (Roje, 2006); (ii) the SMM cycle, which operates by the activities of SAM-Met *S*-methyltransferase (MMT) and homocysteine *S*-methyltransferase (HMT) (Ranocha et al., 2001); and (iii) the Yang cycle (also termed the MTA cycle), where ethylene, biotin, and polyamines are synthesized, and the methylthio moieties are recycled to methionine via methylthiobutyrate (Fig. 16–1) (Yang et al., 1990).

Although the importance of these pathways is accepted, little is known about their regulation and contribution to the methionine pool. In addition, the nature of the competition for methionine between protein synthesis and SAM synthesis in plants is not fully clarified yet. Studies of metabolic fates of methionine using the aquatic plant *Lemna paucicostata* Hegelm. indicated that the synthesis and turnover of SAM accounts for 80% of the methionine metabolism, whereas the synthesis of proteins drives about 20% of the methionine metabolism (Giovanelli, 1987). In mature *Arabidopsis* rosette leaves, however, Ranocha et al. (2001), who use radioactive methionine and in silico modeling, suggest that about half of the soluble methionine converts to SAM and SMM and half to protein synthesis (Ranocha et al., 2001). They also suggest that the rate of release of methionine from the storage pool and from protein turnover together have a similar magnitude to SAM synthesis (Ranocha et al., 2001). A quantitative analysis of the methionine metabolism in *Lemna*, which does not produce ethylene, showed that the Yang cycle (used for polyamine synthesis) accounts for 6% of the methionine, whereas the de novo synthesis of methionine contributed 19%, and recycling from SAM-dependent methylations accounted for 75% of the methionine pool (Giovanelli et al., 1985). By means of radioactive methionine, it was suggested that SAM is used predominantly for transmethylation, whose major product is phosphatidylcholine, with progressively smaller amounts directed to the synthesis of methyl groups of pectin and chlorophyll methyl esters (Mudd and Datko, 1986). The role of the methyl cycle in methionine content is far from being resolved, and further studies are required to reveal the role of this cycle.

Questions also remain about the role of the Yang cycle in methionine homoeostasis, especially in plants that produce ethylene and in climacteric fruits where bursts of ethylene occur during the ripening stages. To further study the role of the Yang cycle in SAM production and its homoeostasis, Bürstenbinder et al. (2007) used an *mtk* mutant that has a disruption of the Yang cycle. The researchers conclude, on the basis of their data that the Yang cycle contributes to SAM homeostasis, especially when de novo SAM synthesis is limited, such as the sulfur limitation that affects SAM synthesis (Nikiforova et al., 2005, 2006). The data also showed that this cycle is required to sustain a high level of ethylene synthesis. However, additional evidence suggests that in addition to the recycling of the methionine moieties via the Yang cycle, the de novo synthesis of methionine is required when high rates of ethylene production are induced (Katz et al., 2006). This was suggested on the basis of the observation that the transcript expression level of CGS is positively correlated to ethylene during the ethylene burst of the climacteric ripening of tomato fruit and during leaf wounding when ethylene was emitted (Katz et al., 2006). The level of methionine also increased accordingly under these conditions (Katz et al., 2006).

The role of the third cycle, the SMM cycle, which is unique to plants, is also unclear. A number roles were proposed for this cycle in plants, including the following: (i) SMM mediates the long-distance transport of labile methyl moieties and reduced sulfur; this is suggested on the basis of the observation that SMM is the major sulfur compound in the phloem of plants (Giovanelli et al., 1980); (ii) SMM functions as a storage form of methionine-to-protein synthesis and keeps the free methionine pool from being depleted by an overshoot in SAM synthesis (Mudd and Datko, 1990; Pimenta et al., 1998); and (iii) SMM controls the SAM level in plant tissue. Using radioactive tracer measurements, Ranocha et al. (2001) suggested that the cycle consumes half the SAM produced and that the cycle serves to stop the accumulation of SAM rather than prevent the depletion of free Met. Controlling the level of SAM is crucial to many methyl transfer reactions, since the ratio between SAM and the *S*-adenosylhomocysteine metabolite in the methyl cycle, which is a potent inhibitor of methyl transferases, determined the activity of these essential enzymes (Chiang et al., 1996). This assumption was recently strengthened by the observation that the *Arabidopsis mmt* mutant, which lacks the SMM cycle, had a significantly higher level of SAM and a lower level of *S*-adenosylhomocysteine than wild-type plants and, consequently, a higher methylation ratio (SAM to *S*-adenosylhomocysteine ratio). These results also support the hypothesis that the SMM cycle contributes to the regulation of SAM levels

(Kocsis et al., 2003). Further support came recently from studies showing that feeding with methionine affected the SAM level, which increased 10-fold, but mainly contributed to the SMM level whose level increased followed by a steady value of SAM (Rebeille et al., 2006).

Taken together, the results described above show that the role of the three recycling pathways in the methionine pool, their relationship to methionine synthesis, and their role in plant metabolism are not fully known. Thus, to reveal the network of the methionine metabolism, additional studies are required.

Methionine Catabolism

SMM production has been proposed to function as a storage or transport form of methionine when the level of this amino acid is present in excess (Bourgis et al., 1999; Kocsis et al., 2003). The accumulation of SMM, however, cannot deal with a large excess of Met, possibly because of the energy cost of its synthesis (one ATP per SMM molecule) (Mudd and Datko, 1990), and thus, methionine must be removed through other catabolic pathways.

Several studies have reported that plants accumulate soluble methionine-emitting volatile sulfur-methyl containing compounds, such as methanethiol, dimethyl disulfide, or dimethyl sulphide. Therefore, it was suggested that these metabolites are methionine catabolized products (Hacham et al., 2002; Schmidt et al., 1985; Boerjan et al., 1994). The process responsible for the production of these volatiles and their physiological significance is still unknown. It was previously shown that bacteria can produce methanethiol by the activity of methionine γ-lyase (Dias and Weimer, 1998); therefore, it was suggested that plants have a similar enzyme with a similar function. A similar sequence to the bacterial gene was detected in the *Arabidopsis* genome, and two groups of researchers recently cloned the cDNA encoding the methionine γ-lyase from this plant (Rebeille et al., 2006; Goyer et al., 2007). The researchers found that this cytosolic enzyme is abundant in all plant organs, except in the seeds, and it catalyzes the conversion of methionine into methanethiol, α-ketobutyrate, and ammonia. Western blot studies have indicated that this gene is expressed under standard growth conditions and was strongly induced when the cells accumulated methionine (Rebeille et al., 2006). The enzyme has a relatively high K_m level for methionine (~10 mM), indicating that this pathway operates preferentially when methionine has accumulated above a certain value in the cytoplasm. Knocking out the methionine γ-lyase gene in *Arabidopsis* significantly increased leaf methionine content (9-fold) and SMM content under sulfate starvation but did not affect methionine level under normal growth conditions (Goyer et al., 2007). This finding suggests that this catabolic pathway plays a role during sulfate starvation, but since the level of methionine is not altered under this stress (Nikiforova et al., 2005), the situation is not clear and further studies are required.

Questions still remain regarding the role of methanethiol in the plant metabolism. Rebeille et al. (2006) suggest that 75% is released from cells as a volatile compound and 25% react with *O*-acetylserine (metabolites from the cysteine biosynthesis pathway) to produce *S*-methylcysteine. However, the authors could not identify metabolites of *S*-methylcysteine, and they suggest that this compound, which is produced in the cytoplasm, is rapidly transferred to vacuole and could play a storage role. However, Goyer et al. (2007) suggest, on the basis of their observations, that under low-sulfur content, about 9% of the radioactive methionine appeared as cysteine in the proteins and that methanethiol can convert to cysteine. They, therefore, assume that the methanethiol pathway is alternative to the reverse transsulfuration pathway operating in certain bacteria (Hacham et al., 2003; Saint-Girons et al., 1988) in which methionine converts to cysteine. In addition to methanethiol, the second catabolic product of methionine, α-ketobutyrate, produced in the cytosol can be transported to plastids and integrate the isoleucine biosynthesis pathway, which enhances its level in the methionine feeding plants (Rebeille et al., 2006).

Toward Improving Methionine Content in Plants for Enhanced Nutritional Quality

The Importance of Methionine in Human and Livestock Nutrition

Methionine is one of the 10 protein amino acids that humans and other animals are unable to synthesize and, hence, belongs to the essential amino acids. It belongs to those essential amino acids that are found in low content in plants and thus limits the nutritional quality of protein for humans and livestock worldwide (Tabe and Higgins, 1998). Among these nutritional-limiting essential amino acids, methionine is found at a low level in most crop plants but is particularly deficient in legume plants [e.g., soybean—*Glycine max* (L.) Merr., pea—*Pisum sativa* L., bean—*Phaseolus vulgaris* L., chickpea—*Cicer arietinum* L., alfalfa, lentil—*Lens culinaris* Medikus, clover—*Trifolium* spp.), which are among the most important nutritional sources of protein for humans and livestock (Galili et al., 2005). The biological value of a plant-based diet with limited methionine content can be equivalent to only 50 to 75% of that of a diet with balanced, essential amino acids. In cultures having a primarily vegetarian diet, or in developing countries in which plant-derived foods are predominant, this can lead to nonspecific signs of protein deficiencies in humans such as lowered resistance to disease, decreased blood proteins, and retarded mental and physical development in young children (Waterlow, 1975). This syndrome is referred to as Protein-Energy Malnutrition (PEM). The World Health Organization (WHO) estimates that around 30% of the populations in the developing world suffer from PEM. In addition, because methionine is one of the four main dietary sources of methyl groups, its deficiency can be associated with methylation-related disorders such as fatty liver, atherosclerosis, neurological disorders, and tumorigenesis (Poirier, 2002; Fukagawa, 2006; Fukagawa and Galbraith, 2004). Methionine through SAM also influences DNA synthesis and repairs the expression of genes; its deficiency has been associated with DNA fragmentation and strand breaks (Lertratanangkoon et al., 1996). In animals, methionine depletion lowers the threshold of chemical-induced toxicity, suggesting that this may be significant in carcinogenesis processes (Lertratanangkoon et al., 1996).

Methionine levels are generally not limited in human foods in western countries because of the significant consumption of livestock products, meat, eggs, and milk, which generally contain adequate levels of this essential amino acid. However, methionine deficiencies in plant-derived feeds for farm animals limit animal growth as well as animal products, such as reduced wool growth in sheep, milk production by dairy animals, and meat quality (Pickering and Reis, 1993;

Tabe et al., 1995; Xu et al., 1998). Broadbean (*Vicia faba* L.) and soybean seed proteins, for example, contain only 0.7 to 1.3 (g/100 g protein) methionine and 1.1 to 1.3 (g/100 g protein) cysteine, respectively, falling short of human and animal growth requirements, which are in the range of 3.5 (g/100 g protein) for both sulfur amino acids, methionine and cysteine (Shewry, 2000). Hence, the demand to improve pasture and forage legumes as sources of animal feed has grown recently (Habben and Larkins, 1995). In contrast to plants, animals can convert methionine to cysteine, but not conversely; hence, this defines methionine as an essential amino acid that can supply the complete requirement for sulfur amino acids in the diet (Tabe and Higgins, 1998).

To meet the requirements of monogastric animal diets, methionine has recently been added in a synthetic form to an animal-based diet in many western countries. Furthermore, methionine has also recently been added to processed soybean products (e.g., soybean milk and tofu) for human consumption. For ruminant animals, however, methionine must be supplied in the form of proteins that are resistant to rumen proteolysis, since unprotected dietary proteins are rapidly degraded by bacteria in the rumen and converted to bacterial proteins (Khan et al., 1996). For this reason, the sulfur-rich protein candidate for expression in transgenic forage legumes should be resistant to proteolytic degradation in the rumen (Bagga et al., 2004).

Because of the importance of methionine in human food and animal feed, many efforts have been made to produce plants having higher methionine content. However, traditional plant breeding methods have not been successful in increasing the level of sulfur amino acids. Recent developments in recombinant DNA technology, plant tissue culture, and in vitro regeneration are proposing new ways of increasing the level of essential amino acids, including Met, by manipulating existing genes and/or introducing foreign genes into plants.

Two main strategies have been used to increase methionine content by gene technology: through manipulation of the methionine biosynthetic pathway and the creation of additional protein "sinks" for methionine storage. Both molecular approaches have provided new insights into the control of the methionine level in plants and, in some cases, have yielded significantly higher levels of methionine and thus improved plant nutritional value. However, to further enhance methionine levels in different plant tissues, the methionine metabolism must be studied intensively. The point of its regulation along its biosynthesis and catabolism must be elucidated to tailor its optimal level in different tissues without causing an abnormal phenotype.

Factors Regulating Methionine Synthesis in Seeds

Analyses of *Arabidopsis* transgenic plants overexpressing AtCGS and *Arabidopsis* mutants, *mto1–1* and *mto2–1*, have shown that a relatively high level of methionine can be found in young rosette leaves. However, this accumulation is strongly dependent on development stage and organs (Inba et al., 1994; Bartlem et al., 2000; Kim et al., 2002). As the mutant plants began to flower, the levels of methionine and SMM declined gradually in the leaves while a high level of methionine was found in the reproductive tissues (Inba et al., 1994; Bartlem et al., 2000; Kim et al., 2002). These findings suggest that the soluble methionine accumulation in leaves during the vegetative growth period was translocated to the sink organs at the onset of reproductive growth and that methionine is most probably not synthesized in situ in these organs. Experiments using radioactive methionine have indeed shown that methionine is exported from leaves in the form of SMM, which is the major sulfur-metabolite in the phloem of different plants, including *Arabidopsis* (Bourgis et al., 1999). Moreover, the results suggested that SMM is the major donor for methionine required for the synthesis of proteins in wheat seeds (Bourgis et al., 1999).

SMM as described above is formed from methionine via the activity of SAM: Met S-methyltransferase (MMT) and can be reconverted to methionine by donating a methyl group to homocysteine in a reaction catalyzed by homocysteine S-methyltransferase (HMT) (Mudd and Datko, 1990) (Fig. 16–1). Each turn of this cycle consumes and then regenerates two methionines while hydrolyzing ATP, and thus this cycle is considered to be futile. Hence, the assumption proposed by Bourgis et al. (1999) that SMM is the major donor for methionine in wheat seeds demands a complete separation in space and time between the activities of MMT and HMT. Indeed, measurements and calculations have further suggested that in wheat (*Triticum aestivum* L.), the flux in leaves is mainly from methionine to SMM and from SMM to methionine in seeds (Bourgis et al., 1999). However, it has been shown that in *Arabidopsis* and maize (*Zea mays* L), MMT and HMT are coexpressed in leaves, roots. and developing seeds with a very similar activity ratio (Ranocha et al., 2001). Further studies have shown that the lack of SMM cycle in the *mmt* mutant of *Arabidopsis* does not alter the level of methionine, thiols, and seed sulfur content compared with wild-type seeds (Kocsis et al., 2003). This suggests that other sulfur sources such as glutathione or sulfate can replace SMM as a long-distance transporter.

While the debate about the role of SMM cycle in methionine accumulation in seeds is still ongoing, Tabe and Droux (2001) have demonstrated that lupine seeds were able to transfer the sulfur atom from radioactive sulfate into seed proteins, showing that seeds have the ability to perform all of the steps of sulfur assimilation and synthesis of methionine and cysteine. Accordingly, the activity measurements of enzymes that synthesized these amino acids were found to be relatively high in developing seeds when compared with leaves and sufficient to account for all the sulfur amino acids stored in proteins of the mature seeds (Tabe and Droux, 2001, 2002). Moreover, it was further found that an increase in sink for methionine in seeds in the form of methionine-rich seed storage proteins increases CGS activity in the transgenic lupine (Tabe and Droux, 2001). Indeed, the level of methionine doubled in these lupine seeds relative to nontransformed seeds (Molvig et al., 1997). These findings provide evidence that cysteine and methionine synthesized in seeds, at least in the case of grain legumes. Methionine is most probably synthesized in tobacco and narbon bean (*Vicia narbonensis* L.) seeds through the CGS activity since the enhanced flux of the carbon–amino skeleton toward its biosynthesis pathway, caused by the expression of bacterial AK under the seed-specific promoter, possesses small but significant increases of methionine content compared with wild-type seeds (Karchi et al., 1993; Demidov et al., 2003). This implies that the soluble methionine content limits the synthesis of methionine-rich storage proteins, as was previously suggested for transgenic rice (*Oryza sativa* L.), soybean, and chickpea (Townsend and Thomas, 1994; Tabe et al., 2002; Chiaiese et al., 2004).

Although it was previously suggested that methionine limits the synthesis of proteins, only minor efforts were invested to increase the soluble methionine content in seeds. As far as we know, the only results showing a dramatic elevation in soluble methionine content in seeds (up to 50-fold compared with wild-type seeds) are those found in *Arabidopsis* seeds, which have a significantly higher level of lysine (Zhu and Galili, 2003, 2004). In general, the results described here present current knowledge, which is far from being resolved, and it is clear that further studies are required to reveal the role of the SMM cycle and HMT in addition to the role of CGS in the methionine synthesis in seeds.

Accumulation of Methionine in Seed Proteins

In higher plants, the compositions of seed storage proteins depend on the availability of nutrients in the soil (Tabe et al., 2002). While nitrogen supply plays a dominant role in determining the total amount of proteins stored in seeds, the sulfur availability plays a fine-tuned role in the composition of these proteins. The effects of sulfur on the storage protein profile have been documented, particularly for legumes and cereal grains (Higgins et al., 1986; Tabe et al., 2002). The changes in sulfur-rich or poor-protein accumulation are attributed to the altered expression of genes encoding the storage protein response to signals, indicating the relative availability of nitrogen and sulfur in the plants. O-Acetylserine, a metabolite of the cysteine assimilation pathway, is thought to be a signal that changes gene expression and harmonizes the assimilation and utilization of nitrogen and sulfur in plants (Kim et al., 1999). The availability of methionine is also found to affect the expression level and stability of some other proteins (Holowach et al., 1984; Naito et al., 1994; Hirai et al., 1995). Accumulating evidence has shown that the signals affected the transcription process in some sulfur-rich proteins or posttranscription processes in the other genes, such as the pea albumin1 protein (Higgins et al., 1986; Morton et al., 1998). It was recently shown that posttranslational regulation also occurs in the storage proteins of *Arabidopsis* in the 12S globulin group, whereby members from this group are sulfur-rich proteins (Higashi et al., 2006). In addition, it was found that the accumulation of two sulfur-rich proteins, one of which belongs to the 12S globulin and the other to the 2S albumin, was suppressed under sulfur-deficient conditions and their regulation occurs at a posttranslation modification (Higashi et al., 2006).

These results demonstrated that plants sense environmental conditions, and in sulfur-deficient conditions, the sulfur-rich proteins decrease their level and the sulfur-poor proteins increase their accumulation (Hirai et al., 1995; Tabe et al., 2002). This mechanism was studied mainly in soybean seeds that contain approximately 35 to 40% protein, 70% of which are salt-soluble globulins, sulfur-poor 7S β-conglycinins, and sulfur-rich 11S glycinins. The 7S is comprised mainly of three subunits, whereby the β-subunit is regulated by the levels of sulfur and methionine through the repression of the transcription process when the methionine level increases (Holowach et al., 1984; Naito et al., 1994; Hirai et al., 1995). Similar to soybean proteins, the accumulations of the major seed proteins of pea were also affected by the sulfur level in plants (Beach et al., 1985). During recovery from sulfur deficiency, it was shown that the mRNA level of legumin, a sulfur-rich protein, increased 20-fold, while the transcript level of vicillin, a sulfur-poor protein, significantly decreased. The authors conclude that the regulation of legumin occurs at the posttranscriptional level, whereas that of vicillin occurs on the transcriptional level (Beach et al., 1985).

Recent data have suggested that the regulation between sulfur-rich and sulfur-poor proteins can be more complex, since by using different nitrogen fertilization regimes it was found that seeds having the lowest protein concentration had the highest sulfur amino acid content, and vice versa, seeds having a high nitrogen content have the lowest sulfur content. Notably, the researchers found that this phenomenon is not due to the sulfur-rich or poor storage protein expression but to the expression of the gene encoded for the Bowman-Birke protease inhibitor, which is a sulfur-rich protein (Krishnan et al., 2005). The researchers found that nitrogen had a negative influence on the expression of this gene, indicating that the negative correlation between total protein and sulfur amino acid content is mediated by the different accumulations of this inhibitor (Krishnan et al., 2005).

Taken together, these results show that the ratio between nitrogen and sulfur in the soil and the level of compounds such as O-acetylserine and methionine can modulate the accumulation of seed storage proteins and enable plants to respond to sulfur enrichment or sulfur deficiency.

Improving Methionine Levels in Seeds

To increase the nutritional value of plants by increasing their methionine content and to study the nature of the competition for methionine between SAM and its metabolites and protein synthesis in seeds, the researchers expressed genes encoding methionine-rich storage proteins in seeds. These attempts were performed in a number of plant species using a variety of genes. Such attempts were discussed in detail in several reviews (see, for example, Higgins et al., 1986; Altenbach and Simpson, 1990; Muntz, 1997; Amir and Tabe, 2006). In these studies, gene expression was almost exclusively directed by storage protein gene promoters, which enabled the seed-specific formation of the foreign protein during the period of storage-protein deposition of seed development. In most cases, genes for methionine-rich 2S albumin storage proteins were used, particularly Brazil nut 2S albumin, which contains 18% methionine, and a 2S albumin from sunflower with 16% methionine residues.

Expression of the 2S albumin Brazil nut protein under the control of the seed-specific promoter increased the methionine content of tobacco and canola (*Brassica napus* L.) seeds by 33%, in which the transgenic proteins accumulated to levels of 1.7 to 7% of total seed protein (Altenbach et al., 1989, 1992; Altenbach and Simpson, 1990). Attempts were also made to express this gene in legume seeds suffering from low content of sulfur amino acids. Similar results were obtained, showing that in narbon bean and soybean, the methionine level increased to 50 and 100% of total soluble seed protein, respectively (Pickard et al., 1995; Saalbach et al., 1995). This improvement brought the proportion of methionine in seed protein close to that required by animals for optimal growth. Seed-specific expression of 2S albumin of sunflower (*Helianthus annuus* L.) also significantly increased methionine content in the seeds of transgenic lupin, an important legume grain crop for animal feed in Australia (Molvig et al., 1997). The 2S albumin from sunflower seeds is a promising donor of sulfur amino acids because it has been shown to be resistant to microbial degradation in the rumen of sheep, thereby raising the possibility of increasing wool growth in sheep given supplementary feeding with transgenic lupins during periods of poor pasture quality (Molvig et al., 1997). Unfortunately, this Brazil nut protein and the 2S albumin of sunflower was subsequently found to be allergenic in some people, reducing the usefulness of this protein as a target for increasing plant nutritional quality (Bartolome et al., 1997; Kelly and Hefle, 2000).

Efforts were also made to increase methionine content in cereal grains. A 10-kDa high-methionine zein, which originates from maize kernels, was overexpressed in transgenic maize resulting in increases in total methionine in individual, first-generation kernels of up to 30% (Anthony et al., 1997). Similarly, expression of the sulfur-rich sesame 2S albumin in transgenic rice was reported to increase total seed methionine up to 75% compared with wild-type seed (Lee et al., 2003). In contrast, the high-level expression of sunflower 2S albumin in transgenic rice did not increase the methionine level in these seeds (Hagan et al., 2003). Instead, endogenous seed protein composition was markedly altered in a way that resembled the responses of seed protein composition to sulfur nutritional stress (Hagan et al., 2003). This means that sulfur-poor proteins became overabundant, while endogenous sulfur-rich proteins were strongly downregulated in these transgenic rice grains. The results suggested that the transgenic rice grains were unable to accumulate any more methionine in response to the added demand from the expression of the transgene. The reasons for the differences in the responses of rice to transgenic expression of the very similar 2S albumins from sunflower and sesame (*Sesamum indicum* L.) are not clear. However, compensatory changes in endogenous pools of sulfur were repeatedly observed when many different kinds of transgenic seeds expressing foreign, sulfur-rich proteins are closely examined (reviewed by Tabe et al., 2002). For example, endogenous sulfur-rich proteins were downregulated in transgenic soybean expressing the Brazil nut 2S protein and in transgenic lupins expressing the sunflower albumin (Jung, 1997; Tabe et al., 2002). In the case of transgenic lupins, levels of oxidized sulfur in mature transgenic seeds that accumulated the sunflower 2S protein decreased compared with wild-type seeds grown in similar conditions (Hagan et al., 2003). Correspondingly, the expression of this gene in chickpea enhanced the sulfur-assimilation rate but also appeared to downregulate the synthesis of endogenous sulfur-containing seed proteins (Chiaiese et al., 2004).

Taken together, the increases in seed methionine content have certainly resulted from the expression of foreign, methionine-rich proteins in some transgenic seeds, implying that the protein synthesis can compete with methionine–SAM metabolites for soluble methionine content (Tabe et al., 2002). However, in other transgenic seeds, the expression of methionine-rich protein affects the total seed sulfur pool, and the reallocation of methionine from endogenous proteins to the transgenes was found. This suggests that the level of soluble methionine limits the accumulation of foreign protein in seeds. To further study this point, Demidov et al. (2003) produced a double transformant of *Vicia narbonensis* that expressed the Brazil nut protein with plants having higher methionine content because of the expression of bacterial AK, both in a seed-specific manner. The double transformants exhibited additive effects on seed methionine content, showing that in mature seeds the protein-bound seeds reached levels 2.4 times higher than the wild-type seeds. These results suggest that in the future it will be worthwhile to express enzyme(s) that lead to the overaccumulation of soluble methionine together with methionine-rich storage protein to increase methionine content in the seeds.

Factors Regulating Methionine Levels in Vegetative Tissues: Increasing Methionine Content in Forage Crops

In addition to improving the nutritional quality of seeds, efforts were also made to improve the quality of vegetative tissues, with particular attention to forage legumes. To this end, the researchers used genes encoding methionine-rich seed storage proteins fused to a constitutive promoter. A different fate for methionine is likely to be in vegetative tissues than in seeds, since the primary difference between seeds and vegetative tissues is the commitment of the seeds to the synthesis and accumulation of seed-storage proteins, and hence less catabolism of methionine via SAM is expected in seeds (Giovanelli et al., 1985).

Sulfur-rich proteins may also be less stable in vegetative tissues than in seeds because in seeds they can accumulate in protein bodies derived from the endoplasmic reticulum (ER) or in storage vacuole-derived protein bodies, which may protect them from proteolysis. Whereas the ER-accumulating storage proteins are retained in the ER of vegetative cells, vacuolar proteins that targeted to vacuoles may enter the protease-rich vegetative vacuole and risk being degraded. This is probably why Brazil nut 2S albumin constitutively expressed in *Vicia narbonensis* and in tobacco plants could not accumulate to high levels in the vacuoles of mesophyll leaf cells (Bagga et al., 1995; Habben and Larkins, 1995). Proteins retained in the ER may thus be more stable in nonseed tissue simply because they are not exposed to vacuolar proteases. This hypothesis is supported by the fact that engineering an ER retention signal (KDEL) to the pea vicillin storage protein retained this vacuolar protein within the ER and stabilized it 100 times more than unmodified vicillin in leaves of transgenic plants (Wandelt et al., 1992).

Consistent with these findings are those obtained in transgenic plants (alfalfa, lotus— *Lotus corniculatus* L., tobacco and white clover—*Trifolium repens* L.) overexpressing the β-zein, γ-zein, and/or δ-zein. These methionine-rich storage proteins that originate from maize accumulate naturally in seed ER-derived protein bodies (Bagga et al., 2004; Sharma et al., 1998; Bellucci et al., 2000, 2002, 2005). Notably tobacco and alfalfa plants overexpressing the β-zein and the δ-zein produced novel ER-derived protein bodies in leaves that apparently protect them from degradation (Bagga et al., 1995, 2004). Coexpression of these two proteins together significantly increased the level and stability of the δ-zein (Bagga et al., 1995, 2004), implying that interactions between different zeins may be important for their accumulation (Kim et al., 2002). Using this knowledge about the ability of the γ-zein to produce protein bodies within leaves and thus protect the protein from degradation, Mainieri et al. (2004) designed a chimeric protein composed of phaseolin, the major seed protein of bean, and 89 amino acids of γ-zein. Unlike wild-type phaseolin, the protein, which they called zeolin, accumulates to very high amounts in leaves of transgenic tobacco. They conclude that the γ-zein portion is sufficient to induce the formation of protein bodies also when fused to another protein. Expression of this protein in other organelles such as chloroplasts led to an unstable protein, demonstrating the role of the ER in protecting the forging protein (Bellucci et al., 2007). Since the storage proteins of cereals and legumes nutritionally complement one other, zeolin can be used as a starting point for such manipulations in the future (Mainieri et al., 2004).

Expressing methionine-rich storage proteins in leaves, demonstrated, as in some of the transgenic seeds, that the elevation of foreign proteins was at the expense of the other sulfur-rich endogenous proteins as well as of other sulfur compounds in the cells. To further test whether the level of free methionine limits the production of sulfur-rich proteins in vegetative tissues, the transgenic alfalfa plants constitutively expressing the β-zein were crossed with those constitutively expressing the AtCGS that exhibit significantly higher levels of methionine (Avraham et al., 2005). Compared with plants expressing only the β-zein, those co-expressing both transgenes showed significantly enhanced levels of the β-zein, concurrently with a reduction in the level of soluble methionine when compared with plants expressing the AtCGS alone, implying that more soluble methionine was incorporated into the β-zein in the crossed plants (Bagga et al., 2005; Golan et al., 2005). The elevation of the β-zein in alfalfa plants is of particular nutritional importance to ruminant animals because this protein is resistant to rumen proteolysis (Bagga et al., 2004). Similar phenomena also occur in tobacco expressing these two genes, but they are considerably less pronounced (Bagga et al., 2005; Golan et al., 2005). The results showed that the accumulation of the β-zein is regulated in a species-specific manner and that soluble methionine plays a major role in the accumulation of the β-zein in some plant species but less so in others (Bagga et al., 2005; Golan et al., 2005). These studies also show that the soluble methionine content may limit the accumulation of methionine-rich storage proteins. Therefore, it is further suggested that to enhance the nutritional quality of forage plants by increasing their methionine content in proteins, one should express both a enzyme like CGS that lead to high methionine synthesis and methionine-rich storage proteins in the same plant tissues (Golan et al., 2005).

Future Prospects

Although major progress has been made in recent years to understand the factors that regulate methionine level in leaves and seeds, some control points of these regulations are still unclear. Since methionine is comprised of three different moieties, and thus three different pathways contribute to its synthesis, and since methionine is required for the production of various essential metabolites in addition to its role in protein synthesis, its metabolism is quite complicated. Some regulatory points of methionine metabolic network are still unknown. For example, it is not yet known what the major pathway that contributes to methionine synthesis is in seeds, whether it is the SMM cycle and the activity of HMT, or whether methionine is synthesized in seeds through the aspartate family by the CGS activity. In addition, in the regulatory role of the methionine recycling pathways, it is not yet clear how these pathways contribute to the content of soluble Met and how their rate is changed as a result of various conditions such as sulfur starvation, biotic and abiotic stresses, how they are altered in different organs, as well as their role during plant development. The knowledge accumulated of the methionine metabolism can contribute to our general knowledge of plant metabolism since Met-associated metabolites regulate different processes in the metabolism of plants and in their growth and development. This knowledge can be achieved by using transgenic plants and mutants that have higher and lower expression levels of genes encoding key enzymes in methionine metabolism pathways regulated by their own promoters or by different tissue-specific and stress-induced promoters. Genomics approaches, such as gene expression profiling in microarrays, proteomics, metabolic profiling, and flux analysis measurements, will greatly contribute to our knowledge in the future. Such approaches are already being used extensively with respect to the metabolism of sugars, lipids, and secondary metabolites, and it is expected that they will strongly penetrate into the metabolism of amino acids and sulfur metabolites. In addition to furthering our knowledge of plant metabolism, this knowledge will help in manipulating the methionine metabolism and in crop plants having higher levels of methionine, thus, improving nutritional quality.

References

Abdel-Kader, Z. 2000. Enrichment of Egyptian 'Balady' bread. Part 1. Baking studies, physical and sensory evaluation of enrichment with decorticated cracked broadbeans flour (*Vicia faba* L.). Nahrung 44:418–421.

Altenbach, S.B., B. Pearson, K.W. Meeker, G. Staraci, and L.C. Samuel. 1989. Enhancement of the methionine content of seed proteins by expression of a chimeric gene encoding a methionine-rich protein in transgenic plants. Plant Mol. Biol. 13:513–522.

Altenbach, S.B., and R.B. Simpson. 1990. Manipulation of methionine-rich protein genes in plant seeds. Trends Biotechnol. 8:156–160.

Altenbach, S.B., C.C. Kuo, L. Staraci, C.K.W. Pearson, C. Wainwright, and A. Georgescu. 1992. Accumulation of Brazil nut albumin in seeds of transgenic canola results in enhanced levels of seed protein methionine. Plant Mol. Biol. 18:235–245.

Amir, R., Y. Hacham, and G. Galili. 2002. Cystathionine g-synthase and threonine synthase operate in concert to regulate carbon flow towards methionine in plants. Trends Plant Sci. 7:153–156.

Amir, R., and L. Tabe. 2006. Molecular approaches to improving plant methionine content. p. 1–26. *In* K.J. Pawan et al. (ed.) Plant genetic engineering Vol. 8: Metabolic engineering and molecular farming II. Studium Press, Houston, TX.

Anthony, J., W. Brown, D. Buhr, G. Ronhovde, D. Genovesi, T. Lane, R. Yingling, K. Aves, M. Rosato, and P. Anderson. 1997. Transgenic maize with elevated 10 kD zein and methionine. p. 295–297. *In* W.J. Cram et al. (ed.) Sulfur metabolism in higher plants: Molecular, ecophysiological and nutritional aspects. Backhuys Publishers, Leiden, the Netherlands.

Avraham, T., and R. Amir. 2005. Methionine and threonine regulate the branching point of their biosynthesis pathways and thus controlling the level of each other. Transgenic Res. 14:299–311.

Avraham, T., H. Badani, S. Galili, and R. Amir. 2005. Enhanced levels of methionine and cysteine in transgenic alfalfa (*Medicago sativa* L.) plants over expressing the *Arabidopsis* cystathionine g-synthase gene. Plant Biotechnol. J. 3:71–80.

Azevedo, R.A., M. Lancien, and P.J. Lea. 2006. The aspartic acid metabolic pathway, an exciting and essential pathway in plants. Amino Acids 30:143–162.

Bagga, S. H. Adams, J.D. Kemp, and C. Sengupta-Gopalan. 1995. Accumulation of 15-kilodalton zein in novel protein bodies in transgenic tobacco. Plant Physiol. 107: 13–23.

Bagga, S., A. Armendaris, N. Klypina, I. Ray, S. Ghoshroy, M. Endress, D. Dutton, J.D. Kemp, and C. Sengupta-Gopalan. 2004. Genetic engineering ruminal stable high methionine protein in the foliage of alfalfa. Plant Sci. 166:273–283.

Bagga, S., C. Ponteza, J. Ross, M.N. Martin, T. Luestek, and C. Sengupta-Gopalan. 2005. A transgene for high methionine protein is post-transcriptionally regulated by methionine. In Vitro Cell. Dev. Biol. Plant 41:731–741.

Bartlem, D.L., I. Okamoto, T. Itaya, A. Uda, Y. Kijima, Y. Tamaki, E. Nambara, and S. Naito. 2000. Mutation in the threonine synthase gene results in an over-accumulation of soluble methionine in *Arabidopsis*. Plant Physiol. 123:101–110.

Bartolome, B., J. Mendez, A. Armentia, A. Vallverdu, and R. Palacios. 1997. Allergens from Brazil nut: Immunochemical characterization. Allergol. Immunopathol. (Madr.) 25:135–144.

Beach, L.R., D. Spencer, P.J. Randall, and T.J.V. Higgins. 1985. Transcriptional and post-transcriptional regulation of storage protein gene expressing in sulfur-deficient pea seeds. Nucleic Acids Res. 13:999–1013.

Bellucci, M., A. Alpini, and A. Arcioni. 2002. Zein accumulation in forage species (*Lotus coriculatus* and *Medicago sativa*) and co-expression of the gamma-zein:KDEL and beta-zein:KDEL polypeptides in tobacco leaf. Plant Cell Rep. 20:848–856.

Bellucci, M., A. Alpini, F. Paolocci, L. Cong, and S. Arcioni. 2000. Accumulation of maize gamma-zein and gamma-zein: KDEL to high levels in tobacco leaves and differential increase of Bip synthesis in transformants. Theor. Appl. Genet. 101:796–804.

Bellucci, M., M. De Marchis, R. Mannucci, R. Bock, and S. Arcioni. 2005. Cytoplasm and chloroplasts are not suitable subcellular locations for beta-zein accumulation in transgenic plants. J. Exp. Bot. 56:1205–1212.

Bellucci, M., M. De Marchis, I. Nicoletti, and S. Arcioni. 2007. Zeolin is a recombinant storage protein with different solubility and stability properties according to its localization in the endoplasmic reticulum or in the chloroplast. J. Biotechnol. 131:97–105.

Ben Tzvi-Tzchori, I., A. Perl, and G. Galili. 1996. Lysine and threonine metabolism are subject to complex patterns of regulation in *Arabidopsis*. Plant Mol. Biol. 32:727–734.

Blaszczyk, A., R. Brodzik, and A. Sirko. 1999. Increased resistance to oxidative stress in transgenic tobacco plants over expressing bacterial serine acetyltransferase. Plant J. 20:237–243.

Boerjan, W., M. Bauw, M.V. Montagu, and D. Inze. 1994. Distinct phenotypes generate by over expression and suppression of S-adenosyl-l-methionine synthetase reveal developmental patterns of gene silencing in tobacco. Plant Cell 6:1401–1414.

Bourgis, F., S. Roje, M.L. Nuccio, D.B. Fisher, M.C. Tarczynski, C. Li, C. Herschbach, H. Rennenberg, M.J. Pimenta, T.L. Shen, D.A. Gage, and A.D. Hanson. 1999. S-methylmethionine plays a major role in phloem sulfur transport and is synthesized by a novel type of methyltransferase. Plant Cell 11:1485–1498.

Brinch-Pedersen, H., G. Galili, S. Knudsen, and P.B. Holm. 1996. Engineering of the aspartate family biosynthetic pathway in barley (*Hordeum vulgare* L.) by transformation with heterologous genes encoding feed-back-insensitive aspartate kinase and dihydrodipicolinate synthase. Plant Mol. Biol. 32:611–620.

Bürstenbinder, K., G. Rzewuski, M. Wirtz, R. Hell, and M. Sauter. 2007. The role of methionine recycling for ethylene synthesis in *Arabidopsis*. Plant J. 49:238–249.

Cassol, T., and A. Mattoo. 2003. Do polyamines and ethylene interact to regulate plant growth, development and senscence. p. 22–32. *In* P. Nath et al. (ed.) Molecular insights in plant biology. Oxford-IBH, New Delhi.

Chiaiese, P., M. Ohkama-Ohtsu, L. Molvig, R. Godfree, H. Dove, C. Hocart, T. Fujiwara, T.J. Higgins, and L.M. Tabe. 2004. Sulphur and nitrogen nutrition influence the response of chickpea seeds to an added, transgenic sink for organic sulphur. J. Exp. Bot. 55:1889–1901.

Chiang, P.K., R. Gordon, J. Tal, G.C. Zeng, B.P. Doctor, K. Pardhasaradhi, and P.P. McCann. 1996. S-Adenosylmethionine and methylation. FASEB J. 10:471–480.

Chiba, Y., M. Ishikawa, F. Kijima, R.H. Tyson, J. Kim, A. Yamamoto, E. Nambara, T. Leustek, R.M. Wallsgrove, and S. Naito. 1999. Evidence for autoregulation of cystathionine gamma-synthase mRNA stability in *Arabidopsis*. Science 286:1371–1374.

Chiba, Y., R. Sakurai, M. Yoshino, K. Ominato, M. Ishikawa, H. Onouchi, and S. Naito. 2003. *S*-adenosyl-L-methionine is an effector in the post-transcriptional autoregulation of the cystathionine gamma-synthase gene in *Arabidopsis*. Proc. Natl. Acad. Sci. USA 100:10225–10230.

Curien, G., R. Dumas, S. Ravanel, and R. Douce. 1996. Characterization of an *Arabidopsis thaliana* cDNA encoding an *S*-asenosylmethionine-sensitive threonine synthase from higher plants. FEBS Lett. 390:85–90.

Curien, G., D. Job, R. Douce, and R. Dumas. 1998. Allosteric activation of *Arabidopsis* threonine synthase by S-adenosylmethionine. Biochemistry 37:13212–13221.

Curien, G., R. Ravanel, and R. Dumas. 2003. A kinetic model of the branch-point between the methionine and threonine biosynthesis pathways in *Arabidopsis thaliana*. Eur. J. Biochem. 270:4615–4627.

Demidov, D., C. Horstmann, M. Meixner, T. Pickard, I. Saalbach, G. Galili, and K. Muntz. 2003. Additive effects of the feed-back insensitive bacterial aspartate kinase and the Brazil nut 2S albumin on the methionine content of transgenic narbon bean (*Vicia narbonensis* L.). Mol. Breed. 11:187–201.

Di, R., J. Kim, M.N. Martin, T. Leustek, J. Jhoo, C.T. Ho, and N.E. Tumer. 2003. Enhancement of the primary flavor compound methional in potato by increasing the level of soluble methionine. J. Agric. Food Chem. 51:5695–5702.

Dias, B., and B. Weimer. 1998. Conversion of methionine to thiols by lactococci, lactobacilli, and brevibacteria. Appl. Environ. Microbiol. 64:3320–3326.

Droux, M. 2004. Sulfur assimilation and the role of sulfur in plants metabolism: A survey. Photosyn. Res. 79:331–348.

Droux, M., B. Gakiere, L. Denis, S. Ravanel, L. Tabe, A.G. Lappartient, and D. Job. 2000. Methionine biosynthesis in plants: Biochemical and regulatory aspects. Paul Haupt, Bern, Switzerland.

Eckermann, C., J. Eichel, and J. Schroder. 2000. Plant methionine synthase: New insights into properties and expression. Biol. Chem. 381:695–703.

Eichel, J., J.C. Gonzalez, M. Hotez, R.G. Matthews, and J. Schroder. 1995. Vitamin-B_{12} independent methionine synthase from higher plant (*Catharanthus roseus*). Eur. J. Biochem. 230:1053–1058.

Frankard, V., G. Ghislain, I. Negrutiu, and M. Jacobs. 1991. High threonine producer mutant of *Nicotiana sylvestris* (Spegazzini. and Comes). Theor. Appl. Genet. 82:273–282.

Fukagawa, N.K. 2006. Sparing of methionine requirements: Evaluation of human data takes sulfur amino acids beyond protein. J. Nutr. 136:1676S–1681S.

Fukagawa, N.K., and R. Galbraith. 2004. Advancing age and other factors influencing the balance between amino acid requirements and toxicity. J. Nutr. 134:1569S–1574S.

Gakiere, B., L. Denis, M. Droux, S. Ravanel, R. Douce, and D. Job. 2000. Methionine synthesis in higher plants: Sense strategy applied to cystathionine gamma-synthase and cystathionine beta-lyase in *Arabidopsis thaliana*. Paul Haupt, Bern, Switzerland.

Gakiere, B., S. Ravanel, M. Droux, R. Douce, and D. Job. 2000. Mechanisms to account for maintenance of the soluble methionine pool in transgenic *Arabidopsis* plants expressing antisense cystathionine gamma-synthase cDNA. C. R. Acad. Sci. III 323:841–851.

Galili, G., R. Amir, R. Hoefgen, and H. Hesse. 2005. Improving the levels of essential amino acids and sulfur metabolites in plants. Biol. Chem. 386:817–831.

Galili, S., D. Guenoune, S. Wilinger, and Y. Kapulnic. 2000. Enhanced levels of free and protein bound threonine in transgenic alfalfa (*Medicago sativa* L.) expressing a bacterial feed back insensitive aspartate kinase gene. Transgenic Res. 9:137–144.

Gigolashvili, T., R. Yatusevich, B. Berger, C. Müller, and U.I. Flügge. 2007. The R2R3-MYB transcription factor HAG1/MYB28 is a regulator of methionine-derived glucosinolate biosynthesis in *Arabidopsis thaliana*. Plant J. 51:247–261.

Giovanelli, J.G. 1987. Sulfur amino acids in plants: An overview. Methods Enzymol. 143:419–428.

Giovanelli, J.G., S.H. Mudd, and A.H. Datko. 1980. Sulfur amino acids in plants. p. 329–357. *In* B.J. Miflin (ed.) The biochemistry of plants. Academic Press, New York.

Giovanelli, J., S.H. Mudd, and A.H. Datko. 1985. Quantitative analysis of pathways of methionine metabolism and their regulation in *Lemna*. Plant Physiol. 78:555–560.

Giovanelli, J., G. Mudd, and H. Datko. 1985. *In vivo* regulation of de novo methionine biosynthesis in a higher plant (*Lemna*). Plant Physiol. 77:450–455.

Golan, A., T. Avraham, I. Matityahu, H. Badani, S. Galili, and R. Amir. 2005. Soluble methionine enhanced the accumulation of 15 kD zein, a methionine rich storage protein, in BY2 cells and in alfalfa transgenic plants but not in transgenic tobacco plants. J. Exp. Bot. 56:2443–2452.

Goto, D.B., M. Ogi, F. Kijima, T. Kumagai, F.V. Werven, H. Onouchi, and S. Naito. 2002. A single-nucleotide mutation in a gene encoding S-adenosylmethionine synthetase is associated with methionine over-accumulation phenotype in *Arabidopsis thaliana*. Genes Genet. Syst. 77:89–95.

Goyer, A., E. Collakova, Y. Shachar-Hill, and A.D. Hanson. 2007. Functional characterization of a methionine gamma-lyase in *Arabidopsis* and its implication in an alternative to the reverse trans-sulfuration pathway. Plant Cell Physiol. 48:232–242.

Habben, J.E., and B. Larkins. 1995. Improving protein quality in seeds. Marcel Dekker, New York.

Hacham, Y., T. Avraham, and R. Amir. 2002. The N-terminal region of *Arabidopsis* cystathionine gamma-synthase plays an important regulatory role in methionine metabolism. Plant Physiol. 128:454–462.

Hacham, Y., U. Gofna, and R. Amir. 2003. In vivo analysis of various substrates utilized by cystathionine {gamma}-synthase and O-acetylhomoserine sulfhydrylase in methionine biosynthesis. Mol. Biol. Evol. 20:1513–1520.

Hacham, Y., I. Matityahu, G. Schuster, and R. Amir. 2008. Overexpression of mutated forms of aspartate kinase and cystathionine γ-synthase in tobacco leaves resulted in the high accumulation of methionine and threonine. Plant J. 54:260–271.

Hacham, Y., G. Schuster, and R. Amir. 2006. An *in vivo* internal deletion in the n-terminus of cystathionine g-synthase in *Arabidopsis* results with decreased modulation of expression by methionine. Plant J. 45:955–967.

Hacham, Y., L. Song, G. Schuster, and R. Amir. 2007. Lysine enhances methionine content by modulating the expression of *S*-adenosylmethionine synthase. Plant J. 51:850–861.

Hagan, N.D., N. Upadhyaya, L.M. Tabe, and T.J. Higgins. 2003. The redistribution of protein sulfur in transgenic rice expressing a gene for a foreign, sulfur-rich protein. Plant J. 34:1–11.

Hanson, A.D., J. Rivoal, L. Paquet, and D.A. Gage. 1994. Biosynthesis of 3-dimethylsulfoniopropionate in *Wollastonia biflora* (L.) DC. Evidence that S-methylmethionine is an intermediate. Plant Physiol. 105:103–110.

Harms, K., P. von Ballmoos, C. Brunold, R. Hofgen, and H. Hesse. 2000. Expression of a bacterial serine acetyltransferase in transgenic potato plants leads to increased levels of cysteine and glutathione. Plant J. 22:335–343.

Hesse, H., and R. Hofgen. 2003. Molecular aspects of methionine biosynthesis. Trends Plant Sci. 8:259–262.

Hesse, H., O. Kerft, S. Maimann, M. Zeh, and R. Hoefgen. 2004. Current understanding of the regulation of methionine biosynthesis in plants. J. Exp. Bot. 55:1799–1808.

Higashi, Y., M. Hirai, T. Fujiwara, S. Naito, M. Noji, and K. Saito. 2006. Proteomic and transcriptomic analysis of *Arabidopsis* seeds: Molecular evidence for successive processing of seed proteins and its implication in the stress response to sulfur nutrition. Plant J. 48:557–571.

Higgins, T.J., P. Chandler, P.J. Randall, D. Spencer, L.R. Beach, R.J. Blagrove, A.A. Kortt, and A.S. Inglis. 1986. Gene structure, protein structure, and regulation of the synthesis of a sulfur-rich protein in pea seeds. J. Biol. Chem. 261:11124–11130.

Hirai, M.Y., T. Fujiwara, M. Chino, and S. Naito. 1995. Effects of sulfate concentrations on the expression of a soybean seed storage protein gene and its reversibility in transgenic *Arabidopsis thaliana*. Plant Cell Physiol. 36:1331–1339.

Hirai, M.Y., K. Sugiyama, Y. Sawada, T. Tohge, T. Obayashi, A. Suzuki, R. Araki, N. Sakurai, H. Suzuki, K. Aoki, H. Goda, O.I. Nishizawa, D. Shibata, and K. Saito. 2007. Omics-based identification of *Arabidopsis* Myb transcription factors regulating aliphatic glucosinolate biosynthesis. Proc. Natl. Acad. Sci. USA 104:6478–6483.

Holowach, L.P., J. Thompson, and J.T. Madison. 1984. Storage protein composition of soybean cotyledons grown in vitro in media of various sulfate concentrations in the presence and absence of exogenous l-methionine. Plant Physiol. 74:584–589.

Inba, K., T. Fujiwara, H. Hayashi, M. Chino, Y. Komeda, and S. Naito. 1994. Isolation of an *Arabidopsis thaliana* mutant, mto1, that over accumulates soluble methionine. Temporal and spatial patterns of soluble methionine accumulation. Plant Physiol. 104:881–887.

Jung, R. 1997. Expression of a 2S albumin from *Bertholletia excelsain* Soybean. The 39th NIBB Conference: Dynamic aspects of seed maturation and germination, National Institute for Basic Biology Okazaki, Japan.

Kagan, R.M., and S. Clarke. 1994. Widespread occurrence of three sequence motifs in diverse S-adenosylmethionine-dependent methyltransferases suggests a common structure for these enzymes. Arch. Biochem. Biophys. 310:417–427.

Karchi, H., O. Shaoul, and G. Galili. 1993. Seed specific expression of a bacterial desensitized aspartate kinase increases the production of seed threonine and methionine in transgenic tobacco. Plant J. 3:721–727.

Katz, Y.S., G. Galili, and R. Amir. 2006. Regulatory role of cystathionine-gamma-synthase and de novo synthesis of methionine in ethylene production during tomato fruit ripening. Plant Mol. Biol. 61:255–268.

Kaur-Sawhney, R., T. Albabella, A.F. Tiburcio, and W. Galston. 2003. Plyamines in plants: An overview. J. Cell. Mol. Biol. 2:1–12.

Kelly, J.D., and S. Hefle. 2000. 2S methionine-rich protein (SSA) from sunflower seed is an IgE-binding protein. Allergy 55:556–560.

Khan, M.R., A. Ceriotti, L. Tabe, A. Aryan, W. McNabb, A. Moore, S. Craig, D. Spencer, and T.J. Higgins. 1996. Accumulation of a sulphur-rich seed albumin from sunflower in the leaves of transgenic subterranean clover (*Trifolium subterraneum* L.). Transgenic Res. 5:179–185.

Kim, H., M. Hirai, H. Hayashi, M. Chino, S. Naito, and T. Fujiwara. 1999. Role of O-acetyl-l-serine in the coordinated regulation of the expression of a soybean seed storage-protein gene by sulfur and nitrogen nutrition. Planta 209:282–289.

Kim, J., M. Lee, R. Chalam, M.N. Martin, T. Leustek, and W. Boerjan. 2002. Constitutive over expression of cystathionine g-synthase in *Arabidopsis thaliana* leads to accumulation of soluble methionine and S-methylmethionine. Plant Physiol. 128:95–107.

Kim, J., and T. Leustek. 1996. Cloning and analysis of the gene for cystathionine g-synthase from *Arabidopsis thaliana*. Plant Mol. Biol. 32:1117–1124.

Kim, C.S., Y. Woo, A.M. Clore, R.J. Burnett, N.P. Carneiro, and B.A. Larkins. 2002. Zein protein interactions, rather than the asymmetric distribution of zein mRNAs on endoplasmic reticulum membranes, influence protein body formation in maize endosperm. Plant Cell Physiol. 14:655–672.

Kocsis, M.G., P. Nolte, D.A. Gage, E.S. Simon, D. Rhodes, G.J. Peel, S. Mellema, K. Saito, M. Awazuhara, C. Li, R.B. Meeley, M.C. Tarczynski, C. Wagner, and A.D. Hanson. 2003. Insertional inactivation of the methionine S-methyltransferase gene eliminates the S-methylmethionine cycle and increases the methylation ratio. Plant Physiol. 131:1808–1815.

Kreft, O., R. Hofgen, and H. Hesse. 2003. Functional analysis of cystathionine gamma-synthase in genetically engineered potato plants. Plant Physiol. 131:1843–1854.

Krishnan, H.B., J. Bennett, W.S. Kim, A.H. Krishnan, and T.P. Mawhinney. 2005. Nitrogen lowers the sulfur amino acid content of soybean [*Glycine max* (L.) Merr.] by regulating the accumulation of Bowman-Birk protease inhibitor. J. Agric. Food Chem. 53:6347–6354.

Lee, M., M. Martin, A.O. Hudson, J. Lee, M.J. Muhitch, and T. Leustek. 2005. Methionine and threonine synthesis are limited by homoserine availability and not the activity of homoserine kinase in *Arabidopsis thaliana*. Plant J. 41:685–696.

Lee, T.T., M. Wang, R.C. Hou, L.J. Chen, R.C. Su, C.S. Wang, and J.T. Tzen. 2003. Enhanced methionine and cysteine levels in transgenic rice seeds by the accumulation of sesame 2S albumin. Biosci. Biotechnol. Biochem. 67:1699–1705.

Lertratanangkoon, K., R. Orkiszewski, and J.M. Scimeca. 1996. Methyl-donor deficiency due to chemically induced glutathione depletion. Cancer Res. 56:995–1005.

Loizeau, K., B. Gambonnet, G.F. Zhang, G. Curien, S. Jabrin, D. Van Der Straeten, W.E. Lambert, F. Rébeillé, and S. Ravanel. 2007. Regulation of C1 metabolism in *Arabidopsis*: The N-terminal regulatory domain of cystathionine {gamma} synthase is cleaved in response to folate starvation. Plant Physiol. 145:491–503.

Maimann, S., R. Hofgen, and H. Hesse. 2001. Enhanced cystathionine beta-lyase activity in transgenic potato plants does not force metabolite flow towards methionine. Planta 214:163–170.

Mainieri, D., M. Rossi, M. Archinti, M. Bellucci, F. De Marchis, S. Vavassori, A. Pompa, S. Arcioni, and A. Vitale. 2004. Zeolin. A new recombinant storage protein constructed using maize gamma-zein and bean phaseolin. Plant Physiol. 136:3447–3456.

Matilla, A.J. 2000. Ethylene in seed formation and germination. Seed Sci. Res. 6:111–126.

Matityahu, I., L. Kachan, I. Bar Ilan, and R. Amir. 2005. Transgenic tobacco plants over expressing the Met25 gene of *Saccharomyces cerevisiae* exhibit enhanced levels of cysteine and glutathione and increased tolerance to oxidative stress. Amino Acids 30:185–194.

Mills, W.R. 1980. Photosynthetic formation of the aspartate family of amino acids in isolated chloroplasts. Plant Physiol. 65:1166–1172.

Miyazaki, J., and S.F. Yang. 1987. The methionine salvage pathway in relation to ethylene and polyamine biosynthesis. Physiol. Plant. 69:366–370.

Molvig, L., L. Tabe, B.O. Eggum, A. Moore, S. Craig, D. Spencer, and T.J. Higgins. 1997. Enhanced methionine level and increased nutritive value of seeds of transgenic lupins (*Lupinus angustifolius* L.) expressing a sunflower seed albumin gene. Proc. Natl. Acad. Sci. USA 94:8393–8398.

Morton, R.L., A. Ellery, and T.J. Higgins. 1998. Downstream elements from the pea albumin 1 gene confer sulfur responsiveness on a reporter gene. Mol. Gen. Genet. 259:309–316.

Mudd, H.D., and A.H. Datko. 1990. The S-methylmethionine cycle in *Lemna paucicostata*. Plant Physiol. 93:623–630.

Mudd, S.H., and A.H. Datko. 1986. Methionine methyl group metabolism in *Lemna*. Plant Physiol. 81:103–114.

Muntz, K. 1997. How does the seed's sulphur metabolism react on high level formation of foreign methionine rich proteins in transgenic narbon bean (*Vicia narbonensis* L.)? p. 14–15. *In* The 39th NIBB Conference: Dynamic aspects of seed maturation and germination. National Institute for Basic Biology, Okazaki, Japan.

Naito, S., M. Hirai, M. Chino, and Y. Komeda. 1994. Expression of a soybean (*Glycine max* [L.] Merr.) seed storage protein gene in transgenic *Arabidopsis thaliana* and its response to nutritional stress and to abscisic acid mutations. Plant Physiol. 104:497–503.

Naito, S., K. Inaba-Higano, T. Kumagai, T. Kanno, E. Nambara, T. Fujiwara, M. Chino, and Y. Komeda. 1994. Maternal effects of *mto1* mutation, that causes over accumulation of soluble methionine, on the expression of soybean beta-conglycinin gene promoter-GUS fusion in transgenic *Arabidopsis thaliana*. Plant Cell Physiol. 35:1057–1063.

Nikiforova, V., J. Freitag, S. Kempa, M. Adamik, H. Hesse, and R. Hoefgen. 2003. Transcriptome analysis of sulfur depletion in *Arabidopsis thaliana*: Interlacing of biosynthetic pathways provides response specificity. Plant J. 33:633–650.

Nikiforova, V., S. Kempa, M. Zeh, S. Maimann, O. Kreft, A.P. Casazza, K. Riedel, E. Tauberger, R. Hoefgen, and H. Hesse. 2002. Engineering of cysteine and methionine biosynthesis in potato. Amino Acids 22:259–278.

Nikiforova, V.J., M. Bielecka, B. Gakière, S. Krueger, J. Rinder, S. Kempa, R. Morcuende, W.R. Scheible, H. Hesse, and R. Hoefgen. 2006. Effect of sulfur availability on the integrity of amino acid biosynthesis in plants. Amino Acids 30:173–183.

Nikiforova, V.J., C.O. Daub, H. Hesse, L. Willmitzer, and R. Hoefgen. 2005. Integrative gene-metabolite network with implemented causality deciphers informational fluxes of sulphur stress response. J. Exp. Bot. 56: 1887–1896.

Nikiforova, V.J., V. Tolstikov, O. Fiehn, L. Hopkins, M.J. Hawkesford, H. Hesse, and R. Hoefgen. 2005. Systems rebalancing of metabolism in response to sulfur deprivation, as revealed by metabolome analysis of *Arabidopsis* plants. Plant Physiol. 138:304–318.

Noji, M., M. Saito, M. Nakamura, M. Aono, H. Saji, and K. Saito. 2001. Cysteine synthase over expression in tobacco confers tolerance to sulfur-containing environmental pollutants. Plant Physiol. 126:973–980.

Ohkama-Ohtsu, N., P. Zhao, C. Xiang, and D.J. Oliver. 2007. Glutathione-conjugates in the vacuole are degraded by glutamyl transpeptidase GGT3 in *Arabidopsis*. Plant J. 49:878–888.

Onouchi, H., I. Lambein, R. Sakurai, A. Suzuki, Y. Chiba, and S. Naito. 2004. Autoregulation of the gene for cystathionine gamma-synthase in *Arabidopsis*: Post-transcriptional regulation induced by S-adenosylmethionine. Biochem. Soc. Trans. 32:597–600.

Onouchi, H., Y. Nagami, Y. Haraguchi, M. Nakamoto, Y. Nishimura, R. Sakurai, N. Nagao, D. Kawasaki, Y. Kadokura, and S. Naito. 2005. Nascent peptide-mediated translation elongation arrest coupled with mRNA degradation in the CGS1 gene of *Arabidopsis*. Genes Dev. 19:1799–1810.

Pickard, T., I. Saalbach, D.R. Waddell, M. Meixner, K. Muntz, and O. Schieder. 1995. Seed specific expression of the 2S albumin gene from Brazil nut (*Bertholletia excelsa* H.B.K.) in transgenic *Vicia narbonensis*. Mol. Breed. 1:295–301.

Pickering, F.S., and P. Reis. 1993. Effects of abomasal supplements of methionine on wool growth of grazing sheep. Aust. J. Exp. Agric. 33:7–12.

Pimenta, M., K. Kaneta, T. Larondelle, Y. Dohmae, and N. Kamiya. 1998. S-adenosyl-L-methionine:L-methionine S-methyltransferase from germinating barley. Purification and localization. Plant Physiol. 118:431–438.

Poirier, L. 2002. The effects of diet, genetics and chemicals on toxicity and aberrant DNA methylation: An introduction. J. Nutr. 132:2336S–2339S.

Ranocha, P., S. McNeil, M.J. Ziemak, C. Li, M.C. Tarczynski, and A.D. Hanson. 2001. The S-methylmethionine cycle in angiosperms: Ubiquity, antiquity and activity. Plant J. 25:575–584.

Ravanel, S., M. Block, P. Rippert, S. Jabrin, G. Curien, F. Rebeille, and R. Douce. 2004. Methionine metabolism in plants: Chloroplasts are autonomous for de novo methionine synthesis and can import S-adenosylmethionine from the cytosol. J. Biol. Chem. 279:22548–22557.

Ravanel, S., M. Droux, and R. Douce. 1995. Methionine biosynthesis in higher plants. I. Purification and characterization of cystathionine gamma-synthase from spinach chloroplasts. Arch. Biochem. Biophys. 316:572–584.

Ravanel, S., B. Gakiere, D. Job, and R. Douce. 1998a. Cystathionine gama-synthase from *Arabidopsis thaliana*: purification and biochemical characterization of the recombinant enzyme over expressed in *Escherichia coli*. Biochem. J. 331:639–648.

Ravanel, S., B. Gakiere, D. Job, and R. Douce. 1998b. The specific features of methionine biosynthesis and metabolism in plants. Proc. Natl. Acad. Sci. USA 95:7805–7812.

Rebeille, F., S. Jabrin, R. Bligny, K. Loizeau, B. Gambonnet, V. Van Wilder, R. Douce, and S. Ravanel. 2006. Methionine catabolism in *Arabidopsis* cells is initiated by a gamma-cleavage process and leads to S-methylcysteine and isoleucine syntheses. Proc. Natl. Acad. Sci. USA 103:15687–15692.

Roje, S. 2006. S-Adenosyl-l-methionine: Beyond the universal methyl group donor. Phytochemistry 67:1686–1698.

Saalbach, I., T. Pickard, D.R. Waddell, S. Hillmer, O. Schieder, and K. Muntz. 1995. The sulphur-rich Brazil nut 2S albumin is specifically formed in transgenic seeds of the grain legume *Vicia narbonensis*. Euphytica 85:181–192.

Saalbach, I., D. Waddell, and D. Schieder. 1995. Stable expression of the sulphur rich 2S albumin gene in transgenic *Vicia narbonensis* increases the methionine content of seeds. J. Plant Physiol. 145:674–681.

Saint-Girons, I., C. Parsot, M.M. Zakin, O. Barzu, and G.N. Cohen. 1988. Methionine biosynthesis in enterobacteriaceae: Biochemical, regulatory, and evolutionary aspects. CRC Crit. Rev. Biochem. 23(Suppl. 1):S1–S42.

Saito, K., M. Kurosawa, K. Tatsuguchi, Y. Takagi, and I. Murakoshi. 1994. [O-acetylserine(thiol)-lyase] Modulation of cysteine biosynthesis in chloroplasts of transgenic tobacco over expressing cysteine synthase. Plant Physiol. 106:887–895.

Schmidt, A., H. Rennenberg, L.G. Wilson, and P. Filner. 1985. Formation of methanethiol from methionine by leaf tissue. Phytochemistry 24:1181–1185.

Sharma, S.B., K. Hancock, P.M. Ealing, and D.W. White. 1998. Expression of a sulfur-rich maize seed storage protein, delta-zein, in white clover (*Trifolium repens*) to improve forage quality. Mol. Breed. 4:435–448.

Shaul, O., and G. Galili. 1992. Threonine overproduction in transgenic tobacco plants expressing a mutant desensitized aspartate kinase from *Escherichia coli*. Plant Physiol. 100:1157–1163.

Shen, B., C. Li, and M.C. Tarczynski. 2002. High free-methionine and decreased lignin content result from a mutation in the *Arabidopsis* S-adenosyl-L-methionine synthetase 3 gene. Plant J. 29:371–380.

Shewry, P.R. 2000. Seed p. p. 42–84. *In* M.A.B. Black and J.D. Sheffield (ed.) Seed technology and its biological basis. Academic Press, Sheffield, UK.

Stiller, I., G. Dancs, H. Hesse, R. Hoefgen, and Z. Bánfalvi. 2007. Improving the nutritive value of tubers: Elevation of cysteine and glutathione contents in the potato cultivar White Lady by marker-free transformation. J. Biotechnol. 128:335–343.

Tabe, L., N. Hagan, and T.J. Higgins. 2002. Plasticity of seed protein composition in response to nitrogen and sulfur availability. Curr. Opin. Plant Biol. 5:212–217.

Tabe, L.M., and T.J. Higgins. 1998. Engineering plant protein composition for improved nutrition. Trends Plant Sci. 3:282–286.

Tabe, L.M., T. Wardley-Richardson, A. Ceriotti, A. Aryan, W. McNabb, A. Moore, and T.J. Higgins. 1995. A biotechnological approach to improving the nutritive value of alfalfa. J. Anim. Sci. 73:2752–2759.

Tabe, M., and M. Droux. 2001. Sulfur assimilation in developing lupin cotyledons could contribute significantly to the accumulation of organic sulfur reserves in the seed. Plant Physiol. 126:176–187.

Tabe, M., and M. Droux. 2002. Limits to sulfur accumulation in transgenic lupin seeds expressing a foreign sulfur-rich protein. Plant Physiol. 128:1137–1148.

Thompson, G., A.H. Datko, and H. Mudd. 1982. Methionine synthesis in *Lemna*. Inhibition of cystathionine gamma synthase by proparglyglycine. Plant Physiol. 70:1347–1352.

Townsend, J.A., and L. Thomas. 1994. Factors which influence the *Agrobacterium*-mediated transformation of soybean. J. Cell. Biochem. Suppl. 18A:78.

Wandelt, C., R. Rafiqul, I. Khan, S. Craig, H.E. Schroeder, D. Spencer, and T.J. Higgins. 1992. Vicilin with carboxy terminal KDEL is retained in the endoplasmic reticulum and accumulates to high levels in the leaves of transgenic plants. Plant J. 2:181–192.

Waterlow, J.C. 1975. The protein gap. Nature 258:113–117.

Wirtz, M., and R. Hell. 2003. Production of cysteine for bacterial and plant biotechnology: Application of cysteine feedback-insensitive isoforms of serine acetyltransferase. Amino Acids 24:195–203.

Xu, S., J. Harrison, W. Chalupa, C. Sniffen, W. Julien, H. Sato, T. Fujieda, K. Watanabe, T. Ueda, and H. Suzuki. 1998. The effect of ruminal bypass lysine and methionine on milk yield and composition of lactating cows. Dairy Sci. 81:1062–1077.

Yang, S.F., and N.E. Hoffman. 1984. Ethylene biosynthesis and its regulation in higher plants. Annu. Rev. Plant Physiol. 35:155–189.

Yang, S.F., W.K. Yip, and J.G. Dong. 1990. Mechanism and regulation of ethylene biosynthesis. Am. Soc. Plant Physiol. 5:24–35.

Zeh, M., A. Casazza, O. Kreft, U. Roessner, K. Bieberich, L. Willmitzer, R. Hoefgen, and H. Hesse. 2001. Antisense inhibition of threonine synthase leads to high methionine content in transgenic potato plants. Plant Physiol. 127:792–802.

Zeh, M., G. Leggewie, R. Hoefgen, and H. Hesse. 2002. Cloning and characterization of a cDNA encoding a cobalamin-independent methionine synthase from potato (*Solanum tuberosum* L.). Plant Mol. Biol. 48:255–265.

Zhu, X., and G. Galili. 2003. Increased lysine synthesis coupled with a knockout of its catabolism synergistically boosts lysine content and also transregulates the metabolism of other amino acids in *Arabidopsis* seeds. Plant Cell 15:845–853.

Zhu, X., and G. Galili. 2004. Lysine metabolism is concurrently regulated by synthesis and catabolism in both reproductive and vegetative tissues. Plant Physiol. 135:129–136.

17

Plant Sulfur Compounds and Human Health

Joseph M. Jez
Donald Danforth Plant Science Center, St. Louis, Missouri

Naomi K. Fukagawa
University of Vermont College of Medicine, Burlington

Abstract

The contribution of sulfur to plant growth, the synthesis of sulfur-containing compounds by plants, and the consumption of plant foodstuffs by humans ultimately affect human health. This chapter provides an overview of the nutritional requirements and dietary sources of the sulfur amino acids, methionine and cysteine, for the human body. Multiple diseases and disorders have components associated with sulfur amino acid metabolism. In addition, the allyl sulfur compounds and the isothiocyanates in plants are a rich source of chemically diverse molecules with potential therapeutic activities, including cancer preventative effects.

Although sulfur is the seventh most abundant element in the human body, and the sulfur amino acids (methionine and cysteine) are required in the diet, compendiums of human nutrition and food composition often overlook this element and its role in health and nutrition (Ingenbleek, 2006). Recent insights on the connections between sulfur-containing compounds and human health, including cancer prevention, proper functioning of the immune system, and development of cardiovascular disease, have stirred a resurgence of interest in sulfur amino acid metabolism in humans.

Mammals do not produce either methionine or cysteine de novo and require the intake of both amino acids, or at least methionine, from dietary sources (Griffith, 1987). Thus, methionine is one of the nine essential amino acids in human health. Cysteine is considered a "conditionally" essential amino acid since under most circumstances the metabolism of methionine in the body can provide it in the necessary amounts. Because the body can convert methionine to cysteine, these two amino acids are considered as nutritionally equivalent compounds with respect to sulfur. Plants, which synthesize both methionine and cysteine, are the ultimate source of these amino acids in the food chains of animals and humans.

Plants are also a rich source of other sulfur-containing compounds with possible health benefits. Researchers have explored the allylsulfur compounds found in garlic as antithrombotic and anticancer treatments (Milner, 2001). Likewise, the molecular diversity of the glucosinolates and isothiocyanates found in cruciferous vegetables, including broccoli (*Brassica oleracea* var. *botrytis* L.) and Brussels sprouts (*Brassica oleracea* var. *gemmifera* DC), may have benefits for prevention of multiple types of cancer (Fahey et al., 2001).

The Sulfur Amino Acids: Methionine and Cysteine

Nutritional Requirements and Dietary Sources

Eating plant and animal foodstuffs provides sulfur amino acids in the human diet. Consumed plant and animal food proteins are broken down into component amino acids in the gastrointestinal tract by exposure to digestive acids and hydrolysis of peptide bonds by proteases. The typical Western diet, containing a balance of meats, vegetables, and fruits, fulfills the recommended dietary allowances for both methionine and cysteine (Table 17–1) to replace losses that result from catabolism (Fukagawa and Galbraith, 2004; Fukagawa, 2006). For an average 70 kg adult male, the average protein intake level is 0.75 g per kg body weight (or 52 g per day). Of this, the recommended minimum sulfur amino acid intake is 13 to 16 mg per kg body weight (or 910–1120 mg per day). An average Western diet includes 3.6 g per day of the sulfur amino acids (roughly two-thirds methionine and one-third cysteine) (Ingenbleek, 2006). Absorption of methionine and cysteine in the gastrointestinal tract is rapid and efficient (Shoveller et al., 2005; Burrin and Stoll, 2007). Examination of the sulfur amino acid content of various foods shows that methionine and cysteine are often limiting essential amino acids in many plant-derived foods (Table 17–2). Thus, issues of sulfur amino acid deficiency arise in the diets of strict vegans and in the developing world where the intake of cereals lacking methionine and cysteine is common.

The recommended dietary intake of methionine and cysteine (Table 17–1) is based on the diet and activities of an "average" adult male. An unclear question is whether these requirements change with age. With aging, changes in body composition, decreases in energy needs and physical activity, increased morbidity, and disease likely contribute to alterations in how the human body uses

Table 17–1. Adult essential amino acid requirements.†

Amino acid	Requirement
	mg kg^{-1} d^{-1}
Methionine, cysteine	13–16
Phenylalanine, tyrosine	14–39
Leucine	14–40
Lysine	12–30
Isoleucine	10–23
Valine	10–20
Threonine	7–16
Tryptophan	3.2–6
Histidine	8–12

† Data show the range of recommended daily intake from studies by the FAO/WHO/UNU (1985), Fereday et al. (1997), Millward et al. (1997), and Young and Borgonha (2000).

Table 17–2. Sulfur and sulfur amino acid contents in plant and meat food sources.†

Food	Met	Cys	Sulfur
		mg per 100 g	
Avocado (fruit)	43	6	11
Banana (fruit)	9	2	2
Beans (white)	234	188	100
Beef	650	280	214
Carrot	14	12	6
Cassava (meal)	12	9	5
Cheese, cheddar	390	110	113
Chicken, meat	710	330	195
Chicken, egg	470	290	178
Corn (grain)	120	97	51
Fish, cod	600	250	189
Fish, salmon	700	290	227
Fish, herring	621	243	198
Lamb	600	280	203
Lentils	194	221	100
Milk, bovine	86	28	26
Pork	720	310	237
Rice (grain)	133	96	54
Potato	26	12	9
Spinach (leaf)	43	36	19
Soy (bean)	580	590	280
Sweet potato	222	14	8
Tomato	7	7	3
Wheat	94	159	62

† Data compiled from the FAO (1970), McCance and Widdowson (1991), Souci et al. (1994), and Ingenbleek (2006).

dietary amino acids (Fukagawa and Galbraith, 2004). Currently, an incomplete understanding of the interplay of sulfur amino acid metabolism and diet make it difficult to determine the optimal intake with changes in age. By contrast, the need for methionine and cysteine in the diet during pregnancy and early childhood is even more important than in later life, as the demands of growth and development are higher earlier in life. The direct metabolic link between methionine metabolism and the synthesis of folic acid, an essential nutrient during pregnancy (Rees, 2002), can have severe consequences. Lack of sulfur amino acids leads directly to decreased folic acid production, which results in an increased incidence of neural tube defects (Rees et al., 2006). Adequate sulfur amino acid nutrition is essential at all stages of life because insufficient intake causes deficiencies; however, high plasma levels may have detrimental effects (van de Poll et al., 2006). Ultimately, a well-balanced diet containing a mix of meat, fruits, and vegetables meets the nutritional needs of a healthy body.

Uses of Sulfur Amino Acids in the Body

Methionine and cysteine, along with other amino acids obtained from digestion, are the major source of materials for protein synthesis in the body. Dietary

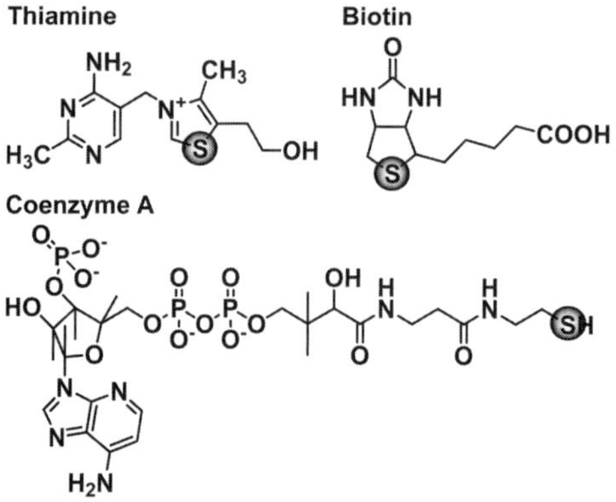

Fig. 17–1. Structures of sulfur-containing vitamins—thiamin, biotin, coenzyme A. The location of sulfur in each structure is highlighted.

consumption supplies these building blocks of proteins. Dietary availability of amino acids reduces the energy demand for de novo production of the nonessential amino acids in the body. Food protein with a composition that allows for resynthesis in the gastrointestinal tract and the gradual release of amino acids to the liver and the body is of higher value than those with a faster degradation. Slower amino acid release from the gut results in more efficient metabolism with a decreased loss of nutritional components to excretion (van de Poll et al., 2006).

The sulfur amino acids are the primary route for the body to incorporate sulfur into a vast array of molecules. They serve as metabolic precursors in production of the glycosamineglycans (chondroitin sulfate, dermata sulfate, and hyaluronic acid), which are important components in connective tissues like cartilage and skin (Komarnisky et al., 2003). Likewise, the synthesis of vitamins containing sulfur, including thiamin, biotin, and coenzyme A (Fig. 17–1), depends on available sulfur amino acid supplies. The metabolism of methionine and cysteine also links into pathways involved in the synthesis of glutathione (an important antioxidant and xenobiotic detoxification molecule), S-adenosylmethionine (a universal methyl group donor), and folic acid. In addition to serving as amino acids in protein synthesis, methionine and cysteine offer chemically reactive side-chains that are the site of assorted modifications, including methylation, oxidation, acylation, nitrosylation, and glutathionylation, that modulate protein function in the cell (Hoshi and Heinemann, 2001; Kiley and Storz, 2004). Sulfur amino acids also play a key role in the regulation of protein metabolism. The first step in the initiation of mRNA translation involves the binding of the initiator methionine-tRNA to the 40S ribosomal subunit to form the 43S preinitiation complex; methionine deficiency can inhibit this step. The capacity to influence cellular redox status and to protect certain proteins from irreversible oxidative damage also contributes to specific roles in the maintenance of cellular homeostasis.

Metabolism of Sulfur Amino Acids in Humans

Dietary methionine and cysteine enter a metabolic cycle that feeds into multiple other pathways (Fig. 17–2) (Finkelstein and Martin, 1986; Selhub, 1999; Stipanuk, 2004). As described above, both sulfur amino acids are raw materials for protein synthesis. In addition, conversion of methionine into S-adenosylmethionine (SAM) by L-methionine-S-adenosyltransferase provides a molecule that serves as the universal methyl group donor in the body. Transfer of the methyl

Plant Sulfur Compounds and Human Health

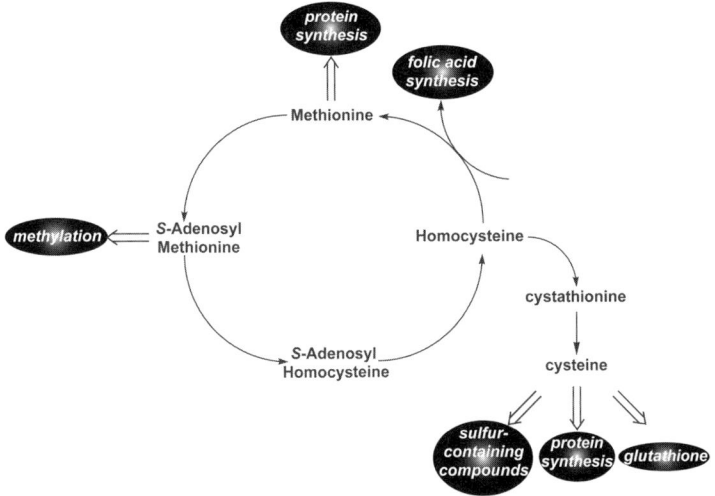

Fig. 17–2. Metabolic relationship of methionine, homocysteine, and cysteine and their biological functions in humans.

group of SAM to target molecules yields S-adenosylhomocysteine. Subsequent hydrolysis of S-adenosylhomocysteine by adenosylhomocysteine hydrolase generates homocysteine, which functions at a branchpoint in the methionine cycle.

Remethylation of homocysteine to methionine by the folate-dependent methionine synthase (methyltetrahydrofolate homocysteine methyltransferase) links sulfur amino acid metabolism to folic acid metabolism. An alternative pathway that regenerates methionine from homocysteine uses betaine as the methyl donor in a reaction catalyzed by betaine-homocysteine methyltransferase. Alternatively, homocysteine may be catabolized through transsulfuration in which a vitamin B6-dependent enzyme (cystathionine-β-synthase) condenses serine and homocysteine to produce cystathionine. This molecule may be hydrolyzed to form cysteine and α-ketobutyrate. Cysteine is then available for synthesis of proteins, glutathione, and other sulfur-containing compounds in the body.

Regulation of the methionine cycle depends on the dietary intake of methionine, as well as the availability of folate, vitamin B12, and vitamin B6. Under conditions of high methionine intake, rapid conversion to SAM occurs. The resulting increase in intracellular SAM concentrations inhibits key enzymes in folic acid synthesis and activates cystathionine-β-synthase. These changes in enzyme activities enhance transsulfurylation of homocysteine into cysteine and reduce remethylation to methionine. Conversely, when methionine intake is low, decreased SAM levels release these inhibitory and activation effects on pathway enzymes. This results in remethylation of homocysteine and increased de novo recycling to compensate for a lack of dietary methionine. Low availability of SAM may, in turn, affect key methylation reactions and the production of polyamines. Importantly, because this pathway converts methionine into cysteine, providing additional dietary cysteine can "spare" or replace a portion of the requirement for dietary methionine by 50 to 80% in mammals and birds (Finkelstein et al., 1986 and 1988, Fukagawa, 2006, Ball et al., 2006).

Diseases and Disorders Associated with Methionine and Cysteine

Given the roles that methionine and cysteine play in the human body, deficiencies of either sulfur amino acid can result in severe health problems.

Low dietary intake of methionine impairs protein synthesis and alters the availability of methyl groups for other synthetic pathways. Deficiency of this essential amino acid in the diet consequently affects folic acid synthesis and the use of SAM in methylation processes. Chronic lack of methionine and folic acid decrease the body's threshold to handle chemical toxicity and may be related to the development of colorectal and breast cancers (Shrubsole et al., 2001; Jiang et al., 2005). Likewise, reductions in intracellular SAM concentrations resulting from methionine deficiency have consequences in health and disease. Because SAM is the primary source of methyl groups in the body, lowered methionine intake directly alters DNA methylation, a process that influences both DNA synthesis and gene expression (Jaenisch and Bird, 2003). Changes in the methylation state of DNA are essential for normal cell growth and development. Global DNA hypomethylation resulting from inadequate dietary supplies of methionine (or folic acid and choline) are associated with DNA fragmentation and strand scission, which can result in mutations and carcinogenesis (Ross, 2003). This has raised interest in "methylation diets" in the prevention and treatment of a number of conditions (e.g., neural tube defects and aberrant gene expression).

The effects of cysteine deficiency in the human diet are unclear. Multiple animal studies show that cysteine adequately fulfills part of the total sulfur amino acid dietary requirement in poultry, pigs, cats, and dogs; however, conclusions from human studies are equivocal (Finkelstein et al., 1988, 1986; Finkelstein and Martin, 1986; Griffith, 1987; Stipanuk, 2004; Baker, 2006). Nonetheless, a decrease in either cysteine intake or transsulfurylation of methionine into cysteine may lead to changes in glutathione metabolism. Because glutathione is a key antioxidant in the body, alterations of cysteine intake may change cellular redox state and thereby influence key redox-sensitive signaling pathways.

Inherited disease due to the genetically determined deficiency of enzymes involved in the methionine cycle is relatively rare but has been reported. The terms hypermethioninemia, homocystinuria (or -emia), and cystathioninuria (or -emia) refer to the biochemical abnormalities found in patients but not the specific disease entities. Each of these may be the result of one or more specific and distinct genetic lesions or defective clearance or catabolism. Perhaps the most commonly recognized is hyperhomocysteinemia that can be the result of defective synthesis of folic acid, homocysteine remethylation, or homocysteine transsulfuration (Selhub, 1999). Increased levels of homocysteine in blood are associated with an increased risk for cardiovascular disease in adults (Refsum et al., 1998), ischemic and hemorrhagic stroke in newborns and children (van Beynum et al., 1999), Alzheimer's disease in adults (Selhub, 1999; Stipanuk, 2004), dementia and schizophrenia (Christensen and Ueland, 1993; Stead et al., 2001), and inflammatory bowel disease (Danese et al., 2005). For example, studies show increased homocysteine levels in plasma and mucosal biopsies from patients with Crohn's disease and ulcerative colitis or in renal disease because of reduced clearance into the urine (Romagnuolo et al., 2001; Morgenstern et al., 2003; Danese et al., 2005). It appears that high blood levels of homocysteine are a marker of disease rather than the cause of disease. The impact of the fortification of foods, especially grain products, with folate in the USA on the prevalence of hyperhomocysteinemia remains to be determined.

Bioactive Sulfur Compounds from Plants

Beyond methionine and cysteine, numerous plants also contain a myriad of sulfur-containing molecules with an array of chemical structures. Certain classes of compounds, i.e., allylsulfur molecules and the glucosinolate-derived isothiocyanates, are widely examined for their potential therapeutic properties. Consumption of plants containing these bioactive compounds may have benefits for human health.

Allylsulfur Compounds from Garlic

The organic sulfur compounds in garlic, *Allium sativa* L., (and onion, *Allium cepa* L.) provide a protective function against bacteria, fungi, and animals (Jones et al., 2004). Before crushing of a garlic clove, the precursors of the allylsulfur molecules are in storage cells, and the enzyme alliinase is in bundle sheath cells. Damage that breaks these cells allows mixing of their contents and subsequent production of allylsulfur compounds by alliinase. Certain compounds, like allicin, are antibacterial. In general, the pungency of these molecules acts as a repellant for animals; however, this same aroma is the reason for using garlic in cooking. Writings from ancient Egypt, Greece, China, and India extol the medicinal properties of garlic, and modern research suggests that the allylsulfur compounds (Fig. 17–3A) found in this plant have potential value as antithrombotic and anticancer agents (Milner, 2001).

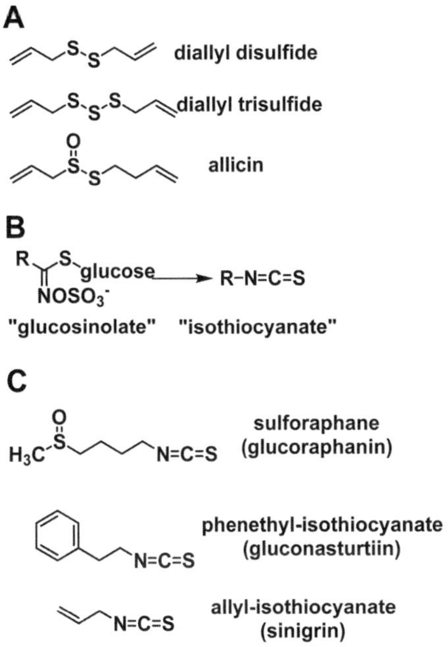

A

diallyl disulfide

diallyl trisulfide

allicin

B

$R-S-glucose \longrightarrow R-N=C=S$
$NOSO_3^-$

"glucosinolate"   "isothiocyanate"

C

sulforaphane (glucoraphanin)

phenethyl-isothiocyanate (gluconasturtiin)

allyl-isothiocyanate (sinigrin)

Fig. 17–3. Allylsulfur and isothiocyanates. (A) Representative allylsulfur compounds. (B) Overall conversion of glucosinolates into isothiocyanates by myrosinase. (C) Representative isothiocyanates. Starting glucosinolates are indicated in parentheses.

The formation of plaques on the arterial wall (atherosclerosis) and the accumulation of blood clots at ruptured lesions in arteries (thrombosis) contribute to cardiovascular disease. Although some allylsulfur compounds, such as methyl allyl trisulfide, are active inhibitors of platelet aggregation (Ariga and Seki, 2006), the origin of this antithrombotic activity is unclear.

Allylsulfur compounds from garlic are reported to reduce the incidence of breast, colon, skin, uterine, and lung cancers and to depress proliferation of tumor cells (reviewed in Milner, 2001). Experiments suggest that allylsulfur compounds affect a number of molecular processes, all of which likely contribute to their effect on cancer. Garlic-derived allylsulfur compounds inhibit formation of nitrosamines, which are suspected carcinogens and potential environmental factors in carcinogenesis (Brown, 1999; Ferguson, 1999). Moreover, allylsulfur molecules effectively block bioactivation of other nonnitrosamine chemical carcinogens, including polycyclic aromatic hydrocarbons and aflatoxins (Milner, 2001). Although the suppression of carcinogen-activation likely contributes to their anticancer activity, allylsulfur compounds also alter cell proliferation and induce apoptosis (Sundaram and Milner, 1996; Knowles and Milner, 1998). Both of these processes are important in cancer progression.

Glucosinolates and Isothiocyanates: Broccoli and Cancer Prevention

Plants of the Brassicaceae family, which includes nearly 3000 species such as broccoli, Brussels sprouts, cabbage (*Brassica oleracea* var. *capitata* L.), bok choy (*Brassica chinensis* L.), cauliflower (*Brassica oleracea* var. *botrytis* L.), collard greens [*Brassica juncea* (L.) Czernj, & Cosson var. *juncea*], kale (*Brassica oleracea* var. *acephala* DC), watercress (*Nasturtium officinale* R. Br.), mustard(*Brassica oleracea* var. *acephala* DC), and turnips (*Brassica rapa* L. var. *rapa*), contain up to 1% dry weight of a chemically diverse class of sulfur-containing compounds called glucosinolates or β-thioglucoside-N-hydroxysulfates (Fahey et al., 2001; Higdon et al., 2007). More than 120 glucosinolates (and their corresponding isothiocyanates) are known (Fahey et al., 2001). These molecules function as fungicidal, bactericidal, nematicidal, and allelopathic compounds in plants but require activation for their effects. Within a plant, sequestration of the enzyme myrosinase in vacuoles prevents conversion of glucosinolates into isothiocyanates. Upon damage to the plant, breakage of cells releases myrosinase into tissues containing glucosinolates. The activity of myrosinase converts glucosinolates into isothiocyanates (Fig. 17–3B). The cancer preventative property of many isothiocyanates, such as sulforaphane (Fig. 17–3C), has spurred intense research on this class of plant sulfur-containing compounds.

Isothiocyanates modify carcinogen metabolism in laboratory animals and in cell culture (Hecht, 1999; Bianchini and Vainio, 2004). Environmental and dietary carcinogens, such as polycyclic aromatic hydrocarbons, require transformation and activation in the body (Fig. 17–4). This process begins with the cytochrome P450 enzymes involved in "Phase 1" metabolism of xenobiotics. By introducing electrophilic oxygens into target molecules, Phase 1 metabolism prepares chemical carcinogens for modification by "Phase 2" enzymes, including glutathione-S-transferase and sulfotransferases. Excretion of the resulting modified molecules detoxifies the body. Problems occur when the chemical reactive intermediates formed by cytochrome P450 enzymes react with DNA, proteins, and

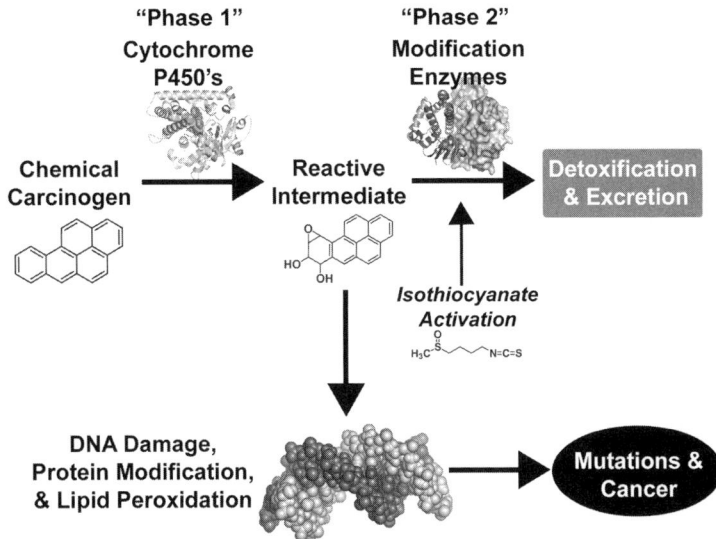

Fig. 17–4. Role of isothiocyanates in cancer prevention. "Phase 1" and "Phase 2" enzymes detoxify potential chemical carcinogens for excretion. Accumulation of reactive metabolites can cause DNA damage, protein modifications, or lipid peroxidation that lead to mutations and carcinogenesis. Isothiocyanates activate "Phase 2" enzymes and accelerate detoxification.

lipids, causing mutations and abnormal cellular function that can lead to carcinogenesis. The cancer preventative effects of isothiocyanates result from blocking carcinogen activation (Phase 1) and/or enhancing detoxification by modification enzymes (Phase 2). Numerous isothiocyanates are potent inducers of the Phase 2 detoxification enzymes (Zhang and Talalay, 1994; Hecht, 1999; Fimognari and Hrelia, 2007). Although studies suggest that intake of cruciferous vegetables (and isothiocyanates) is associated with decreased cancer risk, and that isothiocyanates alter the metabolism of chemical carcinogens (Verhoeven et al., 1996; Kristal and Lampe, 2002; Higdon et al., 2007), these protective effects may also be related to other nutritional factors associated with diets high in vegetables.

Acknowledgments

J.M.J. acknowledges support from a U.S. Department of Agriculture Presidential Early Career Award for Scientists and Engineers (PECASE) (NRI-2005-02518) grant and a grant from the Illinois–Missouri Biotechnology Alliance. N.K.F. receives support from the National Institutes of Health (Grant #AG01106-01A1 and #M01 RR00109) and the Gustavus and Louise Pfeiffer Research Foundation.

References

Ariga, T., and T. Seki. 2006. Antithrombotic and anticancer effects of garlic-derived sulfur compounds: A review. Biofactors 26:93–103.

Baker, D.H. 2006. Comparative species utilization and toxicity of sulfur amino acids. J. Nutr. 136:1670S–1675S.

Ball, R.O., G. Courtney-Martin, and P.B. Pencharz. 2006. The in vivo sparing of methionine by cysteine in sulfur amino acid requirements in animal models and adult humans. J. Nutr. 136:1682S–1693S.

Bianchini, F., and H. Vainio. 2004. Isothiocyanates in cancer prevention. Drug Metab. Rev. 36:655–667.
Brown, J.L. 1999. N-nitrosamines. Occup. Med. 14:839–848.
Burrin, D.G., and B. Stoll. 2007. Emerging aspects of gut sulfur amino acid metabolism. Curr. Opin. Clin. Nutr. Metab. Care 10:63–68.
Christensen, B., and P.M. Ueland. 1993. Methionine synthase inactivation by nitrous oxide during methionine loading of normal human fibroblasts: Homocysteine remethylation as determinant of enzyme inactivation and homocysteine export. J. Pharmacol. Exp. Ther. 267:1298–1303.
Danese, S., A. Sgambato, A. Papa, F. Scaldaferri, R. Pola, M. Sans, M. Lovecchio, G. Gasbarrini, A. Cittadini, and A. Gasbarrini. 2005. Homocysteine triggers mucosal microvascular activation in inflammatory bowel disease. Am. J. Gastroenterol. 100:886–895.
Fahey, J.W., A.T. Zalcmann, and P. Talalay. 2001. The chemical diversity and distribution of glucosinolates and isothiocyanates among plants. Phytochemistry 56:5–51.
Fereday, A., N.R. Gibson, M.D. Cox, P.J. Pacy, and D.J. Millward. 1997. Protein requirements and ageing: Metabolic demand and efficiency of utilization. Br. J. Nutr. 77:685–702.
Ferguson, L.R. 1999. Nature and man-made mutagens and carcinogens in the human diet. Mutat. Res. 443:1–10.
Fimognari, C., and P. Hrelia. 2007. Sulforaphane as a promising molecule for fighting cancer. Mutat. Res. 635:90–104.
Finkelstein, J.D., and J.J. Martin. 1986. Methionine metabolism in mammals: Adaptation to methionine excess. J. Biol. Chem. 261:1582–1587.
Finkelstein, J.D., J.J. Martin, and B.J. Harris. 1986. Effect of dietary cystine on methionine metabolism in rat liver. J. Nutr. 116:985–990.
Finkelstein, J.D., J.J. Martin, and B.J. Harris. 1988. Methionine metabolism in mammals. The methionine-sparing effect of cystine. J. Biol. Chem. 263:11750–11754.
Food and Agriculture Organization. 1970. Amino-acid content of foods and biological data on proteins. FAO, Rome.
Food and Agriculture Organization/World Health Organization/United Nations University. 1985. Energy and protein requirements. WHO Tech. Rep. ser. no. 724.
Fukagawa, N.K. 2006. Sparing of methionine requirements: Evaluation of human data takes sulfur amino acids beyond protein. J. Nutr. 136:1676S–1681S.
Fukagawa, N.K., and R.A. Galbraith. 2004. Advancing age and other factors influencing the balance between amino acid requirements and toxicity. J. Nutr. 134:1569S–1574S.
Griffith, O.W. 1987. Mammalian sulfur amino acid metabolism: An overview. Methods Enzymol. 143:366–376.
Hecht, S.S. 1999. Chemoprevention of cancer by isothiocyanates, modifiers of carcinogen metabolism. J. Nutr. 129:768S–774S.
Higdon, J.V., B. Delage, D.E. Williams, and R.H. Dashwood. 2007. Cruciferous vegetables and human cancer risk: Epidemiologic evidence and mechanistic basis. Pharm. Res. 55:224–236.
Hoshi, T., and S. Heinemann. 2001. Regulation of cell function by methionine oxidation and reduction. J. Physiol. 531:1–11.
Ingenbleek, Y. 2006. The nutritional relationship linking sulfur to nitrogen in living organisms. J. Nutr. 136:1641S–1651S.
Jaenisch, R., and A. Bird. 2003. Epigentic regulation of gene expression: How the genome integrates intrinsic and environmental signals. Nat. Genet. 33:245–254.
Jiang, Q., K. Chen, X. Ma, Q. Li, W. Yu, G. Shu, and K. Yao. 2005. Diets, polymorphisms of methylenetetrahydrofolate reductase, and the susceptibility of colon cancer and rectal cancer. Cancer Detect. Prev. 29:146–154.
Jones, M.G., J. Hughes, A. Tregova, J. Milne, A.B. Tomsett, and H.A. Collin. 2004. Biosynthesis of the flavour precursors of onion and garlic. J. Exp. Bot. 55:1903–1918.
Kiley, P.J., and G. Storz. 2004. Exploiting thiol modifications. PLoS Biol. 2:e400.

Knowles, L.M., and J.A. Milner. 1999. Depressed p34cdc2 kinase activity and G2/M phare arrest induced by diallyl disulfide in HCT-15 cells. Nutr. Cancer 30:169–174.

Komarnisky, L.A., R.J. Christopherson, and T.K. Basu. 2003. Sulfur: Its clinical and toxicologic aspects. Nutrition 19:54–61.

Kristal, A.R., and J.W. Lampe. 2002. Brassica vegetables and prostate cancer risk: A review of the epidemiological evidence. Nutr. Cancer 42:1–9.

McCance, R.A., and E.M. Widdowson. 1991. The composition of foods, 5th ed. Royal Society of Chemistry, Cambridge.

Millward, D., A. Fereday, N. Gibson, and P. Pacy. 1997. Aging, protein requirements, and protein turnover. Am. J. Clin. Nutr. 66:774–786.

Milner, J.A. 2001. A historical perspective on garlic and cancer. J. Nutr. 131:1027S–1031S.

Morgenstern, I., M.T. Raijmakers, W.H. Peters, H. Hoensch, and W. Kirch. 2003. Homocysteine, cysteine, and glutathione in human colonic mucosa: Elevated levels of homocysteine in patients with inflammatory bowel disease. Dig. Dis. Sci. 48:2083–2090.

Rees, W.D. 2002. Manipulating the sulfur amino acid content of the early diet and its implications for long-term health. Proc. Nutr. Soc. 61:71–77.

Rees, W.D., F.A. Wilson, and C.A. Maloney. 2006. Sulfur amino acid metabolism in pregnancy: The impact of methionine in the maternal diet. J. Nutr. 136:1701S–1705S.

Refsum, H., P.M. Ueland, O. Nygard, and S.E. Vollset. 1998. Homocysteine and cardiovascular disease. Annu. Rev. Med. 49:31–62.

Romagnuolo, J., R.N. Fedorak, V.C. Dias, F. Bamforth, and M. Teltscher. 2001. Hyperhomocysteinemia and inflammatory bowel disease: Prevalence and predictors in a cross-sectional study. Am. J. Gastroenterol. 96:2143–2149.

Ross, S.A. 2003. Diet and DNA methylation interactions in cancer prevention. Ann. N. Y. Acad. Sci. 983:197–207.

Selhub, J. 1999. Homocysteine metabolism. Annu. Rev. Nutr. 19:217–246.

Shoveller, A.K., B. Stoll, R.O. Ball, and D.G. Burrin. 2005. Nutritional and functional importance of intestinal sulfur amino acid metabolism. J. Nutr. 135:1609–1612.

Shrubsole, M.J., F. Jin, Q. Dai, X.O. Shu, J.D. Potter, J.R. Hebert, Y.T. Gao, and W. Zheng. 2001. Dietary folate intake and breast cancer risk: Results from the Shanghai Breast Cancer Study. Cancer Res. 61:7136–7141.

Souchi, S.W., W. Fachman, and H. Kraut. 1994. Food composition and nutrition tables, 5th ed. CRC Press, Baton Rouge, LA.

Stead, L.M., K.P. Au, R.L. Jacobs, M.E. Brosnan, and J.T. Brosnan. 2001. Methylation demand and homocysteine metabolism: Effects of dietary provision of creatine and guanidinoacetate. Am. J. Physiol. Endocrinol. Metab. 281:1095–1100.

Stipanuk, M.H. 2004. Sulfur amino acid metabolism: Pathways for production and removal of homocysteine and cysteine. Annu. Rev. Nutr. 24:539–577.

Sundaram, S.G., and J.A. Milner. 1996. Diallyl disulfide induces apoptosis in human colon tumor cells in culture. Carcinogenesis 17:669–673.

van Beynum, I.M., J.A. Smeitink, M. den Heijer, M.T. te Poele Pothoff, and H.J. Blom. 1999. Hyperhomocysteinemia: A risk factor for ischemic stroke in children. Circulation 99:2070–2072.

van de Poll, M.C., C.H. Dejong, and P.B. Soeters. 2006. Adequate range for sulfur-containing amino acids and biomarkers for their excess: Lessons from enteral and parenteral nutrition. J. Nutr. 136:1694S–1700S.

Verhoeven, D.T., R.A. Goldbohm, G. van Poppel, H. Verhagen, and P.A. van den Brandt. 1996. Epidemiological studies on Brassica vegetables and cancer risk. Cancer Epidemiol. Biomarkers Prev. 5:735–748.

Young, V.R., and S. Borgonha. 2000. Nitrogen and amino acid requirements: The Massachusetts Institute of Technology amino acid requirement pattern. J. Nutr. 130:1841S–1849S.

Zhang, Y., and P. Talalay. 1994. Anticarcinogenic activities of organic isothiocyanates: Chemistry and mechanisms. Cancer Res. 54:1976s–1981s.

18

A Future Crop Biotechnology View of Sulfur and Selenium

Muhammad Sayyar Khan and **Rüdiger Hell**
Heidelberg Institute of Plant Sciences, University of Heidelberg, Germany

Abstract

Sulfur is an essential nutrient for plants, animals, and humans. In plants, it also is required for the synthesis of compounds for defense against herbivores and pathogens. Optimized sulfur supply to plants is thus important for quality and yield of many crops. Breeding and biotechnology aim at the improvement of sulfur relations at several levels but most importantly in sulfur-rich seed storage proteins. Apart from lysine and some other amino acids, cysteine and especially methionine often limit the nutritional value of food and feed. Selenium is the uneven sister of sulfur, and although essential in human nutrition, it is potentially toxic at elevated concentrations. Phytoremediation of selenium-contaminated soils and biofortification of foods are two major goals of plant biotechnology. Because of their similar chemical and physical properties, sulfate and selenate are believed to share the initial route for uptake and assimilation. Recently, the regulation of some of the potentially health-promoting compounds of sulfur and selenium in the Brassicaceae family has received considerable interest. Improved crop production appears to be hampered by the joint pathway of uptake and assimilation followed by a bias between reduced sulfur- or selenium-containing secondary compounds, such as glucosinolates and methylselenocysteine, preventing a simultaneous increase of both metabolite groups. A comprehensive understanding of the key regulatory steps of sulfur and selenium metabolism is indispensible to this end.

Sulfur and selenium have received little attention with respect to biotechnology-based crop improvement, at least when compared with nitrogen or phosphorus. Plant nutritional aspects may be the major reason for this lack of interest. Sulfate is the least required among the six macronutrients and is often sufficiently available in soils of arable land. Its mineral fertilization is relatively affordable, mostly combined with chemically reduced nitrogen (ammonium sulfate), and even sulfur contaminations of nitrogen and phosphate mineral fertilizers may be sufficient to support crop growth in some cases (Pasricha and Abrol, 2003). Selenium, on the other hand, still has not been identified as an essential nutrient for plants (not regarding algae) and only plays a role as a potentially deleterious component in small agricultural areas with high selenate content in soil. However, many reduced sulfur compounds, such as methionine and several vitamins (Hell, 1997), are essential in the human diet as is selenide in a steadily increasing number of specialized enzymatic functions (Sors et al., 2005b). In pursuit of

---

Copyright © 2008. American Society of Agronomy, Crop Science Society of America, Soil Science Society of America, 677 S. Segoe Rd., Madison, WI 53711, USA. *Sulfur: A Missing Link between Soils, Crops, and Nutrition.* Agronomy Monograph 50.

higher yield, better nutritional value, and quality in combination with sustainable plant management, several biotechnological approaches have attempted to improve crop plants in recent years.

Biotechnological approaches in this respect made use of new results from basic plant biology research and has been applied to quite different aspects of sulfur and selenium in crop plants. First, enhanced contents of reduced primary sulfur and selenium metabolites have been a goal that will be covered in this chapter. Second, sulfur-based defense compounds have been recognized as important for yield. Molecular biology has finally unraveled the biosynthesis of glucosinolates, potentially toxic secondary sulfur compounds of the Brassicaceae family. Third, selenium hyperaccumulator plants provided important insight in the detoxification of selenate. While these aspects will be covered in this chapter, several approaches can only be briefly reported because they are less advanced and space is limited.

Approaches for Improvement of Seed Sulfur Contents

The importance of sulfur in crop seeds is based on quantity as well as quality. A balanced fertilization regime of nitrogen and sulfur is required to optimize nutrient use efficiency, defined as the amount of grains obtained per amount of fertilizer. Only if vegetative growth enhanced by nitrogen fertilizers is matched by sulfur availability can the genetically defined nitrogen/sulfur (N/S) ratio in proteins of a plant be sustained (Oenema and Pastma, 2003). The nutritional situation during the vegetative or generative growth phase strongly affects not only grain yield but also composition and thus quality. Quality can be defined either as biological value or technological value for processing. As an example, barley (*Hordeum vulgare* L.) varieties for beer brewing have a higher carbon/nitrogen (C/N) ratio and sulfur content compared with food and feed barley to optimize flavor development after fermentation. Sulfur supply and concomitant sulfur content of storage proteins are known to be important for baking quality of bread wheat, *Triticum aestivum* L. (Zhao et al., 1999). Low glucosinolates and erucic acid content in oilseed rape (such as canola, *Brassica napus* L.) are required to reduce antinutritional effects since, after removal of oil, the pressed meal contains about 37% of protein that is used as a low-grade animal feed.

Improvement of nutritional value is largely based on the requirement of non-ruminant animals and humans for the essential amino acid methionine. Cysteine in a strict sense is only essential for infants up to about 2 years of age and can serve as substrate for the transsulfurylation pathway via cystathionine to methionine, thus complementing up to 50% of methionine requirement in adults (Müntz et al., 1998). In animal production (poultry and monogastric animals), the diets mainly consist of grains of the legume and the cereal families. The conversion factor of feed protein nitrogen to body protein nitrogen is higher when the amino acid composition of the feed protein is the more balanced. Leguminosae [e.g., soybeans, *Glycine max* (L.) Merr.] are known to be relatively low in methionine and cysteine but high in lysine. In contrast, cereals (e.g., corn, *Zea mays* L.) are notoriously low in lysine but comparatively high in sulfur amino acids; although, they are still regarded as target for improvement in this respect (Wang et al., 2003). Adequate mixtures of dietary components complement these shortcomings but are still insufficient to achieve maximal animal production, giving rise to supplementation by methionine, lysine, and several other amino acids to enhance the biological value of the feed (Müntz et al., 1998). Amino acids for this purpose have been chemically produced in the past and are beginning to be replaced by products of fermentation (Leinfelder and Heinrich, 1995). Most desirable would be grains with enhanced contents of nutritionally valuable factors, in particular cysteine and methionine.

Such fortified food and feed is proposed to be provided by plant biotechnology in genetically modified crops using so-called push and pull approaches (Galili et al., 2005). Push approaches try to enhance the biosynthetic production of cysteine and methionine. This usually results in the accumulation of free amino acids that are supposed to be incorporated into storage proteins. In seeds, 20 to 40 times more sulfur is bound in proteins than free sulfur amino acids (Chiaiese et al., 2004). These approaches so far have mostly been performed in model species such as tobacco (*Nicotiana tabacum* L.) and *Arabidopsis* or in comparatively easily transformable crops (potato, *Solanum tuberosum* L.; rice, *Oryza sativa* L.) to avoid trial-and-error experiences in costly transformation processes of major grain crops (soybean, corn). With respect to primary sulfur metabolism, several targets can be envisaged. To begin with, no successful push approach leading to increased sulfur amino acid content has been reported for sulfate uptake or allocation. Interestingly, modification of mineral uptake as target was obviously not satisfactory in nitrogen metabolism (Hell and Hillebrand, 2001) and may thus be a general problem. The next possible target in the pathway is the assumed key regulatory step in the assimilatory sulfate reduction pathway that is catalyzed by adenosine 5'-phosphosulfate reductase (APR). Overexpression of a functionally equivalent APR gene from *Pseudomonas aeruginosa* in corn enhanced flux through the sulfate assimilation pathway and increased free levels of sulfur amino acids and glutathione. However, the approach seemed not to be applicable for crop improvement because overexpression caused deleterious side effects because of toxicity of intermediates of the reduction pathway (Martin et al., 2005).

Methionine is a member of the aspartate kinase protein family, together with lysine and threonine. Aspartate kinase and dihydrodipicolinate dehydrogenase play important roles in feedback control of the pathway (Hesse et al., 2004), and overexpression of feedback-insensitive forms of both enzymes from *E. coli* resulted in accumulation of lysine, threonine, and methionine in transgenic tobacco (Galili et al., 2005) and of lysine in tubers of transgenic potato (Sevenier et al., 2002). An alternative approach is the diversion of flux at the branching point of the biosynthesis of threonine and methionine. Antisense expression to downregulate expression of the threonine synthase gene in potato resulted in an accumulation of methionine. The participation of methionine in the S-adenosylmethionine cycle, however, increases the risk of unwanted side effects, in particular because the latter compound has allosteric effects on several steps upstream in the aspartate kinase pathway (Hesse and Hoefgen, 2008).

Instead of changing flux of the carbon–nitrogen backbone, the availability of cysteine for transsulfurylation via cystathionine was attempted in *Arabidopsis* and tobacco. The rate-limiting step of cysteine synthesis is catalyzed by serine acetyltransferase, and plant and *E. coli* orthologs of the enzyme, either feedback sensitive to cysteine or not, were expressed in cytosol and/or plastids. Accumulation of free cysteine in leaves was 2- to 20-fold and of glutathione, a transient reservoir for excess cysteine, was 1.5- to 3-fold, but methionine levels only rose

1.5- to 2.5-fold (Galili et al., 2005; Hell and Wirtz, 2008; Sirko et al., 2004). This suggested that increased pressure of the reduced sulfur donor cysteine had little effect on free methionine contents and that the regulatory networks of sulfate assimilation–cysteine synthesis on one side and of the aspartate family–methionine synthesis on the other, operate largely independently of each other. It might be noteworthy that enhanced cysteine contents could be advantageous to form complexes with iron, a micronutrient with strong underrepresentation in many staple crops, especially rice. Overexpression of the cysteine-rich protein metallothionine in rice would be such an approach as a complementary aspect of the "Golden Rice" project (Lucca et al., 2006). Enhanced cysteine levels would also be a prerequisite to improve glutathione contents. Glutathione is generally acknowledged as important because of its redox properties in cells and as a dietary component (Galili et al., 2005; Maughan and Foyer, 2006).

More direct feed-forward approaches are aimed at overexpression of cystathionine-γ-synthase (CGS) and cystathionine-β-lyase. Overexpression of the latter resulted in no appreciable increases in methionine contents in potato (Hesse et al., 2004), suggesting that this step is not rate limiting for methionine synthesis. Overexpression of CGS and analysis of mutant or transgenic plants of *Arabidopsis*, tomato (*Lycopersicon esculentum* Mill.), and potato revealed an interesting regulatory difference between species. The *Arabidopsis* mutant *mto1* was found to contain very high levels of free methionine and CGS activity because of a mutation in the first exon of the *CGS* mRNA (Onouchi et al., 2004). A domain within this exon in the mRNA determines the stability of the message that is further controlled by the amount of translated CGS protein and S-adenosylmethionine. This effect was also observed on expression in transgenic tobacco and tomato. The elimination of this mRNA domain alone resulted in more CGS protein and accumulation of methionine in leaf (Hacham et al., 2006; Katz et al., 2006). Remarkably, the CGS mRNA of potato does not contain this domain and is not regulated at the translational level according to overexpression of potato CGS, indicating different regulatory mechanisms of methionine synthesis in plants (Kreft et al., 2003). Taken together, push approaches so far were highly elucidating with respect to biosynthesis, cellular compartmentation, regulation, and transport of sulfur amino acids. Significant increases of free cysteine and methionine were achieved but were never anywhere near the recommended FAO standard for nutritionally balanced food proteins. Nevertheless, enhanced flux through the biosynthetic pathways of both amino acids as indicated by their elevated steady-state levels may be necessary to achieve accumulation of reduced sulfur in end products of nutritional interest such as glutathione, glucosinolates, and storage proteins.

The pull approach employs sulfur-rich storage proteins as sinks for cysteine and methionine. Naturally occurring proteins of the water soluble 2S albumin fraction with high contents of methionine or both amino acids have been identified, such as 8-kDa Gm2S-1 from soybean [7.8% methionine and 7.8% cysteine (de Lumen et al., 1999, corn 21-kDa zein (37% of residues), and 12-kDa sunflower (*Helianthus annuus* L.) seed albumin (SSA) that contains 7.5% cysteine and 15.5% methionine and has been employed in numerous transgenic approaches (Tabe et al., 2002). The Brazil nut protein is also sulfur rich (23% methionine) and has been applied early on (Müntz et al., 1998) but was found to be a highly allergenic potential for human nutrition (Nordlee et al., 1996). Ideally, seed storage proteins are equipped with a KDEL targeting signal for localization to the endoplasmic reticulum and deposition in protein bodies (Vitale and Hinz, 2005) and expressed under the control of developmentally controlled seed-specific promoters that are preferably active in the embryo (legume crops) or the endosperm (cereals).

Using the seed-specific legumin B4 promoter from *Vicia faba* L. and the phaseolin promoter from *Phaseolus vulgaris* L. the nonfood legume *Vicia narbonensis* L. was transformed with genes encoding a bacterial feedback-insensitive aspartate kinase and the Brazil nut 2S albumin. Additive but not synergistic effects were observed, reaching methionine content in mature seeds up to 2.4-fold higher than in wild type, while contents of sulfate and free sulfur compounds were decreased. This approach yielded approximately the FAO standard for methionine in a balanced protein diet for humans or monogastric animals (Demidov et al., 2003). A pea vicillin promoter and sunflower 2S seed albumin were combined for transformation of lupin (*Lupinus angustifolius* L.), a major grain legume in Australia (Molvig et al., 1997). Methionine contents almost doubled in cost of total cysteine and sulfate contents and feeding trails with rats that showed positive results. Expression of sunflower seed albumin gene in rice under the control of a wheat glutenin promoter resulted in accumulation of the transgenic protein to 7% of total seed protein (Hagan et al., 2003). However, the total sulfur amino acid content remained unchanged and gave rise to the conclusion that reduced sulfur had been allocated from endogenous proteins to the new sulfur sink. In addition, enhancement of sulfur amino acid contents has been attempted in forage crop such as clover (*Trifolium repens* L.) with the aim to improve diet of sheep and wool growth (Christiansen et al., 2000). Using a leaf-specific promoter and the sunflower seed albumin with endoplasmic reticulum sorting signal an accumulation of up to 0.2% of total leaf protein was achieved, corresponding, according to calculations, to an amount that could improve wool growth by 4%.

These examples demonstrate the potential of biotechnological approaches compared with rather modest improvements based on natural variation. However, so far the potentials have not been realized in commercial products (Wang et al., 2003). Major reasons, apart from public acceptance, may be the insufficient increases in sulfur amino acids that remain below FAO recommendations and requirements for commercial animal production. In addition, lack of understanding of seed biology and thus development of still not optimized transgenic approaches are also important. Several considerations have to be taken into account in more successful future approaches. For one, viable and evolutionary successful seeds are dry. This reduces weight to allow easier spreading and increase durability in time and against pathogens. Desiccation is achieved by a complex developmentally controlled dehydration process and structured packing of storage compounds (starch, oils, proteins) in microenvironments with very low free water contents (Borisjuk et al., 2004). Storage proteins end up in paracrystalline protein bodies after complex posttranslational processing events in the different types of globulins and albumins (Shutov et al., 2003). These evolutionarily optimized and densely packed structures often are limited in their capacity to incorporate foreign storage proteins. The seed developmental program apparently allows little flexibility between the major groups of storage compounds within the limited space of the seed organs.

Second, most nutritionally relevant seeds consist of variable combinations of sulfur-poor and sulfur-rich storage proteins. An example for a sulfur-poor protein is the bean β-conglycinin protein whose promoter is upregulated in response to sulfate deficiency (Kim et al., 1999). Alternatively, mRNA stability may be lowered during sulfate deficiency as shown for the gene, encoding the sulfur-rich pea albumin PA1 (Morton et al., 1998). Expression of the underlying genes thus often is controlled by the sulfur and nitrogen status of the plant, resulting in induction of genes encoding sulfur-poor proteins and repression of sulfur-rich protein genes (Chiaiese et al., 2004; Tabe and Droux, 2002); sulfate deficiency caused a complex program of transcriptional and posttranslational processes as shown for seeds of *Arabidopsis thaliana* (L.) Heynh.(Higashi et al., 2006). Furthermore, the biosynthetic capacity for synthesis of cysteine and methionine in the seed may become limiting with the introduction of artificial sinks (Tabe and Droux, 2002), depending on the contribution of import of reduced sulfur from vegetative parts of the plant (Bourgis et al., 1999).

A third mechanism that needs to be regarded is nutrition and allocation. To avoid situations of sulfur deficiency during grain filling, exact knowledge of the developmentally controlled allocation of nutrients is required. Major grain crops can differ substantially in their grain-filling mechanisms. In a strongly reduced view, cereals such as barley tend to allocate most of the sulfur (oxidized and reduced) for the developing seed from flag leaf or below and incorporate most sulfur in the vegetative phase. In contrast, legumes such as soybean take up most of the sulfur required for seed filling directly from soil during the generative phase (Anderson and Fitzgerald, 2001). These different developmental patterns need to be considered in fertilization treatments of crops. Interference with these complex programs that have been evolved over a long time under various environmental conditions to ensure survival of the next generation is consequently difficult and requires detailed biological knowledge of the different growth and seed types of crops. Push and pull approaches may be just the beginning of improvement of nutritionally important seed crops.

Sulfur And Selenium: Uneven Twins in Plant and Human Nutrition

Selenium is a Group V1A element with chemical properties similar to sulfur (Broadley et al., 2006; White et al., 2007). It displays metalloid characteristics and occurs in several oxidation states as selenide (Se^{2-}), elemental selenium (Se^0), selenite (Se^{4+}), and selenate (Se^{6+}) (White et al., 2004). Plant species have been divided into three groups on the basis of their ability to accumulate selenium: nonaccumulator, selenium-indicator, and selenium-accumulator (Brown and Shrift, 1982; Dhillon and Dhillon, 2003; Ellis and Salt, 2003; Ihnat, 1989; Shrift, 1969; Wu, 1998). Selenium is toxic for nonaccumulator plants at tissue concentrations as low as 10 to 100 µg g^{-1} dry matter (White et al., 2004), whereas for selenium-indicator plants, this limit is around 1000 µg selenium g^{-1} dry matter (Moreno Rodriguez et al., 2005). On the other hand, selenium-accumulator plants can accumulate up to 20 to 40 mg selenium g^{-1} dry matter (Brown and Shrift, 1982). The chemical form of selenium and its concentration in the soil solution, soil redox conditions, pH of the rhizosphere, and the presence of competing anions such as sulfate and phosphate are some of the factors affecting the uptake of selenium from soil by plant roots (Blaylock and James, 1994; Dhillon and Dhillon, 2003).

Because of their similar chemical properties, selenium and sulfur are thought to share the initial steps for the uptake and assimilation. Plants acquire selenium

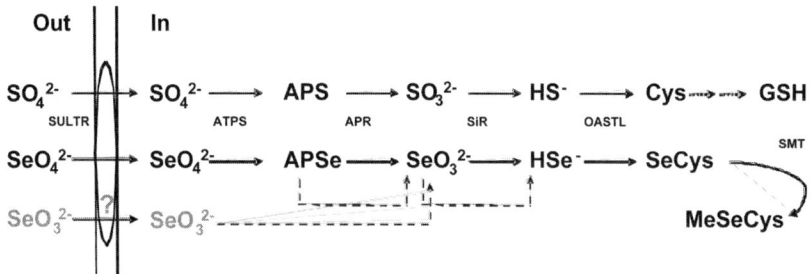

Fig. 18–1. Assimilatory sulfate–selenate reduction pathway in higher plants. The dotted lines in the lower row indicate the possible nonenzymatic steps. SULTR: sulfate transporters, ATPS: ATP sulfurylase, APR: APS reductase, SiR: sulfite reductase, OASTL: O-acetylserine (thiol) lyase, SMT: selenocysteine methyltransferase.

primarily as selenate (SeO_4^{2-}), which enters root cells through high affinity sulfate transporters (Sors et al., 2005b; Terry et al., 2000; White et al., 2007, 2004), but selenite and organic selenium compounds are also taken up readily (Asher et al., 1977; Martin et al., 1971; White et al., 2004). The sulfate transporters differ in their selectively between sulfate and selenate, and several of these appear to contribute to selenate uptake and accumulation (White et al., 2007). The high affinity sulfate transporter *SULTR1;2* has been shown to be a major contributor in the acquisition of selenate (El Kassis et al., 2007). It has been hypothesized that the dominant high affinity sulfate transporters of selenium-accumulator plants are selective for selenate, whereas those in the other species are more selective for sulfate (Broadley et al., 2006; Sors et al., 2005a; White et al., 2004).

The proposed assimilatory sulfate–selenate reduction pathway in higher plants is outlined in Fig. 18–1. On uptake, selenate needs to be activated for further assimilation. ATP sulfurylase, the enzyme that activates sulfate, has been shown in vitro to activate selenate as well (Wilson and Bandurski, 1958), forming adenosine 5′-phosphoselenate (APSe). APSe can then be further reduced to selenite by APS reductase (Sors et al., 2005b); however, there is biochemical evidence that APSe can be reduced nonenzymatically by glutathione (Dilworth and Bandurski, 1977). Moreover, there is evidence that the reduction of selenite to selenide can also occur nonenzymatically with glutathione functioning as electron donor (Ng and Anderson, 1978). *Escherichia coli* mutants lacking sulfite reductase activity have been shown to reduce selenite, strengthening the argument for nonenzymatic reduction (Müller et al., 1997). The nonenzymatic reduction of selenite has been suggested (de Souza et al., 1998) as a possible explanation of why selenite is more easily assimilated into seleno amino acids than selenate. It appears that the last step of assimilation of selenide to form selenocysteine (SeCys) proceeds in a similar way as specified for sulfide assimilation to form cysteine (Cys), by the cysteine synthase complex (Sors et al., 2005b). A more elaborate view of selenium uptake, assimilation, and metabolic fate can be found elsewhere (Sors et al., 2005a).

Biological Roles and Significance of Selenium Metabolites and Proteins

Selenium is an essential element for animals, including humans (Dhillon and Dhillon, 2003; Rayman, 2000, 2002). The number of proteins that are found to require selenide as cofactor increases steadily (Kryukov et al., 2003). Selenium has been shown to prevent cancer in many animal model systems when fed at levels exceeding the nutritional requirement (Combs and Gray, 1998; Ip, 1998; Medina and Morrison, 1988). During the course of metabolism of selenium, a wide array of products are formed (Ganther and Lawrence, 1997; Ganther, 1986). The amount and the chemical form of selenium that are produced during the course of metabolism are the determinants of its biological activity as an essential nutrient, cancer preventive agent, or toxicant (Ganther, 1999). Methylation is a major pathway for selenium metabolism in microbes, plants, and animals that can produce less toxic forms (Ganther, 1999). After testing the role of a number of methylated forms of selenium for cancer prevention (Ip and Ganther, 1991; Ip et al., 1991), the monomethylated forms of selenium were found to have strong effects on carcinogenesis. These metabolites are also lacking some of the toxic effects that are associated with other forms such as selenite (Ganther and Lawrence, 1997; Ip, 1998). Among different monomethylated forms of selenium, methylselenocysteine (MeSeCys) serves as a reservoir (Ganther, 1999). In addition to cancer prevention, selenium has also been reported to play a role in the prevention of cardiovascular diseases (Beckett et al., 2004; Rayman, 2002) and viral infection (Beck, 2001). Moreover, its essential role for optimal endocrine and immune function and moderating the inflammatory response has been demonstrated by different researchers (Arthur et al., 2003; McKenzie et al., 2002). Many of the biological actions ascribed to selenium are also thought to be mediated in most cases through the action of different selenoproteins. The selenoproteins incorporate selenium cotranslationally as a selenocysteine residue that is fully ionized at physiological pH and acts as a very efficient redox catalyst (Beckett and Arthur, 2005). Up to 30 selenoproteins have been identified bioinformatically so far, out of which six are gluthathione peroxidase, three are iodothyronine deiodinases, and three are thioredoxin reductases (Kryukov et al., 2003).

Future Prospects and Goals in Selenium Metabolism

Interest in selenium metabolism is derived from two primary areas: environmental remediation and human nutrition (Sors et al., 2005a). Although selenium is an essential micronutrient for humans and animals at very low doses, it can be very toxic at high doses (Vinceti et al., 2001), and the observation that selenium bioaccumulation is toxic to wildlife has sparked further interest in the phytoremediation of selenium (Banuelos, 2001; Berken et al., 2002; Wu, 2004). Selenium in the environment can occur because of several reasons such as natural geological processes or human activities (Sors et al., 2005a). Being a worldwide problem, there is a demand for the cleanup of selenium-contaminated soils. Phytoremediation is a promising technology that makes use of suitable plants to stabilize, remove, or detoxify pollutants (Terry et al., 2000). Phytoremediation of selenium can be achieved either through phytoextraction or phytovolatilization. In the phytoextraction strategy, selenium taken up by plants from the soil and water is removed by harvesting these plants (LeDuc et al., 2004).

Phytovolatilization is particularly attractive for phytoremediation of selenium-contaminated environments because it completely removes selenium from the local food chain (Atkinson et al., 1990). In this process, plants metabolize inorganic selenium to relatively nontoxic, volatile forms such as dimethyl selenide (DMSe) and dimethyl diselenide (DMDSe), which then escapes to the atmosphere to be diluted and degraded (Lewis et al., 1966; Terry et al., 2000). Phytoremediation of selenium has been achieved under field conditions by planting fast-growing plant species, such as Indian mustard, *Brassica juncea* (L.) Czernj, & Cosson var. *juncea* (Banuelos et al., 1997), which accumulates selenium to several hundred micrograms per gram (Banuelos and Schrale, 1989). Overexpression of the gene encoding selenocysteine methyltransferase (SMT) from selenium hyperaccumulator *Astragalus bisulcatus* (Hook.) Gray in *Arabidopsis* and Indian mustard resulted in increased selenium tolerance, accumulation, and volatilization (LeDuc et al., 2004). However, in this study the advantage conferred by SMT was less pronounced when selenium was supplied as selenate instead of selenite. However, the double transgenic lines overexpressing both, ATP sulfurylase (ATPS) and SMT resulted in substantial accumulation of selenium in Indian mustard when selenium was supplied as selenate (LeDuc et al., 2006), underpinning the importance of this step in selenium metabolism.

The anticarcinogenic properties of various organic forms of selenium against certain types of cancer have been comprehensively documented (Combs and Gray, 1998; Ip, 1998; Medina and Morrison, 1988; Reid et al., 2002; Whanger, 2004, 2002) and promoted research interest in the development of anticarcinogenic selenium-enriched nutritional supplements (Orser et al., 1999). MeSeCys is the major form of selenium found in onions, *Allium cepa* L. (Cai et al., 1995; Uden et al., 1998), broccoli florets, *Brassica oleracea* var. *botrytis* L. (Cai et al., 1995), broccoli sprouts (Finley et al., 2001), and *Astragalus* and selenium enriched garlic, *Allium sativa* L. (Ip et al., 2000; Kotrebai et al., 1999; Uden et al., 1998). Although selenium-enriched yeast is the most popular selenium supplement presently available to the general public, some other biofortified sources of selenium have been shown to be even more effective. For example selenium-enriched garlic was twice as effective as selenium-enriched yeast in prevention of mammary cancer (Ip et al., 2000). Similarly, selenium-enriched broccoli florets (Davis et al., 2002; Finley et al., 2000, 2001) and sprouts (Finley et al., 2001) have been shown to reduce colon tumors in rats.

To develop crops with enhanced selenium content, one useful strategy is to improve the husbandry of the crops. To this end it is important to determine the potential for different crops to accumulate selenium by characterizing the responses of their growth and selenium content to the application of selenium and sulfur fertilizers. Addition of selenium to agricultural fertilizers in Finland had increased the human dietary intake of selenium by Finish people from 20 to 30 $\mu g\ d^{-1}$ (in 1986) to 80 to 90 $\mu g\ d^{-1}$ (in 1989), with the primary food source being wheat (Makela et al., 1993). Some other examples of producing selenium-enriched food through selenium fertilization includes potatoes (Poggi et al., 2000), tomatoes, strawberries (*Fragaris vesca* L. subsp. *vesca*) , radish (*Raphanis sativus* L.), lettuce (*Lactuca sativa* L.)(Carvalho et al., 2003), and soybeans (Yang et al., 2003). The second strategy is to develop crop genotypes with improved selenium accumulation and tolerance traits through screening of existing germplasm, conventional breeding, or genetic modification approaches. The latter strategy is likely to benefit from knowledge of the genes that affect selenium accumulation and tolerance. While pursuing any strategies for developing selenium-enriched crops, the unintended interaction of selenium with other compounds has to be taken into consideration. Given the fact that the chemical properties of sulfur and selenium are quite similar, unintended interactions are most likely to happen. The interaction between selenium and glucosinolates in broccoli provides a good example of the unintended consequences of manipulation of a single bioactive compound. The use of a commercially available broccoli variety enriched in selenium through selenium fertilization in animal cancer trials (Finley, 2003; Finley et al., 2000) revealed that as a result of selenium fertilization the production of sulforaphane, which itself is a major candidate for cancer prevention, was inhibited by about 75% compared with unfertilized controls (Charron et al., 2001). In view of these considerations, the production of the so-called selenium-enriched functional foods without significantly compromising plant health and other equally important bioactive compounds is a major challenge for plant scientists. To overcome such challenges a comprehensive knowledge of the genes affecting sulfur and selenium uptake, assimilation, and metabolism coming from model organisms is indispensable. It has been recently reported that the high affinity sulfate transporter *SULTR1;2* is a major contributor in the acquisition of not only sulfate but also of selenate in *Arabidopsis* roots (El Kassis et al., 2007). Moreover, transgenic *Arabidopsis* lines overexpressing ATPS and APR showed a significant enhancement of selenium reduction as a proportion of total selenium, whereas overexpression of the first step of cysteine biosynthesis catalyzed by serine acetyltransferase resulted in only a slight increase in selenate reduction to organic forms (Sors et al., 2005a). These results suggest that ATPS and APR are major contributors of selenate reduction in planta.

Biotechnological Potential of the Glucosinolates

Glucosinolates (GSs) are nitrogen and sulfur containing natural plant products mainly found in the order Capparales, which includes agriculturally important crop plants of the Brassicaceae family. The GS contents of *Brassica* species range from 1.7 to 8.0% of the total sulfur (Blake-Kalff et al., 1998; Fieldsend and Milford, 1994); however, the proportion could be higher in other species. To date, more than 120 different glucosinolates have been detected in hundreds of plant species of the order Capparales and in the genus *Drypetes* (Fahey et al., 2001). All GSs have a common core structure consisting of a β-D-thioglucose group linked to a sulfonated aldoxime moiety and a variable side chain derived from amino acids. Generally, they are grouped into aliphatic, aromatic, and indole glucosinolates on the basis of their origin from aliphatic amino acids (methionine, alanine, valine, leucine, isoleucine), aromatic amino acids (tyrosine, phenylalanine), or tryptophan.

GSs are stored in vacuoles of specialized cells and are hydrolyzed by endogenous myrosinases (β-glucosidases) to primarily nitrils and isothiocyanates (ITCs) on tissue disruption that might be caused by pathogen attack or wounding (Rask et al., 2000). This GS–myrosinase system has been termed the "mustard oil bomb" and plays an important role in the defense of plants against generalist herbivores and possibly against pathogens (Brader et al., 2001; Rask et al., 2000). The wide range of biological activities of the products derived from the GS-myrosinase system have intrigued scientists for decades. On the one hand, these products play an important role in mediating the interaction of plants with their biotic environment. On the other hand, they influence the flavor and/or health characteristics of agriculturally important vegetable, oil, and fodder crops (Kliebenstein et al., 2005). Breeding efforts during the middle of the last century provided the low GS varieties of oilseed rape, that, together with low erucic acid contents, are described as "00" varieties and were first marketed under the name canola for Canadian oil. While these varieties are based on mutations that lead to low GS contents in seeds but not in the leaves, there is a considerable interest in manipulating the individual GSs tissue-specifically to improve the nutritional value of seed meal and pest resistance of the crops (Wittstock and Halkier, 2002).

Several studies demonstrated that GSs or rather their degradation products are involved in plant defense against insects and pathogens. The ITCs in particular have been reported to play important roles as repellents against insects (Agrawal and Kurashige, 2003; Rask et al., 2000). The inhibition of fungal and bacterial pathogens by different GSs have been reported in several in vitro studies (Brader et al., 2001; Manici et al., 1997; Mithen et al., 1986; Tierens et al., 2001). *Arabidopsis* plants with high levels of novel GSs as a result of the introduction of single CYP79 genes exhibited altered disease resistance (Brader et al., 2006). Transgenic *Arabidopsis* expressing the sorghum CYP79A1 or overexpressing endogenous CYP79A2 accumulated *p*-hydroxybenzyl or benzyl GS, respectively and were more resistant to the bacterial pathogen *Pseudomonas syringae*, whereas the expression of CYP79D2 from cassava (*Manihot esculenta* Crantz) in *Arabidopsis* resulted in the accumulation of aliphatic, isopropyl, and methylpropyl GSs accompanied by enhanced resistance against bacterial soft-rot pathogen *Erwinia carotovora* (Brader et al., 2006).

In addition to GSs, other sulfur-containing compounds have been shown to contribute significantly to innate plant immunity against fungal and microbial pathogens. This includes the phytoalexin group related to camalexin, sulfur-rich peptides of the thionin and defensin families and elemental sulfur (Kruse et al., 2007). Targeted expression of thionins and defensins provide a particularly promising approach to enhance resistance (Kruse et al., 2005). These observations together with others (Bloem et al., 2007; Burow et al., 2008; Rausch and Wachter, 2005) led to the concept of sulfur-enhanced resistance that suggests a quantitative trait based on optimized sulfur nutrition and sulfur-containing defense compounds (Kruse et al., 2007).

GSs are among one the most studied bioactive compounds of the crucifers associated with cancer protection (Fenwick et al., 1983). Several complementary pieces of evidence have reported that isothiocyanates affect many steps of cancer development including modulation of Phase I and II detoxification enzymes (Bogaards et al., 1994; Jiao et al., 1996; Rabot et al., 1993; Talalay and Fahey, 2001), induction of apoptosis (Chiao et al., 2002; Yu et al., 1998), and control of cell cycle (Wang et al., 2004; Yu et al., 1998). Sulforaphane, phenethyl isothiocyanate, allyl isothiocynate, and indole-3-carbinol are the most characterized GS compounds (Hecht, 1999), but many other isothiocyanates that are present in lower quantities may also contribute to the anticarcinogenic properties of crucifers (Finley, 2005). Sulforaphane has been shown to reduce not only the incidence, but also delayed the appearance of, and reduce the size of tumors in a rat mammary tumor model (Fahey et al., 1997). It has also been shown to induce cell cycle arrest and apoptosis in HT29 human colon cancer cells in vitro (Gamet-Payrastre et al., 2000). Phenethyl isothiocyanate has been shown to inhibit induction of esophageal and lung cancer in both rat and mouse tumor models (Hecht et al., 1996; Stoner et al., 1999).

Prospects for Glucosinolate Engineering

Recently, major progress has been made in the identification of genes responsible for biosynthesis of the core GS structure (Kliebenstein et al., 2005; Mikkelsen et al., 2004, 2002; Wittstock and Halkier, 2002). During the past few years, the extensive use of *Arabidopsis* as a model plant has contributed a wealth of information toward understanding the biosynthesis of GSs. Mutant screens, mapping of the genetic loci determining GS profiles, functional genomics, and reverse genetics have been the driving forces responsible for rapid progress (Wittstock and Halkier, 2002). The biosynthesis of GSs takes place in three independent stages. First, elongation of some amino acids by one or more methylene groups take place, then the precursors amino acids are converted to the parent glucosinolates, and finally secondary modifications of parent glucosinolate takes place (Wittstock and Halkier, 2002). During biosynthesis of GSs, the conversion of precursor amino acids to their corresponding aldoximes by substrate-specific cytochrome P450 enzymes belonging to the CYP79 family is the committed step (Brader et al., 2006). Identification of the CYP79 homologs now provides tools for metabolic engineering of GS profiles. Modification of endogenous CYP79 homologs and overexpression of exogenous CYP79 homologs from cyanogenic plants resulted in dramatically changed levels of the corresponding GSs and opened the possibility of developing transgenic crops with custom-designed GS profiles. Although at the stage of basic research, the expression of the sorghum CYP79A1 or overexpression of the endogenous CYP79A2 in *Arabidopsis* resulted in accumulation of *p*-hydroxy benzyl or benzyl GS. Transgenic lines were more resistant to *Pseudomonas syringae*, whereas the expression of CYP79D2 from cassava in *Arabidopsis* resulted in the accumulation of aliphatic isopropyl and methylpropyl GSs. The latter was accompanied by improved resistance against the bacterial soft-rot pathogen *Erwinia carotovora* (Brader et al., 2006). These results highlight the importance of the transgenic plants with altered GSs profile for evaluating the exact biological role(s) of individual GSs. At the same time, they offer potential as biotechnological tools for introducing tailor-made disease resistance traits in plants. The identification of genes involved in GS biosynthesis and genetic engineering of GS profiles should improve the nutritional quality and pest resistance of crop plants in next steps.

Summary

In conclusion, engineering of sulfur and selenium metabolism for crop improvement aims at highly valuable traits but is still at its beginning stage. Better understanding of physiological processes is required to address the complex traits associated with sulfur-containing compounds in crop plants.

References

Agrawal, A., and N. Kurashige. 2003. A role for isothiocyanates in plant resistance against the specialist herbivore *Pieris rapae*. J. Chem. Ecol. 29:1403–1415.

Anderson, J.W., and M.A. Fitzgerald. 2001. Physiological and metabolic origin of sulfur for the synthesis of seed storage proteins. J. Plant Physiol. 158:447–456.

Arthur, J.R., R.C. McKenzie, and G.J. Beckett. 2003. Selenium in the immune system. J. Nutr. 133:1457S–1459S.

Asher, C.J., G.W. Bulter, and P.J. Peterson. 1977. Selenium transport in root systems of tomato. J. Exp. Bot. 28:279–291.

Atkinson, R., S. Aschmann, D. Hasegawa, E. Thompson-Eagle, and W.J. Frankenberger. 1990. Kinetics of the atmospherically important reactions of dimethylselenide. Environ. Sci. Technol. 24:1326–1332.

Banuelos, G. 2001. The green technology of selenium phytoremediation. Biofactors 14:255–260.

Banuelos, G., H. Ajwa, L. Wu, X. Guo, S. Akohoue, and S. Zambrzuski. 1997. Selenium-induced growth reduction in Brassica land races considered for phytoremediation. Ecotoxicol. Environ. Saf. 36:282–287.

Banuelos, G., and G. Schrale. 1989. Plants that remove selenium from soils. Calif. Agric. 43:19–20.

Beck, M. 2001. Selenium as a antiviral agent. p. 235–247. In D.L. Hatfield. (ed.) Selenium. Its molecular biology and role in human health. Kluwer Academic Publishers, Boston, MA.

Beckett, G.J., and J.R. Arthur. 2005. Selenium and endocrine systems. J. Endocrinol. 184:455–465.

Beckett, G., J. Arthur, S. Miller, and R. McKenzie. 2004. Minerals and Immune responses-selenium. p. 217–240. In D.A. Hughes et al. (ed.) Diet and human immune function. Humana Press, Totowa, NJ.

Berken, A., M.M. Mulholland, D.L. LeDuc, and N. Terry. 2002. Genetic engineering of plants to enhance selenium phytoremediation. Crit. Rev. Plant Sci. 21:567–582.

Blake-Kalff, M.M.A., K.R. Harrison, M.J. Hawkesford, F.J. Zhao, and S.P. McGrath. 1998. Distribution of sulfur within oilseed rape leaves in response to sulfur deficiency during vegetative growth. Plant Physiol. 118:1337–1344.

Blaylock, M., and B. James. 1994. Redox transformations and plant uptake of selenium resulting from root-soil interactions. Plant Soil 158:1–12.

Bloem, E., S. Haneklaus, I. Salac, P. Wickenhauser, and E. Schnug. 2007. Facts and fiction about sulfur metabolism in relation to plant-pathogen interactions. Plant Biol. (Stuttg.) 9:596–607.

Bogaards, J.J., H. Verhagen, M.I. Willems, G. van Poppel, and P.J. van Bladeren. 1994. Consumption of brussels sprouts results in elevated alpha-class glutathione S-transferase levels in human blood plasma. Carcinogenesis 15:1073–1075.

Borisjuk, L., H. Rolletschek, R. Radchuk, W. Weschke, U. Wobus, and H. Weber. 2004. Seed development and differentiation: A role for metabolic regulation. Plant Biol. (Stuttg.) 6:375–386.

Bourgis, F., S. Roje, M.L. Nuccio, D.B. Fisher, M.C. Tarczynski, C. Li, C. Herschbach, H. Rennenberg, M.J. Pimenta, T.L. Shen, D.A. Gage, and A.D. Hanson. 1999. S-methylmethionine plays a major role in phloem sulfur transport and is synthesized by a novel type of methyltransferase. Plant Cell 11:1485–1498.

Brader, G., M.D. Mikkelsen, B.A. Halkier, and E. Tapio Palva. 2006. Altering glucosinolate profiles modulates disease resistance in plants. Plant J. 46:758–767.

Brader, G., E. Tas, and E.T. Palva. 2001. Jasmonate-dependent induction of indole glucosinolates in Arabidopsis by culture filtrates of the nonspecific pathogen *Erwinia carotovora*. Plant Physiol. 126:849–860.

Broadley, M., P. White, R. Bryson, M. Meacham, H. Bowen, S. Johnson, M. Hawkesford, S. McGrath, F. Zhao, N. Breward, M. Harriman, and M. Tucker. 2006. Biofortification of UK food crops with selenium. Proc. Nutr. Soc. 65:169–181.

Brown, T., and A. Shrift. 1982. Selenium: Toxicity and tolerance in higher plants. Biol. Rev. 57:59–84.

Burow, M., U. Wittstock, and J. Gershenzon. 2008. Sulfur-containing secondary metabolites and their role in plant defense p. 205–226. In R. Hell et al. (ed.) Sulfur metabolism in phototrophic organisms. Springer Publisher, Dordrecht, the Netherlands.

Cai, X.-J., E. Block, P.C. Uden, B.D. Quimby, and J.J. Sullivan. 1995. Allium chemistry: Identification of natural abundance organoselenium compounds in human breath after ingestion of garlic using gas chromatography with atomic emission detection. J. Agric. Food Chem. 43:1751–1753.

Carvalho, K.M., M.T. Gallardo-Williams, R.F. Benson, and D.F. Martin. 2003. Effects of selenium supplementation on four agricultural crops. J. Agric. Food Chem. 51:704–709.

Charron, C., D. Kopsell, W. Randle, and C. Sams. 2001. Sodium selenate fertilisation increases selenium accumulation and decreases glucosinolate concentration in rapid-cycling Brassica oleracea. J. Sci. Food Agric. 81:962–966.

Chiaiese, P., N. Ohkama-Ohtsu, L. Molvig, R. Godfree, H. Dove, C. Hocart, T. Fujiwara, T.J. Higgins, and L.M. Tabe. 2004. Sulfur and nitrogen nutrition influence the response of chickpea seeds to an added, transgenic sink for organic sulfur. J. Exp. Bot. 55:1889–1901.

Chiao, J.W., F.L. Chung, R. Kancherla, T. Ahmed, A. Mittelman, and C.C. Conaway. 2002. Sulforaphane and its metabolite mediate growth arrest and apoptosis in human prostate cancer cells. Int. J. Oncol. 20:631–636.

Christiansen, P., J.M. Gibson, A. Moore, C. Pedersen, L. Tabe, and P.J. Larkin. 2000. Transgenic *Trifolium repens* with foliage accumulating the high sulfur protein, sunflower seed albumin. Transgenic Res. 9:103–113.

Combs, G., and W. Gray. 1998. Chemopreventive agents. Selenium. Pharmacol. Ther. 79:179–192.

Davis, C.D., H. Zeng, and J.W. Finley. 2002. Selenium-enriched broccoli decreases intestinal tumorigenesis in multiple intestinal neoplasia mice. J. Nutr. 132:307–309.

de Lumen, B.O., A.F. Galvez, M.J. Revilleza, and D.C. Krenz. 1999. Molecular strategies to improve the nutritional quality of legume proteins.p.117–127. *In* F. Shahidi et al. (ed.) Chemicals via higher plant bioengineering. Kluwer Academic/Plenum Publishers, New York.

de Souza, M.P., E.A. Pilon-Smits, C.M. Lytle, S. Hwang, J. Tai, T.S. Honma, L. Yeh, and N. Terry. 1998. Rate-limiting steps in selenium assimilation and volatilization by Indian mustard. Plant Physiol. 117:1487–1494.

Demidov, D., C. Horstmann, M. Meixner, T. Pickardt, I. Saalbach, G. Galili, and K. Müntz. 2003. Additive effects of the feed-back insensitive bacterial aspartate kinase and the Brazil nut 2S albumin on the methionine content of transgenic narbon bean *(Vicia narbonensis* L.). Mol. Breed. 11:187–201.

Dhillon, K., and S. Dhillon. 2003. Distribution and management of seleniferous soils. Adv. Agron. 79:119–184.

Dilworth, G., and R. Bandurski. 1977. Activation of selenate by adenosine 5′-triphosphate sulfurylase from *Saccharomyces cerevisiae*. Biochem. J. 163:521–529.

El Kassis, E., N. Cathala, H. Rouached, P. Fourcroy, P. Berthomieu, N. Terry, and J.-C. Davidian. 2007. Characterization of a selenate-resistant Arabidopsis mutant. root growth as a potential target for selenate toxicity. Plant Physiol. 143:1231–1241.

Ellis, D.R., and D.E. Salt. 2003. Plants, selenium and human health. Curr. Opin. Plant Biol. 6:273–279.

Fahey, J.W., A.T. Zalcmann, and P. Talalay. 2001. The chemical diversity and distribution of glucosinolates and isothiocyanates among plants. Phytochemistry 56:5–51.

Fahey, J.W., Y. Zhang, and P. Talalay. 1997. Broccoli sprouts: An exceptionally rich source of inducers of enzymes that protect against chemical carcinogens. Proc. Natl. Acad. Sci. USA 94:10367–10372.

Fenwick, G., R. Heaney, and W. Mullin. 1983. Glucosinolates and their breakdown products in food and food plants. Crit. Rev. Food Sci. Nutr. 18:123–201.

Fieldsend, J., and G.F.J. Milford. 1994. Changes in glucosinolates during crop development in single- and double-low genotypes of winter oilseed rape *(Brassica napus)*: Production and distribution in vegetative tissues and developing pods during development and potential role in the recycling of sulfur within the crop. Ann. Appl. Biol. 124:531–542.

Finley, J. 2003. Reduction of cancer risk by consumption of selenium-enriched plants: Enrichment of broccoli with selenium increases the anticarcinogenic properties of broccoli. J. Med. Food 6:19–26.

Finley, J.W. 2005. Proposed criteria for assessing the efficacy of cancer reduction by plant foods enriched in carotenoids, glucosinolates. polyphenols and selenocompounds. Ann. Bot. 95:1075–1096.

Finley, J.W., C.D. Davis, and Y. Feng. 2000. Selenium from high selenium broccoli protects rats from colon cancer. J. Nutr. 130:2384–2389.

Finley, J.W., C. Ip, D.J. Lisk, C.D. Davis, K.J. Hintze, and P.D. Whanger. 2001. Cancer-protective properties of high-selenium broccoli. J. Agric. Food Chem. 49:2679–2683.

Galili, G., R. Amir, R. Hoefgen, and H. Hesse. 2005. Improving the levels of essential amino acids and sulfur metabolites in plants. Biol. Chem. 386:817–831.

Gamet-Payrastre, L., P. Li, S. Lumeau, G. Cassar, M.A. Dupont, S. Chevolleau, N. Gasc, J. Tulliez, and F. Terce. 2000. Sulforaphane, a naturally occurring isothiocyanate, induces cell cycle arrest and apoptosis in HT29 human colon cancer cells. Cancer Res. 60:1426–1433.

Ganther, H.E. 1986. Parhway of selenium metabolism including respiratory excretory products. JAMA 5:1–5.

Ganther, H.E. 1999. Selenium metabolism, selenoproteins and mechanisms of cancer prevention: Complexities with thioredoxin reductase. Carcinogenesis 20:1657–1666.

Ganther, H., and J. Lawrence. 1997. Chemical transformations of selenium in living organisms. Improved forms of selenium for cancer prevention. Tetrahedron 53:12229–12310.

Hacham, Y., G. Schuster, and R. Amir. 2006. An in vivo internal deletion in the N-terminus region of Arabidopsis cystathionine gamma-synthase results in CGS expression that is insensitive to methionine. Plant J. 45:955–967.

Hagan, N.D., N. Upadhyaya, L.M. Tabe, and T.J.V. Higgins. 2003. The redistribution of protein sulfur in transgenic rice expressing a gene for a foreign, sulfur-rich protein. Plant J. 34:1–11.

Hecht, S.S. 1999. Chemoprevention of cancer by isothiocyanates, modifiers of carcinogen metabolism. J. Nutr. 129:768S–774S.

Hecht, S.S., N. Trushin, J. Rigotty, S.G. Carmella, A. Borukhova, S. Akerkar, and A. Rivenson. 1996. Complete inhibition of 4-(methylnitrosamino)-1-(3-pyridyl)-1-butanone-induced rat lung tumorigenesis and favorable modification of biomarkers by phenethyl isothiocyanate. Cancer Epidemiol. Biomarkers Prev. 5:645–652.

Hell, R. 1997. Molecular physiology of plant sulfur metabolism. Planta 202:138–148.

Hell, R., and H. Hillebrand. 2001. Plant concepts for mineral acquisition and allocation. Curr. Opin. Biotechnol. 12:161–168.

Hell, R., and M. Wirtz. 2008. Metabolism of cysteine in plants and phototrophic bacteria. p. 61–94. In R. Hell et al. (ed.) Sulfur metabolism in phototrophic organisms. Springer Publisher, Dordrecht, the Netherlands.

Hesse, H., and R. Hoefgen. 2008. Metabolism of methionine in plants and phototrophic bacteria. p. 95–112. In R. Hell et al. (ed.) Sulfur metabolism in phototrophic organisms. Springer Publisher, Dordrecht, the Netherland.

Hesse, H., O. Kreft, S. Maimann, M. Zeh, and R. Hoefgen. 2004. Current understanding of the regulation of methionine biosynthesis in plants. J. Exp. Bot. 55:1799–1808.

Higashi, Y., M.Y. Hirai, T. Fujiwara, S. Naito, M. Noji, and K. Saito. 2006. Proteomic and transcriptomic analysis of Arabidopsis seeds: Molecular evidence for successive processing of seed proteins and its implication in the stress response to sulfur nutrition. Plant J. 48:557–571.

Ihnat, M. 1989. Plants and agricultural materials. p. 33–104. In M. Ihnat (ed.) Occurrence and distribution of selenium. CRC Press, Boca Raton, FL.

Ip, C. 1998. Lessons from Basic Research in Selenium and Cancer Prevention. J. Nutr. 128:1845–1854.

Ip, C., M. Birringer, E. Block, M. Kotrebai, J.F. Tyson, P.C. Uden, and D.J. Lisk. 2000. Chemical speciation influences comparative activity of selenium-enriched garlic and yeast in mammary cancer prevention. J. Agric. Food Chem. 48:2062–2070.

+Ip, C., and H.E. Ganther. 1991. Combination of blocking agents and suppressing agents in cancer prevention. Carcinogenesis 12:365–367.

Ip, C., C. Hayes, R.M. Budnick, and H.E. Ganther. 1991. Chemical form of selenium, critical metabolites, and cancer prevention. Cancer Res. 51:595–600.

Jiao, D., C.C. Conaway, M.H. Wang, C.S. Yang, W. Koehl, and F.L. Chung. 1996. Inhibition of N-nitrosodimethylamine demethylase in rat and human liver microsomes by isothiocyanates and their glutathione, L-cysteine, and N-acetyl-L-cysteine conjugates. Chem. Res. Toxicol. 9:932–938.

Katz, Y.S., G. Galili, and R. Amir. 2006. Regulatory role of cystathionine-gamma-synthase and de novo synthesis of methionine in ethylene production during tomato fruit ripening. Plant Mol. Biol. 61:255–268.

Kim, H., M.Y. Hirai, H. Hayashi, M. Chino, S. Naito, and T. Fujiwara. 1999. Role of O-acetyl-L-serine in the coordinated regulation of the expression of a soybean seed storage-protein gene by sulfur and nitrogen nutrition. Planta 209:282–289.

Kliebenstein, D.J., J. Kroymann, and T. Mitchell-Olds. 2005. The glucosinolate-myrosinase system in an ecological and evolutionary context. Curr. Opin. Plant Biol. 8:264–271.

Kotrebai, M., M. Birringer, J. Tyson, E. Block, and P. Uden. 1999. Identification of the principal selenium compounds in selenium-enriched natural sample extracts by ion-pair liquid chromatography with inductively coupled plasma- and electrospray ionization-mass spectrometric detection. Anal. Commun. 36:249–252.

Kreft, O., R. Hoefgen, and H. Hesse. 2003. Functional analysis of cystathionine gamma-synthase in genetically engineered potato plants. Plant Physiol. 131:1843–1854.

Kruse, C., R. Jost, and R. Hell. 2005. Sulfur-rich proteins and their agrobiotechnological potential for resistance to plant pathogens. FAL Agric. Res. 283:73–80.

Kruse, C., R. Jost, M. Lipschis, B. Kopp, M. Hartmann, and R. Hell. 2007. Sulfur-enhanced defence: Effects of sulfur metabolism, nitrogen supply, and pathogen lifestyle. Plant Biol. (Stuttg) 9:608–619.

Kryukov, G.V., S. Castellano, S.V. Novoselov, A.V. Lobanov, O. Zehtab, R. Guigo, and V.N. Gladyshev. 2003. Characterization of mammalian selenoproteomes. Science 300:1439–1443.

LeDuc, D., M. AbdelSamie, M. Montes-Bayon, C. Wu, S. Reisinger, and N. Terry. 2006. Overexpressing both ATP sulfurylase and selenocysteine methyltransferase enhances selenium phytoremediation traits in Indian mustard. Environ. Pollut. 144:70–76.

LeDuc, D.L., A.S. Tarun, M. Montes-Bayon, J. Meija, M.F. Malit, C.P. Wu, M. AbdelSamie, C.-Y. Chiang, A. Tagmount, M. deSouza, B. Neuhierl, A. Bock, J. Caruso, and N. Terry. 2004. Overexpression of selenocysteine methyltransferase in Arabidopsis and Indian mustard increases selenium tolerance and accumulation. Plant Physiol. 135:377–383.

Leinfelder, W., and P. Heinrich. 1995. Process for preparing O-acetylserine, L-cysteine-related products. German patent WO97/15673.

Lewis, B.G., C.M. Johnson, and C.C. Delwiche. 1966. Release of volatile selenium compounds by plants. collection procedures and preliminary observations. J. Agric. Food Chem. 14:638–640.

Lucca, P., S. Poletti, and C. Sauter. 2006. Genetic engineering approaches to enrich rice with iron and vitamin A. Physiol. Plant. 126:291–303.

Makela, A.L., V. Nanto, P. Makela, and W. Wang. 1993. The effect of nationwide selenium enrichment of fertilizers on selenium status of healthy Finnish medical students living in south western Finland. Biol. Trace Elem. Res. 36:151–157.

Manici, L.M., L. Lazzeri, and S. Palmieri. 1997. In vitro fungitoxic activity of some glucosinolates and their enzyme-derived products toward plant pathogenic fungi. J. Agric. Food Chem. 45:2768–2773.

Martin, J., A. Shrift, and M. Gerlach. 1971. Use of 75Se-selenite for the study of selenium metabolism in Astragalus. Biochemistry 10:945–952.

Martin, M.N., M.C. Tarczynski, B. Shen, and T. Leustek. 2005. The role of 5'-adenylylsulfate reductase in controlling sulfate reduction in plants. Photosynth. Res. 86:1–15.

Maughan, S., and C. Foyer. 2006. Engineering and genetic approaches to modulating the glutathione network in plants. Physiol. Plant. 126:382–397.

McKenzie, R., J. Arthur, S. Miller, T. Rafferty, and G. Beckett. 2002. Selenium and immune function. p. 229–250. In P.C. Calder et al. (ed.) Nutrition and immune function. CABI Publ.,Wallingford, UK.

Medina, D., and D.G. Morrison. 1988. Current ideas on selenium as a chemopreventive agent. Pathol. Immunopathol. Res. 7:187–199.

Mikkelsen, M., P. Naur, and B. Halkier. 2004. Arabidopsis mutants in the C-S lyase of glucosinolate biosynthesis establish a critical role for indole-3-acetaldoxime in auxin homeostasis. Plant J. 37:770–777.

Mikkelsen, M.D., B.L. Petersen, C.E. Olsen, and B.A. Halkier. 2002. Biosynthesis and metabolic engineering of glucosinolates. Amino Acids 22:279–295.

Mithen, R.F., B.G. Lewis, and G.R. Fenwick. 1986. In vitro activity of glucosinolates and their products against *Leptosphaeria maculans*. Trans. Br. Mycol. Soc. 87:433–440.

Molvig, L., L.M. Tabe, B.O. Eggum, A.E. Moore, S. Craig, D. Spencer, and T.J.V. Higgins. 1997. Enhanced methionine levels and increased nutritive value of seeds of transgenic lupins (*Lupinus angustifolius* L.) expressing a sunflower seed albumin gene. Proc. Natl. Acad. Sci. USA 94:8393–8398.

Moreno Rodriguez, M., V. Cala Rivero, and R. Jiménez Ballesta. 2005. Selenium distribution in topsoils and plants of a semi-arid Mediterranean environment. Environ. Geochem. Health 27:513–519.

Morton, R.L., A.J. Ellery, and T.J. Higgins. 1998. Downstream elements from the pea albumin 1 gene confer sulfur responsiveness on a reporter gene. Mol. Gen. Genet. 259:309–316.

Müller, S., J. Heider, and A. Bock. 1997. The path of unspecific incorporation of selenium in Escherichia coli. Arch. Microbiol. 168:421–427.

Müntz, K., V. Christov, G. Saalbach, I. Saalbach, D. Waddell, T. Pickardt, O. Schieder, and T. Wustenhagen. 1998. Genetic engineering for high methionine grain legumes. Nahrung 42:125–127.

Ng, B.H., and J.W. Anderson. 1978. Synthesis of selenocysteine by cysteine synthase from selenium accumulator and nonaccumulator plants. Phytochemistry 17:2069–2074.

Nordlee, J.A., S.L. Taylor, J.A. Townsend, L.A. Thomas, and R.K. Bush. 1996. Identification of a Brazil-nut allergen in transgenic soybeans. N. Engl. J. Med. 334:688–692.

Oenema, O., and R. Pastma. 2003. Managing sulfur in agroecosystems. p. 45–70. In Y.P. Arbol and A. Ahmad (ed.) Sulfur in plants. Kluwer Academic Press, Dordrecht, the Netherlands.

Onouchi, H., I. Lambein, R. Sakurai, A. Suzuki, Y. Chiba, and S. Naito. 2004. Autoregulation of the gene for cystathionine gamma-synthase in Arabidopsis: Post-transcriptional regulation induced by S-adenosylmethionine. Biochem. Soc. Trans. 32:597–600.

Orser, C.S., D.E. Salt, I.J. Pickering, R. Prince, A. Epstein, and B.D. Ensley. 1999. Brassica plants to provide enhanced human mineral nutrition: Selenium phytoenrichment and metabolic transformation. J. Med. Food 1:253–261.

Pasricha, N.S., and Y.P. Abrol. 2003. Food production and plant nutrient sulfur. p. 29–44. In Y.P. Abrol and A. Ahmad (ed.) Sulfur in plants. Kluwer Academic Press, Dordrecht, the Netherlands.

Poggi, V., A. Arcioni, P. Filippini, and P. Pifferi. 2000. Foliar application of selenite and selenate to potato (*Solanum tuberosum*): Effect of a ligand agent on selenium content of tubers. J. Agric. Food Chem. 48:4749–4751.

Rabot, S., L. Nugon-Baudon, and O. Szylit. 1993. Alterations of the hepatic xenobiotic-metabolizing enzymes by a glucosinolate-rich diet in germ-free rats: Influence of a pre-induction with phenobarbital. Br. J. Nutr. 70:347–354.

Rask, L., E. Andreasson, B. Ekbom, S. Eriksson, B. Pontoppidan, and J. Meijer. 2000. Myrosinase: Gene family evolution and herbivore defence in Brassicaceae. Plant Mol. Biol. 42:93–113.

Rausch, T., and A. Wachter. 2005. Sulfur metabolism: A versatile platform for launching defence operations. Trends Plant Sci. 10:503–509.

Rayman, M. 2000. The importance of selenium to human health. Lancet 356:233–241.

Rayman, M. 2002. The argument for increasing selenium intake. Proc. Nutr. Soc. 61:203–215.

Reid, M.E., A.J. Duffield-Lillico, L. Garland, B.W. Turnbull, L.C. Clark, and J.R. Marshall. 2002. Selenium supplementation and lung cancer incidence: An update of the nutritional prevention of cancer trial. Cancer Epidemiol. Biomarkers Prev. 11:1285–1291.

Sevenier, R., I.M. van der Meer, R. Bino, and A.J. Koops. 2002. Increased production of nutriments by genetically engineered crops. J. Am. Coll. Nutr. 21:199S–204S.

Shrift, A. 1969. Aspects of selenium metabolism in higher plants. Annu. Rev. Plant Physiol. 20:475–494.

Shutov, A.D., H. Baumlein, F.R. Blattner, and K. Muntz. 2003. Storage and mobilization as antagonistic functional constraints on seed storage globulin evolution. J. Exp. Bot. 54:1645–1654.

Sirko, A., A. Blaszczyk, and F. Liszewska. 2004. Overproduction of SAT and/or OASTL in transgenic plants: A survey of effects. J. Exp. Bot. 55:1881–1888.

Sors, T.G., D.R. Ellis, G.N. Na, B. Lahner, S. Lee, T. Leustek, I.J. Pickering, and D.E. Salt. 2005a. Analysis of sulfur and selenium assimilation in Astragalus plants with varying capacities to accumulate selenium. Plant J. 42:785–797.

Sors, T.G., D.R. Ellis, and D.E. Salt. 2005b. Selenium uptake, translocation, assimilation and metabolic fate in plants. Photosynth. Res. 86:373–389.

Stoner, G.D., L.A. Kresty, P.S. Carlton, J.C. Siglin, and M.A. Morse. 1999. Isothiocyanates and freeze-dried strawberries as inhibitors of esophageal cancer. Toxicol. Sci. 52:95–100.

Tabe, L., N. Hagan, and T.J. Higgins. 2002. Plasticity of seed protein composition in response to nitrogen and sulfur availability. Curr. Opin. Plant Biol. 5:212–217.

Tabe, L.M., and M. Droux. 2002. Limits to sulfur accumulation in transgenic lupin seeds expressing a foreign sulfur-rich protein. Plant Physiol. 128:1137–1148.

Talalay, P., and J.W. Fahey. 2001. Phytochemicals from cruciferous plants protect against cancer by modulating carcinogen metabolism. J. Nutr. 131:3027S–3033S.

Terry, N., A.M. Zayed, M.P. De Souza, and A.S. Tarun. 2000. Selenium in higher plants. Annu. Rev. Plant Physiol. Plant Mol. Biol. 51:401–432.

Tierens, K.F., B.P. Thomma, M. Brouwer, J. Schmidt, K. Kistner, A. Porzel, B. Mauch-Mani, B.P. Cammue, and W.F. Broekaert. 2001. Study of the role of antimicrobial glucosinolate-derived isothiocyanates in resistance of Arabidopsis to microbial pathogens. Plant Physiol. 125:1688–1699.

Uden, P., S. Bird, M. Kotrebai, P. Nolibos, J. Tyson, E. Block, and E. Denoyer. 1998. Analytical selenoamino acid studies by chromatography with interfaced atomic mass spectrometry and atomic emission spectral detection. Fresenius J. Anal. Chem. 362:447–456.

Vinceti, M., E. Wei, C. Malagoli, M. Bergomi, and G. Vivoli. 2001. Adverse health effects of selenium in humans. Rev. Environ. Health 16:233–251.

Vitale, A., and G. Hinz. 2005. Sorting of proteins to storage vacuoles: How many mechanisms? Trends Plant Sci. 10:316–323.

Wang, L., D. Liu, T. Ahmed, F. Chung, C. Conaway, and J. Chiao. 2004. Targeting cell cycle machinery as a molecular mechanism of sulforaphanein prostate cancer prevention. Int. J. Oncol. 24:187–192.

Wang, T.L., C. Domoney, C.L. Hedley, R. Casey, and M.A. Grusak. 2003. Can we improve the nutritional quality of legume seeds? Plant Physiol. 131:886–891.

Whanger, P. 2004. Selenium and its relationship to cancer: An update dagger. Br. J. Nutr. 91:11–28.

Whanger, P.D. 2002. Selenocompounds in plants and animals and their biological significance. J. Am. Coll. Nutr. 21:223–232.

White, P.J., H.C. Bowen, B. Marshall, and M.R. Broadley. 2007. Extraordinarily high leaf selenium to sulfur ratios define 'Se-accumulator' plants. Ann. Bot. 100:111–118.

White, P.J., H.C. Bowen, P. Parmaguru, M. Fritz, W.P. Spracklen, R.E. Spiby, M.C. Meacham, A. Mead, M. Harriman, L.J. Trueman, B.M. Smith, B. Thomas, and M.R. Broadley. 2004. Interactions between selenium and sulfur nutrition in Arabidopsis thaliana. J. Exp. Bot. 55:1927–1937.

Wilson, L.G., and R.S. Bandurski. 1958. Enzymatic reactions involving sulfate, sulfite, selenate, and molybdate. J. Biol. Chem. 233:975–981.

Wittstock, U., and B. Halkier. 2002. Glucosinolate research in the Arabidopsis era. Trends Plant Sci. 7:263–270.

Wu, L. 1998. Selenium accumulation and uptake by crop and grassland plant species.p. 657–686. *In* W.T. Frankenberger and R.A. Engberg (ed.) Environmental chemistry of selenium. Marcel Dekker, New York.

Wu, L. 2004. Review of 15 years of research on ecotoxicology and remediationof land contaminated by agricultural drainage sediment rich in selenium. Ecotoxicol. Environ. Saf. 57:257–269.

Yang, F., L. Chen, Q. Hu, and G. Pan. 2003. Effect of the application of selenium on selenium content of soybean and its products. Biol. Trace Elem. Res. 93:249–256.

Yu, R., S. Mandlekar, K.J. Harvey, D.S. Ucker, and A.N. Kong. 1998. Chemopreventive isothiocyanates induce apoptosis and caspase-3-like protease activity. Cancer Res. 58:402–408.

Zhao, F., S.E. Salmon, P.J.A. Withers, J.M. Monaghan, E.J. Evans, P.R. Shewry, and S.P. McGrath. 1999. Variation in the breadmaking quality and rheological properties of wheat in relation to sulfur nutrition under field conditions. J. Cereal Sci. 30:19–31.

Index

11S glycinin, 237
7S globulins, 239

AA, formation of, 162
Absorption, direct, S gasses by soil and plants, 65–66
Accretion, rainfall studies, 206
Acid rain, 45
Acid soils, S components, 62
Acid sulfate soils, 63, 202–203
Acid, Sulfuric, 66
Acidity, oxidation of S, 5
Acrylamide formation, 161–166
Adsorption, sulfate, 31, 70–71
Aeration of soils, rice production, 200
Aerobic soils, rice production, 199–200
Akiochi, 204
Alfalfa
   methionine, increasing levels, 270–271
   predicting response, S, 74
   S critical values, at growth stages, 19, 77
   S removal, 70
   S shoot uptake, 126
Allergens, 240, 242, 268–269
Alliin, 185, 191
Alliums, 184, 189–192
Allysulfur, diagrams, 287
Amino acids
   biosynthetic production, 295
   content, improving in soybeans, 237–239
   forms of, 3
   humans, 282–287
   improving in soybeans for animal feeds, 235–244
   onions, 184–185
   plants, 205
   S in foods, 283
   soybeans, 121–122
Ammonium phosphate sulfate, application management, 20
Ammonium sulfate
   application management, 20
   common scab, potatoes, 174–176
   concentrations, 67
   excessive fertilization, 32
   fertilizer response in peas, 109
   fertilizers, S form analysis, 6
   N-P-K nutrient concentrations, 67
   production, 66, 68
   S removal and, 117–118
Ammonium thiosulfate, 6, 67
Animal feed, soybeans, importance of, 235–236
Animal health, 46–47
Anion exchange resin strip, rice testing, 213
Anthocyanines, formation, 53
Anthocyanins, formation, 52
Antibacterials, 287–288
Applications
   based on S fertility, 131–132
   canola sulfate effectiveness and growth stage, 112
   dosages, 46
   fertilizer, methods, 21
   fertilizers, timing of, 21
   nitrogen on corn, 147
   S, dose recommendations, 20
   S, IGA, 16
   S, time and frequency, corn, 149
   soybean production, 125–129
   wet ammonium sulfate and thiosulfate, 175–176
APR, pathway in sulfate reduction, 89
APS reductase, 244
APS, sulfate assimilation, plants, 88
Aquoll, nature of organic S, 4
Arabidopsis, 93
Argentina, 128, 130
Arid conditions, 8, 62
Aridic boroll, 4, 8
Ashed rice straw, 202
Asia, S deficit area, 11
ASP1, 242–243
Asparagine, concentrations in plants, 163
Asparagine models, 162
Aspartate kinase, 257
Assimilation, S, 84–88, 266
Atlantic Coastal Plains, corn yields, 147
Atmospheric S, 63
ATP-bound sulfate, 88–91
ATP sulfurylase, 244
ATPS isoforms, 89–90
Available S, plants, 59–78
Avocado, 283

Bagels, acrylamide concentrations, 161
Baking quality, bread (wheat), 47, 158–159

Bananas, 127, 283
Bangladesh
    irrigation studies, rice, 206
    response to S in rice, 16–17
    S application and topsoil type, 128
    S deficiencies, 11–12
Barley, 77, 109, 126–127
Beans, 126, 283
Beets, S shoot uptake, 126–127
Benzyl-isothiocyanate, bacterial infections and, 51
Bermuda, coastal, S shoot uptake, 126–127
Berseem, yield increase due to S, 15
Bioactive S compounds, humans, 287–289
Biochemical activation, sulfate, 88
Biochemical mineralization, 29–30
Biological mineralization, 29–30
Biomass S, 28
Biosolid sources, N-P-K nutrient concentrations, 67
Biotechnology, S and selenium, 293–303
Biotin, 3, 205, 284
Biscuits, acrylamide concentrations, 161
Black scurf, potatoes, 172
Blackgram, 14–15, 17, 20
BOLIDES evaluation, mathematical assessment S, 50, 221–228
Boralf, 4, 8
Boron, optimum ranges, 225–227
Boundary line development systems, 221–228
Boundary step function, 223
Bowman-Birk protease inhibitor, 267–268
Brazil, 128, 134
Brazil nut protein, 268, 296–297
Bread, 110, 121, 161
Brigets, UK, S, wheat yields, 156
Brinjal, S concentrations in dry matter, 19
Broken stick method, 220–221
Build-up, studies, 34–35

C/S ratio, 3, 36, 61
Cabbage, S shoot uptake, 126–127
Calcium, 224, 228
Calcium, potatoes and, 176, 225–227
California, 188
Canada, S application and topsoil type, 128
Canada, western, 106, 108
Cancer, prevention, humans, 282, 289, 300
Canola
    demand for S, 111
    glucosinolate content, 47–51
    S critical values, at growth stages, 77
    S deficiencies, characteristics, 52–53
    S deficiency susceptibility, 110
    S removal, 70
    sulfate effectiveness and growth stage, 112
    symptoms, S deficiency, 110
    yields, Rock Lake, ND, 107
Carbo-amino skeleton, 253, 259–260
Carbon-amino flow into methionine, 257
Carbon backbone, cysteine biosynthesis, 91–93
Carbon bonding, S, 26
Carbon, organic ratios in soils, 61
Carrots, 283
CAS enzymes, function, 93
Cassava, 127, 283
Castorbean, S shoot uptake, 126–127
Catabolic pathways, methionine, 261–264
Cauliflower, S concentrations, dry matter, 19
Cell proliferation, 88
Cereal grains, methionine content, increasing, 269
Cereals
    acrylamide concentrations, 161
    optimum nutrient ranges, 226
    S deficiency, 53
    threshold values, concentrations, S, 228
CGS levels, regulation, 253–255
Cheese, 121, 283
Chemical extraction, organic S in soils, 26–27
Chicken, 283
Chickpeas
    methionine contents, 269
    protein content increase due to S, 18
    S application rates, 20
    S concentrations in dry matter, 19
    S, N, P, K uptakes, IGA of Asia, 14
    S shoot uptake, 126–127
    yield increase due to S, 15, 17
China, rainfall studies, 206
Chips, corn and potato, acrylamide concentrations, 161
Chives, 185
Chlorine, optimum ranges, 225–227
Citrus, S shoot uptake, 127
Clean air acts, 45–47, 63
Clinton, NC, 150
Clovers, 126, 270–271, 297
Cocoa, S shoot uptake, 126–127
Coconut, S shoot uptake, 126
Coenzyme A., 284
Coffee, S shoot uptake, 126–127
Columbia basin region, WA, 188
Common scab, potatoes, 172–177
Composts, N-P-K nutrient concentrations, 67
Concentrations, S, 13–15, 59, 119–120, 228
Copper, optimum ranges, 225–227
Corn
    nitrogen application, 147
    S critical values, at growth stages, 77

Index

Corn *continued*
   S fertilization, 109, 145–150
   S needs prediction, 143–144
   S removal, 70
   silage, quality, 149–150
   yields, 147–148
Corn chips, acrylamide concentrations, 161
Cotton, 70, 77, 126
Cowpea, S concentrations in dry matter, 19
Critical nutrient value, 220
Critical S concentrations in plants, 77, 220–221
Critical values, S in plants, 18–20
Crop residues, 208
Crops
   demand for S, 107–110
   determining S needs, 112–113
   managing S content, 18–21
   predicting response available S, 72–78
   predicting S requirements, 50
   productivity variables, 83–84
   removal of S, northern great plains, 108
   response to S application, 15–17
   S management, 21–22
   S nutrition and quality, IGA south Asia, 17–18
   S requirements and responsiveness, 108–110
   S uptake and removal, 69–72
Cucumber, S concentrations in dry matter, 19
Cycle, S, 1, 32
Cycling, S, model, temperate soils, 31–32
Cystathionine, regulatory, plants, 253–255
Cysteine
   7S β-conglycinin, 237
   animal feeds, 236
   biosynthesis regulation, 96
   content, mungbeans and S application, 18
   diseases and disorders, humans, 286–287
   enhanced contents, 295–296
   flavor, 184–187
   forms of, 3
   free, 296
   function of, 107
   genetic manipulation, 96–97
   human nutrition, 282–287, 294
   Hyperhomocysteinemia, humans, 286–287
   metabolic relationships and functions, humans, 285
   metabolism, S, 83–98
   production of, 88–89
   reductive pathway to, 88–91
   S flow regulation to methionine synthesis, 257–258
   S role in plants, 205
   SAT relationship, 92
   sinks, 296–297
   soybeans, 121–122
   structure, 183
   sulfoxides, 185
   synthase, metabolic regulation of, 95
   synthesis process, 91–97
   synthesized in seeds, 266
   transsulfurylation, 295–296
   uses of, human bodies, 283–284
Cysteine synthase complex (CSC), 94–97

Davis, CA, 150
Decomposable S sources, 3
Deficiency, protein, human, 264
Deficiency, S
   assessing in soybeans, 129–136
   causes, 1, 60, 84
   characteristics, canola (rapeseed), 52–53, 110
   characteristics, potato, 171–172
   characteristics, soybeans, 124–125
   characteristics, Sugar beets, 54
   critical values, plants and growth stages, 19
   in crops, 45–54
   diagnosing, rice, 210–212
   humans, 286
   IGP of Asia, 12–13
   latent, 129
   levels, wheat, 154
   likelihood, 105–106
   lupines, 87–88
   northern great plains, 105–107
   onions, 189
   origins of, 45–46
   quantifying equation, 74
   ranges, 224
   rice, 87–88, 198–199, 208, 210–212
   soybeans, 124–129
   and sufficiency levels, 19
   symptoms, 108
   testing in plants, 76–78
   visual diagnosis in plants, 52–54
   western Canada, 106
   world soybean production, 117–118
Denmark, 37–38
Deposition, 12, 150
Deprivation, SAT levels, 92
Designer proteins, soybeans, 242–243
Diagnosing S deficiencies in soybeans, 129–136
Diets, humans, S amino acids, 282–287
Digestibility, corn silage, 149–150
Diminishing returns, law of, 228–229
Disorders, nutrition, 47–51

Dissimilatory sulfate reduction, 5
Distribution, S in rice, 204–205
Disulphide bonds, 159
Dixon Springs, IL, 150
DNA hypomethylation, humans, 286
Dough rheology, 159–160
Dry matter accumulation, soybeans, 119
Dry matter, rice production, 207
Dust, S deposition, 65

Eggplants, S shoot uptake, 127
Eggs, protein contents, 121
Elasticity, dough, 160
Elemental S
    advantages, 33
    application management, 20
    fertilizer, 69
    fertilizer response in peas, 109
    fertilizers, S form analysis, 6
    forms, 5
    N-P-K nutrient concentrations, 67
    oxidation, 111
    oxidized to sulfate, 7
    rice, 208–209
Elephant garlic, 185
Emulsions, suspensions, fertilizers, S form analysis, 6
Environment, qualities, S, 47
Enzymes, 29–30, 244
ER protein bodies, 270–271
Ester-bound S, rice uptake, 213

Feed additives, alternative, 51
Fertility, S, influence on, 5
Fertilizer
    alliums, 187–189
    analysis of, 6
    application, 21, 49, 229–231
    corn, programs, 143–150
    crop system responses, 21
    deposition, selected locations, 150
    fluid, 148
    inorganic and organic sources, S, 67
    management, rice, 205–213
    management, S in corn, 145–149
    S amounts, 61
    S inputs, 66–69
    S response, rice, 208–210
    sources, corn yields, 148
    sulfate, effectiveness in canola, 112
Fertilizers
    application management, 20
    N P K, Denmark experiments, 34
    plant-available, 34–36
    S management,NGP, 110–112
    sulfate-containing, 32–37

FGD. *see* Flue gas desulfurization (FGD)
Fish, 121, 283
Flavor, cysteine sulfoxides, 184–187
Flour, 153, 160
Flue gas desulfurization (FGD), 68
Foliar uptake, 65–66, 125, 129
Folic acid, human production, 283–284
Foods, processed, acrylamide concentrations, 161
Foods, protein contents, 121
Forage grasses, S, N, P, K uptakes, IGA of Asia, 14
Forests, S deficiency and, 48
Free sulfate, 85–86
French bean, S concentrations in dry matter, 19
Fungicidal effect, S, 49–50
Fusarium head blight, S applications, 49–50

Garlic
    Alliin, 185
    allysulfur compounds, humans, 287–288
    cysteine sulfoxides, 185
    production, 188
    S shoot uptake, 127
Garrison, ND., 109
Gary, IN, S depositions, 63
Gasses, S, 65–66, 71–72
Genetic manipulation, cysteine, 96–97
Genetic regulation, onions, 192–193
Genotypes, free amino acids, 162
Georgia, 188–189
Gliadin proteins, available S in wheat, 157
Gliadins proteins, 156–157
Global S cycle, 63–65
Glucosinolates, 88, 287, 302–304
Glutathione (GHS), 205
    cysteine synthesis process, 91–97
    enhanced cysteine contents, 296
    PIPPA estimates, 230
    reductant, 89
    rice, 204
    S transportation, 87
Glutenin proteins, wheat qualities, 157–159
Glycinin, 121–122, 241
Glycosamineglycans, 284
Gobhi sarson, S nutrition, 16
Grain filling, 86, 297
Grain, rice S levels, 211
Grain sorghum, S removal, 70
Grain sulfur content, rice, 204
Grains, 33
Grains, deficiencies and S distribution, 135
Grapes, S shoot uptake, 126–127
Grasses, cool-season, S removal, 70
Grasses, forage, 14, 20, 126

Index

Grasses, tropical, S shoot uptake, 126–127
Great Plains, S fertilizer management, 110–112
Green gram, 17–18, 20, 126
Green manures, 37
Ground water, 20, 48–49
Groundnut, 14–21, 126
Gypsum
    application management, 20
    availability, rice, 208
    common scab, potatoes, 176–177
    fertilizer, 66–67
    fertilizers, S form analysis, 6
    N-P-K nutrient concentrations, 67
    sources, 65
    wet limestone forced, process, 68–69

Health, nutraceuticals, 51
Heterologous genes, soybeans, 239–242
Heterotrophic bacteria, 111
Histidine, human requirements, 282
Homocysteine, humans, 285
Honeybees, 47–48
Human health, 281–289, 298–299
Humus fraction, organic S, 3
Hydrogen sulfide assimilation, 85
Hydrogen sulfide toxicity, 204
Hydrogen sulfide uptake, 65–66, 177–178
Hyperhomocysteinemia, humans, 286–287
Hypervariable region, soybeans, 240–241

IGP, 12–13
Immobilization, microbial plant available S, 3
Immobilization, microbial S, 2, 71
India, 60, 128, 134, 211
Industrial deposits, S, 63
Inorganic S, 31, 62–63
Insects, onion sulfur, 192
Iron, optimum ranges, 225–227
Iron pyrites, application management, 20
Iron sulfide mineral formation, 5
Irrigation, 20, 48–49, 206–207
Isoleucine, human requirements, 282
Isothiocyanates, 287

Japanese bunching onions, 185
Japanese over-wintering onions, 190

Kalamazoo, MI, 150

Lamb, 283
Lamberton, MN, 150
Landscape, S qualities and, 48
Law of diminishing returns, 228–229
Law of the Minimum, 220, 228–229
Leaching
    nitrates, 37
    nitrogen, 47
    rice, 207–208
    S, 69–71
    S balance studies, 38
    soil texture, 130
Leaves, 108, 120, 124–125, 132–134, 211
Leeks, cysteine sulfoxides, 185
Legumes, 20, 297
Legumin, regulation in plants, 267
Lentils
    S amino acid in foods, 283
    S application rates, 20
    S, N, P, K uptakes, IGA of Asia, 14
    S shoot uptake, 126–127
Leucine, human requirements, 282
Liebig's Law of the Minimum, 220, 228–229
Lime, common scab, potatoes, 176–177
Linseed, 15, 17–19
Livestock, use of soybeans, 236
Loaf, bread, S influence, 161
Loam, 109
Lunasin, soybeans, 241–242
Lupines, 87–88, 269, 297
Luvisolic soils, sulfate-S concentrations, 7–8
Lysine, 253, 259–260, 282

Magnesium, optimum ranges, 225–227
Magnesium sulfate, N-P-K nutrient
        concentrations, 67
Maillard reaction, 165–166
Maize
    direct and residual benefits, S fertilizer, 21
    S application rates, 20
    S concentrations in dry matter, 19
    S critical values, at growth stages, 19
    S, N, P, K uptakes, IGA of Asia, 14
    S shoot uptake, 126
    yield increase due to S, 15
Malaysia, rainfall studies, 206
Manganes, 225–227
Manganese sulphate, common scab,
        potatoes, 176–177
Manures
    animal, S content, 68
    C/S ratio, 36
    dairy, N-P-K nutrient concentrations, 67
    general animal, N-P-K nutrient
        concentrations, 67
    green, catch crops, 37
    N-P-K nutrient concentrations, 67
    poultry, N-P-K nutrient concentrations, 67
    S contents, 61
    S sources, 33
    S through mineralization, 4, 34–35
    sheep, N-P-K nutrient concentrations, 67
Meat, 121, 283

Mehlich III, soil fertility, 76
Met, 258–259
Metallic sulfates, common scab, potatoes, 176–177
Methionine
 7S β-conglycinin, 237
 accumulation in seed proteins, 267–268
 animal feeds, 236
 aspartate kinase, 257
 AtCGS levels and, 254
 biosynthesis pathway, plants, 252–253
 catabolism, 261–264
 content, 255
 content, mungbeans and S application, 18
 content, S nutrition, IGA south Asia, 17–18
 cycle and regulation, humans, 285
 depletion, animals, 264
 diseases and disorders, humans, 286–287
 feedback-insensitive forms, 295
 forms of, 3
 free, 296
 function of, 107
 human nutrition, 282–287
 improving levels in seeds, 268–271
 increasing levels, forage crop tissues, 270–271
 isoleucine, regulation, 260
 metabolic relationships and functions, humans, 285
 metabolism, human, 281
 metabolism, plants, 251–272
 metabolism, synthesis of proteins, 261
 nutritional quality, 264–271
 plants, importance of, 251–252
 recycling pathways, 253, 261–263
 S role in plants, 205
 SAM and, 256
 sinks, 296–297
 soybeans, 121–122
 structure, 183
 synthesis, 265–267, 296
 through SAM, 264
 TS levels, 255
 uses of, human bodies, 283–284
Methyl cycle, 261
Methyl cysteine sulfoxide, 191
Methyl, methionine and, 253, 259
Microbes, 5–7, 28
Microbial mineralization and immobilization, defined, 2
Microbial oxidation, of S, 1
Milk, 121, 283
Millet, 19, 126–127

Mineralization
 biological and biochemical, 29–30
 catch crop S, 37
 microbial S, 2–5
 mineralization in humus, 4
 of S, 1
 S in straw, 202
 S, particle size, 26
 soil organic matter, 60–62
 soil organic S, 28–31
 soil tests, S availability, 72
 sulphatase enzymes, 29–30
Mineralization-immobilization, factors controlling, 3–4
Minerals, essentials, 219
Minimum, law of the, 220, 228–229
Minnesota, S fertilizer response in corn, 109
Mitscherlich's law of diminishing returns, 228–229
Molecular weight fractionation, organic S, soils, 27
Molybdenum, optimum ranges, 225–227
Mono-ammonium phosphate, rice, 209–210
MOPS, 50
MTA cycle, 261
MTO1 domain, 253–254
Muleshoe, TX, 150
Mungbean, 13–14, 16, 18–19
Mustard, 18–19, 21
Mustard oil bomb, GS-myrosinase system, 302

N/S ratio, 49, 155–156, 294
Nebraska, S, corn yields, 147
New Mexico, 188
New York, 188
Nitrogen
 application, corn, 147
 availability, S and, 77–78
 flooded soils, 202
 leaching, 47
 optimum ranges, 225–227
 organic ratios in soils, 61
 regulation of S, 122
 S and metabolism, 155–156
 S deficiency and, 130
North Dakota, fertilizer studies, 109
Northern great plains, 105–107, 110–112
Nutraceuticals, health and, 51
Nutrient supply, algorithms for deducting, 220–221
Nutrition, S
 humans, 281–289, 298–299
 major disorder, S, 47–51
 optimum ranges, 225–227
 optimum ranges, oilseed rape, 225

Nutrition, S *continued*
  plant stress, 133, 135
  S, IGA south Asia, 17–18
  S in crops, Indo-Gangetic plains, South Asia, 11–22
  S in plants, 105
  S in rice, 203–205
  threshold markers, supply, 219

O-phosphohomoserine, 255–256
OAS formation, SAT and, 92
OAS, function, 93
Oats, S critical values, at growth stages, 77
Ohio
  S deficiencies, 60
  S depositions, 65
  soil S availability, 74
Oilpalm, S shoot uptake, 126–127
Oils, content, S nutrition, IGA south Asia, 17–18
Okra, S shoot uptake, 127
Onions
  bulb quality, 190–192
  cysteine sulfoxides, 185
  S deficiency, 189
  S role, 183–193
  S shoot uptake, 126–127
  yield increase, 15
Optimum nutrient ranges, 225–227
optimum ranges, 225–227
Oranges, S removal, 70
Ordinary superphosphate, N-P-K nutrient concentrations, 67
Organic carbons, S build-up, studies, 34–35
Organic farms, 50
Organic materials, mineralization rates, 36
Organic matter, 106–107, 199–200
Organic S, 61
Organic S pools in soils, 26–30, 34–35
Organic sulfates, 3
OSTIL, function, 93–94
Outliers, identification of, 221–222
Overexpression, cysteine, methionine, 296
Oxidation
  S, 27–28, 62
  S fertilizers, 33
  submerged conditions, 202
  Thiobacillus, 69
  wet limestone forced, process, 68–69
Oxidation rate, plant-available S, 5
Oxidized rhizosphere, rice, 203, 208

Papays, S shoot uptake, 127
Particles, S, 6–7, 26
Peanuts, 70, 77
Pearling index, 160

Peas, S fertilizer, response and soil type, 109
PEM. *see* Protein-Energy Malnutrition (PEM)
Penn State, PA, 150
Petal fall, 48
Petals, S deficiency, 53
PH levels, 7, 70–71, 172–173, 201–202
Philippines, 206
Phloem transport, sulfate-sulfur, rice, 204
Phosphorus, optimum and deficiency ranges, 130, 225–227
Physical protection, S pools in soil, 27
Phytochelatin, compounds, 205
Phytoremediation, 300
Phytovolatilization, 300–301
Pigeonpea, 14, 126–127
Pineapples, S shoot uptake, 127
PIPPA, fertilizer application recommendations, 228–231
PIPPA3, S routine, 230
Plant-available S, 5, 34–36
Plant tissues, S concentration and uptake, 13–15
Plants
  analysis of nutrition, 48, 144–145, 210–212
  available S in soil, equations for in and outputs, 73–74
  bioactive S compounds, humans, 287–289
  diagnostics, S availability, 72
  direct absorption, S gasses, 65–66
  importance of S, 59–60
  metabolism, 83
  methionine, 251–253
  methionine metabolism, 251–272
  N/S ratio, 113
  S amino acid in foods, 283
  S biochemical role, 205
  S fertility, assessing, 130
  S roles in, 107–108
Pod development, S deficiency, 53
Polypeptides, S function, 205
Pork, 283
Potassium magnesium sulfate, 67
Potassium, optimum ranges, 225–227
Potassium sulfate
  application management, 20
  concentrations, 67
  fertilizers, S form analysis, 6
  N-P-K nutrient concentrations, 67
  S removal and, 117–118
Potassium thiosulfate, fertilizers, S form analysis, 6
Potatoes
  high methionine, 254
  S amino acid in foods, 283

Potatoes *continued*
    S and marketable yield, 15, 171–178
    S concentrations in dry matter, 19
    S removal, 70
    S shoot uptake, 127
Precipitation, S depositions, 63–65
Professional interpretation program for plant analysis, 228–229
Propyl cysteine sulfoxide, 191
Protein-Energy Malnutrition (PEM), 264
Proteins
    biosynthesis, 87
    components, 3
    components, analysis, soybeans, 136
    composition, improving in soybeans, 237–239
    consumption, soybeans, 122
    content, S nutrition, IGA south Asia, 17–18
    content, soybean animal feeds, 235–236
    function, humans, 284
    impairment, humans, 286
    plant, S regulation, 267
    salt-soluble, 237
    stored in soybeans, 121
    structure, 183
    synthesis, 59
    synthesis, humans, 283–284
    synthetic, amino acids in soybeans, 242–243
    wheat, composition, S and, 156–157
Pullman, WA, 150
Pungency testing, alliums, 186–7
Push and pull approaches, biotechnology, 295–297
Pyruvate testing, 186–7, 191

Quincy, FL, 150

Radial oxygen loss (ROL), rice, 204
Rainfall, S dissolution, 106
Rakkyo, 185
Rapeseed canola
    demand for S, 111–112
    fertilizer application PIPPA calculations, 228–231
    glucosinolate content, 47–51
    S critical values, at growth stages, 77
    S deficiencies, 52–53, 110
    S removal, 70
    S shoot uptake, 126
    sulfate effectiveness and growth stage, 112
    threshold values, concentrations, S, 228
    yields, Rock Lake, ND, 107
Rapeseed-mustard
    protein content increase due to S, 18
    S application rates, 20
    S critical values, at growth stages, 19
    S, N, P, K uptakes, IGA of Asia, 14
    yield increase due to S, 15–17
Raya, S critical values, at growth stages, 19
Reactivity, reducing agents, 26
Redox potential, sulfide formation, 201
Reduced soils, S components in acid, 62
Reduction, S, 84–88
Reliability, S testing, 76
Removal, S in crops, S removal, 70
Requirements, S in crops, 14, 125, 205
Residuals, response, S fertilizers, 21
Residues, crops, N-P-K nutrient concentrations, 67
Resistance, S induced (SIR), 50, 177–178
Rhizosphere, sulfate-sulfur, rice, 201
Rice
    aerenchyma formation, 200
    aerobic soils, production, 199–200
    deficiencies and diagnosing, 87–88, 210–212
    direct and residual benefits, S fertilizer, 21
    glutathione, 204
    grain sulfur content, 204
    irrigation studies, 207
    phosphate fertilizers, 209–210
    production, wet and drylands, 197–214
    protein contents, 121
    quality of, 205
    recovery S from fertilizer, 209
    S amino acid in foods, 283
    S application rates, 20
    S budget and cycling, 205–208, 211
    S concentrations in dry matter, 19
    S critical values, at growth stages, 19, 77
    S distribution, 204
    S fertilizer response, 208–210
    S from rain, 206
    S, N, P, K uptakes, IGA of Asia, 14
    S normal and critical levels, 211
    S nutrition, 203–205
    S remobilization, 204–205
    S removal, 70
    S shoot uptake, 127
    submerged soils, 198
    wetland soils, 200–203
    yield increase due to S, 15–17
ROL. *see* Radial oxygen loss (ROL)
Roots, submerged rice, 201
Ryegrass, S critical values, at growth stages, 77

S-alk(en)yl cysteine sulfoxides, 184–185
S coating fertilizer, rice, 209–210

S-methylmethionine (SMM), S transportation, 87
Salt-soluble proteins, 237
Salts, sulfate, 1, 5
SAM synthase, 261–262, 284–285
Sampling procedures, PIPPA, 228–229
Sand, loamy fine, 109
Saskatchewan, Canada, 4, 6–8
SAT, OAS, 92, 94
Sediment, S rice, 207
Seed-protein production, Soybeans, 120, 122
Seeds
	critical S concentrations, 134
	deficiencies, 135
	improving S contents, 294–298
	protein storage, 267–268
	quality, S and, 120–124
Selenium, 191, 299–302
Serine, function of, 91–93
Sesame, 14–15, 126–127
Shallots, cysteine sulfoxides, 185
Shoots, critical S concentrations, 134
Silage, corn, quality, 149–150
Single superphosphate, 6, 20, 117–118
SiR, 90–91
SIR, Sulfur induced Resistance, 50
Slurry, C/S ratio, 36
Slurry, S contents, 33, 33–34
SMM cycle, 261–263, 266
Sodium chloride levels, 191–192
Soil organic S, 28–31
Soil water
	S A leaching, 70
	S storage, 48–49
Soils
	analysis, 131–132, 144, 212–213
	available S, removal of, 69–72
	clay, leaching, 70
	direct absorption, S gasses, 65–66
	extractants test, S availability, 72
	fractions tests, S availability, 72
	humus, S mineralization, 4
	inorganic S, 31
	landscape position, S and, 107
	luvisolic, sulfate-S concentrations, 7–8
	microbes, S gas uptake, 66
	moisture, S gas uptake, 65–66
	onions, 188, 192
	organic S, 2–3, 26–30
	S accumulation and losses, 37–38
	S amendments, 32–37
	S and soybean growth, 128
	S build-up, studies, 34–35
	S components in acid, 62
	S concentration, Saskatchewan, Canada, 8
	S cycling, temperate systems, 25–39
	S deficiencies, 60, 106
	S fertility, assessing, 75–76, 130
	S in, 25–31
	S in submerged, 201
	S microbial biomass, 28
	S pool studies, 38
	sandy, leaching, 70
	selenium contaminated, 300
	sulfate status, 75
	temperate, model S cycling, 31–32
	texture and response to S fertilizer in corn, 109
	texture, S content and, 106
	wetland, 200–203
Sorghum, 20, 126–127
Sorption, aerobic soils, 200
Sources, 187
	available S, 60–69
	organic S, 33–36
	S aerobic soils, 199–200
	and sinks, plant available S, 73
	sulfate or elemental S, 111
Soybeans
	amino acid, improvements, 238–239
	animal feeds demand, 235
	assessment of S deficiency, 129–136
	composition and improvement, 236–243
	critical S concentrations, 134
	dry matter accumulation, 119
	heterologous genes, expression of, 239–242
	hypervariable region, S insertion, 240–241
	improving S amino acids nutrition, animal feeds, 235–244
	latent deficiency, 129
	major producers, 118
	oil content, S and, 18
	pods, S accumulation, 120
	protein content increase due to S, 18
	protein contents, 121
	S amino acid in foods, 283
	S application and topsoil type, 128
	S concentrations, 119
	S concentrations in dry matter, 19
	S critical values, at growth stages, 19, 77
	S deficiency, 124–129
	S management, 117–137
	S/N ratio, 122–123
	S nutrition of, 118–120
	S removal, 70
	S shoot uptake, 126
	S utilization, 118–120
	storage proteins, 121
	transgenic, 241
	transgenic improvements to, 123

Soybeans *continued*
　　wild, quality, 239
　　world production and deficiencies, 117–118
　　yield and quality traits, 123
　　yield increase due to S, 15–16
Soymeal. *see* Soybeans
Spatial-temporal variation of soils, 75
Spinach, 19, 283
Sri Lanka, rainfall studies, 206
Storage, internal, 87
Straighthead, 204
Straw, C/S ratio, 36
Straw, rice, S levels, 211
Stubble, S balance, rice, 207
Stylosanthes, S shoot uptake, 126
Submerged rice, 198
Submerged soils, S, 201
Submergence, soil oxygen supply, 200
Subsoil S pools, rice, 213
Sugar beets
　　optimum ranges, 225–227
　　S and Mn content BOLIDES analysis, 223
　　S deficiency, 54
　　S demands, 49
　　S removal, 70
　　S shoot uptake, 126–127
　　threshold values, concentrations, S, 228
Sugarcane, 14–15, 77, 126–127
Sulfate, 2
　　adsorption, 7–8, 31, 62
　　assimilation, subcellular organization, 85
　　cells, importation of, 88–89
　　concentrations, rice, 203–204
　　formation reactions, 63
　　leaching, 7–8
　　precipitation, 7–8
　　reduction, ATPS pathway, 89–90
　　reduction to sulfite, 89–90
　　retention of, 31
　　in soils, 62
　　sorption, submerged soils, 202
　　submerged soils, 201
　　uptake, 85–88
　　wed deposition, 64
Sulfate ion deposition, Northern plains, 106
Sulfate-selenate reduction pathway, 299
Sulfate-sulfur movement, 129–130
Sulfide, reduced from sulfite, 90–91
Sulfite, 89–91
Sulfolipids, 205
Sulfotransferases, 88
Sulfur-bentonites, granular fertilizers, S form analysis, 6
Sulfur dioxide, 68
Sulfur induced Resistance (SIR), 50, 177–178

Sulfuric acid, 66
Sulphatase enzymes, 29–30
Sunflower
　　methionine content, 268
　　oil content, S and, 18
　　protein content increase due to S, 18
　　S application rates, 20
　　S concentrations in dry matter, 19
　　S critical values, at growth stages, 19
　　S, N, P, K uptakes, IGA of Asia, 14
　　S removal, 70
　　S shoot uptake, 126–127
　　yield increase due to S, 15, 17
Superphosphate, 67–68
Sweet potatoes, 283
Symptomatological value, 219
Synthetic genes, soybeans, 242–243

Teas, S shoot uptake, 126
Temperate climates, S cycling in soil, 25–39
Temperate regions, S input, 46
Temperature, 4, 163–164, 192
Tests, S, 72–73, 75–76,
Texas, 188
Thailand, 206–207, 211
Thiamine, 3, 205, 284
Thiosulfate, 2
　　fertilizers, 6, 67
　　oxidation, 5
　　structure, 185
Threonine, human requirements, 282
Threonine synthase (TS), 255–257, 259–260
Tissues, vegetative, storage, 88
Tobacco, 126–127, 270–271
Tomatoes, 70, 126–127, 283
Topsoils, S and soybean growth, 128
Toria, protein content increase due to S, 18
Toxicity, sulfate, 68
Toxicological value, 220
Transformation, S in upland agricultural systems, 2
Transgenic seeds, 269, 295–298
Transgenic soybeans, 241
Transition phase crops, IGA of Asia, 13
Transporters, 86, 203
Treasure Valley, ID, 188
Triple superphosphate, rice, 209–210
Tryptophan, human requirements, 282
TS. *see* Threonine synthase (TS)
Tubers, diseases in potato, 172
Turnips, S shoot uptake, 126–127
Turnover rate, solid organic S, 61

Ultisols, water soluble sulfate-sulfur, 71
Upper boundary line functions, 224

Uptakes, S
  nitrogen, phosphorus and potassium, IGA, 14
  regulation, 87
  soils, and removal, 69–72
Urea, di-ammonium phosphate, rice, 209–210
USA, 128, 134

Valine, human requirements, 282
Variables, BOLIDES, 221–228
Variables, soil samples, 75
Vector analysis, 220–221
Vegetative tissues, methionine contents, 270–271
Vidalia onions and region, 189–190
Vietnam, S application and topsoil type, 128
Vitamins, 3, 205, 284

Weather, growing season, 163–164
Weather, S availability, 72
Wet limestone forced oxidation, 68–69
Wet scrubber sludge, 68
Wetland soils, 200–203
Wheat, 16–17
  direct and residual benefits, S fertilizer, 21
  methionine, SMM, methionine flux, 266
  protein composition effected by S, 156–157
  quality, and S nutrition, 153–167
  S amino acid in foods, 283
  S and Nitrogen fertilizer, effects, 154–155
  S application rates, 20
  S concentrations in dry matter, 19
  S critical values, at growth stages, 19, 77
  S deficiency, 47
  S, N, P, K uptakes, IGA of Asia, 14
  S removal, 70
  S shoot uptake, 126
  yields, increase due to S, 15–17, 109, 156
Wheat, winter, S uptake, 155
White blooming, flowers, 53
White clover, S critical values, at growth stages, 77
Whole shoot, tillering, rice, S levels, 211
Wooster, OH, 150

XANES spectroscopy, S oxidation states, 27–28

Y leaf-flowering, rice S levels, 211
Yang cycle, 261
Yields
  corn, irrigated and non-irrigated S, 147
  increase due to S, 15
  S sources and, 111
  wheat, increased, 15–17, 109, 155–156

Zinc, 225–227